A First Course on
WAVELETS

Studies in Advanced Mathematics

Series Editor

STEVEN G. KRANTZ
Washington University in St. Louis

Editorial Board

R. Michael Beals
Rutgers University

Dennis de Turck
University of Pennsylvania

Ronald DeVore
University of South Carolina

Lawrence C. Evans
University of California at Berkeley

Gerald B. Folland
University of Washington

William Helton
University of California at San Diego

Norberto Salinas
University of Kansas

Michael E. Taylor
University of North Carolina

A First Course on
WAVELETS

Eugenio Hernández
Universidad Autónoma de Madrid

Guido Weiss
Washington University in St. Louis

CRC Press
Boca Raton   New York

Library of Congress Cataloging-in-Publication Data

Hernández, Eugenio, 1954–
    A first course on wavelets / Eugenio Hernández, Guido Weiss.
           p. cm. — (Studies in advanced mathematics)
     Includes bibliographical references and indexes.
     ISBN 0-8493-8274-2 (alk. paper)
     1. Wavelets (Mathematics). I. Weiss, Guido L., 1928– II. Title.
     III. Series.
QA403.3.H47 1996
515′.2433—dc20   96–27111
                                                      CIP

No claim to original U.S. Government works
International Standard Book Number 0-8493-8274-2
Library of Congress Card Number 96-27111
Printed in the United States of America   2 3 4 5 6 7 8 9 0
Printed on acid-free paper

About the Authors

Eugenio Hernández graduated from the Universidad Complutense de Madrid in 1977 and obtained his Ph.D. degree from Washington University in St. Louis in 1981. He has been a member of the faculty of the Universidad Autónoma de Madrid since 1982 and is now Professor of Mathematics. During the academic year 1987–88, Dr. Hernández was a Fulbright Fellow and a member of the Mathematical Sciences Research Institute in Berkeley, California.

Guido Weiss (left) and Eugenio Hernández

He was a visiting professor at Washington University in St. Louis in 1994–95. His research interests lie in the areas of the theory of interpolation of operators, weighted inequalities, and most recently, in the theory of wavelets.

Guido Weiss obtained his undergraduate and graduate degrees from the University of Chicago, receiving his Ph.D. degree in 1956. He served on the faculty at DePaul University from 1955 to 1960, and joined the faculty of Washington University in 1960 where he is now the Elinor Anheuser Professor of Mathematics. During the past 35 years he has had leaves of absence that have allowed him to be visiting professor in several different institutions: the Sorbonne, the University of Geneva, the University of Paris in Orsay, the Mathematical Sciences Research Institute in Berkeley, California (in each case for an academic year). He also was visiting professor during semester academic leave at the Universidad de Buenos Aires, Peking University, Beijing Normal University, and the Universidad Autónoma de Madrid.

His research involves a broad area of mathematical analysis, particularly harmonic analysis. Some of his work, especially his contributions to the atomic and molecular characterizations of certain function spaces (particularly the Hardy spaces), is closely related to the theory of wavelets, a subject that has commanded his attention during the last few years. He has been awarded several honors—among them the Chauvenet Prize and honorary degrees from Beijing Normal University, the University of Milano, and the University of Barcelona.

To **Barbara** and **Jody**

Contents

Foreword

by

Yves Meyer

Membre de l'Institut (Académie des Sciences)

Foreign Honorary Member of the American Academy of Arts and Sciences

Wavelet analysis can be defined as an alternative to the classical windowed Fourier analysis. In the latter case the goal is to measure the local frequency content of a signal, while in the wavelet case one is comparing several magnifications of this signal, with distinct resolutions. The building blocks of a windowed Fourier analysis are sines and cosines (waves) multiplied by a sliding window. They are usually referred to as time-frequency atoms. In a wavelet analysis, the window is already oscillating and is called a mother wavelet. This mother wavelet is no longer multiplied by sines or cosines. Instead it is translated and dilated by arbitrary translations and dilations. That is the way the mother wavelet generates the other wavelets which are the building blocks of a wavelet analysis. These dilations are precisely the magnifications we alluded to, and the building blocks are called time-scale atoms.

Fourier analysis, windowed Fourier analysis, and wavelet analysis are based on an identical recipe. In the three cases, the analysis of a function amounts to computing all the correlations between this function and the time-frequency or time-scale atoms which are being used. The synthesis is obtained exactly as if these building blocks were an orthonormal basis.

A common wisdom among numerical analysts and image processing people is that the inverse of a scale is a frequency: small scales correspond to large frequencies and large scales to small frequencies. Moreover, very distinct scales should provide independent (i.e., non-redundant) information. Wavelet analysis could be defined as an attempt to give a very precise meaning to this folk belief.

Wavelets were implicit in mathematics, physics, signal or image processing, and numerical analysis long before they were given the status of a unified scientific field.

In pure mathematics, three algorithms have been created to overcome some drawbacks of standard Fourier series expansions. These difficulties appear when one is facing the problem of measuring the size or the smoothness of a function. For example, the simplest norms, based on quadratic estimates, can easily be extracted from Fourier coefficients. But as soon as L^p or H^p estimates are addressed, Fourier coefficients do not answer the problem, while the algorithms that do answer it involve the **Haar basis** (1909), the **Franklin orthonormal system** (1927), or the **Littlewood-Paley theory** (1930); these have, in the past, proven to be the correct tools.

Later, **Calderón's reproducing identity** (1960) and **atomic decompositions** (1972) were widely used in other functional settings (Hardy spaces, for example). Both the Littlewood-Paley theory and atomic decompositions play a key role in a branch of operator theory created by Calderón, Zygmund, and their school which is known as the Calderón-Zygmund Theory. Just before wavelets became popular, J.O. Strömberg used this precise tool for solving a celebrated problem in the geometry of Banach spaces: the existence of a specific unconditional basis for the Hardy space $H^1(\mathbb{R})$.

In signal or image processing a similar and parallel evolution started from the standard windowed Fourier analysis and culminated in some discrete versions of Calderón's reproducing identity. Indeed, D. Gabor (1946) introduced **time-frequency atoms** in speech signal processing; Croisier, Esteban, and Galand developed **subband coding** in signal processing (1975); and only a little later Burt and Adelson described **pyramidal algorithms** in image processing (1982). D. Marr was convinced that both human vision and computer vision were based on similar algorithms which should be, in some sense, independent of the "wires" used in their realizations. These specific algorithms involve the **zero-crossings** of the wavelet transform of a two-dimensional signal (1982). In numerical analysis, wavelets are related to spline approximation. Before wavelets became fashionable, V. Rokhlin created the so-called **multipole algorithms**: refinement schemes that play a key role in computer graphics.

Finally, let us turn to mathematical physics. **Coherent states** are fundamental in quantum mechanics. **Renormalization in quantum field theory** is needed for extracting finite numbers from divergent integrals. It is based on some variants of Littlewood-Paley techniques which were mainly developed by K. Wilson, K. Gawedzki and A. Kupiainen, J. Glimm and A. Jaffe, G. Battle and P. Federbush.

Therefore wavelets were implicit in several scientific fields but nobody knew that, for instance, Littlewood-Paley theory and the Burt & Adelson pyramidal algorithms were telling the same story. The great unification was a shock, and many people still do not accept it. This unification was a fairy tale come true, which explains why the subject became immediately popular. The great unification meant a scientific status incorporating the heuristics and the wisdom of the distinct fields where protowavelets were already used. This unification was made possible through the efforts of several people. Let me especially mention Alex Grossmann and Stephane Mallat.

I have a vivid and nostalgic memory of many discussions with Antoni Zygmund. He used to test me on whatever problem he was dreaming about. He silently waited for my answer. Then he listened with a smile to my often stupid comments. Finally he often tried to correct my erroneous viewpoints. This happened when R.R. Coifman and G. Weiss and their collaborators launched the so-called **atomic decompositions** program. Zygmund asked my opinion about what Guido Weiss was doing. Zygmund immediately recognized the relevance of this endeavour, while it took me a slightly longer time.

But it is hard to believe that Zygmund would have guessed that atomic decompositions are also relevant in signal processing. He would have been surprised to learn that the celebrated composer and conductor Pierre Boulez and his collaborators decided to find a compact atomic decomposition for an aria by Mozart interpreted by Rita Streich. P. Boulez and his collaborators were indeed using (time-frequency) waveforms instead of (time-scale) wavelets.

We now come to the present book. It is not just one more book about wavelets. This unique book is distinct, since it is co-authored by one of the pioneers of atomic decompositions. Who else is more appropriate to talk about wavelets? Indeed atomic decompositions are at the heart of signal and image processing.

The careful writing of the authors, Eugenio Hernández and Guido Weiss, is well known and this book reflects their desire to make this subject most accessible. It will be applauded by all lovers of the precise, powerful, and elegant mathematics which Guido Weiss and his school have promoted.

This book contains many new and impressive results. Nowadays, there is a tendency to derive wavelets from the multiresolution analysis construction. By this method one cannot address basic issues like the ones that are discussed in this book and are, indeed, crucial. For example, the Fourier localization of a wavelet is discussed in full detail. This has been neglected by other authors. I hope the reader will enjoy this remarkable contribution as much as I did, and I thank the authors for letting me read the manuscript.

Preface

Introduction

Wavelets were introduced relatively recently, in the beginning of the 1980s. They attracted considerable interest from the mathematical community and from members of many diverse disciplines in which wavelets had promising applications. A consequence of this interest is the appearance of several books on this subject and a large volume of research articles. In order to explain why we have written this book, describe where it might play a useful role in this field and to whom it is addressed, we find it necessary to state what we mean by the word "wavelet" and mention some of its properties. Let us do this for wavelets defined on the real line $\mathbb{R}$.

The real line is endowed with two basic algebraic operations, addition and multiplication. From these two operations we obtain two families of operators acting on functions defined on $\mathbb{R}$: the **translations** and the **dilations**. More precisely, translation by $h \in \mathbb{R}$ is the operator τ_h that maps a function f into the function whose value at $x \in \mathbb{R}$ is $(\tau_h f)(x) = f(x - h)$. The dilation ρ_r, $r > 0$, is defined by the equality $(\rho_r f)(x) = f(rx)$. Many of the important linear operators acting on functions defined on $\mathbb{R}$ have simple relations with these two families. For example, differentiation commutes with the translations. More generally, in the setting of tempered distributions, the class of convolution operators are characterized by this property of commuting with translations (differentiation is obtained by convolving with the distribution that is the derivative of the "Dirac-delta function"). Similar observations can be made about the family of dilations. A most important operator acting on functions (or, more generally, on tempered distributions) is the **Fourier Transform**, which maps f into $\hat{f}$, where

$$\hat{f}(\xi) = \int_{\mathbb{R}} e^{-i\xi x} f(x) \, dx.$$

It is well known that convolution operators are converted, via the Fourier transform, into multiplication operators. This is a consequence of the formula $(f * g)^\wedge = \hat{f} \, \hat{g}$. In particular, $(\tau_h f)^\wedge(\xi) = e^{-ih\xi} \hat{f}(\xi)$; that is, translation by h corresponds to multiplication by the exponential $e^{-ih\xi}$. All these properties are particularly natural if we consider them in the context of $L^2(\mathbb{R})$: the Fourier transform can then be expressed in terms of a unitary operator, and this allows one to study many convolution operators in terms of particularly simple multiplier operators.

In view of these observations, it is only natural to look for bases of $L^2(\mathbb{R})$ having properties that reflect the importance of translations, dilations and the Fourier transform. For example, in the analogous periodic case, the "trigonometric" system, $\{\frac{1}{\sqrt{2\pi}} e^{inx} : n \in \mathbb{Z}\}$, is an orthonormal basis for $L^2(0, 2\pi)$ that simultaneously diagonalizes all the bounded operators on this space that commute with translations. This property makes this system a most important basis for $L^2(0, 2\pi)$ and is of fundamental importance to the study of Fourier series. The various wavelets provide us with orthonormal bases for $L^2(\mathbb{R})$ that are particularly natural when dealing with the analysis that involves the action of translations, dilations and the Fourier transform (that is, Harmonic Analysis). We see that this is most plausible from their definition: a function $\psi \in L^2(\mathbb{R})$ is an **orthonormal wavelet** provided the system $\{\psi_{j,k} : j, k \in \mathbb{Z}\}$ is an orthonormal basis for $L^2(\mathbb{R})$, where

$$\psi_{j,k}(x) = 2^{\frac{j}{2}} \psi(2^j x - k) \qquad \text{for all } j, k \in \mathbb{Z}.$$

That is, this system is generated from one function, ψ, by translating it by the integers and applying the dyadic dilations ρ_r, where $r = 2^j$, to these translates. The multiplication by the factor $2^{j/2}$ is forced upon us if we require each member of this system to have L^2-norm equal to one; moreover, it renders the action of the Fourier transform on this system particularly simple: if $\gamma = \hat{\psi}$, then the Fourier transform of $\psi_{j,k}$ is

$$(\psi_{j,k})^{\wedge}(\xi) = e^{-i2^{-j}k\xi} 2^{-\frac{j}{2}} \gamma(2^{-j}\xi).$$

That is, we still have dyadic dilations and the translations are converted into "modulations" (which is a term that means multiplication by exponentials).

The philosophy of the book

The purpose of this book is to show how such wavelets can be constructed, illustrate why they provide us with a particularly powerful tool in mathematical analysis, and indicate how they can be used in applications. The title of the book reflects our hope that it can be read by those who are familiar with the Fourier transform and its basic properties; we feel that this amount of knowledge suffices for the understanding of the material presented. Let us explain in more detail what we mean by this. We shall show that wavelets can be applied to a large variety of mathematical subjects. For example, they can be used to characterize several function spaces: the Lebesgue, Hardy, Sobolev, Besov and Triebel-Lizorkin spaces are some of these. The Lebesgue spaces are easily defined, and some of their basic properties are not hard to explain. This is not the case for all these spaces. For example, the Hardy spaces have many different, but equivalent, definitions. Originally they were introduced as spaces of holomorphic functions in the domain in the complex plane that lies above the x-axis. About twenty-five years ago it was discovered that they can be identified as functions (really, distributions) on $\mathbb{R}$ having an appropriate maximal function. A few years later their "atomic" characterization was discovered. This approach involves certain "building blocks" called atoms, which are particularly simple functions, that can be used to express the general element of the Hardy space. It would carry us way beyond the scope of this book if, before discussing these spaces, we were to present all the material that is necessary to establish the equivalence of these various versions. It is not difficult, however, to present clear statements of those properties that are most relevant to the use of wavelets; when we do this, we do give appropriate references. In this sense this book is not "self-contained," but this does not mean that more is demanded from the reader in order to appreciate the roles that are played by wavelets in these applications.

Wavelets can be defined on other domains. For example, we can introduce a natural extension of the definition of the function $\psi_{j,k}$ by considering ψ to be defined in $\mathbb{R}^n$, n-dimensional Euclidean space, by letting $k = (k_1, \cdots, k_n) \in \mathbb{Z}^n$ be an n-tuple of integers and replacing $2^{j/2}$ by $2^{nj/2}$ (so that $\|\psi_{j,k}\|_2 = \|\psi\|_2$). The situation, in this case, is more complicated: if one makes certain natural assumptions, it can be shown that one cannot obtain an orthonormal basis of $L^2(\mathbb{R}^n)$ by such a construction; in fact, $2^n - 1$ such generating functions are needed if one wants to obtain such a basis. Other domains can be considered where the roles played by the translations and dilations need to be played by different actions on the domain. We decided in this "first course on wavelets" not to present the theory of wavelets in these more complicated settings and to concentrate on the one-dimensional case. We felt that a good understanding of the one-dimensional theory provides a good background for its extensions to other domains.

Let us make a few comments about some of the other books on wavelets. Perhaps the two most important treatises on the subject are Y. Meyer's three-volume set ([Me1], [Me2], and [CM1] – the third one is co-authored with R. Coifman) and I. Daubechies "Ten lectures on wavelets" ([Da1]). Both are excellent presentations, and we recommend them with enthusiasm. They are more advanced than this book and cover much more material. Since they were written, however, the theory has advanced considerably (partly due to their contributions). Some of the original constructions have been simplified and extended. We hope that this book can serve as an introduction to these two treatises. The book by C. Chui ([Chu]) should also be mentioned. It is a good complement to the ones by Daubechies and Meyer (as, we hope, is ours). We cite it often, particularly when we discuss spline wavelets.

Description of the book

It is, perhaps, useful to describe this book in more detail and give some advice about how to read it. The first four chapters, together with Chapter 7, make up a "natural" inter-related group. They are devoted to the construction of wavelets. We feel that Chapter 7 is the most important one in the book. There are two simple equations that completely characterize all orthonormal wavelets. They are

$$\sum_{j\in\mathbb{Z}}|\hat{\psi}(2^j\xi)|^2 = 1 \qquad \text{for } a.e. \; \xi \in \mathbb{R}, \tag{1}$$

and for every odd integer m,

$$\sum_{j=0}^{\infty}\hat{\psi}(2^j\xi)\,\overline{\hat{\psi}(2^j(\xi+2m\pi))} = 0 \qquad \text{for } a.e. \; \xi \in \mathbb{R}. \tag{2}$$

More precisely, $\psi \in L^2(\mathbb{R})$ is an orthonormal wavelet if and only if ψ satisfies (1) and (2), provided $\|\psi\|_2 = 1$. The proof of this is elementary but it is not simple, and we present it in the seventh chapter. These equations are known and have been used by many investigators working with wavelets. The proof of this characterization in full generality, however, did not appear in the published literature until recently. It can be found in a paper by G. Gripenberg ([Gri1]), the Ph.D. thesis of one of our students, X. Wang ([Wan]), and will appear in an expository article we wrote with him ([HWW3]). It has been one of our goals to study the properties of wavelets by examining their Fourier transforms. One of the principal features of this book, in fact, is the important role played by the Fourier transform.

The first four chapters are devoted to different ways of constructing wavelets. Chapter 1 deals with the local sine and cosine bases that were discovered by R. Coifman and Y. Meyer. We show how they lead us to bases for $L^2(\mathbb{R})$ that have the important features described in the beginning of this introduction; that is, they enjoy particularly simple relations with the basic operators: translations, dilations and the Fourier transform. We use these bases to construct the wavelets of Lemarié and Meyer, the first class of orthonormal wavelets that were introduced and that includes ones such that they and their Fourier transform are smooth.

In the second chapter we develop a general method that was introduced by Mallat and Meyer for constructing wavelets: the **multiresolution analysis** (MRA). We apply this method to obtain the compactly supported wavelets introduced by Daubechies. The third chapter is devoted to the "band-limited" wavelets (the ones having compactly supported Fourier transforms). We show that the elements of this class have some surprising properties; for example, their Fourier transforms vanish in a neighborhood of the origin. Perhaps one of the best reasons for studying this class separately is that the basic equations (1) and (2) are particularly easy to study. Among other things, the series involved have only a finite number of non-zero terms and we do not need to worry about their convergence. This allows us to pave the way for the technically more difficult analysis involved in the seventh chapter. The fourth chapter introduces the reader to the "spline wavelets." This class appears to be particularly important in the various applications of wavelet theory to signal and image analyses. We also explain in this fourth chapter how one can construct periodic wavelets.

By the end of the first four chapters we have enough examples and have obtained sufficiently many properties of wavelets to introduce the reader to some of the uses of wavelets and their connection to other parts of Analysis. We therefore interrupt our program of characterizing all wavelets in terms of their Fourier transform and show how they provide us with tools for the study of the important scales of function spaces we mentioned above. In addition to providing us with orthonormal bases for the Hilbert space $L^2(\mathbb{R})$, some wavelets give us natural bases for these other topological linear spaces as well. Let us illustrate this with the Lebesgue spaces $L^p(\mathbb{R})$, $1 < p < \infty$, of all those measurable functions f such that

$$\|f\|_p = \left(\int_{-\infty}^{\infty} |f(x)|^p \, dx \right)^{\frac{1}{p}} < \infty. \tag{3}$$

When $p = 2$ the finiteness of this norm $\|f\|_2$ is equivalent to the finiteness of the norm

$$\|\mathfrak{c}\|_2 = \left(\sum_{j \in \mathbb{Z}} \sum_{k \in \mathbb{Z}} |c_{j,k}|^2 \right)^{\frac{1}{2}}$$

of the coefficient sequence $\mathfrak{c} = \{c_{j,k}\} = \{\langle f, \psi_{j,k} \rangle\}$ that provides us with the representation

$$f = \sum_{j \in \mathbb{Z}} \sum_{k \in \mathbb{Z}} c_{j,k} \, \psi_{j,k} \, .$$

Thus, $L^2(\mathbb{R})$ can be represented as the space $\ell^2(\mathbb{Z} \times \mathbb{Z})$ of all sequences $\mathfrak{c}$ such that $\|\mathfrak{c}\|_2 < \infty$. Appropriate wavelet bases provide us with the characterization of $L^p(\mathbb{R})$ in terms of a sequence space for the other indices $p \in (1, \infty)$. It can be shown that f belongs to $L^p(\mathbb{R})$ if and only if

$$\|\mathfrak{c}\|^{(p)} = \|\{c_{j,k}\}\|^{(p)} = \left(\int_{-\infty}^{\infty} \left\{ \sum_{j \in \mathbb{Z}} \sum_{k \in \mathbb{Z}} 2^j |c_{j,k}|^2 \chi_{j,k}(x) \right\}^{\frac{p}{2}} dx \right)^{\frac{1}{p}} < \infty,$$

where $\chi_{j,k}$ is the characteristic function of the interval $[2^{-j}k, 2^{-j}(k+1)]$ and $\mathfrak{c}$ is the sequence of coefficients of f associated with $\{\psi_{j,k}\}$. Observe that the finiteness of $\|\mathfrak{c}\|^{(p)}$ is a condition on the size (or absolute value) of the coefficients $c_{j,k}$. This provides us with the ability to study $L^p(\mathbb{R})$ in terms of a corresponding sequence space in a way that is analogous to the reduction of properties of $L^2(\mathbb{R})$ to properties of $\ell^2(\mathbb{Z} \times \mathbb{Z})$. Note that $\|\mathfrak{c}\|^{(2)} = \|\mathfrak{c}\|_2$. Thus, an operator that is diagonalized by the basis $\{\psi_{j,k}\}$ can be analyzed in terms of its proper values, as is the case in Hilbert space theory. Many important operators are "essentially" diagonalized by wavelet bases. It is this circle of ideas that is presented in Chapter 5 and Chapter 6. More specifically, we present a brief treatment of bases in Banach spaces, with an emphasis on the notion of unconditionality, in Chapter 5. In Chapter 6 we give the characterizations described above. This treatment uses properties of **Calderón-Zygmund operators**; consequently, we have an opportunity to see how wavelets are associated with the study of these important operators.

In Chapter 7 we resume the study of wavelets in $L^2(\mathbb{R})$. We not only give a characterization of all wavelets, as described above, but we also characterize all wavelets that arise from an MRA and the basic functions (the scaling functions and low-pass filters) involved in this method. These characterizations allow us to construct several other classes of wavelets as well.

Though most of the bases discussed in the first seven chapters are orthonormal, we do mention some other types of bases. In Chapter 8 we present a more thorough treatment of systems that are more general, with particular attention to **frames** and their importance to wavelets. We pay special attention to the way they can be used to analyze and reconstruct functions; we also extend the Balian-Low theorem to frames.

The last chapter is devoted to certain topics that are important and relevant to the applications of the theory of wavelets. We indicate how the mathematical theory is transformed when it is applied to "discrete" signals. We develop the Discrete Fourier (and Cosine) Transform in what is, probably, a manner that is different from the usual treatment but has some features that are adaptable to programming for computers. We also describe the decomposition and reconstruction algorithms for wavelets and we end the chapter with a treatment of "wavelet packets."

One of our colleagues, M.V. Wickerhauser, has recently written a book, "Adapted wavelet analysis from theory to software" ([Wi2]), that treats the subject we just mentioned, and many more applications, in great detail. We believe that his book will prove to be most useful. We found no need, therefore, to go further than we did in this direction. As we stated about the books by Daubechies and Meyer, we hope that our book makes a good companion to, and complements, the book by Wickerhauser.

Some advice to the reader

The background we assume of the reader is a "good undergraduate" preparation in mathematics. We do use the language of measure theory; for example, we talk about "measurable functions." One should not be discouraged if he/she only knows the ordinary Riemann integral. Substituting the Lebesgue integral for the latter will, in general, not affect the meaning or the validity of most statements. Some notions in elementary functional analysis are used; again, the results or statements involving these notions can almost always be understood by ignoring unfamiliar language.

It is our hope that graduate students in mathematics, the sciences and engineering can profit from our presentation. We advise the reader not to be discouraged by the **few** somewhat technical notions we introduce at times (distributions, maximal functions, vector-valued inequalities, etc). If it's "too much," just skip it at first; there is enough material that can be understood with the background mentioned in the previous paragraph. It is our experience that even those whose main interest is in the applications can profit by learning about the theory we present.

In each section we number results consecutively; that is, we do **not** form separate lists of theorems, propositions, corollaries, lemmas, formulae and inequalities. These items are listed as ordered pairs n.m, where "n" denotes the section (in the chapter) and "m" the mth item so numbered in the section. If we need to refer to something in another chapter, we mention the chapter and the relevant ordered pair. The sections in each chapter are also assigned an ordered pair, n.m; in this case, "n" denotes the number of the chapter and "m" the mth section.

We do not present a list of exercises at the end of each chapter. In many cases we leave certain calculations to be worked out by the reader. This is particularly true of the comments made in the last section of each chapter, which is labelled "Notes and references."

We also feel that we should state quite clearly that, though the bibliography we include is quite large, it is far from a list that comes anywhere close to exhausting what has been published in the theory of wavelets during the relatively short period of its existence. We have tried to give proper credits; however, since some of the material we discuss is quite new, we realize that it is very likely we omitted some references that should have been included.

Acknowledgments

The idea of writing a book on wavelets was suggested by F. Soria when he asked the second author to give a course in this subject in the Universidad Autónoma de Madrid during the spring semester of 1993. This course was repeated and extended considerably during the academic year 1994 – 95 when both authors were at Washington University.

Many individuals helped us in the preparation of these two courses, before and during the time when the lectures were given, and while we were writing this book: P. Auscher, A. Bonami, M.J. Carro, J. Dziubañski, X. Fang, G. Garrigós, Y. Han, A. Ho, A. Hulanicki, E. Laeng, Y. Meyer, M. Paluszyñski, P.M. Soardi, F. Soria, J. Soria, M. Taibleson, A. Trgo, X. Wang and D. Weiland. Some on this list were particularly important to us and we feel that special thanks are in order.

Y. Meyer, during many years, continuously provided us with the many manuscripts he wrote, not only for publication, but for private circulation among his many students and friends. We learned a lot from him. P. Auscher, A. Bonami and F. Soria were collaborators on several projects with one of us and were constant consultants. The "logo" for this book represents the method described in section 3.3, for obtaining Lemarié-Meyer wavelets as approximations to the Shannon wavelet. This graphic design was done by J. Soria.

Special thanks are due to X. Wang, who obtained his Ph.D. with G. Weiss at Washington University. His thesis and collaborations with us contain much that is novel in this book. In addition, he prepared most of our manuscript for the publisher. The first three chapters, and some later material, was originally typed by J. Doran.

We are also most grateful to the Southwestern Bell Telephone Company, the Air Force Office of Scientific Research (U.S.A.), the National Science Foundation, and the Ministerio de Educación y Ciencia (Spain) for giving us the financial support that allowed us to work together for the creation of this book.

Eugenio Hernández, Universidad Autónoma de Madrid
Guido Weiss, Washington University in St. Louis

1

Bases for $L^2(\mathbb{R})$

Classical systems of orthonormal bases for $L^2([0,1))$ include the exponentials $\{e^{2\pi imx} : m \in \mathbb{Z}\}$ and various appropriate collections of trigonometric functions. (See Theorem 4.1 below.) The analogs of these bases for $L^2([\alpha,\beta))$, $-\infty < \alpha < \beta < \infty$, are obtained by appropriate translations and dilations of the ones above. To find an orthonormal basis for $L^2(\mathbb{R})$ we can cover $\mathbb{R}$ with a disjoint union of intervals

$$[\alpha_j, \alpha_{j+1}), \quad j \in \mathbb{Z}, \quad -\infty < \cdots < \alpha_j < \alpha_{j+1} < \cdots < \infty,$$

and consider one of these bases for each space $L^2([\alpha_j, \alpha_{j+1}))$, multiply the basis elements by the characteristic function of $[\alpha_j, \alpha_{j+1})$, and take the totality of the functions so obtained. This orthonormal basis, however, produces "undesirable edge effects" at the endpoints α_j when we try to represent a function in terms of it.

In order to remedy this situation one is led to consider smooth functions that replace the characteristic function of $[\alpha_j, \alpha_{j+1})$ for $j \in \mathbb{Z}$. In the case of complex exponentials and the simple partition

$$\mathbb{R} = \bigcup_{n \in \mathbb{Z}} [n, n+1)$$

we examine systems of the form

$$\{g_{m,n}(x) = e^{2\pi imx} g(x - n) : m, n \in \mathbb{Z}\}.$$

For a system of this type (often called a **Gabor basis**) to be an orthonormal basis for $L^2(\mathbb{R})$ g cannot be "too smooth" or "very localized." This is made precise by the Balian-Low theorem presented in section 1.2. If appropriate bases of sine (or cosine) functions, however, are used, a much more general family of functions g, arbitrarily smooth and "very localized," can be used to obtain orthonormal bases of $L^2(\mathbb{R})$.

This is done in section 1.3 where we present a theory of **smooth projections**, introduced by Coifman and Meyer, that allows us to "join" appropriate bases associated with the intervals $[\alpha_j, \alpha_{j+1})$. Several examples of this construction are given, but the most relevant for our purpose are the ones that produce **orthonormal wavelets**: $\psi \in L^2(\mathbb{R})$ such that

$$\psi_{j,k}(x) = 2^{\frac{j}{2}} \psi(2^j x - k), \qquad j, k \in \mathbb{Z},$$

is an orthonormal basis for $L^2(\mathbb{R})$. It is in this way that, in section 1.4, we construct the wavelets of Lemarié and Meyer.

In section 1.5 we describe the smooth projections presented in section 1.3 in terms of certain unitary "folding operators." Some theoretical results can be obtained in an elegant manner by using these operators; however, it is, perhaps, more important that they provide some simple ways for programming the uses of these local bases. Furthermore, this approach does lend itself to extending the theory to higher dimensions.

1.1   Preliminaries

We assume that the reader is familiar with the basic notions of Lebesgue measure and integration theory, Hilbert space theory and Functional Analysis. We begin by introducing some notation and a few results that we shall assume. $\mathbb{R}$ refers to the real line; $\mathbb{T}$ will denote the unit circle in the complex plane which can be identified with the interval $[-\pi, \pi)$, though sometimes we use the interval $[-\frac{1}{2}, \frac{1}{2})$ or $[0, 1)$; and $\mathbb{Z}$ will denote the collection of integers. The inner product of functions f and g defined on either of these two spaces is

$$<f, g> = \int f \overline{g},$$

where the integral is taken over $\mathbb{R}$ or over $\mathbb{T}$, as the case may be. We have Schwarz's inequality

$$|<f, g>| \leq \|f\|_2 \|g\|_2,$$

where

$$\|f\|_2 = \left(\int |f|^2 \right)^{\frac{1}{2}}$$

is the L^2-norm of f. Schwarz's inequality allows us to prove Minkowski's inequality

$$\|f + g\|_2 \leq \|f\|_2 + \|g\|_2.$$

We say that two functions f and g are **orthogonal**, and write $f \perp g$, when $<f, g> = 0$. A sequence of functions $\{f_n\}_{n \in \mathbb{Z}}$ is an **orthonormal sequence** if $<f_m, f_n> = \delta_{m,n}$, where

$$\delta_{m,n} = \begin{cases} 1, & \text{if } n = m, \\ 0, & \text{if } n \neq m. \end{cases}$$

A well known example of an orthonormal sequence on $\mathbb{T} = [-\pi, \pi)$ is $\{ \frac{1}{\sqrt{2\pi}} e_n \}_{n \in \mathbb{Z}}$, where $e_n(x) = e^{inx}$.

Given an orthonormal system $\{f_n : n \in \mathbb{Z}\}$ and a function f, we define the Fourier coefficients of f with respect to $\{f_n : n \in \mathbb{Z}\}$ to be

$$c_k = <f, f_k>, \qquad k \in \mathbb{Z}.$$

A basic question that we shall study is to determine when, and in what sense, it is true that

$$f = \sum_{k \in \mathbb{Z}} c_k f_k. \tag{1.1}$$

When $f_k(x) = e^{ikx}$, $k \in \mathbb{Z}$, and $f \in L^2(\mathbb{T})$, the representation (1.1) is valid in the L^2-norm sense. In general, when this is the case we say that $\{f_k : k \in \mathbb{Z}\}$ is an **orthonormal basis** for $L^2(\mathbb{T})$. Equality (1.1) is a reconstruction formula and it is the basis for many applications of the theory of wavelets. Given a function f (a signal or a sound), we can encode it by means of the coefficients $\{c_k\}_{k \in \mathbb{Z}}$. Equality (1.1) allows us to reconstruct the signal from the numbers c_k and the basis used in the encoding. Some bases, in particular wavelet bases, perform this job more efficiently than others. For any orthonormal system $\{f_n : n \in \mathbb{Z}\}$ we have Bessel's inequality

$$\sum_{k \in \mathbb{Z}} |c_k|^2 \leq \|f\|_2^2.$$

Moreover, if the system is a basis, we have equality. Conversely, if an orthonormal system $\{f_n : n \in \mathbb{Z}\}$ satisfies

$$\sum_{k \in \mathbb{Z}} |c_k|^2 = \|f\|_2^2 \tag{1.2}$$

for all $f \in L^2(\mathbb{T})$, the system is a basis for $L^2(\mathbb{T})$.

In $\mathbb{R}$ we have an "analogous" theory. The Fourier transform of a function $f \in L^1(\mathbb{R}) \cap L^2(\mathbb{R})$ is defined by

$$\hat{f}(\xi) = \int_{-\infty}^{\infty} f(x) e^{-i\xi x} \, dx.$$

We will often say that x is the "time" variable and ξ will be referred as the "frequency" variable.

The inverse Fourier transform is

$$\check{g}(x) = \frac{1}{2\pi} \int_{-\infty}^{\infty} g(\xi) e^{i\xi x} \, d\xi,$$

and if we apply it to $g = \hat{f}$ we obtain f; that is $(\hat{f})^{\vee} = f$. With this definition the Plancherel theorem asserts that

$$<f,g> = \frac{1}{2\pi} <\hat{f}, \hat{g}> . \tag{1.3}$$

The Fourier transform extends to all $f \in L^2(\mathbb{R})$ and the operator $f \mapsto \frac{1}{\sqrt{2\pi}}\hat{f}$ is unitary. When f' exists in the L^2 sense, then

$$\widehat{f'}(\xi) = i\xi\hat{f}(\xi). \tag{1.4}$$

It can be proved that the integration by parts formula

$$\int_{-\infty}^{\infty} f'(x)g(x) \, dx = -\int_{-\infty}^{\infty} f(x)g'(x) \, dx \tag{1.5}$$

is valid when $f, g \in L^2(\mathbb{R})$ and $f'g, fg' \in L^1(\mathbb{R})$. In the case $f, g, f', g' \in L^2(\mathbb{R})$ this can be proved using (1.3) and (1.4).

A notion, which will be used in several proofs, is that of a Lebesgue point. Suppose f is a measurable function which is locally integrable, then a point

x_o is called a **Lebesgue point** for f whenever

$$\lim_{\delta \to 0+} \frac{1}{2\delta} \int_{x_o-\delta}^{x_o+\delta} |f(x) - f(x_o)| \, dx = 0.$$

It follows from the Lebesgue Differentiation Theorem that almost every x_o is a Lebesgue point. The reader may consult [Rud] for this particular theorem as well as for other results in measure theory.

Three simple operators on functions defined on $\mathbb{R}$ play an important role in our theory: **translation by** h, τ_h, defined by $(\tau_h f)(x) = f(x - h)$, **dilation by** $r > 0$, ρ_r, defined by $(\rho_r f)(x) = f(rx)$ and **multiplication by** e^{imx}. (Sometimes referred to as a **modulation** operator.) One of our main goals is to construct orthonormal bases of $L^2(\mathbb{R})$ by applying some of these operators to a single function in $L^2(\mathbb{R})$.

Of particular interest to us are the wavelet bases for which the first two operators are applied to an appropriate function. More precisely, an **orthonormal wavelet on** $\mathbb{R}$ is a function $\psi \in L^2(\mathbb{R})$ such that $\{\psi_{j,k} : j, k \in \mathbb{Z}\}$ is an orthonormal basis of $L^2(\mathbb{R})$, where

$$\psi_{j,k}(x) = 2^{\frac{j}{2}} \psi(2^j x - k), \qquad j, k \in \mathbb{Z}.$$

Observe that the $\psi_{j,k}$ are normalized so that $\|\psi_{j,k}\|_2 = \|\psi\|_2 = 1$ for all $j, k \in \mathbb{Z}$.

EXAMPLE A: If

$$\psi(x) = \begin{cases} 1, & \text{if } 0 \le x < \frac{1}{2}, \\ -1, & \text{if } \frac{1}{2} \le x < 1, \\ 0, & \text{elsewhere,} \end{cases}$$

then ψ is an orthonormal wavelet for $L^2(\mathbb{R})$. This is called the **Haar wavelet**. It is easy to prove that $\{\psi_{j,k} : j, k \in \mathbb{Z}\}$ is an orthonormal system in $L^2(\mathbb{R})$. It is also a basis for $L^2(\mathbb{R})$, a fact that will become obvious when we develop the theory of "multiresolution analysis" in Chapter 2.

EXAMPLE B: Let ψ be such that $\hat{\psi}(\xi) = \chi_I(\xi)$, where

$$I = [-2\pi, -\pi] \cup [\pi, 2\pi].$$

We shall show that ψ is an orthonormal wavelet for $L^2(\mathbb{R})$. A simple calculation shows

$$(\psi_{j,k})^{\wedge}(\xi) = 2^{-\frac{j}{2}}\hat{\psi}(2^{-j}\xi)e^{-i2^{-j}k\xi}.$$

For $j \neq \ell$ this equality shows that the intersection of the supports of $(\psi_{j,k})^{\wedge}$ and $(\psi_{\ell,m})^{\wedge}$ has measure zero; hence,

$$<\psi_{j,k},\psi_{\ell,m}> = \frac{1}{2\pi}<(\psi_{j,k})^{\wedge},(\psi_{\ell,m})^{\wedge}> = 0 \qquad \text{for } j \neq \ell.$$

When $j = \ell$ we can write

$$<\psi_{j,k},\psi_{j,m}> = \frac{1}{2\pi}2^{-j}\int_{\mathbb{R}}|\hat{\psi}(2^{-j}\xi)|^2 e^{i2^{-j}(m-k)\xi}\,d\xi$$

$$= \frac{1}{2\pi}\left\{\int_{-2\pi}^{-\pi}e^{i(m-k)\eta}\,d\eta + \int_{\pi}^{2\pi}e^{i(m-k)\eta}\,d\eta\right\} = \delta_{k,m}.$$

To prove that the system is a basis we use (1.2). The Plancherel theorem and a change of variables allow us to write

$$\sum_{j,k\in\mathbb{Z}}|<f,\psi_{j,k}>|^2 = \sum_{j,k\in\mathbb{Z}}\frac{2^{-j}}{4\pi^2}\left|\int_{\mathbb{R}}\hat{f}(\xi)\,\overline{\hat{\psi}(2^{-j}\xi)}\,e^{i2^{-j}k\xi}\,d\xi\right|^2$$

$$= \sum_{j\in\mathbb{Z}}\frac{2^j}{2\pi}\sum_{k\in\mathbb{Z}}\left|\int_I \hat{f}(2^j\mu)\frac{e^{ik\mu}}{\sqrt{2\pi}}\,d\mu\right|^2.$$

We now use the fact that the system $\{\frac{1}{\sqrt{2\pi}}e^{ik\mu} : k \in \mathbb{Z}\}$ is an orthonormal basis of $L^2(I)$ (a fact that is equivalent to the orthonormality of the same system on $[0,2\pi]$) to write

$$\sum_{j,k\in\mathbb{Z}}|<f,\psi_{j,k}>|^2 = \sum_{j\in\mathbb{Z}}\frac{2^j}{2\pi}\int_I|\hat{f}(2^j\mu)|^2\,d\mu$$

$$= \frac{1}{2\pi}\sum_{j\in\mathbb{Z}}\int_{\mathbb{R}}\chi_I(2^{-j}\xi)|\hat{f}(\xi)|^2\,d\xi = \frac{1}{2\pi}\|\hat{f}\|_2^2 = \|f\|_2^2,$$

since

$$\sum_{j\in\mathbb{Z}}\chi_I(2^{-j}\xi) = 1 \qquad \text{for a.e. } \xi \in \mathbb{R}.$$

This shows that ψ is an orthonormal wavelet for $L^2(\mathbb{R})$. This is related to the **Shannon wavelet** which will be described in Example C of Chapter 2.

1.2 Orthonormal bases generated by a single function; the Balian-Low theorem

Another way of producing an orthonormal basis from a single function involves translations and modulations. For example, a basis for $L^2(\mathbb{R})$ is the following: let $g = \chi_{[0,1]}$ and

$$g_{m,n}(x) = e^{2\pi i m x} g(x-n) \qquad \text{for } m,n \in \mathbb{Z}.$$

It is not difficult to see that $\{g_{m,n} : m,n \in \mathbb{Z}\}$ is an orthonormal basis for $L^2(\mathbb{R})$. D. Gabor ([Gab]) considered this type of system in 1946 and proposed its use for communication purposes. For a general $g \in L^2(\mathbb{R})$ the following theorem gives conditions that g must satisfy if the system $\{g_{m,n} : m,n \in \mathbb{Z}\}$ is an orthonormal basis.

THEOREM 2.1 (Balian-Low) *Suppose $g \in L^2(\mathbb{R})$ and*

$$g_{m,n}(x) = e^{2\pi i m x} g(x-n), \quad m,n \in \mathbb{Z}.$$

If $\{g_{m,n} : m,n \in \mathbb{Z}\}$ is an orthonormal basis for $L^2(\mathbb{R})$, then either

$$\int_{-\infty}^{\infty} x^2 |g(x)|^2 \, dx = \infty \qquad or \qquad \int_{-\infty}^{\infty} \xi^2 |\hat{g}(\xi)|^2 \, d\xi = \infty.$$

PROOF : We introduce the operators Q and P, defined on, say, the space $\mathcal{S}'$ of tempered distributions, given by

$$(Qf)(x) = xf(x) \qquad \text{and} \qquad (Pf)(x) = -if'(x).$$

The relevance of these operators to the theorem is that

$$\int_{-\infty}^{\infty} |Qg(x)|^2 dx = \int_{-\infty}^{\infty} x^2 |g(x)|^2 dx$$

and

$$\int_{-\infty}^{\infty} |Pg(x)|^2 \, dx = \frac{1}{2\pi} \int_{-\infty}^{\infty} \xi^2 |\hat{g}(\xi)|^2 \, d\xi,$$

where the last formula is a consequence of (1.3) and (1.4). Hence, we need to show that both (Qg) and (Pg) **cannot** belong to $L^2(\mathbb{R})$.

Suppose that both Qg and Pg belong to $L^2(\mathbb{R})$. We will show that this leads to a contradiction, and this proves the theorem. We claim that

$$<Qg, Pg> = \sum_{m,n \in \mathbb{Z}} <Qg, g_{m,n}> <g_{m,n}, Pg>, \qquad (2.2)$$

$$<Qg, g_{m,n}> = <g_{-m,-n}, Qg> \quad \text{for all} \ \ m, n \in \mathbb{Z}, \qquad (2.3)$$

and

$$<Pg, g_{m,n}> = <g_{-m,-n}, Pg> \quad \text{for all} \ \ m, n \in \mathbb{Z}. \qquad (2.4)$$

Equalities (2.2), (2.3) and (2.4) imply

$$<Qg, Pg> = <Pg, Qg> . \qquad (2.5)$$

But (2.5) cannot hold if Pg and Qg belong to $L^2(\mathbb{R})$. If this were the case we could apply the integration by parts formula (1.5) to obtain

$$<Qg, Pg> = \int_{-\infty}^{\infty} x g(x) \left\{ \overline{-i g'(x)} \right\} dx$$

$$= -i \int_{-\infty}^{\infty} \left\{ g(x) + x g'(x) \right\} \overline{g(x)} \, dx$$

$$= -i <g, g> + <Pg, Qg> .$$

Since $<g, g> = \|g\|_2^2 = \|g_{0,0}\|_2^2 = 1$ we obtain

$$<Qg, Pg> = -i + <Pg, Qg>,$$

which contradicts (2.5).

Hence, the theorem is proved if we establish (2.2), (2.3) and (2.4). Since $Qg, Pg \in L^2(\mathbb{R})$ and $\{g_{m,n} : m, n \in \mathbb{Z}\}$ is an orthonormal basis we have

$$<Qg, Pg> = < \sum_{m,n \in \mathbb{Z}} <Qg, g_{m,n}> g_{m,n}, Pg >$$

$$= \sum_{m,n \in \mathbb{Z}} <Qg, g_{m,n}> <g_{m,n}, Pg>,$$

which proves (2.2). To prove (2.3) observe that $n < g, g_{m,n} >= 0$ for all $m, n \in \mathbb{Z}$; this obviously holds for $n = 0$ and if $n \neq 0$, $g = g_{0,0}$ is orthogonal to $g_{m,n}$. Thus,

$$<Qg, g_{m,n}> = <Qg, g_{m,n}> -n<g, g_{m,n}>$$

$$= \int_{-\infty}^{\infty} g(x)(x - n)\, \overline{g(x - n)}\, e^{-2\pi i m x}\, dx$$

$$= \int_{-\infty}^{\infty} g(y + n)\, y\, \overline{g(y)}\, e^{-2\pi i m(y+n)}\, dy = <g_{-m,-n}, Qg>,$$

which gives us (2.3). To prove (2.4) we use the integration by parts formula (1.5) to obtain

$$<Pg, g_{m,n}> = -i \int_{-\infty}^{\infty} g'(x)\, \overline{g(x - n)}\, e^{-2\pi i m x}\, dx$$

$$= i \int_{-\infty}^{\infty} g(x)\{-2\pi i m\, \overline{g(x - n)} + \overline{g'(x - n)}\}\, e^{-2\pi i m x}\, dx$$

$$= 2\pi m \delta_{m,0} \delta_{0,n} + \int_{-\infty}^{\infty} g(y + n)\{\overline{-i g'(y)}\}\, e^{-2\pi i m y}\, dy$$

$$= <g_{-m,-n}, Pg>.$$

■

EXAMPLE C: For $g = \chi_{[0,1)}$, $\{g_{m,n} : m, n \in \mathbb{Z}\}$, as we have seen, is an orthonormal basis of $L^2(\mathbb{R})$; in this case the first integral in the announcement of the Balian-Low theorem is finite, but the second is infinite since

$$\xi^2 |(\chi_{[0,1)})^{\wedge}(\xi)|^2 = [2 \sin(\tfrac{\xi}{2})]^2.$$

EXAMPLE D: For $g(x) = \frac{\sin(\pi x)}{\pi x} \equiv \mathrm{sinc}(x)$, $\{g_{m,n} : m, n \in \mathbb{Z}\}$ is an orthonormal basis of $L^2(\mathbb{R})$; observe that

$$(\chi_{[0,1)})^{\wedge}(\xi) = e^{-i\frac{\xi}{2}} \frac{\sin(\xi/2)}{(\xi/2)} = e^{-i\frac{\xi}{2}} \mathrm{sinc}(\tfrac{\xi}{2\pi}).$$

In this case the first integral in the announcement of the Balian-Low theorem is infinite.

If $g \in L^2(\mathbb{R})$ and

$$g_{m,n}(x) = e^{imw_o x} g(x - nt_o) \qquad (2.6)$$

with $w_o t_o = 2\pi$, Theorem 2.1 is still true; to see this observe that the operator U defined by $Ug(x) = (2\pi w_o^{-1})^{\frac{1}{2}} g(2\pi w_o^{-1} x)$ is unitary in $L^2(\mathbb{R})$ and

$$Ug_{m,n}(x) = e^{2\pi i m x} Ug(x - n)$$

since $2\pi w_o^{-1} = t_o$. This theorem tells us that if $w_o t_o = 2\pi$, the basis given by (2.6) does not have good time and frequency localization simultaneously.

In particular, if $b(x)$ is sufficiently smooth and compactly supported the Balian-Low theorem tells us that the system

$$\left\{ b_m(x) \right\}_{m \in \mathbb{Z}} = \left\{ e^{2\pi i m x} b(x) \right\}_{m \in \mathbb{Z}}$$

will not produce an orthonormal basis by translating the elements of the system by the integers. This is easy to see due to the decay at infinity of the Fourier transform of b, that is a consequence of the smoothness of b.

If we consider a more local situation, however, we can find a smooth and compactly supported "bell" function $b(x)$ for which

$$\left\{ b_m(x) \right\}_{m \in \mathbb{Z}} = \left\{ e^{2\pi i m x} b(x) \right\}_{m \in \mathbb{Z}}$$

is an orthonormal system. For example, suppose that b is a function defined on $\mathbb{R}$ with $\operatorname{supp}(b) \subseteq [-\varepsilon, 1 + \varepsilon']$, where $\varepsilon + \varepsilon' \leq 1$, $\varepsilon, \varepsilon' > 0$ and $b(x) \geq 0$. It is easy to find conditions on b so that $\{ b_m : m \in \mathbb{Z} \}$ is an orthonormal system. The idea is to use a "folding argument" to write the orthonormal relations $<e^{2\pi i m (\cdot)} b, e^{2\pi i n (\cdot)} b> = \delta_{m,n}$ on the interval $[0, 1]$:

$$\delta_{m,n} = <e^{2\pi i m (\cdot)} b, e^{2\pi i n (\cdot)} b> = \int_{-\varepsilon}^{1+\varepsilon'} b^2(x)\, e^{2\pi i (m-n) x}\, dx$$

$$= \left(\int_{-\varepsilon}^0 + \int_0^{\varepsilon'} + \int_{\varepsilon'}^{1-\varepsilon} + \int_{1-\varepsilon}^1 + \int_1^{1+\varepsilon'} \right) \left\{ b^2(x)\, e^{2\pi i (m-n) x}\, dx \right\}.$$

In the first integral we perform the change of variables $y = 1 + x$; in the last integral we use the change of variables $y = x - 1$. We therefore obtain

$$\delta_{m,n} = \int_0^{\varepsilon'} \left[b^2(x) + b^2(1 + x) \right] e^{2\pi i (m-n) x}\, dx$$

$$+ \int_{\varepsilon'}^{1-\varepsilon} b^2(x)\, e^{2\pi i (m-n)x}\, dx$$

$$+ \int_{1-\varepsilon}^{1} \left[b^2(x) + b^2(x-1) \right] e^{2\pi i (m-n)x}\, dx.$$

That is, the function f having values $b^2(x) + b^2(1+x)$ on $[0, \varepsilon']$, $b^2(x)$ on $[\varepsilon', 1-\varepsilon]$ and $b^2(x) + b^2(x-1)$ on $[1-\varepsilon, 1]$ has Fourier coefficients $\hat{f}(k) = 0$ if $k \neq 0$ and $\hat{f}(0) = 1$. It follows easily that, if these orthonormal relations are to hold, b must satisfy

$$\left. \begin{aligned}
b^2(x) + b^2(1+x) &= 1 && \text{if } x \in [0, \varepsilon'], \\
b^2(x) &= 1 && \text{if } x \in [\varepsilon', 1-\varepsilon], \\
b^2(x) + b^2(x-1) &= 1 && \text{if } x \in [1-\varepsilon, 1].
\end{aligned} \right\} \qquad (2.7)$$

It follows that (2.7) is a necessary and sufficient condition for

$$\left\{ e^{2\pi i m x} b(x) \right\}_{m \in \mathbb{Z}}$$

to be an orthonormal system in $L^2(\mathbb{R})$. The Balian-Low theorem tells us that if we choose such a smooth bell function, translations by integers will not produce an orthonormal basis for $L^2(\mathbb{R})$. In the next two sections we shall show that if the exponentials $e^{2\pi i m x}$ are replaced by appropriate sines and cosines we can obtain such bases.

1.3   Smooth projections on $L^2(\mathbb{R})$

We will show that we can construct a smooth "bell" function associated with the interval $[0, 1]$, in such a way that the system

$$\sqrt{2} b(x - k) \sin\left(\tfrac{2j+1}{2} \pi (x - k) \right), \qquad j, k \in \mathbb{Z},$$

is an orthonormal basis for $L^2(\mathbb{R})$. In fact, we will see that for each fixed $k \in \mathbb{Z}$, the family

$$\left\{ \sqrt{2} b(x - k) \sin\left(\tfrac{2j+1}{2} \pi (x - k) \right) : j \in \mathbb{Z} \right\}$$

is an orthonormal basis for a closed subspace H_k of $L^2(\mathbb{R})$, and that $L^2(\mathbb{R})$ is the direct sum of these H_k. More generally, we shall construct smooth "bell" functions associated with a general finite interval $I = [\alpha, \beta)$ that can be multiplied by appropriate sines and cosines to obtain an orthonormal basis of a subspace H_I of $L^2(\mathbb{R})$ in such a way that if we have

$$-\infty < \cdots < \alpha_{k-1} < \alpha_k < \alpha_{k+1} < \cdots < \infty,$$

the H_{I_k}'s ($I_k = [\alpha_k, \beta_k)$) form a complete system of mutually orthogonal subspaces of $L^2(\mathbb{R})$. This does not have the form of a wavelet system, but it can be used for analyzing general functions in L^2 and, moreover, we will see how it can be used for constructing wavelets.

We start with the special case $I = [0, \infty)$ and our goal is to construct a smooth "bell" function that "approximates" $\chi_{[0,\infty)}$. Since any projection is idempotent, multiplication by a function gives a projection only if the function takes the values 0 or 1 almost everywhere on $\mathbb{R}$; this shows that the projection we are looking for cannot be given simply by multiplication by a smooth function.

Let us pose the problem of finding a non-negative bounded function $\rho \in C^\infty$ such that supp $(\rho) \subseteq [-\varepsilon, \infty)$ for an $\varepsilon > 0$ and, like $\chi_{[0,\infty)}$, satisfies $\rho(x) + \rho(-x) = 1$, $x \neq 0$, and a real-valued function t so that

$$(Pf)(x) = \rho(x)f(x) + t(x)f(-x)$$

is a projection. A simple calculation, based on the fact that P has to be idempotent and self-adjoint, leads us to the equality $t(x) = \pm\sqrt{\rho(x)\rho(-x)}$. Writing $s = \sqrt{\rho}$, we are led to the formula

$$(Pf)(x) = s(x)[s(x)f(x) \pm s(-x)f(-x)].$$

In fact, more generally, no longer assuming s to be real valued, if we introduce the operator $P = P_{0,\varepsilon}$ defined by

$$(Pf)(x) \equiv (P_{0,\varepsilon}f)(x) = \overline{s(x)}\left[s(x)f(x) \pm s(-x)f(-x)\right] \qquad (3.1)$$

with

$$|s(x)|^2 + |s(-x)|^2 = 1, \qquad (3.2)$$

it is easy to show that it is an orthogonal projection. To see this it is enough to show that P is idempotent ($P^2 = P$) and self-adjoint ($P^* = P$). In fact,

$$(P^2 f)(x) = \overline{s(x)}\left[s(x)(Pf)(x) \pm s(-x)(Pf)(-x)\right]$$

$$= \overline{s(x)} \left[|s(x)|^2 s(x)f(x) \pm |s(x)|^2 s(-x)f(-x) \right.$$
$$\left. \pm |s(-x)|^2 s(-x)f(-x) + |s(-x)|^2 s(x)f(x) \right]$$
$$= \overline{s(x)} \left[s(x)f(x) \pm s(-x)f(-x) \right] = (Pf)(x),$$

and

$$<P^*f,g> = <f,Pg> = \int_{-\infty}^{\infty} f(x)s(x) \overline{[s(x)g(x) \pm s(-x)g(-x)]} \, dx$$

$$= \int_{-\infty}^{\infty} \left(f(x)s(x)\overline{s(x)g(x)} \pm f(-x)s(-x)\overline{s(x)g(x)} \right) dx$$

$$= \int_{-\infty}^{\infty} \overline{s(x)} \left[s(x)f(x) \pm s(-x)f(-x) \right] \overline{g(x)} \, dx = <Pf,g>.$$

For the moment we shall suppose that s is a real-valued function. Let us construct a smooth function $s(x)$ that satisfies (3.2). Choose ψ to be an even C^∞ function on $\mathbb{R}$ supported on $[-\varepsilon, \varepsilon]$, $\varepsilon > 0$, such that $\int_{-\varepsilon}^{\varepsilon} \psi(x) \, dx = \pi/2$. Let $\theta(x) = \int_{-\infty}^{x} \psi(t) \, dt$ and observe that

$$\theta(x) + \theta(-x) = \int_{-\infty}^{x} \psi(t) \, dt + \int_{-\infty}^{-x} \psi(t) \, dt$$

$$= \int_{-\infty}^{x} \psi(t) \, dt + \int_{x}^{\infty} \psi(-t) \, dt$$

$$= \int_{-\infty}^{x} \psi(t) \, dt + \int_{x}^{\infty} \psi(t) \, dt = \frac{\pi}{2}.$$

Putting $s(x) \equiv s_\varepsilon(x) = \sin(\theta(x))$ and $c(x) \equiv c_\varepsilon(x) = \cos(\theta(x))$ we have $s(-x) = \sin(\theta(-x)) = \sin(\frac{\pi}{2} - \theta(x)) = \cos(\theta(x)) = c(x)$. Hence,

$$s^2(x) + s^2(-x) = \sin^2(\theta(x)) + \cos^2(\theta(x)) = 1,$$

and (3.2) is satisfied.

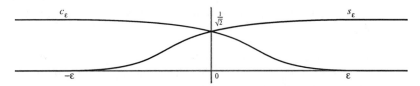

Figure 1.1: The functions s_ε and c_ε.

We have thus obtained two projections, $P^+_{0,\varepsilon}$ and $P^-_{0,\varepsilon}$, associated with the interval $[0,\infty)$ corresponding to the choice $+$ or $-$ in (3.1). We also have the analogous projections

$$(P^{0,\varepsilon'}_{+,-} f)(x) = c_{\varepsilon'}(x)\big[c_{\varepsilon'}(x)f(x) \pm c_{\varepsilon'}(-x)f(-x)\big]$$

associated with the interval $(-\infty, 0]$, where $\varepsilon' > 0$.

Now we wish to construct smooth projections on a general interval

$$I = [\alpha, \beta], \qquad -\infty < \alpha < \beta < \infty.$$

We do this by using the translation operator $\tau_h f(x) = f(x-h)$, introduced at the beginning of section 1.2, and defining

$$P_\alpha = \tau_\alpha P_0 \tau_{-\alpha} \qquad \text{and} \qquad P^\beta = \tau_\beta P^0 \tau_{-\beta},$$

where we have suppressed, for the moment, the subindices and superscripts $\varepsilon, \varepsilon', +, -$. Each one of these operators is idempotent and self-adjoint since P_0 and P^0 are orthogonal projections. Thus P_α and P^β are also orthogonal projections. Using (3.1), we obtain the formulae

$$(P_\alpha f)(x) = (\tau_\alpha P_0 \tau_{-\alpha} f)(x) = (P_0 \tau_{-\alpha} f)(x - \alpha)$$

$$= s_\varepsilon(x-\alpha)\big[s_\varepsilon(x-\alpha)f(x) \pm s_\varepsilon(\alpha - x)f(2\alpha - x)\big] \quad (3.3)$$

and, similarly,

$$(P^\beta f)(x) = (\tau_\beta P^0 \tau_{-\beta} f)(x) = (P^0 \tau_{-\beta} f)(x - \beta)$$

$$= c_{\varepsilon'}(x-\beta)\big[c_{\varepsilon'}(x-\beta)f(x) \pm c_{\varepsilon'}(\beta - x)f(2\beta - x)\big]. \quad (3.4)$$

Observe that $2\alpha - x$ and x are symmetric with respect to α. (That is, they lie on opposite sides and are equidistant to α.) This motivates the following definition. We say that a function g is **even with respect to** $\gamma \in \mathbb{R}$ if $g(x) = g(2\gamma - x)$ for all $x \in \mathbb{R}$.

If g is an even function with respect to α, it is easily seen that $P_\alpha(gf) = g(P_\alpha f)$ when $g \in L^\infty(\mathbb{R})$ and $f \in L^2(\mathbb{R})$; that is, multiplication by g commutes with P_α. Similarly, if g is even with respect to β, from (3.4) we see that $P^\beta(gf) = g(P^\beta f)$.

For a general interval $I = [\alpha, \beta]$ we choose $\varepsilon, \varepsilon' > 0$ such that $\alpha + \varepsilon \le \beta - \varepsilon'$ and observe that

$$P_\alpha P^\beta f = \chi_{[\alpha-\varepsilon,\alpha+\varepsilon]} P_\alpha f + \chi_{[\alpha+\varepsilon,\beta-\varepsilon']} f + \chi_{[\beta-\varepsilon',\beta+\varepsilon']} P^\beta f$$

$$= P^\beta P_\alpha f. \tag{3.5}$$

To obtain this, observe that

$$P_\alpha f = P_\alpha \chi_{[\alpha-\varepsilon,\alpha+\varepsilon]} f + P_\alpha \chi_{[\alpha+\varepsilon,\infty)} f$$

$$= \chi_{[\alpha-\varepsilon,\alpha+\varepsilon]} P_\alpha f + \chi_{[\alpha+\varepsilon,\infty)} f, \tag{3.6}$$

where we have used the fact that $\chi_{[\alpha-\varepsilon,\alpha+\varepsilon]}$ is even with respect to α, and, hence, commutes with P_α. Similarly, we have

$$P^\beta f = P^\beta \chi_{(-\infty,\beta-\varepsilon']} f + P^\beta \chi_{[\beta-\varepsilon',\beta+\varepsilon']} f$$

$$= \chi_{(-\infty,\beta-\varepsilon']} f + \chi_{[\beta-\varepsilon',\beta+\varepsilon']} P^\beta f. \tag{3.7}$$

Now apply P^β to the first equality and P_α to the second to obtain the desired result.

Since P_α and P^β commute, the operator

$$P_I f \equiv P_{[\alpha,\beta]} f = P_\alpha P^\beta f = P^\beta P_\alpha f \tag{3.8}$$

is a bounded, orthogonal projection on $L^2(\mathbb{R})$.

Observe that $P_I \equiv P_{[\alpha,\beta]}$ depends on $\alpha, \beta, \varepsilon, \varepsilon'$ and the signs we choose at α and β. Thus, if $\alpha, \beta, \varepsilon$ and ε' are fixed, the choice of signs gives us four projections.

An expression for $P_I \equiv P_{[\alpha,\beta]}$ that is different from the one given in (3.8) is obtained by introducing the function

$$b(x) = s_\varepsilon(x - \alpha)\, c_{\varepsilon'}(x - \beta).$$

We refer to $b = b_I$ as a "bell" function associated with the interval $[\alpha, \beta]$. Observe that b depends on $\alpha, \beta, \varepsilon$ and ε'. By translating the graphs of s_ε and c_ε given in Figure 1.1 we obtain the graph of a bell function associated with $[\alpha, \beta]$:

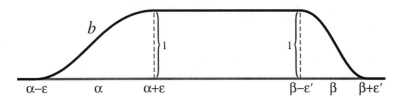

Figure 1.2: *Graph of the bell function b associated with* $[\alpha, \beta]$.

It is easy to prove the following basic properties of $b(x)$:

i)   supp $(b) \subseteq [\alpha - \varepsilon, \beta + \varepsilon']$;

on $[\alpha - \varepsilon, \alpha + \varepsilon]$

     ii) $b(x) = s_\varepsilon(x - \alpha)$,

     iii) $b(2\alpha - x) = s_\varepsilon(\alpha - x) = c_\varepsilon(x - \alpha)$,

     iv) $b^2(x) + b^2(2\alpha - x) = 1$;

v)   supp $(b(\cdot)b(2\alpha - \cdot)) \subseteq [\alpha - \varepsilon, \alpha + \varepsilon]$;

vi)   on $[\alpha + \varepsilon, \beta - \varepsilon']$, $b(x) = 1$;

on $[\beta - \varepsilon', \beta + \varepsilon']$

     vii) $b(x) = c_{\varepsilon'}(x - \beta)$,

     viii) $b(2\beta - x) = c_{\varepsilon'}(\beta - x) = s_{\varepsilon'}(x - \beta)$,

     ix) $b^2(x) + b^2(2\beta - x) = 1$;

x)   supp $(b(\cdot)b(2\beta - \cdot)) \subseteq [\beta - \varepsilon', \beta + \varepsilon']$;

xi) $b^2(x) + b^2(2\alpha - x) + b^2(2\beta - x) = 1$ on   supp (b).

(3.9)

Not all these properties are independent. For example, iv) follows from ii) and iii). The reader will find it instructive to compare these conditions with (2.7).

Using (3.5), the definition of P_α and P^β given in (3.3) and (3.4) and these properties, we can easily derive the following new formula for P_I in terms of the bell function b:

$$(P_I f)(x) = b(x)\{b(x)f(x) \pm b(2\alpha - x)f(2\alpha - x) \pm b(2\beta - x)f(2\beta - x)\}. \quad (3.10)$$

Observe that we have four choices for such a projection. The choice of $\pm$ associated with α is referred to as the **polarity** of $P_{[\alpha, \beta]}$ at α, and the choice of $\pm$ associated with β is referred to as the **polarity** of $P_{[\alpha, \beta]}$ at

β. Thus, if we choose "+" before the second summand in the bracket in (3.10), we say that the projection has **positive polarity at** α.

DEFINITION 3.11 *Suppose $I = [\alpha, \beta]$ and $J = [\beta, \gamma]$ are adjacent; we say that they have* **compatible bell functions** b_I *and* b_J *if*

$$\alpha - \varepsilon < \alpha < \alpha + \varepsilon \leq \beta - \varepsilon' < \beta < \beta + \varepsilon' \leq \gamma - \varepsilon'' < \gamma < \gamma + \varepsilon''$$

and

$$b_I = s_\varepsilon(x - \alpha)\, c_{\varepsilon'}(x - \beta), \qquad b_J = s_{\varepsilon'}(x - \beta)\, c_{\varepsilon''}(x - \gamma).$$

If $I = [\alpha, \beta]$ and $J = [\beta, \gamma]$ are intervals with compatible bell functions, we have

$$b_I(x) = b_J(2\beta - x), \qquad \text{if } x \in [\beta - \varepsilon', \beta + \varepsilon']; \qquad (3.12)$$

$$b_I^2(x) + b_J^2(x) = 1, \qquad \text{if } x \in [\beta - \varepsilon', \beta + \varepsilon']; \qquad (3.13)$$

$$b_I^2(x) + b_J^2(x) = b_{I \cup J}^2(x) \qquad \text{for all } x \in \mathbb{R}. \qquad (3.14)$$

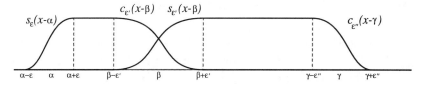

Figure 1.3: *Compatible bell functions on $[\alpha, \beta]$ and $[\beta, \gamma]$.*

These properties follow easily from (3.9). The next result establishes the main property of these projections that will allow us to decompose $L^2(\mathbb{R})$ as a direct sum of orthogonal subspaces.

THEOREM 3.15 *Let $I = [\alpha, \beta]$ and $J = [\beta, \gamma]$ be adjacent intervals with compatible bell functions and suppose P_I and P_J have opposite polarities at β. Then*

$$P_I + P_J = P_{I \cup J}, \qquad (3.16)$$

$$P_I P_J = 0 = P_J P_I. \qquad (3.17)$$

PROOF : According to (3.5), letting I also denote the identity operator,

$$P_I + P_J = \chi_{[\alpha-\varepsilon,\alpha+\varepsilon]} P_\alpha + \chi_{[\alpha+\varepsilon,\beta-\varepsilon']} I + \chi_{[\beta-\varepsilon',\beta+\varepsilon']} P^\beta$$

$$+ \chi_{[\beta-\varepsilon',\beta+\varepsilon']} P_\beta + \chi_{[\beta+\varepsilon',\gamma-\varepsilon'']} I + \chi_{[\gamma-\varepsilon'',\gamma+\varepsilon'']} P^\gamma$$

with P^β and P_β chosen with opposite polarity at β. This last property allows us to prove that the two middle terms in the above formula add up to $\chi_{[\beta-\varepsilon',\beta+\varepsilon']} I$, and, hence, the result equals $P_{I \cup J}$ according to (3.5). Thus, $P_I + P_J = P_{I \cup J}$.

Formula (3.17) is a consequence of a general result about projections. In fact, if P and Q are orthogonal projections on a Hilbert space such that $P + Q$ is an orthogonal projection, then $PQ = QP = 0$. To see this, observe that $(P + Q)^2 = P + Q$ implies $PQ = -QP$; from this we deduce $PQ = P^2 Q = P(PQ) = -P(QP) = QP^2 = QP$; these two results give us $PQ = QP = 0$.

∎

If H is a Hilbert space and $\{H_k : k \in \mathbb{Z}\}$ is a sequence of mutually orthogonal closed subspaces we let

$$V = \bigoplus_{k=-\infty}^{\infty} H_k$$

denote the closed subspace consisting of all $f = \sum_{k \in \mathbb{Z}} f_k$ with $f_k \in H_k$ and $\sum_{k \in \mathbb{Z}} \|f_k\|^2 < \infty$. We call V the **orthogonal direct sum of the spaces** H_k. If the family consist of two spaces H_1 and H_2 we write $H_1 \oplus H_2$.

The above theorem allows us to decompose $L^2(\mathbb{R})$ as an orthogonal direct sum

$$L^2(\mathbb{R}) = \bigoplus_{k=-\infty}^{\infty} H_k, \tag{3.18}$$

where $H_k = P_k(L^2(\mathbb{R}))$, $P_k = P_{[\alpha_k,\alpha_{k+1}]}$ with

$$-\infty < \cdots < \alpha_{k-1} < \alpha_k < \alpha_{k+1} < \cdots < \infty;$$

moreover, adjacent intervals, $[\alpha_k, \alpha_{k+1}]$ and $[\alpha_{k+1}, \alpha_{k+2}]$, have compatible bell functions, and P_k and P_{k+1} have opposite polarity at α_{k+1}. The orthogonality of the H_k's follows from (3.17); formula (3.16) gives us the decomposition of $L^2(\mathbb{R})$.

Another orthogonal decomposition of $L^2(\mathbb{R})$, that we shall show is pertinent to wavelets, can be achieved as follows. Let $I = [\pi, 2\pi]$ and $J = -I = [-2\pi, -\pi]$. Choose $\varepsilon > 0$ such that $0 < \varepsilon \le \frac{1}{3}\pi$ and $\varepsilon' = 2\varepsilon$; put $I_k = 2^k I$, $J_k = 2^k J$ for $k \in \mathbb{Z}$. Then, associating $\varepsilon_k = 2^k \varepsilon$, $\varepsilon_{k+1} = 2\varepsilon_k$ with I_k, the adjacent intervals I_k and I_{k+1} have compatible bell functions (similarly for J_k and J_{k+1}) and we have

$$L^2(\mathbb{R}) = \left\{ \bigoplus_{k=-\infty}^{\infty} H_{J_k} \right\} \oplus \left\{ \bigoplus_{k=-\infty}^{\infty} H_{I_k} \right\}, \tag{3.19}$$

if we choose the appropriate polarities for P_{I_k}, P_{J_k} and denote the images $P_{I_k}(L^2(\mathbb{R}))$ and $P_{J_k}(L^2(\mathbb{R}))$ by H_{I_k} and H_{J_k}.

Let us now characterize the subspace $H_I = P_I(L^2(\mathbb{R}))$. We say that f is **even** with respect to α on $[\alpha - \varepsilon, \alpha + \varepsilon]$ if $f(2\alpha - x) = f(x)$ on this interval. Similarly, a function g is said to be **odd** with respect to α on $[\alpha - \varepsilon, \alpha + \varepsilon]$ if $g(2\alpha - x) = -g(x)$ on this interval.

By (3.10) we can write

$$(P_I f)(x) = b_I(x) S(x),$$

where $S(x) = b_I(x)f(x) \pm b_I(2\alpha - x)f(2\alpha - x) \pm b_I(2\beta - x)f(2\beta - x)$. Observe that there are four choices for $S(x)$ depending on the signs considered, which give us the four functions $S_+^+(x)$, $S_-^+(x)$, $S_+^-(x)$, and $S_-^-(x)$. S_+^+ is even with respect to α on $[\alpha - \varepsilon, \alpha + \varepsilon]$ and even with respect to β on $[\beta - \varepsilon', \beta + \varepsilon']$; S_-^+ is odd with respect to α on $[\alpha - \varepsilon, \alpha + \varepsilon]$ and even with respect to β on $[\beta - \varepsilon', \beta + \varepsilon']$. The obvious similar statements apply to S_+^- and S_-^-.

THEOREM 3.20 *Let $I = [\alpha, \beta]$; then $f \in H_I = P_I(L^2(\mathbb{R}))$ if and only if $f = b_I S$, where $S \in L^2(\mathbb{R})$, b_I is the bell function associated with I, and S is even or odd on $[\alpha - \varepsilon, \alpha + \varepsilon]$ according to the choice of polarity at α, and even or odd on $[\beta - \varepsilon', \beta + \varepsilon']$ according to the choice of polarity at β.*

PROOF : If $f \in H_I$, there exists $g \in L^2(\mathbb{R})$ such that $f = P_I g$; then $f = P_I g = b_I S$ by (3.10) where, clearly, S has the same polarity at α and β as P_I. Observe that $S \in L^2(\mathbb{R})$.

Suppose now that $f = bS$ with $S \in L^2(\mathbb{R})$, locally even at α and locally odd at β. (The other cases are handled similarly.) It is enough to show that if P_I has the same polarity as S, then $P_I(bS) = bS$, since then $P_I(f) = P_I(bS) = bS = f$. To show $P_I(bS) = bS$ we use (3.10), and the properties of b_I given in iv), vi) and ix) of (3.9). We leave the details to the reader. ∎

1.4 Local sine and cosine bases and the construction of some wavelets

In this section we shall introduce orthonormal bases for the subspaces $H_I = P_I(L^2(\mathbb{R}))$, where P_I are the projection operators defined in the previous section. As we shall see, these bases are closely allied to certain trigonometric systems and consistent with the polarity of P_I. That is, if P_I is chosen with negative polarity at the left endpoint of the interval I and with positive polarity at the right endpoint of I, the elements of the basis will be locally odd at the left endpoint and locally even at the right endpoint. In addition, the bases for these subspaces will be expressed in terms of trigonometric functions and the associated bell function. (As explained at the beginning of the last section.)

Let us first consider I to be the interval $[0, 1]$, and suppose that P_I has polarities $-$ and $+$ at 0 and 1, respectively. (We tacitly assume that P_I is associated with positive ε and ε' such that $\varepsilon + \varepsilon' \leq 1$. As we did on several occasions in the previous section, we do not indicate the dependence of P_I on ε and ε'.) Let $f \in L^2([0, 1])$ and extend f to a function F on $[-2, 2]$ so that F is even with respect to 1 and odd with respect to 0; this is consistent with the choice of the polarities for P_I. (See Figure 1.4 below.)

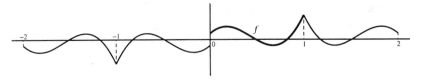

Figure 1.4: Extension F of f to $[-2, 2]$.

On $[-2, 2]$ we have the usual cosine and sine basis

$$\left\{ \frac{1}{2}, \frac{1}{\sqrt{2}} \sin \frac{\pi k x}{2}, \frac{1}{\sqrt{2}} \cos \frac{\pi \ell x}{2} \right\}, \quad k, \ell = 1, 2, \cdots$$

Since F is odd on $[-2,2]$, the cosines are not involved in the Fourier expansion of F. Moreover, the functions $\sin(\frac{2k+1}{2}\pi x)$, $k = 0,1,2,\cdots$, are even with respect to 1 and the functions $\sin(k\pi x)$, $k = 1,2,3,\cdots$, are odd with respect to 1. Therefore, we only need $\sin(\frac{2k+1}{2}\pi x)$, $k = 0,1,2,\cdots$, to represent F. That is,

$$F(x) = \sum_{k=0}^{\infty} c_k \sin\left(\tfrac{2k+1}{2}\pi x\right),$$

where

$$c_k = \frac{1}{2}\int_{-2}^{2} F(x)\sin\left(\tfrac{2k+1}{2}\pi x\right)\,dx,$$

and the above series converges in the norm of $L^2([-2,2])$. Observe that the convergence is also true in the pointwise almost everywhere sense by a deep theorem of L. Carleson concerning almost everywhere convergence of Fourier series. (See [Car1].)

If we restrict ourselves to $[0,1]$, and use the appropriate normalization, we find that $\{\sqrt{2}\sin(\frac{2k+1}{2}\pi x),\ k = 0,1,2,\cdots\}$ is an orthonormal basis for $L^2([0,1])$ with polarities of its elements at 0 and 1 that match the ones of P_I. This provides us with the proof of the first part of the following result.

THEOREM 4.1 *Each one of the systems*

  i) $\{\sqrt{2}\sin(\frac{2k+1}{2}\pi x)\}$, $k = 0,1,2,\cdots$

  ii) $\{\sqrt{2}\sin(k\pi x)\}$, $k = 1,2,3,\cdots$

  iii) $\{\sqrt{2}\cos(\frac{2k+1}{2}\pi x)\}$, $k = 0,1,2,\cdots$

  iv) $\{1,\ \sqrt{2}\cos(k\pi x)\}$, $k = 1,2,3,\cdots$

is an orthonormal basis of $L^2([0,1])$ and the polarities are $(-,+)$ *for* i), $(-,-)$ *for* ii), $(+,-)$ *for* iii) *and* $(+,+)$ *for* iv).

We have already seen how to obtain i); the other three statements are obtained in a similar way.

We use this result to obtain the desired orthonormal bases for $H_I = P_I(L^2(\mathbb{R}))$ when $I = [0,1]$. Let $\varepsilon,\varepsilon' > 0$ with $\varepsilon + \varepsilon' \le 1$ and consider the associated bell function $b(x) = s_\varepsilon(x)c_{\varepsilon'}(x-1)$. Suppose, as before, that the polarities of P_I are $-$ at 0 and $+$ at 1. Thus, (3.10) in this case becomes

$$P_I f(x) = b(x)\{b(x)f(x) - b(-x)f(-x) + b(2-x)f(2-x)\} = b(x)S(x).$$

The function $S(x)$ is odd with respect to 0 and even with respect to 1 because of the properties of b (see (3.9)); hence, S has the right polarity to be represented by the orthonormal basis i) in Theorem (4.1). Therefore, we can write

$$S(x) = \sqrt{2} \sum_{k=0}^{\infty} c_k \sin(\tfrac{2k+1}{2}\pi x),$$

where

$$c_k = \sqrt{2} \int_0^1 S(x) \sin(\tfrac{2k+1}{2}\pi x)\, dx,$$

where the convergence is in $L^2([0,1])$ and, by Carleson's theorem, almost everywhere. Since S and the sine functions we are using have the same polarities at 0 and 1, the expansion is valid on $[-\varepsilon, 1+\varepsilon']$ in the L^2-sense and almost everywhere. Multiplying by $b(x)$ we obtain

$$(P_I f)(x) = b(x)S(x) = \sum_{k=0}^{\infty} c_k \sqrt{2}\, b(x) \sin(\tfrac{2k+1}{2}\pi x)$$

and the convergence is valid in $L^2([-\varepsilon, 1+\varepsilon'])$ and almost everywhere, since b is bounded. This shows that the system

$$\{\sqrt{2}\, b(x) \sin(\tfrac{2k+1}{2}\pi x)\}, \quad k = 0,1,2,\cdots \qquad (4.2)$$

is complete in $H_I = P_I(L^2(\mathbb{R}))$ when P_I has the polarities $(-,+)$. To show that this system is an orthonormal basis, we need to prove the orthonormality relations; if $e_k = \sin(\tfrac{2k+1}{2}\pi x)$, $k = 0,1,2,\cdots$, we have to show

$$2 \int_{-\varepsilon}^{1+\varepsilon'} b^2(x) e_k(x) e_\ell(x)\, dx = \delta_{k\ell}, \quad k,\ell = 0,1,2,\cdots.$$

Since e_k is locally odd with respect to 0, a change of variables together with property iv) of (3.9) gives us

$$\int_{-\varepsilon}^{\varepsilon} b^2(x) e_k(x) e_\ell(x)\, dx = \int_0^{\varepsilon} e_k(x) e_\ell(x)\, dx.$$

Similarly, using that e_k is locally even at 1 and property ix) of (3.9), a change of variables gives us

$$\int_{1-\varepsilon'}^{1+\varepsilon'} b^2(x) e_k(x) e_\ell(x)\, dx = \int_{1-\varepsilon'}^{1} e_k(x) e_\ell(x)\, dx.$$

Finally, since $b \equiv 1$ on $[\varepsilon, 1 - \varepsilon']$, the orthonormality of (4.2) on $[-\varepsilon, 1 + \varepsilon']$ is equivalent to the orthonormality of the system i) on the interval $[0, 1]$ given in Theorem (4.1). Since we know that this is true, we have proved the desired result.

Performing the appropriate translations and dilations and taking into account the different types of polarity, we obtain the following result for the spaces $H_I = P_I(L^2(\mathbb{R}))$ when $I = [\alpha, \beta]$ is an arbitrary finite interval:

THEOREM 4.3 *If $P_I = P_{[\alpha,\beta]}$ has negative polarity at α and positive polarity at β, then*

i) $\left\{ \sqrt{\frac{2}{|I|}}\, b_I(x) \sin\left(\frac{2k+1}{2} \frac{\pi}{|I|}(x - \alpha)\right) \right\}, \quad k = 0, 1, 2, \cdots,$

is an orthonormal basis for $H_I = P_I(L^2(\mathbb{R}))$. If the polarities are $(-, -)$, $(+, -)$ and $(+, +)$ at (α, β) the same is true, respectively, for

ii) $\left\{ \sqrt{\frac{2}{|I|}}\, b_I(x) \sin\left(k \frac{\pi}{|I|}(x - \alpha)\right) \right\}, \quad k = 1, 2, 3, \cdots,$

iii) $\left\{ \sqrt{\frac{2}{|I|}}\, b_I(x) \cos\left(\frac{2k+1}{2} \frac{\pi}{|I|}(x - \alpha)\right) \right\}, \quad k = 0, 1, 2, \cdots,$

iv) $\left\{ \sqrt{\frac{1}{|I|}}\, b_I(x), \ \sqrt{\frac{2}{|I|}}\, b_I(x) \cos\left(k \frac{\pi}{|I|}(x - \alpha)\right) \right\}, \quad k = 1, 2, 3, \cdots.$

This theorem, together with the orthogonal decomposition (3.18), can be used to obtain bases for $L^2(\mathbb{R})$. Choose a strictly increasing sequence of real numbers $\{\alpha_j\}_{j \in \mathbb{Z}}$ such that $\lim_{j \to \infty} \alpha_j = \infty$ and $\lim_{j \to -\infty} \alpha_j = -\infty$; let $\{\varepsilon_j\}_{j \in \mathbb{Z}}$ be a sequence of positive real numbers such that

$$\varepsilon_j + \varepsilon_{j+1} \le \alpha_{j+1} - \alpha_j \equiv \ell_j \qquad \text{for all } j \in \mathbb{Z}.$$

If we choose the polarities $(-, +)$ for each $P_j = P_{[\alpha_j, \alpha_{j+1}]}$ we obtain that the system

$$\theta_{k,j} = \sqrt{\frac{2}{\ell_j}}\, b_{[\alpha_j, \alpha_{j+1}]}(x) \sin\left(\frac{2k+1}{2} \frac{\pi}{\ell_j}(x - \alpha_j)\right), \quad k = 0, 1, 2, ..., \ j \in \mathbb{Z}, \ (4.4)$$

is an orthonormal basis for $L^2(\mathbb{R})$. The convergence of the series expansion of a function $f \in L^2(\mathbb{R})$ with respect to the basis given in (4.4) is valid in $L^2(\mathbb{R})$. A much deeper result is the almost everywhere convergence, which is a consequence of the celebrated theorem of Carleson. More precisely, we have

$$\lim_{N \to \infty} \sum_{|j| \le N} \sum_{k=0}^{\infty} <f, \theta_{k,j}> \theta_{k,j}(x) = f(x)$$

for almost every $x \in \mathbb{R}$, where the second sum indicates the a.e. convergence of the partial sums

$$\sum_{k=0}^{M} <f, \theta_{k,j}> \theta_{k,j}(x)$$

as $M \to \infty$, for each $j \in \mathbb{Z}$.

Combining appropriately the polarities for different intervals $[\alpha_j, \alpha_{j+1}]$ we can obtain, in a similar manner, other bases for $L^2(\mathbb{R})$. Observe that we obtain, by using appropriate sine and cosine functions, a result which is not true in general if we use modulations, that is multiplications by exponentials. (See the Balian-Low theorem, Theorem 2.1.)

The orthogonal decomposition of $L^2(\mathbb{R})$ given in (3.19) can be used to obtain a new orthonormal basis of this space. The elements of this basis are the Fourier transforms of the wavelet basis introduced by Lemarié and Meyer in [LM].

THEOREM 4.5 *The system*

$$\gamma_{j,k}(\xi) = \frac{2^{\frac{j}{2}}}{\sqrt{2\pi}} b(2^j \xi) \, e^{i\frac{2k+1}{2} 2^j \xi}, \qquad j, k \in \mathbb{Z},$$

is an orthonormal basis for $L^2(\mathbb{R})$, where b restricted to $[0, \infty)$ is a bell function for $[\pi, 2\pi]$ associated with $0 < \varepsilon \le \frac{\pi}{3}$, $\varepsilon' = 2\varepsilon$, and b is even on $\mathbb{R}$.

PROOF : Let

$$\left. \begin{array}{l} C_{j,k}(\xi) = \dfrac{2^{\frac{j}{2}}}{\sqrt{2\pi}} b(2^j \xi) \cos\left(\frac{2k+1}{2} 2^j \xi\right), \\[4mm] S_{j,k}(\xi) = \dfrac{2^{\frac{j}{2}}}{\sqrt{2\pi}} b(2^j \xi) \sin\left(\frac{2k+1}{2} 2^j \xi\right), \end{array} \right\} \qquad k \ge 0, \quad j \in \mathbb{Z},$$

so that $\gamma_{j,k}(\xi) = C_{j,k}(\xi) + iS_{j,k}(\xi)$, $k \geq 0$, $j \in \mathbb{Z}$. Observe that $C_{j,k}$ is an even function on $\mathbb{R}$ and $S_{j,k}$ is an odd function on $\mathbb{R}$.

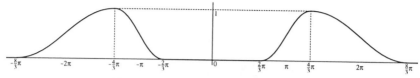

Figure 1.5: The graph of b with $\varepsilon = \frac{\pi}{3}$.

We shall use the trigonometric formulas

$$
\left.
\begin{array}{ll}
\text{i)} & \sin\left(\frac{2k+1}{2}(\xi - \pi)\right) = (-1)^{k+1}\cos\left(\frac{2k+1}{2}\xi\right), \\[2mm]
\text{ii)} & \cos\left(\frac{2k+1}{2}(\xi + 2\pi)\right) = -\cos\left(\frac{2k+1}{2}\xi\right), \\[2mm]
\text{iii)} & \cos\left(\frac{2k+1}{2}(\xi - \pi)\right) = (-1)^{k}\sin\left(\frac{2k+1}{2}\xi\right), \\[2mm]
\text{iv)} & \sin\left(\frac{2k+1}{2}(\xi + 2\pi)\right) = -\sin\left(\frac{2k+1}{2}\xi\right).
\end{array}
\right\}
\qquad (4.6)
$$

Let $b^{-}(\xi) = \chi_{(-\infty,0]}(\xi)b(\xi)$ and $b^{+}(\xi) = \chi_{[0,\infty)}(\xi)b(\xi)$ and define $C_{j,k}^{+}$, $C_{j,k}^{-}$, $S_{j,k}^{+}$ and $S_{j,k}^{-}$ as at the beginning of the proof, replacing b with b^{+} and b^{-}. Observe that $C_{j,k} = C_{j,k}^{+} + C_{j,k}^{-}$ and $S_{j,k} = S_{j,k}^{+} + S_{j,k}^{-}$.

Using formula i) in (4.6) and the basis i) in Theorem 4.3, we deduce that $\{2C_{j,k}^{+} : k \geq 0\}$ is an orthonormal basis for the projection spaces $P_{I_j}^{-,+}(L^2(\mathbb{R}))$, where $I_j = 2^{-j}[\pi, 2\pi]$. In fact, using i) of (4.6), we obtain

$$
2C_{j,k}^{+}(\xi) = \sqrt{\tfrac{2}{|I_j|}}\, b^{+}(2^j\xi)\cos\left(\tfrac{2k+1}{2}2^j\xi\right)
$$

$$
= (-1)^{k+1}\sqrt{\tfrac{2}{|I_j|}}\, b^{+}(2^j\xi)\sin\left(\tfrac{2k+1}{2}(2^j\xi - \pi)\right)
$$

$$
= (-1)^{k+1}\sqrt{\tfrac{2}{|I_j|}}\, b^{+}(2^j\xi)\sin\left(\tfrac{2k+1}{2}\tfrac{\pi}{|I_j|}(\xi - 2^{-j}\pi)\right),
$$

which is the basis i) of Theorem 4.3, except for the factor $(-1)^{k+1}$, which does not affect the orthonormality.

Using formula iii) of (4.6) and the basis iii) of Theorem 4.3, an analogous argument shows that $\{2S_{j,k}^{+} : k \geq 0\}$ is an orthonormal basis for the projection spaces $P_{I_j}^{+,-}(L^2(\mathbb{R}))$.

Similarly, it can be proved that $\{2C_{j,k}^{-} : k \geq 0\}$ and $\{2S_{j,k}^{-} : k \geq 0\}$ are orthonormal bases for $P_{-I_j}^{+,-}(L^2(\mathbb{R}))$ and $P_{-I_j}^{-,+}(L^2(\mathbb{R}))$, respectively, where $-I_j = 2^{-j}[-2\pi, -\pi]$.

Hence, each one of the systems

$$\{2C_{j,k}^+ : k \geq 0, \; j \in \mathbb{Z}\} \quad \text{and} \quad \{2S_{j,k}^+ : k \geq 0, \; j \in \mathbb{Z}\}$$

is an orthonormal basis for $L^2((0,\infty))$, and each one of the systems

$$\{2C_{j,k}^- : k \geq 0, \; j \in \mathbb{Z}\} \quad \text{and} \quad \{2S_{j,k}^- : k \geq 0, \; j \in \mathbb{Z}\}$$

is an orthonormal basis for $L^2((-\infty,0))$.

For $k \geq 0$ and $j \in \mathbb{Z}$, define

$$\alpha_{j,k}(\xi) = C_{j,k}(\xi) + iS_{j,k}(\xi) = \frac{2^{\frac{j}{2}}}{\sqrt{2\pi}} \, b(2^j\xi) \, e^{i\frac{2k+1}{2}2^j\xi}$$

and

$$\beta_{j,k}(\xi) = C_{j,k}(\xi) - iS_{j,k}(\xi) = \frac{2^{\frac{j}{2}}}{\sqrt{2\pi}} \, b(2^j\xi) \, e^{-i\frac{2k+1}{2}2^j\xi}.$$

If $m \leq -1$, $\beta_{j,-(m+1)}(\xi) = \gamma_{j,m}(\xi)$ and if $k \geq 0$, $\alpha_{j,k}(\xi) = \gamma_{j,k}(\xi)$. Hence, the theorem is proved if we show that the system

$$\{\alpha_{j,k} : j \in \mathbb{Z}, \; k \geq 0\} \cup \{\beta_{j,k} : j \in \mathbb{Z}, \; k \geq 0\}$$

is an orthonormal basis of $L^2(\mathbb{R})$.

We start by showing the orthonormality of the system.

$$4 <\alpha_{j,n}, \alpha_{k,\ell}> \; = 4 <C_{j,n}, C_{k,\ell}> + 4 <S_{j,n}, S_{k,\ell}>$$

$$= <2C_{j,n}^+, 2C_{k,\ell}^+> + <2C_{j,n}^-, 2C_{k,\ell}^->$$

$$+ <2S_{j,n}^+, 2S_{k,\ell}^+> + <2S_{j,n}^-, 2S_{k,\ell}^->$$

$$= 4\delta_{j,k}\delta_{n,\ell}.$$

Similarly, $<\beta_{j,n}, \beta_{k,\ell}> \; = \delta_{j,k}\delta_{n,\ell}$. Finally, using the evenness of $C_{j,n}$ and the oddness of $S_{k,\ell}$ we obtain

$$4 <\alpha_{j,n}, \beta_{k,\ell}> \; = 4 <C_{j,n}, C_{k,\ell}> + 4i <C_{j,n}, S_{k,\ell}>$$

$$+ 4i <S_{j,n}, C_{k,\ell}> - 4 <S_{j,n}, S_{k,\ell}>$$

$$= <2C_{j,n}^{+}, 2C_{k,\ell}^{+}> + <2C_{j,n}^{-}, 2C_{k,\ell}^{-}>$$

$$- <2S_{j,n}^{+}, 2S_{k,\ell}^{+}> - <2S_{j,n}^{-}, 2S_{k,\ell}^{-}>$$

$$= 2\delta_{j,k}\delta_{n,\ell} - 2\delta_{j,k}\delta_{n,\ell} = 0.$$

Now we must show completeness. Given $f \in L^2(\mathbb{R})$, let $f^{(e)}$ be the even function $[f(x) + f(-x)]/2$ and $f^{(o)}$ be the odd function $[f(x) - f(-x)]/2$, so that $f = f^{(e)} + f^{(o)}$. Using the evenness of $C_{j,k}$ and the oddness of $S_{j,k}$ we obtain

$$\sum_{j\in\mathbb{Z}}\sum_{k\geq 0} <f,\alpha_{j,k}> \alpha_{j,k} + <f,\beta_{j,k}> \beta_{j,k}$$

$$= 2\sum_{j\in\mathbb{Z}}\sum_{k\geq 0} <f^{(e)}, C_{j,k}> C_{j,k} + <f^{(o)}, S_{j,k}> S_{j,k}$$

$$= 4\sum_{j\in\mathbb{Z}}\sum_{k\geq 0}\{<f^{(e)}, C_{j,k}^{+}> C_{j,k}^{+} + <f^{(e)}, C_{j,k}^{-}> C_{j,k}^{-}$$

$$+ <f^{(o)}, S_{j,k}^{+}> S_{j,k}^{+} + <f^{(o)}, S_{j,k}^{-}> S_{j,k}^{-}\}$$

$$= f^{(e)}\chi_{(0,\infty)} + f^{(e)}\chi_{(-\infty,0)} + f^{(o)}\chi_{(0,\infty)} + f^{(o)}\chi_{(-\infty,0)} = f,$$

where we have used the already observed fact that the systems $\{2C_{j,k}^{+,-}\}$ and $\{2S_{j,k}^{+,-}\}$, $k \geq 0$, $j \in \mathbb{Z}$, form an orthonormal basis of $L^2((0,\infty))$ an $L^2((-\infty,0))$ for the appropriate choice of $+$ and $-$. ∎

COROLLARY 4.7 Let $\gamma(\xi) = \dfrac{1}{\sqrt{2\pi}} e^{i\frac{\xi}{2}} b(\xi)$ be the function $\gamma_{0,0}$ of Theorem 4.5 and define ψ by

$$\hat{\psi}(\xi) = \sqrt{2\pi}\,\gamma(\xi) = e^{i\frac{\xi}{2}} b(\xi).$$

Then, ψ is an orthonormal wavelet.

PROOF: By the Plancherel theorem $\|\psi\|_2^2 = \frac{1}{2\pi}\|\hat{\psi}\|_2^2 = \|\gamma\|_2^2 = 1$. Moreover,

$$(\psi_{j,k})^{\wedge}(\xi) = 2^{-\frac{j}{2}} e^{-i2^{-j}k\xi} \hat{\psi}(2^{-j}\xi) = 2^{-\frac{j}{2}} e^{-i2^{-j}k\xi} b(2^{-j}\xi) e^{i2^{-j}\frac{\xi}{2}}$$

$$= 2^{-\frac{j}{2}} b(2^{-j}\xi) e^{i2^{-j}\frac{1-2k}{2}\xi} = \sqrt{2\pi}\,\gamma_{-j,-k}(\xi).$$

By Theorem 4.5, $\{\psi_{j,k} : j, k \in \mathbb{Z}\}$ is an orthonormal basis for $L^2(\mathbb{R})$.

∎

The orthonormal wavelets obtained in Corollary 4.7 are the ones described by P.G. Lemarié and Y. Meyer in [LM] (see also [Me5]), and will be called the **Lemarié-Meyer wavelets**.

In Figure 1.6 we give the graph of a wavelet ψ whose Fourier tranform is of the form $\hat{\psi}(\xi) = b(\xi)e^{i\frac{\xi}{2}}$ with

$$b(\xi) = \begin{cases} \sin\left(\frac{3}{4}(|\xi| - \frac{2}{3}\pi)\right), & \text{if } \frac{2}{3}\pi < |\xi| \leq \frac{4}{3}\pi, \\ \sin\left(\frac{3}{8}(\frac{8}{3}\pi - |\xi|)\right), & \text{if } \frac{4}{3}\pi < |\xi| \leq \frac{8}{3}\pi, \\ 0 & \text{otherwise.} \end{cases}$$

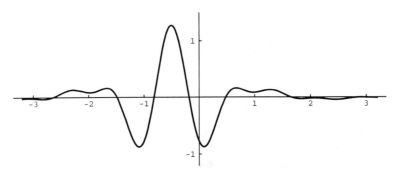

Figure 1.6: The graph of a Lemarié-Meyer wavelet.

Theorem 4.3 together with the orthogonal decomposition (3.18) can be used to obtain other bases for $L^2(\mathbb{R})$. Let

$$\alpha_j = \frac{j}{2}, \quad I_j = [\alpha_j, \alpha_{j+1}] \quad \text{and} \quad \ell_j = |I_j| = \alpha_{j+1} - \alpha_j = \frac{1}{2} \quad \text{for } j \in \mathbb{Z},$$

and choose $0 < \varepsilon \leq \frac{1}{4}$. Let b be the "bell" function associated with $[0, \frac{1}{2}]$ and ε at each endpoint. Observe that

$$b_j \equiv b_{I_j} = b(x - \tfrac{j}{2})$$

if we use the same ε at each endpoint of the interval I_j.

For the interval $I_j = [\frac{j}{2}, \frac{j+1}{2}]$ we choose the polarities indicated in the figure below:

| $-\frac{1}{2}$ | 0 | (+,+) | $\frac{1}{2}$ | (-,-) | 1 | (+,+) | $\frac{3}{2}$ | (-,-) | 2 | $\frac{5}{2}$ |

We now construct an orthonormal basis for $L^2(\mathbb{R})$. If j is even we use the local cosine basis given in iv) of Theorem 4.3 to obtain

$$\left.\begin{array}{l} \sqrt{2}\,b(x-\tfrac{j}{2}), \\[4pt] 2b(x-\tfrac{j}{2})\cos\big(2\pi k(x-\tfrac{j}{2})\big), \quad k=1,2,\cdots. \end{array}\right\} \tag{4.8}$$

If j is odd we use the local sine basis given in ii) of Theorem 4.3 to obtain

$$2b(x-\tfrac{j}{2})\sin\big(2\pi k(x-\tfrac{j}{2})\big), \quad k=1,2,\cdots. \tag{4.9}$$

For j even we have

$$\cos\big(2\pi k(x-\tfrac{j}{2})\big) = \cos(2\pi kx)\cos\big(2\pi k\tfrac{j}{2}\big) + \sin(2\pi kx)\sin\big(2\pi k\tfrac{j}{2}\big)$$

$$= \cos(2\pi kx);$$

and for j odd,

$$\sin\big(2\pi k(x-\tfrac{j}{2})\big) = \sin(2\pi kx)\cos\big(2\pi k\tfrac{j}{2}\big) - \cos(2\pi kx)\sin\big(2\pi k\tfrac{j}{2}\big)$$

$$= (-1)^k \sin(2\pi kx).$$

Thus,

$$\left.\begin{array}{ll} \sqrt{2}\,b(x-\tfrac{j}{2}) & \text{if } j \in 2\mathbb{Z}, \\[4pt] 2b(x-\tfrac{j}{2})\cos(2\pi kx) & \text{if } j \in 2\mathbb{Z},\ k=1,2,\cdots, \\[4pt] (-1)^k 2b(x-\tfrac{j}{2})\sin(2\pi kx) & \text{if } j \in 2\mathbb{Z}+1,\ k=1,2,\cdots, \end{array}\right\} \tag{4.10}$$

is an orthonormal basis for $L^2(\mathbb{R})$. Observe that we have "defeated" the Balian-Low phenomenon by using cosines and sines instead of exponentials, but the product of the "translation step" and the "frequency step" is still 2π (see (2.6)). Observe that if $g_{m,n}(x) = e^{2\pi i m x} g(x-\tfrac{n}{2})$, the family

$$\left.\begin{array}{ll} \sqrt{2}\,g_{0,j} & \text{if } j \in 2\mathbb{Z}, \\[4pt] \big[g_{k,j} + (-1)^j g_{-k,j}\big] & \text{if } j \in \mathbb{Z},\ k=1,2,3,\cdots, \end{array}\right\} \tag{4.11}$$

coincides with (4.10) when $g = b$, except that the factor $(-1)^k$ in the third family of (4.10) is replaced by $2i$.

A basis similar to the one described in (4.10) arises in the work of K. Wilson in quantum mechanics ([Wil]). He observed that for the study of his operators one does not need basis functions that distinguish between positive and negative frequencies of the same order. Instead of having a "peak" function localized around $x = \frac{n}{2}$, he uses functions that arise from the combination of two functions having peaks symmetrically distributed about the origin; this produces a system similar to the one in (4.11). We shall call the basis he uses a **Wilson basis**; more explicitly, using the notation of (4.11), we have

$$\left.\begin{array}{ll} \sqrt{2}\, g_{0,j} & \text{if } j \in 2\mathbb{Z}, \\[2mm] \left[g_{k,j} + (-1)^{k+j} g_{-k,j}\right] & \text{if } j \in \mathbb{Z},\ k = 1,2,3,\cdots, \end{array}\right\} \tag{4.12}$$

(observe the difference between the powers of -1 in (4.11) and (4.12)). This family can be written in the following way:

$$\left.\begin{array}{ll} \sqrt{2}\, g(x-j) & \text{if } k = 0,\ j \in \mathbb{Z}, \\[2mm] 2g\left(x-\frac{i}{2}\right)\cos(2\pi k x) & \text{if } k > 0,\ j+k \text{ even}, \\[2mm] 2g\left(x-\frac{i}{2}\right)\sin(2\pi k x) & \text{if } k > 0,\ j+k \text{ odd}. \end{array}\right\} \tag{4.13}$$

The proof that (4.13) is an orthonormal basis for some function g was simplified in [DJJ]. Here we can give a very simple proof as a consequence of our results on smooth projections and local sine and cosine basis. This was observed independently by P. Auscher ([Au1]) and E. Laeng ([Lae]). What is needed is a simple modification of the scheme developed to obtain (4.10).

Take $\alpha_j = \frac{2j-1}{4}$ for $j \in \mathbb{Z}$ and $0 < \varepsilon < \frac{1}{4}$, $\varepsilon' = \varepsilon$, and use the polarities described below:

$-\frac{3}{4}$	$-\frac{1}{4}$ (+,+)	$\frac{1}{4}$ (-,-)	$\frac{3}{4}$ (+,+)	$\frac{5}{4}$ (-,-)	$\frac{7}{4}$	$\frac{9}{4}$

By using simple trigonometric identities, it is not hard to show that the family

$$\left.\begin{array}{ll} \sqrt{2}\, b\left(x-\frac{i}{2}\right) & \text{if } j \text{ is even and } k = 0, \\[2mm] 2b\left(x-\frac{i}{2}\right)\cos\left(2\pi k\left(x+\frac{1}{4}\right)\right) & \text{if } j \text{ is even and } k > 0, \\[2mm] 2b\left(x-\frac{i}{2}\right)\sin\left(2\pi k\left(x-\frac{1}{4}\right)\right) & \text{if } j \text{ is odd and } k > 0, \end{array}\right\} \tag{4.14}$$

coincides with (4.13) when $b = g$, except for some factors of -1 which do not change the orthonormality of the system.

1.5 The unitary folding operators and the smooth projections

In this section we present another way of defining the projections P_I of section 1.3 and a new proof of Theorem 3.15, which allowed us to obtain orthonormal bases for $L^2(\mathbb{R})$. This section is not necessary to understand the chapters that follow, so that the reader who is interested in the concept of multiresolution analysis can proceed directly to Chapter 2.

We begin with the projections associated with the interval $[0, \infty)$. Recall the definition of $P \equiv P_{0,\varepsilon}^{+,-}$ given in (3.1). Motivated by this definition, we introduce the operator U defined by

$$Uf(x) = \begin{cases} s(x)f(x) + s(-x)f(-x), & x > 0, \\ \overline{s(-x)}f(x) - \overline{s(x)}f(-x), & x < 0, \end{cases} \tag{5.1}$$

where supp $(s) \subseteq [-\varepsilon, \infty)$ and satisfies

$$|s(x)|^2 + |s(-x)|^2 = 1 \qquad \text{for all } x. \tag{5.2}$$

The condition supp $(s) \subseteq [-\varepsilon, \infty)$ is not necessary for the first result we shall prove. In Figure 1.7 we show the graph of Uf for $f(x) = \frac{1}{x^2+1}$ and $\varepsilon = \frac{1}{2}$.

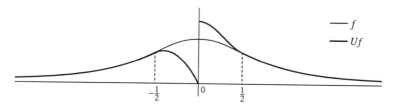

Figure 1.7: The graph of Uf for $f(x) = \frac{1}{x^2+1}$ and $\varepsilon = \frac{1}{2}$.

We consider the space $L^2(\mathbb{R}^+, \mathbb{C}^2)$ of all functions

$$F(t) = \begin{bmatrix} f_1(t) \\ f_2(t) \end{bmatrix},$$

where f_1 and f_2 are complex-valued functions defined on $\mathbb{R}^+$ belonging to $L^2((0, \infty))$. On this space the inner product is defined by

$$[F, G] = \int_0^\infty \left[f_1(t)\overline{g_1(t)} + f_2(t)\overline{g_2(t)} \right] dt.$$

The **folding operator** $\mathcal{F} : L^2(\mathbb{R}, \mathbb{C}) \longrightarrow L^2(\mathbb{R}^+, \mathbb{C}^2)$ is defined by

$$(\mathcal{F}f)(t) = \begin{bmatrix} f(t) \\ f(-t) \end{bmatrix}, \quad t > 0.$$

It is easy to see that $\mathcal{F}$ has an inverse, $\mathcal{F}^{-1}$, given by

$$\left(\mathcal{F}^{-1} \begin{bmatrix} f_1(t) \\ f_2(t) \end{bmatrix} \right)(t) = \begin{cases} f_1(t), & \text{if } t > 0, \\ f_2(-t), & \text{if } t < 0. \end{cases}$$

A simple computation shows that $[\mathcal{F}f, \mathcal{F}g] = <f, g>$, so that $\mathcal{F}$ is a unitary operator.

We can "transfer" U to $L^2(\mathbb{R}^+, \mathbb{C}^2)$ by using the following matrix:

$$A(t) = \begin{pmatrix} s(t) & s(-t) \\ -\overline{s(-t)} & \overline{s(t)} \end{pmatrix}, \quad t > 0.$$

This matrix is unitary due to (5.2); thus, $A(t)A(t)^* = I = A(t)^*A(t)$. We now define the operators $\mathcal{A}, \mathcal{A}^* : L^2(\mathbb{R}^+, \mathbb{C}^2) \longrightarrow L^2(\mathbb{R}^+, \mathbb{C}^2)$ by letting

$$(\mathcal{A}F)(t) = A(t)F(t) \quad \text{and} \quad (\mathcal{A}^*F)(t) = A(t)^*F(t).$$

THEOREM 5.3 $U = \mathcal{F}^{-1}\mathcal{A}\mathcal{F}$ and $U^* = \mathcal{F}^{-1}\mathcal{A}^*\mathcal{F}$, so that U is unitary. Moreover,

$$(U^*f)(x) = \begin{cases} \overline{s(x)}f(x) - s(-x)f(-x), & x > 0, \\ s(-x)f(x) + \overline{s(x)}f(-x), & x < 0, \end{cases}$$

and $U^*\chi_{[0,\infty)}U = P_{0,\varepsilon}^+$, where $P_{0,\varepsilon}^+$ is defined by (3.1).

PROOF : The equality $U = \mathcal{F}^{-1}\mathcal{A}\mathcal{F}$ is easy to check by using the above definitions. That $U^* = \mathcal{F}^{-1}\mathcal{A}^*\mathcal{F}$ follows immediately from the fact that $\mathcal{F}$

is also a unitary operator. The expression for (U^*f) in terms of f follows readily from this last equality.

Finally, let us prove that $U^*\chi_{[0,\infty)}U = P_{0,\varepsilon}^+$: if $x > 0$,

$$(U^*\chi_{[0,\infty)}Uf)(x) = \overline{s(x)}(\chi_{[0,\infty)}Uf)(x) - s(-x)(\chi_{[0,\infty)}Uf)(-x)$$

$$= \overline{s(x)}(Uf)(x) = \overline{s(x)}\left[s(x)f(x) + s(-x)f(-x)\right];$$

if $x < 0$,

$$(U^*\chi_{[0,\infty)}Uf)(x) = \overline{s(x)}(\chi_{[0,\infty)}Uf)(-x)$$

$$= \overline{s(x)}\left[s(-x)f(-x) + s(x)f(x)\right].$$

These two formulae coincide with the definition of $P_{0,\varepsilon}^+$ given in (3.1). ∎

The graph of U^* is shown in Figure 1.8 for $f(x) = \frac{1}{x^2+1}$ and $\varepsilon = \frac{1}{2}$.

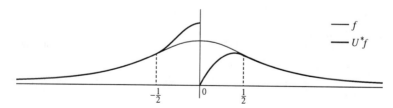

$$— f$$
$$— U^*f$$

Figure 1.8: The graph of U^*f for $f(x) = \frac{1}{x^2+1}$ and $\varepsilon = \frac{1}{2}$.

Observe that f is unchanged under the actions of U^* and U outside the interval $(-\varepsilon, \varepsilon)$. (See Figures 1.7 and 1.8.)

We can translate the point 0 to α as we did in section 1.3. As before, let $\tau_\alpha f(x) = f(x - \alpha)$ be the translation by α operator; we then define

$$U_\alpha = \tau_\alpha U \tau_\alpha^* \quad \text{and} \quad U_\alpha^* = \tau_\alpha U^* \tau_\alpha^*.$$

Observe that $U = U_0$ and $U^* = U_0^*$. With these definitions we have the general formulas:

$$(U_\alpha f)(x) = \begin{cases} \overline{s(x-\alpha)f(x) + s(\alpha-x)f(2\alpha-x)}, & x > \alpha, \\ -\overline{s(x-\alpha)}f(2\alpha-x) + \overline{s(\alpha-x)}f(x), & x < \alpha, \end{cases}$$

and

$$(U_\alpha^* f)(x) = \begin{cases} \overline{s(x-\alpha)} f(x) - \overline{s(\alpha-x)} f(2\alpha - x), & x > \alpha, \\ \overline{s(x-\alpha)} f(2\alpha - x) + s(\alpha-x) f(x), & x < \alpha. \end{cases}$$

PROPOSITION 5.4 *Let* $E = E_\alpha = [\alpha - \varepsilon, \alpha + \varepsilon]$,

$$L^2(E) = \{ f \in L^2(\mathbb{R}) : supp\,(f) \subseteq E \},$$

and suppose that s *satisfies* (5.2) *and, also,* $s(x) = 1$ *if* $x > \varepsilon$. *Then* U_α *and* U_α^* *satisfy:*

(i) $U_\alpha, U_\alpha^* : L^2(E) \longrightarrow L^2(E)$, *and, hence, are unitary on* $L^2(E)$,

(ii) $U_\alpha|_{L^2(E)^\perp} = I = U_\alpha^*|_{L^2(E)^\perp}$, *where* $L^2(E)^\perp = L^2(E^c)$, *and*

(iii) U_α *and* U_α^* *commute with multiplication by* χ_E.

PROOF :   Since $2\alpha - x$ is the point symmetric to x with respect to α, supp $(f(2\alpha - \cdot)) \subseteq [\alpha - \varepsilon, \alpha + \varepsilon]$; this proves that U_α and U_α^* take $L^2(E)$ into $L^2(E)$. (i) is now immediate.

To prove (ii), look at the general formulas for U_α, U_α^* and use the equalities $s(x - \alpha) = 1$, $s(\alpha - x) = 0$ if $x > \alpha + \varepsilon$. Again, examine the general formulas for U_α and U_α^* and observe that χ_E is symmetric with respect to α; this proves (iii).
∎

THEOREM 5.5 *Let* s *satisfy* (5.2) *with support on* $[\varepsilon, \infty)$, *and suppose that* $s \in C^d$, *where* C^d *is the space of all functions with continuous derivatives up to order* d. *Then*

$$U_\alpha : C^d \cap L^2(\mathbb{R}) \longrightarrow S_\alpha \quad and \quad U_\alpha^* : S_\alpha \longrightarrow C^d \cap L^2(\mathbb{R}),$$

and both operators are one-to-one and onto, where

$$S_\alpha = \{ f \in C^d(\mathbb{R} - \{\alpha\}) \cap L^2(\mathbb{R}) : f^{(n)}(\alpha\pm) \ exist \ for \ 0 \leq n \leq d,$$

$$\lim_{x \to \alpha+} f^{(n)}(x) = 0 \ \ if \ n \ \ is \ odd,$$

$$and \ \lim_{x \to \alpha-} f^{(n)}(x) = 0 \ \ if \ n \ \ is \ even \ \}.$$

Proof : Since U_α and U_α^* are unitary (see part (i) of Proposition 5.4), it suffices to show $U_\alpha(C^d \cap L^2(\mathbb{R})) \subseteq S_\alpha$ and $U_\alpha^*(S_\alpha) \subseteq C^d \cap L^2(\mathbb{R})$. Moreover, these inclusions are proved if we can show that $U_0(C^d \cap L^2(\mathbb{R})) \subseteq S_0$ and $U_0^*(S_0) \subseteq C^d \cap L^2(\mathbb{R})$, where $U_0 = U$.

To show the first inclusion, let $h(x) = s(x)f(x)$ so that

$$(Uf)(x) = h(x) + h(-x) \qquad \text{when } x > 0.$$

If $f \in C^d \cap L^2(\mathbb{R})$ we have

$$(Uf)^{(n)}(x) = h^{(n)}(x) + (-1)^n h^{(n)}(-x) \quad \text{for } x > 0,$$

which shows that $(Uf)^{(n)}(0+)$ exists and is zero if n is odd and $0 \leq n \leq d$. Let $g(x) = \overline{s(-x)}f(x)$ so that $(Uf)(x) = g(x) - g(-x)$ for $x < 0$. If $f \in C^d \cap L^2(\mathbb{R})$,

$$(Uf)^{(n)}(x) = g^{(n)}(x) - (-1)^n g^{(n)}(-x) \quad \text{for } x < 0,$$

which shows that $(Uf)^{(n)}(0-)$ exists and is zero if n is even and $0 \leq n \leq d$.

We now show the second inclusion, which is a little more complicated. If $f \in S_0$ it is clear from the formula satisfied by U^*f that $(Uf)^{(n)}(0\pm)$ exists when $0 \leq n \leq d$. It is enough to show

$$\lim_{x \to 0+} \left\{ (U^*f)^{(n)}(x) - (U^*f)^{(n)}(-x) \right\} = 0. \tag{5.6}$$

Let $H(x) = (U^*f)(x) - (-1)^n (U^*f)(-x)$ so that we have to show

$$\lim_{x \to 0+} H^{(n)}(x) = 0.$$

A simple computation using the formula for U^*f shows

$$H(x) = \left[\overline{s(x)} - (-1)^n \overline{s(-x)} \right] f(x) - \left[s(-x) + (-1)^n s(x) \right] f(-x), \; x > 0.$$

Taking derivatives we obtain

$$H^{(n)}(x) = \sum_{k=0}^{n} \binom{n}{k} f^{(n-k)}(x) \left[\overline{s^{(k)}(x)} - (-1)^{n+k} \overline{s^{(k)}(-x)} \right]$$

$$- \sum_{k=0}^{n} \binom{n}{k} (-1)^{n-k} f^{(n-k)}(-x) \left[(-1)^k s^{(k)}(-x) + (-1)^n s^{(k)}(x) \right]$$

$$\equiv \sum_{k=0}^{n} \binom{n}{k} \{G_k(f,s)(x)\}.$$

If $n - k$ is odd, $\lim_{x \to 0+} f^{(n-k)}(x) = 0$ so that

$$\lim_{x \to 0+} G_k(f,s)(x) = -(-1)^{n-k} f^{(n-k)}(0-)\big[(-1)^k s^{(k)}(0) + (-1)^n s^{(k)}(0)\big]$$

$$= 0,$$

since $(-1)^k = -(-1)^n$. If $n - k$ is even, $\lim_{x \to 0+} f^{(n-k)}(-x) = 0$ so that

$$\lim_{x \to 0+} G_k(f,s)(x) = f^{(n-k)}(0+)\big[\overline{s^{(k)}(0)} - (-1)^{n+k} \overline{s^{(k)}(0)}\big] = 0.$$

This finishes the proof of (5.6) and, hence, the theorem. ∎

We shall examine the smooth projection operators defined in section 1.3 in terms of the unitary operators we have defined in this section.

In Theorem 5.3 we have shown that $U^* \chi_{(0,\infty)} U = P_{0,\varepsilon}^+ \equiv P_0^+$ where $(P_0^+ f)(x) = \overline{s(x)}\,[s(x)f(x) \pm s(-x)f(-x)]$ was given in (3.1); we already knew that P_0^+ is an orthogonal projection, but this follows immediately from this equality since U is a unitary operator and the multiplication by $\chi_{(0,\infty)}$ operator is self-adjoint: in fact,

$$(P_0^+)^* = U^* \chi_{(0,\infty)}^* U = U^* \chi_{(0,\infty)} U = P_0^+$$

and

$$(P_0^+)^2 = U^* \chi_{(0,\infty)} U U^* \chi_{(0,\infty)} U = U^* \chi_{(0,\infty)} U = P_0^+.$$

To find the projection P_α^+ corresponding to the interval (α, ∞), we recall the definition $P_\alpha^+ = \tau_\alpha P_0^+ \tau_\alpha^*$ and observe that $\tau_\alpha \chi_{(0,\infty)} \tau_\alpha^* = \chi_{(\alpha,\infty)}$ to obtain

$$P_\alpha^+ = U_\alpha^* \chi_{(\alpha,\infty)} U_\alpha. \tag{5.7}$$

The following, in fact, establishes (5.7):

$$P_\alpha^+ = \tau_\alpha P_0^+ \tau_\alpha^* = \tau_\alpha U^* \chi_{(0,\infty)} U \tau_\alpha^* = \tau_\alpha U^* (\tau_\alpha^* \tau_\alpha \chi_{(0,\infty)} \tau_\alpha^* \tau_\alpha) U \tau_\alpha^*$$

$$= U_\alpha^* (\tau_\alpha \chi_{(0,\infty)} \tau_\alpha^*) U_\alpha = U_\alpha^* \chi_{(\alpha,\infty)} U_\alpha.$$

Observe that equality (5.7) shows immediately that P_α^+ is an orthogonal projection, since U_α is unitary.

We can also define, as we did in section 1.3, the projections associated with the interval $(-\infty, 0)$. Just define $P_-^0 = U^* \chi_{(-\infty,0)} U$ and it is immediate that P_-^0 is an orthogonal projection, since U is unitary. Moreover,

$$(P_-^0 f)(x) = s(-x)\big[\,\overline{s(-x)}f(x) - \overline{s(x)}f(-x)\big].$$

(Observe that this projection has negative polarity at 0.) We can translate this projection to the interval $(-\infty, \beta)$ to obtain

$$P_-^\beta = U_\beta^* \chi_{(-\infty,\beta)} U_\beta$$

as we did when we obtained (5.7).

Let us now show how these unitary operators can be used to obtain the smooth projections P_I associated with the interval $I = [\alpha, \beta]$. Choose the real numbers α, β, ε and ε' with $\varepsilon, \varepsilon' > 0$ and

$$-\infty < \alpha - \varepsilon < \alpha < \alpha + \varepsilon < \beta - \varepsilon' < \beta < \beta + \varepsilon' < \infty.$$

Observe that $U_\alpha f$ and $U_\alpha^* f$ have the same values as f outside the interval $(\alpha - \varepsilon, \alpha + \varepsilon)$, and both $U_\beta f$ and $U_\beta^* f$ coincide with f outside the interval $(\beta - \varepsilon', \beta + \varepsilon')$. As a consequence, we have several commutativity relations; some of them are:

$$\left.\begin{array}{ll}
\text{i)} & U_\alpha U_\beta^* = U_\beta^* U_\alpha \quad \text{and} \quad U_\alpha U_\beta = U_\beta U_\alpha, \\[2mm]
\text{ii)} & \chi_{(\alpha,\infty)} U_\beta^* = U_\beta^* \chi_{(\alpha,\infty)}, \\[2mm]
\text{iii)} & U_\alpha \chi_{(-\infty,\beta)} = \chi_{(-\infty,\beta)} U_\alpha.
\end{array}\right\} \qquad (5.8)$$

These commutativity relations allow us to show $P_\alpha^+ P_-^\beta = P_-^\beta P_\alpha^+$. In fact, using (5.8), we have

$$P_\alpha^+ P_-^\beta = U_\alpha^* \chi_{(\alpha,\infty)} U_\alpha U_\beta^* \chi_{(-\infty,\beta)} U_\beta = U_\alpha^* U_\beta^* \chi_{(\alpha,\beta)} U_\alpha U_\beta,$$

and, similarly,

$$P_-^\beta P_\alpha^+ = U_\alpha^* U_\beta^* \chi_{(\alpha,\beta)} U_\alpha U_\beta.$$

Since $P_\alpha^+ P_-^\beta = P_-^\beta P_\alpha^+$, the operator $P_{(\alpha,\beta)}^{+,-} = U_\alpha^* U_\beta^* \chi_{(\alpha,\beta)} U_\alpha U_\beta$ is an orthogonal projection. Observe that this projection has polarity $+$ at α and polarity $-$ at β. Again, observe that this equality giving us $P_{(\alpha,\beta)}^{+,-}$ also immediately implies that it is a projection.

It is now easy to obtain the version of Theorem 3.15 for $P_{(\alpha,\beta)}^{+,-}$ by using the definition of this projection we have just given in terms of the folding operators. In fact, suppose

$$-\infty < \alpha - \varepsilon < \alpha < \alpha + \varepsilon < \beta - \varepsilon' < \beta < \beta + \varepsilon' < \gamma - \varepsilon'' < \gamma < \gamma + \varepsilon'' < \infty$$

so that the intervals $I = [\alpha, \beta]$ and $J = [\beta, \gamma]$, with these choices of $\varepsilon, \varepsilon', \varepsilon''$, are compatible. Then, if we write

$$P_I = P_{(\alpha,\beta)}^{+,-}, \quad P_J = P_{(\beta,\gamma)}^{+,-} \quad \text{and} \quad P_{I \cup J} = P_{(\alpha,\gamma)}^{+,-},$$

we have

$$\text{i)} \quad P_I + P_J = P_{I \cup J}$$

and

$$\text{ii)} \quad P_I P_J = P_J P_I.$$

Equality i) follows easily from (5.8) once we express P_I, P_J and $P_{I \cup J}$ in terms of the associated folding operators and the characteristic functions χ_I, χ_J and $\chi_{I \cup J}$.

Equality ii) can be proved as we did for Theorem 3.15. But it can also be easily obtained directly by using the fact that $\chi_I \chi_J = 0 = \chi_J \chi_I$ and the commutativity relations (5.8) that the folding operators satisfy.

There are some remarks that should be made about this approach. We have only obtained the special case of Theorem 3.15 for projections with polarities $(+, -)$. To obtain the full statement of this theorem we need to define folding operators related to projections having the other polarities. In addition to the operators U and U^*, we also define V and V^* by

$$(Vf)(x) = \begin{cases} s(x)f(x) - s(-x)f(-x), & x > 0, \\ \overline{s(-x)}f(x) + \overline{s(x)}f(-x), & x < 0, \end{cases}$$

and

$$(V^*f)(x) = \begin{cases} \overline{s(x)}f(x) + s(-x)f(-x), & x > 0, \\ s(-x)f(x) - \overline{s(x)}f(-x), & x < 0. \end{cases}$$

Thus,

$$U^*\chi_{(0,\infty)}U = P_0^+, \qquad U^*\chi_{(-\infty,0)}U = P_-^0,$$

$$V^*\chi_{(0,\infty)}V = P_0^-, \qquad V^*\chi_{(-\infty,0)}V = P_+^0,$$

where

$$(P_0^{+,-}f)(x) = \overline{s(x)}\left[s(x)f(x) \pm s(-x)f(-x)\right],$$

$$(P_{+,-}^0 f)(x) = s(-x)\left[\overline{s(-x)}f(x) \pm \overline{s(x)}f(-x)\right].$$

We can now construct the four projections associated with an interval $I = (\alpha, \beta)$ having chosen appropriate $\varepsilon, \varepsilon' > 0$. They are

$$\left.\begin{aligned}
P_I^{+,-} &= U_\alpha^* U_\beta^* \chi_I U_\alpha U_\beta, \\
P_I^{+,+} &= U_\alpha^* V_\beta^* \chi_I U_\alpha V_\beta, \\
P_I^{-,+} &= V_\alpha^* V_\beta^* \chi_I V_\alpha V_\beta, \\
P_I^{-,-} &= V_\alpha^* U_\beta^* \chi_I V_\alpha U_\beta.
\end{aligned}\right\} \qquad (5.9)$$

The full statement of Theorem 3.15 can then be obtained by using these equalities, as long as we choose compatible projections for adjacent intervals, and provided they have opposite polarities at the common end point.

It is illustrative to present the graphs of $P_0^+ f$, $P_0^- f$, $P_+^0 f$ and $P_-^0 f$ (see Figure 1.9 below) for $f(x) = \frac{1}{(x+1)^2+1}$ and $\varepsilon = \frac{1}{2}$.

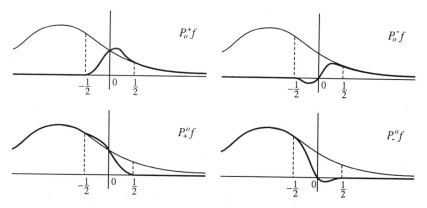

Figure 1.9: Graphs of $P_0^+ f$, $P_0^- f$, $P_+^0 f$ and $P_-^0 f$.

This method for obtaining the projections associated with the interval I by the factorizations presented in (5.9) is particularly useful in applications. The factors used are very simple operators that can be easily represented in a computer program. More generally, the bases we have constructed in terms of a specific function to which we apply certain translation operators, dilation operators and/or modulations are also well suited for applications. The fact that an elementary function is used and the fact that these operators applied to it are particularly simple lead to relatively elegant expressions for the partial sums of the series representing general functions.

1.6   Notes and references

1. Appropriate general assumptions that guarantee the validity of most of the formulas in section 1.1 can be found in [SW]. The Haar function presented in Example A was discovered by A. Haar (see [Haa]) in 1910. More information about Gabor bases can be obtained from [Gab]. Those readers who are not familiar with the space $\mathcal{S}'$ of tempered distributions involved in the proof of Theorem 2.1 can find the definition and relevant properties of $\mathcal{S}'$ in [SW]. Theorem 2.1, referred to as the Balian-Low theorem in this book, was originally proved independently by R. Balian [Bal] and F. Low [Low] in the early 1980s. (See also item 1 in section 8.5.) The proof we presented is due to G. Battle [Bat2]. A good source for the basic properties of orthogonal projections used in section 1.3 is the book written by P. Halmos [Hal]. The almost everywhere convergence of the trigonometric series considered in section 1.4 follows from the Carleson-Hunt theorem (see [Car1] and [Hun]). The detailed construction of the Lemarié-Meyer wavelets introduced in section 1.4 can be found in [LM]. While the local sine and cosine series in Theorem 4.5 were first described by R. Coifman and Y. Meyer in [CM2], they were also introduced by H. Malvar ([Malv]) in connection with the theory of signal processes. A complete account of these facts is also discussed in [AWW]. The Wilson basis mentioned in the same section was introduced by K. Wilson in an unpublished manuscript [Wil]. The proof that the family (4.13) is an orthonormal basis for some function g was presented in [DJJ]. The unitary folding operators of section 1.5 and their application to obtaining smooth localized orthonormal bases were developed in [Wi1] and are also described in [Wi2].

2. Dilation factors other than 2 can be considered to decompose $L^2(\mathbb{R})$ as an orthogonal direct sum in a way similar to (3.19). Consider the intervals $I = [\pi, \lambda\pi]$ and $J = [-\lambda\pi, -\pi]$ for some $\lambda > 1$, and let $I_k = \lambda^{-k}I$ and $J_k = \lambda^{-k}J$ for all $k \in \mathbb{Z}$. Since

$$(0, \infty) = \bigcup_{k\in\mathbb{Z}} \lambda^k I \qquad \text{and} \qquad (-\infty, 0) = \bigcup_{k\in\mathbb{Z}} \lambda^k J,$$

it follows from the theory of smooth projections developed in section 1.3 that

$$L^2(\mathbb{R}) = \left\{\bigoplus_{k\in\mathbb{Z}} P_{J_k}(L^2(\mathbb{R}))\right\} \oplus \left\{\bigoplus_{k\in\mathbb{Z}} P_{I_k}(L^2(\mathbb{R}))\right\}$$

when we choose compatible bell functions for adjacent intervals and appropriate polarities. It then follows, as in the proof that led to Theorem 4.5, that the collection of functions

$$\alpha_{j,k}^\lambda(\xi) = c_{j,k}^\lambda(\xi) + i s_{j,k}^\lambda(\xi) \qquad \text{and} \qquad \beta_{j,k}^\lambda(\xi) = c_{j,k}^\lambda(\xi) - i s_{j,k}^\lambda(\xi),$$

where

$$c_{j,k}^\lambda(\xi) \equiv \frac{\lambda^{\frac{j}{2}}}{\sqrt{2(\lambda-1)\pi}} b(\lambda^j \xi) \cos\left(\frac{2k+1}{2}\frac{1}{\lambda-1}(\lambda^j\xi - \pi)\right)$$

and

$$s_{j,k}^\lambda(\xi) \equiv \frac{\lambda^{\frac{j}{2}}}{\sqrt{2(\lambda-1)\pi}} b(\lambda^j \xi) \sin\left(\frac{2k+1}{2}\frac{1}{\lambda-1}(\lambda^j\xi - \pi)\right),$$

is an orthonormal basis for $L^2(\mathbb{R})$. It is shown in [AWW] that in order for the functions $\alpha_{j,k}^\lambda$ and $\beta_{j,k}^\lambda$ to be generated by a single function, $\alpha_{0,0}^\lambda$, via dilations by λ and multiplications by $e^{in\frac{\xi}{\lambda-1}}$, $n \in \mathbb{Z}$, as was the case for the basis of Theorem 4.5, we must have

$$\lambda = 1 + \frac{1}{m}$$

for some $m \in \mathbb{Z}^+$. With this value of λ one can obtain a wavelet basis of the form $\{\lambda^{\frac{j}{2}}\psi(\lambda^j x - k\frac{1}{\lambda-1}) : j, k \in \mathbb{Z}\}$, where

$$\hat{\psi}(\xi) = \frac{1}{\sqrt{2\pi(\lambda-1)}} e^{i\frac{\xi}{2(\lambda-1)}} b(\xi).$$

The function b we need to use in the above considerations is an even function on $\mathbb{R}$ that, when restricted to $(0, \infty)$, is a bell function associated with the

interval $[\pi, \lambda\pi]$, $\varepsilon = \frac{\lambda-1}{\lambda+1}\pi$ and $\varepsilon' = \frac{\lambda(\lambda-1)}{\lambda+1}\pi = \lambda\varepsilon$. This result is due to
G. David (see [Dav]).

3. For information on the theory of wavelet-like bases with more general
dilation factors see [Au3] and note 2 in section 2.5.

2

Multiresolution analysis and the construction of wavelets

In this chapter we shall present a method for constructing orthonormal wavelets that is based on the existence of a family of subspaces of $L^2(\mathbb{R})$ satisfying certain properties. Such a family is called a **multiresolution analysis**, or, simply, an **MRA**. It will become clear as we present it in the first two sections of this chapter that the main feature of this method is to describe mathematically the process of studying signals at different scales. Furthermore, as we shall see in section 7.3 (Corollary 3.16), those wavelets that do not arise from an MRA are, in a sense, exceptional.

Although it is traditional to define an MRA by specifying five properties that a family of subspaces of $L^2(\mathbb{R})$ must satisfy, in section 2.1 we show that these properties are not independent. The basic construction of a wavelet starting from an MRA is presented in section 2.2, which also includes several examples of wavelets and properties they satisfy.

It was this theory that was used by I. Daubechies to show that for any non-negative integer n there exists an orthonormal wavelet ψ with compact support such that all the derivatives of ψ up to order n exist. This remarkable achievement is presented in section 2.3. Section 2.4 is dedicated to improving the estimates we need to establish the smoothness of compactly supported wavelets.

2.1   Multiresolution analysis

A **multiresolution analysis** (MRA) consists of a sequence of closed subspaces V_j, $j \in \mathbb{Z}$, of $L^2(\mathbb{R})$ satisfying

$$V_j \subset V_{j+1} \qquad \text{for all } j \in \mathbb{Z}; \tag{1.1}$$

$$f \in V_j \text{ if and only if } f(2(\cdot)) \in V_{j+1} \qquad \text{for all } j \in \mathbb{Z}; \tag{1.2}$$

$$\bigcap_{j \in \mathbb{Z}} V_j = \{0\}; \tag{1.3}$$

$$\overline{\bigcup_{j \in \mathbb{Z}} V_j} = L^2(\mathbb{R}); \tag{1.4}$$

$$\left. \begin{array}{l} \text{There exists a function } \varphi \in V_0, \text{ such that} \\ \{\varphi(\cdot - k) : k \in \mathbb{Z}\} \text{ is an orthonormal basis for } V_0. \end{array} \right\} \tag{1.5}$$

The function φ whose existence is asserted in (1.5) is called a **scaling function** of the given MRA.

Sometimes condition (1.5) is relaxed by assuming that $\{\varphi(\cdot - n) : n \in \mathbb{Z}\}$ is a **Riesz basis** for V_0. That is, for every $f \in V_0$ there exists a unique sequence $\{\alpha_n\}_{n \in \mathbb{Z}} \in \ell^2(\mathbb{Z})$ such that

$$f(x) = \sum_{n \in \mathbb{Z}} \alpha_n \varphi(x - n)$$

with convergence in $L^2(\mathbb{R})$ and

$$A \sum_{n \in \mathbb{Z}} |\alpha_n|^2 \le \left\| \sum_{n \in \mathbb{Z}} \alpha_n \varphi(x - n) \right\|_2^2 \le B \sum_{n \in \mathbb{Z}} |\alpha_n|^2$$

with $0 < A \le B < \infty$ constants independent of $f \in V_0$. If this is the case we say that we have an MRA with a Riesz basis. The fact that this weaker version of (1.5) is equivalent to the one announced above is proved at the end of this section.

Observe that (1.5) implies that $\{\varphi(\cdot - n) : n \in \mathbb{Z}\}$ is a Riesz basis for V_0 with $A = B = 1$ (see section 1.1).

Let $\varphi_{j,k}(x) = 2^{\frac{j}{2}}\varphi(2^j x - k)$; since $\varphi_{0,k} = \varphi(x - k)$ we see that $\varphi_{0,k} \in V_0$ for all $k \in \mathbb{Z}$ due to (1.5). Moreover, if $j \in \mathbb{Z}$, condition (1.2) implies that $\{\varphi_{j,n} : n \in \mathbb{Z}\}$ is an orthonormal basis for V_j.

EXAMPLE A : Let V_j be the space of all functions in $L^2(\mathbb{R})$ which are constant on intervals of the form $[2^{-j}k, 2^{-j}(k + 1)]$, $k \in \mathbb{Z}$. Then $\{V_j : j \in \mathbb{Z}\}$ is an MRA and we can take the scaling function to be $\varphi = \chi_{[-1,0]}$. As we shall see, this MRA is related to the Haar basis.

The properties we have used to define an MRA are not independent. Precise formulations of this dependence are contained in the next two theorems.

THEOREM 1.6 *Conditions* (1.1), (1.2) *and* (1.5) *imply* (1.3). *This is the case even if, in* (1.5), *we only assume that* $\{\varphi(\cdot - n) : n \in \mathbb{Z}\}$ *is a Riesz basis.*

PROOF : Suppose that there exists a non-zero f in $\cap_{j \in \mathbb{Z}} V_j$; we can clearly assume $\|f\|_2 = 1$. In particular, $f \in V_{-j}$ for each $j \in \mathbb{Z}$; hence, if we let $f_j(x) = 2^{\frac{j}{2}} f(2^j x)$ we must have $f_j \in V_0$, because of (1.2). Moreover, a change of variables shows $\|f_j\|_2 = \|f\|_2 = 1$. Since we are assuming that $\{\varphi(\cdot - n) : n \in \mathbb{Z}\}$ is a Riesz basis, we can write

$$f_j(x) = \sum_{k \in \mathbb{Z}} \alpha_k^j \varphi(x - k),$$

with convergence in $L^2(\mathbb{R})$, in such a way that

$$A \sum_{k \in \mathbb{Z}} |\alpha_k^j|^2 \leq \|f_j\|_2^2 = 1.$$

Taking Fourier transforms, we can write

$$2^{-\frac{j}{2}} \hat{f}(2^{-j}\xi) = \hat{f}_j(\xi) = \sum_{k \in \mathbb{Z}} \alpha_k^j e^{-ik\xi} \hat{\varphi}(\xi) = m_j(\xi)\hat{\varphi}(\xi),$$

where

$$m_j(\xi) = \sum_{k \in \mathbb{Z}} \alpha_k^j e^{-ik\xi}.$$

Observe that m_j is a 2π-periodic function which belongs to $L^2(\mathbb{T})$ with norm $\leq \sqrt{\frac{2\pi}{A}}$. Thus, $\hat{f}(\xi) = 2^{\frac{j}{2}} m_j(2^j \xi)\hat{\varphi}(2^j \xi)$ and, for $j \geq 1$,

$$\int_{2\pi}^{4\pi} |\hat{f}(\xi)|\, d\xi \leq 2^{\frac{j}{2}} \left(\int_{2\pi}^{4\pi} |\hat{\varphi}(2^j \xi)|^2\, d\xi \right)^{\frac{1}{2}} \left(\int_{2\pi}^{4\pi} |m_j(2^j \xi)|^2\, d\xi \right)^{\frac{1}{2}}$$

$$= 2^{-\frac{j}{2}} \left(\int_{2^{j+1}\pi}^{2^{j+2}\pi} |\hat{\varphi}(\mu)|^2\, d\mu \right)^{\frac{1}{2}} \left(\int_{2^{j+1}\pi}^{2^{j+2}\pi} |m_j(\mu)|^2\, d\mu \right)^{\frac{1}{2}}$$

$$\leq \left(\int_{2^{j+1}\pi}^{\infty} |\hat{\varphi}(\mu)|^2\, d\mu \right)^{\frac{1}{2}} \left(\frac{1}{2^j} \int_{2^{j+1}\pi}^{2^{j+2}\pi} |m_j(\mu)|^2\, d\mu \right)^{\frac{1}{2}}$$

$$= \left(\int_{2^{j+1}\pi}^{\infty} |\hat{\varphi}(\mu)|^2\, d\mu \right)^{\frac{1}{2}} \left(\frac{1}{2^j} \sum_{\ell=0}^{2^j - 1} \int_{2^{j+1}\pi+2\ell\pi}^{2^{j+1}\pi+2(\ell+1)\pi} |m_j(\mu)|^2\, d\mu \right)^{\frac{1}{2}}$$

$$\leq \left(\int_{2^{j+1}\pi}^{\infty} |\hat{\varphi}(\mu)|^2\, d\mu \right)^{\frac{1}{2}} \left(\frac{2\pi}{A} \right)^{\frac{1}{2}}.$$

Hence, letting $j \to \infty$, we obtain $\int_{2\pi}^{4\pi} |\hat{f}(\xi)|\, d\xi = 0$ and we can deduce that $\hat{f}(\xi) = 0$ almost everywhere on $[2\pi, 4\pi]$. We can apply the same argument to $2^{\frac{\ell}{2}} \hat{f}(2^\ell \xi)$, $\ell \in \mathbb{Z}$, to obtain $\hat{f}(\xi) = 0$ almost everywhere on $2^\ell[2\pi, 4\pi]$, $\ell \in \mathbb{Z}$. Hence, $\hat{f}(\xi) = 0$ almost everywhere on $(0, \infty)$. If we apply this argument, with the interval $[-4\pi, -2\pi]$ playing the role of $[2\pi, 4\pi]$, we obtain $\hat{f}(\xi) = 0$ almost everywhere on $(-\infty, 0)$. ∎

THEOREM 1.7 *Let* $\{V_j : j \in \mathbb{Z}\}$ *be a sequence of closed subspaces of* $L^2(\mathbb{R})$ *satisfying* (1.1), (1.2) *and* (1.5); *assume that the scaling function* φ *of condition* (1.5) *is such that* $|\hat{\varphi}|$ *is continuous at* 0. *Then, the following two conditions are equivalent:*

$$\text{i)} \quad \hat{\varphi}(0) \neq 0, \qquad\qquad \text{ii)} \quad \overline{\bigcup_{j\in\mathbb{Z}} V_j} = L^2(\mathbb{R}).$$

Moreover, when either is the case, $|\hat{\varphi}(0)| = 1$.

PROOF : Assume that $\hat{\varphi}(0) \neq 0$. We first claim that

$$W \equiv \overline{\bigcup_{j\in\mathbb{Z}} V_j}$$

is invariant under translations. To prove this we first show that W is invariant under the dyadic translations $\tau_{2^{-\ell}m}$, $\ell, m \in \mathbb{Z}$. Let $f \in W$; given $\varepsilon > 0$, there exists $j_o \in \mathbb{Z}$ and $h \in V_{j_o}$ such that $\|f - h\|_2 < \varepsilon$. From (1.1) we deduce that $h \in V_j$ for all $j \geq j_o$ and, using (1.2) and (1.5), we can write

$$h(x) = \sum_{k \in \mathbb{Z}} c_k^j \varphi(2^j x - k)$$

with convergence in $L^2(\mathbb{R})$. Hence,

$$(\tau_{2^{-\ell}m} h)(x) = h(x - 2^{-\ell}m) = \sum_{k \in \mathbb{Z}} c_k^j \varphi(2^j(x - 2^{-\ell}m) - k).$$

If $j \geq \ell$, $\varphi(2^j(x-2^{-\ell}m) - k) = \varphi(2^j x - 2^{j-\ell}m - k)$ is an element of V_j, since $2^{j-\ell}m \in \mathbb{Z}$. Since $\|\tau_{2^{-\ell}m} f - \tau_{2^{-\ell}m} h\|_2 = \|f - h\|_2 < \varepsilon$ and ε is arbitrarily small, we can conclude that W is invariant under dyadic translations. Now, for a general $x \in \mathbb{R}$, we can find m and ℓ, both integers, such that $2^{-\ell}m$ is arbitrarily close to x; hence, $\|\tau_{2^{-\ell}m} f - \tau_x f\|_2 < \varepsilon$, and it follows that W is invariant under all translations τ_x.

Since $\hat{\varphi}(0) \neq 0$ and $|\hat{\varphi}|$ is continuous at 0, $\hat{\varphi}(\xi) \neq 0$ on $(-\mu, \mu)$ for some $\mu > 0$. Suppose that there exists $g \in W^\perp$; then g is orthogonal to all $f \in W$, and since W is translation-invariant, we have

$$0 = \int_{\mathbb{R}} f(x + t) \overline{g(t)}\, dt$$

for all $x \in \mathbb{R}$ and all $f \in W$. Since $(\tau_{-x} f)^\wedge(\xi) = e^{i\xi x} \hat{f}(\xi)$, this last equality and the Plancherel formula (1.3) in Chapter 1 imply

$$0 = \int_{-\infty}^{\infty} e^{i\xi x} \hat{f}(\xi) \overline{\hat{g}(\xi)}\, d\xi$$

for all $x \in \mathbb{R}$. Since $\hat{f}\overline{\hat{g}} \in L^1(\mathbb{R})$ this shows that $\hat{f}(\xi) \overline{\hat{g}(\xi)} = 0$ for almost every $\xi \in \mathbb{R}$. In particular, let $f(x) = 2^j \varphi(2^j x)$ so that $f \in V_j \subset W$ and $\hat{f}(\xi) = \hat{\varphi}(2^{-j}\xi)$. Hence, $\hat{\varphi}(2^{-j}\xi) \overline{\hat{g}(\xi)} = 0$ for almost every $\xi \in \mathbb{R}$. Since $\hat{\varphi}(2^{-j}\xi) \neq 0$ if $\xi \in (-2^j\mu, 2^j\mu)$, we can conclude that $\hat{g}(\xi) = 0$ for almost all ξ satisfying $|\xi| < 2^j\mu$. Letting $j \to \infty$, we see that $\hat{g} = 0$ a.e. and, therefore, $g = 0$. This shows that $\overline{\cup_{j \in \mathbb{Z}} V_j} = L^2(\mathbb{R})$.

Assume now that $W = \overline{\cup_{j \in \mathbb{Z}} V_j} = L^2(\mathbb{R})$ and let f be such that $\hat{f} = \chi_{[-1,1]}$; then, $\|f\|_2^2 = \frac{1}{2\pi}\|\hat{f}\|_2^2 = \frac{1}{\pi}$. If P_j denotes the orthogonal projection onto V_j we then have $\|f - P_j f\|_2 \to 0$ as $j \to \infty$, due to (1.1) and our assumption. Thus, $\|P_j f\|_2 \to \|f\|_2$ as $j \to \infty$. Therefore, if $\varphi_{j,k}(x) = 2^{\frac{j}{2}} \varphi(2^j x - k)$, we have

$$\|P_j f\|_2^2 = \left\| \sum_{k \in \mathbb{Z}} <f, \varphi_{j,k}> \varphi_{j,k} \right\|_2^2 \longrightarrow \frac{1}{\pi}$$

as $j \to \infty$, since $\{\varphi_{j,k} : k \in \mathbb{Z}\}$ is an orthonormal basis of V_j (because of (1.2) and (1.5)). From the Plancherel theorem (see section 1.1) and the fact that $\hat{f} = \chi_{[-1,1]}$, we have

$$\frac{1}{4\pi^2} \sum_{k \in \mathbb{Z}} \left| \int_{\mathbb{R}} \hat{f}(\xi) \overline{(\varphi_{j,-k})^{\wedge}(\xi)} \, d\xi \right|^2$$

$$= \frac{1}{4\pi^2} \sum_{k \in \mathbb{Z}} \left| \int_{\mathbb{R}} \hat{f}(\xi) 2^{-\frac{j}{2}} e^{-i2^{-j}k\xi} \overline{\hat{\varphi}(2^{-j}\xi)} \, d\xi \right|^2$$

$$= 2^j \sum_{k \in \mathbb{Z}} \left| \frac{1}{2\pi} \int_{-2^{-j}}^{2^{-j}} \overline{\hat{\varphi}(\mu)} \, e^{-ik\mu} d\mu \right|^2 .$$

For j large enough, $[-2^{-j}, 2^{-j}] \subset [-\pi, \pi]$ and the last expression is 2^j times the sum of the squares of the absolute values of the Fourier coefficients of the function $\chi_{[-2^{-j}, 2^{-j}]} \overline{\hat{\varphi}}$; thus, by the Plancherel formula for Fourier series,

$$\frac{2^j}{2\pi} \int_{-2^{-j}}^{2^{-j}} |\hat{\varphi}(\mu)|^2 \, d\mu \longrightarrow \frac{1}{\pi}$$

as $j \to \infty$. But, by the continuity of $|\hat{\varphi}|$ at 0, the last integral expression tends to $\frac{1}{\pi} |\hat{\varphi}(0)|^2$. Therefore, $|\hat{\varphi}(0)| = 1 \neq 0$. ∎

It is easy to see from the proof of this theorem that we can deduce (1.4) if we assume (1.1), (1.2) and (1.5) and that there exists a $\mu > 0$ such that $\hat{\varphi}(\xi) \neq 0$ on $(-\mu, \mu)$. Also, if $\{V_j : j \in \mathbb{Z}\}$ is an MRA, the proof of the above theorem shows that we must have

$$\lim_{j \to \infty} \frac{1}{2^{-j+1}} \int_{-2^{-j}}^{2^{-j}} |\hat{\varphi}(\mu)|^2 \, d\mu = 1.$$

We end this section by showing that if in (1.5) we only assume that $\{\varphi(\cdot - k) : k \in \mathbb{Z}\}$ is a Riesz basis for V_0, we can then find a function $\gamma \in V_0$

such that $\{\gamma(\cdot - k) : k \in \mathbb{Z}\}$ is an orthonormal basis for V_0. This is an easy consequence of the following lemma:

LEMMA 1.8 *Suppose $\varphi \in L^2(\mathbb{R})$ is such that the set of translates $\{\varphi(\cdot - k)\}$, $k \in \mathbb{Z}$, forms a Riesz basis of the closed subspace of $L^2(\mathbb{R})$ that they span; that is,*

$$A \sum_{k \in \mathbb{Z}} |c_k|^2 \leq \left\| \sum_{k \in \mathbb{Z}} c_k \varphi(\cdot - k) \right\|_{L^2(\mathbb{R})}^2 \leq B \sum_{k \in \mathbb{Z}} |c_k|^2, \qquad (1.9)$$

where the constants A and B satisfy $0 < A \leq B < \infty$ and are independent of $\mathfrak{c} = \{c_k\}_{k \in \mathbb{Z}} \in \ell^2(\mathbb{Z})$. Let

$$\sigma_\varphi(\xi) \equiv \left(\sum_{k \in \mathbb{Z}} |\hat{\varphi}(\xi + 2k\pi)|^2 \right)^{\frac{1}{2}}. \qquad (1.10)$$

Then $\sqrt{A} \leq \sigma_\varphi(\xi) \leq \sqrt{B}$ for almost every $\xi \in \mathbb{R}$.

Before presenting the proof of this lemma, let us show why our claim is an easy consequence of it. We define γ by letting $\hat{\gamma} = \hat{\varphi}/\sigma_\varphi$. Since $1/\sigma_\varphi$ is bounded with

$$0 < \frac{1}{\sqrt{B}} \leq \frac{1}{\sigma_\varphi(\xi)} \leq \frac{1}{\sqrt{A}} \qquad \text{for} \quad a.e. \ \xi \in \mathbb{R},$$

$\hat{\gamma}$, and, hence, γ belong to $L^2(\mathbb{R})$. In fact, since σ_φ is 2π-periodic we can find two sequences $\{a_k\}_{k \in \mathbb{Z}}$ and $\{b_k\}_{k \in \mathbb{Z}}$ in $\ell^2(\mathbb{Z})$ such that

$$\frac{1}{\sigma_\varphi(\xi)} = \sum_{k \in \mathbb{Z}} a_k e^{-ik\xi} \qquad \text{and} \qquad \sigma_\varphi(\xi) = \sum_{k \in \mathbb{Z}} b_k e^{-ik\xi}$$

in $L^2(\mathbb{T})$ (or for a.e. $\xi \in \mathbb{T}$). Thus,

$$\hat{\gamma}(\xi) = \hat{\varphi}(\xi) \sum_{k \in \mathbb{Z}} a_k e^{-ik\xi} \qquad \text{and} \qquad \hat{\varphi}(\xi) = \hat{\gamma}(\xi) \sum_{k \in \mathbb{Z}} b_k e^{-ik\xi}.$$

Taking Fourier transforms these equalities become

$$\gamma(x) = \sum_{k \in \mathbb{Z}} a_k \varphi_{0,k}(x) \qquad \text{and} \qquad \varphi(x) = \sum_{k \in \mathbb{Z}} b_k \gamma_{0,k}(x),$$

with convergence in $L^2(\mathbb{R})$. Thus,

$$\gamma \in \overline{\text{span} \{\varphi_{0,k} : k \in \mathbb{Z}\}} \qquad \text{and} \qquad \varphi \in \overline{\text{span} \{\gamma_{0,k} : k \in \mathbb{Z}\}}.$$

Furthermore, it is clear from the definition of $\hat{\gamma}$ and the 2π-periodicity of σ_φ that

$$\sum_{k\in\mathbb{Z}} |\hat{\gamma}(\xi + 2k\pi)|^2 = \frac{1}{\sigma_\varphi^2(\xi)} \sum_{k\in\mathbb{Z}} |\hat{\varphi}(\xi + 2k\pi)|^2 = 1 \quad \text{for a.e. } \xi \in \mathbb{R}.$$

Our claim that $\{\gamma(\cdot - k) : k \in \mathbb{Z}\}$ is an orthonormal basis for V_0 now follows from an easily proved, but most basic, characterization of such orthonormal systems:

PROPOSITION 1.11 *If $g \in L^2(\mathbb{R})$, then $\{g(\cdot - k) : k \in \mathbb{Z}\}$ is an orthonormal system if and only if*

$$\sum_{k\in\mathbb{Z}} |\hat{g}(\xi + 2k\pi)|^2 = 1 \quad \text{for a.e. } \xi \in \mathbb{R}.$$

PROOF : This follows from the following periodization argument and the Plancherel theorem:

$$\delta_{k,0} = \int_{\mathbb{R}} g(x) \overline{g(x-k)} \, dx = \frac{1}{2\pi} \int_{\mathbb{R}} |\hat{g}(\xi)|^2 e^{ik\xi} \, d\xi$$

$$= \frac{1}{2\pi} \sum_{\ell=-\infty}^{\infty} \int_{2\ell\pi}^{2(\ell+1)\pi} |\hat{g}(\xi)|^2 e^{ik\xi} \, d\xi$$

$$= \frac{1}{2\pi} \sum_{\ell=-\infty}^{\infty} \int_{0}^{2\pi} |\hat{g}(\mu + 2\ell\pi)|^2 e^{ik\mu} \, d\mu$$

$$= \frac{1}{2\pi} \int_{0}^{2\pi} \left(\sum_{\ell\in\mathbb{Z}} |\hat{g}(\mu + 2\ell\pi)|^2 \right) e^{ik\mu} \, d\mu.$$

This tells us that the 2π-periodic function $\sum_{k\in\mathbb{Z}} |\hat{g}(\mu + 2k\pi)|^2$ equals 1 a.e. since it has Fourier coefficient 1 at the frequency $k = 0$ and all the other coefficients are zero. The converse is immediate. ∎

We shall now present a proof of Lemma 1.8. Let $\mathfrak{c} = \{c_k\}_{k\in\mathbb{Z}}$ be any sequence in $\ell^2(\mathbb{Z})$ and $S_\mathfrak{c} = \sum_{k\in\mathbb{Z}} c_k \varphi_{0,k}$. Then

$$\widehat{S_\mathfrak{c}}(\xi) = \sum_{k\in\mathbb{Z}} c_k e^{-ik\xi} \hat{\varphi}(\xi) \equiv \vartheta_\mathfrak{c}(\xi) \hat{\varphi}(\xi),$$

where $\vartheta_c(\xi) = \sum_{k \in \mathbb{Z}} c_k e^{-ik\xi}$ is in $L^2(\mathbb{T})$. Moreover,

$$\left\|\widehat{S_c}\right\|^2_{L^2(\mathbb{R})} = \int_{\mathbb{R}} \left|\vartheta_c(\xi)\hat{\varphi}(\xi)\right|^2 d\xi = \sum_{k \in \mathbb{Z}} \int_{2k\pi}^{2(k+1)\pi} \left|\vartheta_c(\xi)\hat{\varphi}(\xi)\right|^2 d\xi$$

$$= \int_{\mathbb{T}} |\vartheta_c(\xi)|^2 \sum_{k \in \mathbb{Z}} |\hat{\varphi}(\xi + 2k\pi)|^2 d\xi = \int_{\mathbb{T}} |\vartheta_c(\xi)\sigma_\varphi(\xi)|^2 d\xi.$$

This and the Plancherel theorem give us

$$\left\|S_c\right\|^2_{L^2(\mathbb{R})} = \frac{1}{2\pi} \int_{\mathbb{T}} |\vartheta_c(\xi)\sigma_\varphi(\xi)|^2 d\xi.$$

This allows us to rewrite the Riesz basis property (1.9) in the form

$$A\left\|\vartheta_c\right\|^2_{L^2(\mathbb{T})} \leq \left\|\vartheta_c\sigma_\varphi\right\|^2_{L^2(\mathbb{T})} \leq B\left\|\vartheta_c\right\|^2_{L^2(\mathbb{T})}. \tag{1.12}$$

Fix $\eta \in \mathbb{T}$ and choose $\vartheta_c(\xi) = \frac{1}{\sqrt{2m\pi}} \sum_{k=0}^{m-1} e^{ik(\xi-\eta)}$ so that

$$|\vartheta_c(\xi)|^2 = \frac{1}{2m\pi} \left[\frac{\sin \frac{m(\xi-\eta)}{2}}{\sin \frac{\xi-\eta}{2}}\right]^2 = K_m(\eta - \xi),$$

where K_m is the m th Fejér kernel. In this case $\left\|\vartheta_c\right\|_{L^2(\mathbb{T})} = 1$ and (1.12) becomes

$$A \leq (K_m * \sigma_\varphi^2)(\eta) \leq B \qquad \text{for} \quad a.e. \ \eta \in \mathbb{T}. \tag{1.13}$$

But $(K_m * \sigma_\varphi^2)(\eta) \to \sigma_\varphi^2(\eta)$ as $m \to \infty$ for a.e. $\eta \in \mathbb{T}$, since the Cesàro means of the Fourier series of a function $f \in L^1(\mathbb{T})$ converge to f almost everywhere. Thus, letting $m \to \infty$ in (1.13) we obtain the conclusion of Lemma 1.8.

$\blacksquare$

We should observe that the function σ_φ defined in Lemma 1.8 is always well defined when $\varphi \in L^2(\mathbb{R})$. In fact,

$$\int_{\mathbb{T}} |\sigma_\varphi(\xi)|^2 d\xi = \sum_{k \in \mathbb{Z}} \int_{2k\pi}^{2(k+1)\pi} |\hat{\varphi}(\xi)|^2 d\xi = \int_{\mathbb{R}} |\hat{\varphi}(\xi)|^2 d\xi = 2\pi \left\|\varphi\right\|^2_{L^2(\mathbb{R})}.$$

This clearly implies $\sigma_\varphi(\xi) < \infty$ a.e. on $\mathbb{T}$. The condition in Proposition 1.11 is clearly a special case (when $A = B = 1$) of the conclusion in Lemma 1.8. This conclusion is easily seen to be equivalent to the property defining a Riesz basis (see (1.12) and the discussion preceding it). This presents another way in which a Riesz basis generalizes an orthonormal basis generated by the integral translates of a function in $L^2(\mathbb{R})$.

2.2   Construction of wavelets from a multiresolution analysis

We now pass to the construction of orthonormal wavelets from an MRA. Let W_0 be the orthogonal complement of V_0 in V_1; that is, $V_1 = V_0 \oplus W_0$. Then, if we dilate the elements of W_0 by 2^j, we obtain a closed subspace W_j of V_{j+1} such that

$$V_{j+1} = V_j \oplus W_j \qquad \text{for each } j \in \mathbb{Z}.$$

Since $V_j \to \{0\}$ as $j \to -\infty$, we see that

$$V_{j+1} = V_j \oplus W_j = \bigoplus_{\ell=-\infty}^{j} W_\ell \qquad \text{for all } j \in \mathbb{Z}.$$

Since $V_j \to L^2(\mathbb{R})$ as $j \to \infty$, we have

$$L^2(\mathbb{R}) = \bigoplus_{j=-\infty}^{\infty} W_j. \tag{2.1}$$

To find an orthonormal wavelet, therefore, all we need to do is to find a function $\psi \in W_0$ such that $\{\psi(\cdot - k) : k \in \mathbb{Z}\}$ is an orthonormal basis for W_0. In fact, if this is the case, then $\{2^{\frac{j}{2}}\psi(2^j \cdot -k) : k \in \mathbb{Z}\}$ is an orthonormal basis for W_j for all $j \in \mathbb{Z}$ due to (1.2) and the definition of W_j. Hence $\{\psi_{j,k} : j, k \in \mathbb{Z}\}$ is an orthonormal basis for $L^2(\mathbb{R})$, which shows that ψ is an orthonormal wavelet on $\mathbb{R}$.

Let us now try to find such a function ψ. Consider $V_0 = W_{-1} \oplus V_{-1}$ and observe that $\frac{1}{2}\varphi(\frac{\cdot}{2}) \in V_{-1} \subset V_0$. By (1.5) we can express this function in terms of the basis $\{\varphi(\cdot + k) : k \in \mathbb{Z}\}$ to obtain

$$\tfrac{1}{2}\varphi(\tfrac{1}{2}x) = \sum_{k \in \mathbb{Z}} \alpha_k \varphi(x + k), \tag{2.2}$$

where $\alpha_k = \tfrac{1}{2}\int_{\mathbb{R}} \varphi(\tfrac{1}{2}x)\,\overline{\varphi(x + k)}\,dx$; the convergence in (2.2) is in $L^2(\mathbb{R})$ and $\sum_{k \in \mathbb{Z}} |\alpha_k|^2 < \infty$. Taking Fourier transforms, we obtain

$$\hat{\varphi}(2\xi) = \hat{\varphi}(\xi)\sum_{k \in \mathbb{Z}} \alpha_k e^{ik\xi} = \hat{\varphi}(\xi)m_0(\xi), \tag{2.3}$$

where

$$m_0(\xi) = \sum_{k \in \mathbb{Z}} \alpha_k e^{ik\xi}$$

is a 2π-periodic function in $L^2(\mathbb{T}) = L^2([-\pi, \pi))$. The function m_0 is called the **low pass filter** associated with the scaling function φ.

We begin by deriving a very simple, but very basic, corollary of Proposition 1.11:

COROLLARY 2.4 *Suppose that $g \in L^2(\mathbb{R})$ and that $\{g(\cdot - k) : k \in \mathbb{Z}\}$ is an orthonormal system; then $|supp\,(\hat{g})| \geq 2\pi$. Equality holds if and only if $|\hat{g}| = \chi_K$ for some measurable set $K \subset \mathbb{R}$ with $|K| = 2\pi$.*

PROOF : Since $\|g\|_2 = 1$, $\|\hat{g}\|_2^2 = 2\pi$; meanwhile, Proposition 1.11 tells us that $|\hat{g}(\xi)| \leq 1$ almost everywhere on $\mathbb{R}$. Hence,

$$|supp\,(\hat{g})| = \int_{supp(\hat{g})} 1\,d\xi \geq \int_{\mathbb{R}} |\hat{g}(\xi)|^2\,d\xi = 2\pi.$$

If $|supp\,(\hat{g})| = 2\pi$ and $0 < |\hat{g}(\xi)| < 1$ on a set E of positive measure we have the contradiction

$$2\pi = \|\hat{g}\|_2^2 = \int_{supp(\hat{g})} |\hat{g}(\xi)|^2\,d\xi < |(supp\,(\hat{g})) - E| + |E| = |supp\,(\hat{g})| = 2\pi;$$

hence, $|\hat{g}(\xi)| = 1$ on the support of $\hat{g}$. Finally, if $\{g(\cdot - k) : k \in \mathbb{Z}\}$ is an orthonormal system and $|\hat{g}| = \chi_K$, then we must have

$$|K| = |supp\,(\hat{g})| = \|\hat{g}\|_2^2 = 2\pi.$$

∎

We continue with the construction of orthonormal wavelets ψ associated with the scaling function φ. An important property of the low pass filter is

$$|m_0(\xi)|^2 + |m_0(\xi + \pi)|^2 = 1 \qquad \text{for } a.e. \ \xi \in \mathbb{R}. \qquad (2.5)$$

To prove this we use Proposition 1.11 with g replaced by φ to obtain

$$\sum_{k \in \mathbb{Z}} |\hat{\varphi}(2\xi + 2k\pi)|^2 = 1 \qquad \text{for } a.e. \ \xi \in \mathbb{R};$$

by using (2.3) we see that this equality is equivalent to

$$\sum_{k \in \mathbb{Z}} |\hat{\varphi}(\xi + k\pi)|^2 |m_0(\xi + k\pi)|^2 = 1 \qquad \text{for } a.e. \ \xi \in \mathbb{R}.$$

We sum the left-hand side of the above formula separately over the even integers and over the odd integers. Using the 2π-periodicity of m_0 and Proposition 1.11, we obtain

$$1 = |m_0(\xi)|^2 \sum_{\ell \in \mathbb{Z}} |\hat{\varphi}(\xi + 2\ell\pi)|^2 + |m_0(\xi + \pi)|^2 \sum_{\ell \in \mathbb{Z}} |\hat{\varphi}(\xi + \pi + 2\ell\pi)|^2$$

$$= |m_0(\xi)|^2 \cdot 1 + |m_0(\xi + \pi)|^2 \cdot 1.$$

This proves (2.5).

To find ψ we examine W_{-1}, which is the orthogonal complement of V_{-1} in V_0; we do this on the Fourier transform side.

LEMMA 2.6 *If φ is a scaling function for an MRA $\{V_j\}_{j \in \mathbb{Z}}$, and m_0 is the associated low-pass filter, then*

$$V_{-1} = \{f : \hat{f}(\xi) = m(2\xi) m_0(\xi) \hat{\varphi}(\xi) \ \text{for some } 2\pi\text{-periodic } m \in L^2(\mathbb{T})\};$$

$$V_0 = \{f : \hat{f}(\xi) = \ell(\xi) \hat{\varphi}(\xi) \ \text{for some } 2\pi\text{-periodic function } \ell \in L^2(\mathbb{T})\}.$$

PROOF: If $f \in V_{-1}$, $f(x) = \dfrac{1}{\sqrt{2}} \displaystyle\sum_{k \in \mathbb{Z}} c_k \varphi(\tfrac{1}{2}x - k)$ with $\displaystyle\sum_{k \in \mathbb{Z}} |c_k|^2 < \infty$.
Hence, using (2.3),

$$\hat{f}(\xi) = \sqrt{2}\hat{\varphi}(2\xi) \sum_{k \in \mathbb{Z}} c_k e^{-i2k\xi} = m(2\xi)\hat{\varphi}(2\xi) = m(2\xi) m_0(\xi) \hat{\varphi}(\xi)$$

where $m(\xi) = \sqrt{2} \sum_{k \in \mathbb{Z}} c_k e^{-ik\xi}$, which is a 2π-periodic function in $L^2(\mathbb{T})$.

Conversely, if $m \in L^2(\mathbb{T})$ and is 2π-periodic, we can retrace the above computation to show that the function f defined by $\hat{f}(\xi) = m(2\xi)m_0(\xi)\hat{\varphi}(\xi)$ belongs to V_{-1}. Observe that, when we do this, we have to prove that $m(2\xi)m_0(\xi)\hat{\varphi}(\xi)$ is a function in $L^2(\mathbb{R})$. This is true for any function of the form $h(\xi)\hat{\varphi}(\xi)$ if h is 2π-periodic and $h \in L^2(\mathbb{T})$ due to the orthonormality of $\{\varphi(\cdot - k) : k \in \mathbb{Z}\}$. In fact,

$$\int_{\mathbb{R}} |h(\xi)|^2 |\hat{\varphi}(\xi)|^2 \, d\xi = \sum_{k \in \mathbb{Z}} \int_0^{2\pi} |h(\mu)|^2 |\hat{\varphi}(\mu + 2k\pi)|^2 \, d\mu = \|h\|_{L^2(\mathbb{T})}^2$$

by Proposition 1.11. Since m_0 is bounded (from (2.5)) and $m \in L^2(\mathbb{T})$, $h(\xi) = m(2\xi)m_0(\xi)$ belongs to $L^2(\mathbb{T})$. This establishes the characterization of V_{-1}.

If $g \in V_0$, $g(x) = \sum_{k \in \mathbb{Z}} d_k \varphi(x - k)$ with $\sum_{k \in \mathbb{Z}} |d_k|^2 < \infty$. Then

$$\hat{g}(\xi) = \hat{\varphi}(\xi) \sum_{k \in \mathbb{Z}} d_k e^{-ik\xi} = \ell(\xi)\hat{\varphi}(\xi),$$

where ℓ is a 2π-periodic function in $L^2(\mathbb{T})$. The proof is finished by arguing as we just did for the characterization of V_{-1}. ∎

Let us continue with the construction of the wavelet ψ. The elements of W_{-1} are those L^2-functions $f \in V_0$ that are orthogonal to V_{-1}. Let $U : V_0 \longrightarrow L^2(\mathbb{T})$ be defined by $U(f) = \ell$, where ℓ is as in Lemma 2.6. It is easy to check that U is linear; moreover, U satisfies

$$\|Uf\|_{L^2(\mathbb{T})}^2 = \|\ell\|_{L^2(\mathbb{T})}^2 = 2\pi \sum_{k \in \mathbb{Z}} |d_k|^2 = 2\pi \|f\|_{L^2(\mathbb{R})}^2.$$

By the polarization identity we obtain

$$<f, g>_{L^2(\mathbb{R})} = \frac{1}{2\pi} <Uf, Ug>_{L^2(\mathbb{T})} \qquad \text{for all } f, g \in V_0.$$

If f is orthogonal to V_{-1}, this last equality and Lemma 2.6 tell us that ℓ must be orthogonal to $m(2\xi)m_0(\xi)$ for all 2π-periodic $m \in L^2(\mathbb{T})$. Hence,

$$0 = \int_0^{2\pi} \ell(\xi) \overline{m(2\xi)m_0(\xi)} \, d\xi$$

$$= \int_0^\pi \overline{m(2\xi)} \left\{ \ell(\xi) \overline{m_0(\xi)} + \ell(\xi + \pi) \overline{m_0(\xi + \pi)} \right\} d\xi$$

for all 2π-periodic functions $m \in L^2(\mathbb{T})$. This tells us that the π-periodic function in the curly brackets is orthogonal to all π-periodic square integrable functions; that is, $\ell(\xi) \overline{m_0(\xi)} + \ell(\xi + \pi) \overline{m_0(\xi + \pi)} = 0$ for almost every $\xi \in \mathbb{T}$. Therefore, we must have

$$(\ell(\xi), \ell(\xi + \pi)) = -\lambda(\xi + \pi) (\overline{m_0(\xi + \pi)}, -\overline{m_0(\xi)}) \tag{2.7}$$

for almost every ξ and an appropriate $\lambda(\xi)$. Letting $\xi = \mu + \pi$ we have

$$(\ell(\mu + \pi), \ell(\mu)) = -\lambda(\mu + 2\pi) (\overline{m_0(\mu)}, -\overline{m_0(\mu + \pi)})$$

because of the 2π-periodicity of m_0 and ℓ. But this equality is equivalent to

$$(\ell(\xi), \ell(\xi + \pi)) = \lambda(\xi + 2\pi) (\overline{m_0(\xi + \pi)}, -\overline{m_0(\xi)}) \tag{2.8}$$

for almost every ξ. We know from (2.5) that the vector

$$(\overline{m_0(\xi + \pi)}, -\overline{m_0(\xi)})$$

has norm 1 for almost every ξ. This fact, together with equalities (2.7) and (2.8), imply $\lambda(\xi + \pi) = -\lambda(\xi + 2\pi)$, or, equivalently, $\lambda(\xi) = -\lambda(\xi + \pi)$ for almost every ξ. Therefore, λ is a 2π-periodic function in $L^2(\mathbb{T})$ satisfying

$$\lambda(\xi) = -\lambda(\xi + \pi) \qquad \text{for} \quad a.e. \ \xi \in \mathbb{T}. \tag{2.9}$$

This means that $\lambda(\xi) = e^{i\xi} s(2\xi)$ where $s \in L^2(\mathbb{T})$ and is 2π-periodic (it is enough to define $s(\xi) = e^{-i\frac{\xi}{2}} \lambda(\frac{1}{2}\xi)$).

From (2.7) and (2.9) we obtain

$$\ell(\xi) = e^{i\xi} s(2\xi) \overline{m_0(\xi + \pi)} \tag{2.10}$$

with $s \in L^2(\mathbb{T})$ and 2π-periodic. But it is a simple exercise to show that the subspace of $L^2(\mathbb{R})$ that contains all functions f for which $\hat{f}(\xi) = \ell(\xi) \hat{\varphi}(\xi)$ with ℓ satisfying (2.10) is in the orthogonal complement of V_{-1} in V_0. This gives us a characterization of W_{-1}:

$$W_{-1} = \left\{ f : \hat{f}(\xi) = e^{i\xi} s(2\xi) \overline{m_0(\xi + \pi)} \hat{\varphi}(\xi) \ \text{ for a } 2\pi\text{-periodic } s \in L^2(\mathbb{T}) \right\},$$

which, in turn, establishes the following characterization of W_0:

LEMMA 2.11 *If φ is a scaling function for an MRA $\{V_j\}_{j\in\mathbb{Z}}$, and m_0 is the associated low-pass filter, then*

$$W_0 = \{f : \hat{f}(2\xi) = e^{i\xi} s(2\xi)\, \overline{m_0(\xi+\pi)}\, \hat{\varphi}(\xi) \quad for\ a\ 2\pi\text{-}periodic\ s \in L^2(\mathbb{T})\}.$$

It follows from this lemma that W_0 is invariant under integral translations. Similarly, we have

$$W_j = \{f : \hat{f}(2^{j+1}\xi) = e^{i\xi} s(2\xi)\overline{m_0(\xi+\pi)}\hat{\varphi}(\xi) \quad for\ a\ 2\pi\text{-}periodic\ s \in L^2(\mathbb{T})\}.$$

If we define ψ by

$$\hat{\psi}(2\xi) = e^{i\xi}\, \overline{m_0(\xi+\pi)}\, \hat{\varphi}(\xi) \tag{2.12}$$

(that is, $s \equiv 1$ in (2.10)), we claim that we have found an orthonormal wavelet we are looking for. In fact, all orthonormal wavelets in W_0 can be characterized as follows:

PROPOSITION 2.13 *Suppose φ is a scaling function for an MRA $\{V_j\}_{j\in\mathbb{Z}}$, and m_0 is the associated low-pass filter; then a function $\psi \in W_0 = V_1 \cap V_0^\perp$ is an orthonormal wavelet for $L^2(\mathbb{R})$ if and only if*

$$\hat{\psi}(2\xi) = e^{i\xi}\nu(2\xi)\, \overline{m_0(\xi+\pi)}\, \hat{\varphi}(\xi) \tag{2.14}$$

a.e. on $\mathbb{R}$, for some 2π-periodic measurable function ν such that

$$|\nu(\xi)| = 1 \qquad a.e.\ on\ \mathbb{T}.$$

PROOF: First of all, it is clear that $\psi \in W_0$, since the last equality ensures that $\nu \in L^2(\mathbb{T})$. Now for any $g \in W_0$, by our characterization of W_0, there is a 2π-periodic $s \in L^2(\mathbb{T})$ such that $\hat{g}(2\xi) = e^{i\xi}s(2\xi)\, \overline{m_0(\xi+\pi)}\, \hat{\varphi}(\xi)$. This gives us

$$\hat{g}(\xi) = \frac{s(\xi)}{\nu(\xi)}\, e^{i\frac{\xi}{2}}\nu(\xi)\, \overline{m_0(\tfrac{1}{2}\xi+\pi)}\, \hat{\varphi}(\tfrac{1}{2}\xi) = \frac{s(\xi)}{\nu(\xi)}\hat{\psi}(\xi) = s(\xi)\, \overline{\nu(\xi)}\, \hat{\psi}(\xi).$$

Since $s\bar{\nu} \in L^2(\mathbb{T})$, we can write $s(\xi)\overline{\nu(\xi)} = \sum_{k\in\mathbb{Z}} c_k e^{-ik\xi}$, for a sequence $\{c_k\}_{k\in\mathbb{Z}} \in \ell^2(\mathbb{Z})$, and obtain

$$g(x) = \sum_{k\in\mathbb{Z}} c_k \psi(x-k),$$

which proves that $\{\psi(\cdot - k) : k \in \mathbb{Z}\}$ generates W_0. The orthonormality of this system can be proved by showing that $\hat{\psi}$ satisfies the equality in Proposition 1.11:

$$\sum_{k \in \mathbb{Z}} |\hat{\psi}(\xi + 2k\pi)|^2 = \sum_{k \in \mathbb{Z}} |\hat{\varphi}(\tfrac{1}{2}\xi + k\pi)|^2 |m_0(\tfrac{1}{2}\xi + k\pi + \pi)|^2$$

$$= \sum_{\ell \in \mathbb{Z}} |\hat{\varphi}(\tfrac{1}{2}\xi + 2\ell\pi)|^2 |m_0(\tfrac{1}{2}\xi + 2\ell\pi + \pi)|^2$$

$$+ \sum_{\ell \in \mathbb{Z}} |\hat{\varphi}(\tfrac{1}{2}\xi + 2\ell\pi + \pi)|^2 |m_0(\tfrac{1}{2}\xi + 2\ell\pi + 2\pi)|^2$$

$$= |m_0(\tfrac{1}{2}\xi + \pi)|^2 + |m_0(\tfrac{1}{2}\xi)|^2 = 1,$$

where we have summed over the even and odd integers separately, used the 2π-periodicity of m_0, Proposition 1.11 for φ and equality (2.5) for m_0.

We have already observed that if $\{\psi(\cdot - k) : k \in \mathbb{Z}\}$ is an orthonormal basis for W_0, then $\{2^{\frac{j}{2}}\psi(2^j \cdot -k) : k \in \mathbb{Z}\}$ is an orthonormal basis for W_j. Hence, (2.1) shows that ψ is, indeed, an orthonormal wavelet for $L^2(\mathbb{R})$.

Now we have to show that all orthonormal wavelets ψ in W_0 are described by (2.14). For any $\psi \in W_0$, by Lemma 2.11, there must be a 2π-periodic function $\nu \in L^2(\mathbb{T})$ such that

$$\hat{\psi}(\xi) = e^{i\frac{\xi}{2}} \nu(\xi) \overline{m_0(\tfrac{1}{2}\xi + \pi)} \, \hat{\varphi}(\tfrac{1}{2}\xi).$$

If ψ is an orthonormal wavelet, then the orthonormality of $\{\psi(\cdot - k) : k \in \mathbb{Z}\}$ gives us

$$1 = \sum_{k \in \mathbb{Z}} |\hat{\psi}(\xi + 2k\pi)|^2 = \sum_{k \in \mathbb{Z}} |\nu(\xi)|^2 |m_0(\tfrac{\xi}{2} + k\pi + \pi)|^2 |\hat{\varphi}(\tfrac{\xi}{2} + k\pi)|^2$$

$$= |\nu(\xi)|^2 \Big(\sum_{\ell \in \mathbb{Z}} |m_0(\tfrac{\xi}{2} + \pi)|^2 |\hat{\varphi}(\tfrac{\xi}{2} + 2\ell\pi)|^2$$

$$+ \sum_{\ell \in \mathbb{Z}} |m_0(\tfrac{\xi}{2})|^2 |\hat{\varphi}(\tfrac{\xi}{2} + 2\ell\pi + \pi)|^2 \Big)$$

$$= |\nu(\xi)|^2 \Big(|m_0(\tfrac{\xi}{2} + \pi)|^2 + |m_0(\tfrac{\xi}{2})|^2 \Big) = |\nu(\xi)|^2 \qquad \text{for } a.e. \ \xi \in \mathbb{T},$$

which finishes our proof.   ∎

This proposition completes the construction of a wavelet from an MRA.

Let us for simplicity consider the wavelet ψ given by (2.12) (in terms of Proposition 2.13 this means $\nu(\xi) \equiv 1$). Since ψ belongs to V_1 it must be a (countable) linear combination of translates of $\varphi(2x)$. In fact, there is a way of expressing ψ as a linear combination of translates of $\varphi(2x)$ with coefficients related to the α_k's that determined $m_0(\xi)$. From (2.12) and (2.3) we obtain

$$\hat{\psi}(2\xi) = \left(\sum_{k\in\mathbb{Z}}(-1)^k \,\overline{\alpha_k}\, e^{-i(k-1)\xi}\right)\hat{\varphi}(\xi).$$

Hence,

$$\hat{\psi}(\xi) = \left(\sum_{k\in\mathbb{Z}}(-1)^k \,\overline{\alpha_k}\, e^{-i(k-1)\frac{\xi}{2}}\right)\hat{\varphi}(\tfrac{1}{2}\xi).$$

Taking the inverse Fourier transform this gives us

$$\psi(x) = 2\sum_{k\in\mathbb{Z}}(-1)^k \,\overline{\alpha_k}\, \varphi(2x - (k-1)). \qquad (2.15)$$

EXAMPLE B: The **Haar wavelet** is constructed (up to a translation) from the MRA generated by the scaling function $\varphi(x) = \chi_{[-1,0)}(x)$ associated with the MRA described in Example A. Since

$$\tfrac{1}{2}\varphi(\tfrac{1}{2}x) = \tfrac{1}{2}\chi_{[-2,0)}(x) = \tfrac{1}{2}\varphi(x) + \tfrac{1}{2}\varphi(x+1),$$

we can use (2.15) to obtain

$$\psi(x) = \varphi(2x+1) - \varphi(2x) = \chi_{[-1,-\frac{1}{2})} - \chi_{[-\frac{1}{2},0)}.$$

This result can also be obtained directly. Since

$$\tfrac{1}{2}\varphi(\tfrac{1}{2}x) = \tfrac{1}{2}\varphi(x) + \tfrac{1}{2}\varphi(x+1),$$

the low-pass filter for the Haar wavelet is $m_0(\xi) = \tfrac{1}{2}(1 + e^{i\xi})$. Since

$$\hat{\varphi}(\xi) = \frac{1 - e^{i\xi}}{-i\xi} = e^{i\frac{\xi}{2}}\,\frac{\sin(\xi/2)}{\xi/2},$$

we obtain

$$\hat{\psi}(2\xi) = e^{i\xi}\,\frac{1 + e^{-i(\xi+\pi)}}{2}\,\frac{e^{i\xi} - 1}{i\xi} = e^{i\xi}\,\frac{(1 - e^{-i\xi})(e^{i\xi} - 1)}{2i\xi}.$$

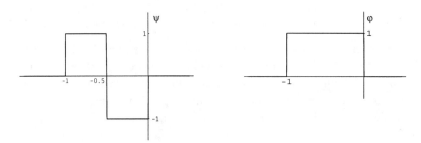

Figure 2.1: The Haar wavelet ψ and its scaling function φ.

Thus,

$$\hat{\psi}(\xi) = e^{i\frac{\xi}{2}} \frac{e^{i\frac{\xi}{2}} - 2 + e^{-i\frac{\xi}{2}}}{i\xi} = ie^{i\frac{\xi}{2}} \frac{\sin^2(\xi/4)}{(\xi/4)}.$$

It is easy to see that the Fourier transform of $\chi_{[-1,-\frac{1}{2})} - \chi_{[-\frac{1}{2},0)}$ coincides with this last expression.

To obtain the expression of the Haar wavelet given in Example A of Chapter 1 we need to take $\nu(\xi) = e^{-i\xi}$ in Proposition 2.13. We leave the details to the reader.

Before giving more examples, we show how we can obtain $|\hat{\varphi}|$ from $|\hat{\psi}|$. That is, we shall show that the modulus of the Fourier transform of a scaling function can be expressed in terms of the modulus of the Fourier transform of the wavelet. From (2.3), (2.12) and (2.5) we obtain

$$|\hat{\varphi}(2\xi)|^2 + |\hat{\psi}(2\xi)|^2 = |\hat{\varphi}(\xi)|^2 \left\{ |m_0(\xi)|^2 + |m_0(\xi + \pi)|^2 \right\} = |\hat{\varphi}(\xi)|^2.$$

Iterating this result we have

$$|\hat{\varphi}(\xi)|^2 = |\hat{\varphi}(2^N \xi)|^2 + \sum_{j=1}^{N} |\hat{\psi}(2^j \xi)|^2 \qquad \text{for all } N \geq 1.$$

Since $|\hat{\varphi}(\xi)| \leq 1$ the sequence $\left\{ \sum_{j=1}^{N} |\hat{\psi}(2^j \xi)|^2 : N = 2, 3, \cdots \right\}$ is an increasing sequence of real numbers bounded by 1 so that its limit exists; hence $\lim_{N \to \infty} |\hat{\varphi}(2^N \xi)|^2$ also exists. Moreover,

$$\int_{\mathbb{R}} |\hat{\varphi}(2^N \xi)|^2 \, d\xi = \frac{1}{2^N} \int_{\mathbb{R}} |\hat{\varphi}(\xi)|^2 \, d\xi \longrightarrow 0, \qquad \text{as } N \to \infty;$$

so that, by Fatou's lemma,

$$\int_{\mathbb{R}} \lim_{N \to \infty} |\hat{\varphi}(2^N \xi)|^2 \, d\xi \leq \lim_{N \to \infty} \int_{\mathbb{R}} |\hat{\varphi}(2^N \xi)|^2 \, d\xi = 0.$$

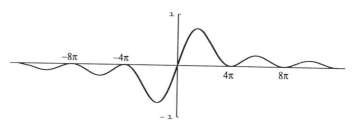

Figure 2.2: Graph of $\frac{\sin^2(\xi/4)}{(\xi/4)}$, related to the Fourier transform of the Haar wavelet ψ.

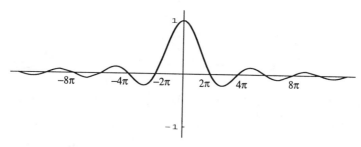

Figure 2.3: Graph of $\frac{\sin(\xi/2)}{(\xi/2)}$, related to the Fourier transform of the scaling function φ for the Haar wavelet.

This shows that

$$\lim_{N\to\infty} |\hat{\varphi}(2^N\xi)|^2 = 0.$$

We have proved

$$|\hat{\varphi}(\xi)|^2 = \sum_{j=1}^{\infty} |\hat{\psi}(2^j\xi)|^2 \qquad \text{for a.e. } \xi \in \mathbb{R}, \qquad (2.16)$$

which expresses $|\hat{\varphi}|$ in terms of the Fourier transform of the wavelet.

EXAMPLE C: The function ψ whose Fourier transform satisfies

$$\hat{\psi}(\xi) = e^{i\frac{\xi}{2}}\chi_I(\xi), \qquad \text{where } I = [-2\pi, -\pi) \cup (\pi, 2\pi],$$

is called the **Shannon wavelet** (see also Example B of Chapter 1). To prove that it is a wavelet we use (2.16) to obtain a scaling function for an MRA that will produce this wavelet. We have $\hat{\psi}(2^j\xi) = e^{i2^{j-1}\xi}\chi_{I_j}(\xi)$, where

$$I_j = [-2^{-j+1}\pi, -2^{-j}\pi) \cup (2^{-j}\pi, 2^{-j+1}\pi].$$

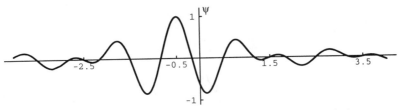

Figure 2.4: The Shannon wavelet $\psi(x) = -2\frac{\sin(2\pi x)+\cos(\pi x)}{\pi(2x+1)}$

for which $\hat{\psi}(\xi) = e^{i\frac{\xi}{2}}\chi_I(\xi)$.

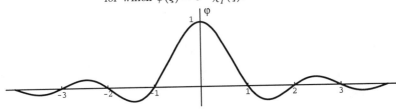

Figure 2.5: The scaling function $\varphi(x) = \frac{\sin(\pi x)}{\pi x}$ for the Shannon wavelet.

Since the I_j's are disjoint and their union, over $j \geq 1$, coincides with $[-\pi, 0) \cup (0, \pi]$ we can take $\hat{\varphi}(\xi) = \chi_{[-\pi,\pi]}(\xi)$. Notice that by Proposition 1.11, $\{\varphi(\cdot - k) : k \in \mathbb{Z}\}$ is an orthonormal system in $L^2(\mathbb{R})$. Choose V_j as the closed span of $\{\varphi_{j,k} = 2^{\frac{j}{2}}\varphi(2^j \cdot -k) : k \in \mathbb{Z}\}$, for all $j \in \mathbb{Z}$. Then $\{V_j : j \in \mathbb{Z}\}$ is a multiresolution analysis if we show that $\frac{1}{2}\varphi(\frac{1}{2}x)$ is an element of V_0; this is equivalent to finding the low pass filter m_0 that satisfies (2.3). Since m_0 has to be a 2π-periodic function in $L^2(\mathbb{T})$, equation (2.3) gives us

$$m_0(\xi) = \begin{cases} 1 & \text{if } -\frac{\pi}{2} \leq \xi < \frac{\pi}{2}, \\ 0 & \text{if } -\pi \leq \xi < -\frac{\pi}{2} \text{ or } \frac{\pi}{2} \leq \xi < \pi, \end{cases}$$

which is then extended periodically from $[-\pi, \pi]$ to $\mathbb{R}$. From (2.12) we can deduce $\hat{\psi}(\xi) = e^{i\frac{\xi}{2}}\chi_I(\xi)$ (we shall see that the factor $e^{i\frac{\xi}{2}}$ is not necessary in this case). Graphs of the Shannon wavelet ψ and the scaling function φ are given in Figures 2.4 and Figure 2.5.

EXAMPLE D: The Lemarié-Meyer family of wavelets was presented in Corollary 4.7 of Chapter 1. This family of wavelets is of the form

$$\hat{\psi}(\xi) = b(\xi)e^{i\frac{\xi}{2}},$$

where b is an even function on $\mathbb{R}$ that, restricted to $[0, \infty)$, is a bell function for $[\pi, 2\pi)$ with $0 < \varepsilon \leq \frac{1}{3}\pi$ and $\varepsilon' = 2\varepsilon$ (see the properties of a bell function described in (3.9) of Chapter 1). By (2.16), if ψ arises from an MRA, the scaling function φ must satisfy

$$|\hat{\varphi}(\xi)| = \begin{cases} 1 & \text{if } \xi \neq 0 \text{ and } |\xi| \leq \pi - \varepsilon, \\ s_{\varepsilon}(\xi + \pi) & \text{if } -\pi - \varepsilon \leq \xi < -\pi + \varepsilon, \\ s_{\varepsilon}(\pi - \xi) & \text{if } \pi - \varepsilon < \xi \leq \pi + \varepsilon, \\ 0 & \text{otherwise.} \end{cases}$$

Let us choose $\hat{\varphi}(\xi) = |\hat{\varphi}(\xi)|$ and let $\hat{\varphi}(0) = 1$; the system

$$\{\varphi(\cdot - k) : k \in \mathbb{Z}\}$$

is orthonormal by Proposition 1.11 of this chapter and (3.9) of Chapter 1. For $j \in \mathbb{Z}$, define V_j to be the closed span of the set

$$\{\varphi_{j,k} = 2^{\frac{j}{2}}\varphi(2^j \cdot -k) : k \in \mathbb{Z}\}.$$

Thus, by Theorem 1.6 and Theorem 1.7, we only have to show that $\frac{1}{2}\varphi(\frac{1}{2}x)$ is an element of V_0, or, equivalently, to find a 2π-periodic function m_0 belonging to $L^2(\mathbb{T})$ such that $\hat{\varphi}(2\xi) = m_0(\xi)\hat{\varphi}(\xi)$ (see (2.3)). This equality and the formula defining $|\hat{\varphi}(\xi)|$ tell us that we can choose

$$m_0(\xi) = \hat{\varphi}(2\xi), \quad \text{if } -\pi \leq \xi < \pi,$$

and then extend m_0 2π-periodically to $\mathbb{R}$. Figure 1.6 of Chapter 1 shows the graph of ψ with $\varepsilon = \frac{1}{3}\pi$. For $0 < \varepsilon \leq \frac{1}{3}\pi$, the graph of $\hat{\varphi}$ is given in Figure 2.6 (see section 3.3 for further details about the phase function associated with m_0 and $\hat{\varphi}$).

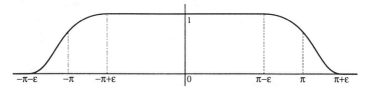

Figure 2.6: The graph of $\hat{\varphi}$ for the Lemarié-Meyer wavelets.

For the Haar wavelet (see Example B) we have

$$\hat{\varphi}(\xi) = e^{i\frac{\xi}{2}}\frac{\sin(\xi/2)}{\xi/2} \quad \text{and} \quad \hat{\psi}(\xi) = ie^{i\frac{\xi}{2}}\frac{\sin^2(\xi/4)}{\xi/4}.$$

Thus, the scaling function satisfies $\hat{\varphi}(2k\pi) = 0$ for all $k \neq 0$, $k \in \mathbb{Z}$, and the wavelet satisfies $\hat{\psi}(4k\pi) = 0$ for all $k \in \mathbb{Z}$ (see Figures 2.2 and 2.3). The same is clearly true for the Lemarié-Meyer wavelets described in Example D. In general, we have the following result:

PROPOSITION 2.17 *If φ is a scaling function for an MRA and $|\hat{\varphi}|$ is continuous, then*

$$\hat{\varphi}(2k\pi) = 0 \qquad \text{for all } k \neq 0, k \in \mathbb{Z}.$$

Furthermore, if ψ is a wavelet constructed from this scaling function via Proposition 2.13 with both $|m_0|$ and $|\nu|$ continuous, then

$$\hat{\psi}(4k\pi) = 0 \qquad \text{for all } k \in \mathbb{Z}.$$

PROOF : We prove the first part by contradiction. If $\hat{\varphi}(2k_o\pi) \neq 0$ for some $k_o \neq 0$, $k_o \in \mathbb{Z}$, say, $|\hat{\varphi}(2k_o\pi)| = a > 0$, we must have

$$|\hat{\varphi}(\xi)|^2 + |\hat{\varphi}(\xi + 2k_o\pi)|^2 > 1 + \tfrac{1}{2}a^2 \qquad \text{when } \xi \in [-\varepsilon, \varepsilon]$$

for some $\varepsilon > 0$, by the continuity of $|\hat{\varphi}|$ and Theorem 1.7. This contradicts Proposition 1.11.

To show the second part we observe from (2.14) that

$$|\hat{\psi}(2\xi)| = |m_0(\xi + \pi)\hat{\varphi}(\xi)| \qquad \text{on } \mathbb{R},$$

by the continuity of $|\hat{\varphi}|$, $|m_0|$ and $|\nu|$. Thus, $|\hat{\psi}(4k\pi)| = |m_0(\pi)\hat{\varphi}(2k\pi)|$ for all $k \in \mathbb{Z}$. But (2.3) and (2.5) together with the continuity of $|\hat{\varphi}|$ and $|m_0|$ give us $m_0(\pi) = 0$. Therefore, $|\hat{\psi}(4k\pi)| = 0$ for all $k \in \mathbb{Z}$. ∎

There are examples of wavelets which do **not** arise from a multiresolution analysis. The first example was given by J.L. Journé (see [Mal1]) and consists of a function ψ which satisfies $|\hat{\psi}| = \chi_K$, where

$$K = [-\tfrac{32}{7}\pi, -4\pi) \cup [-\pi, -\tfrac{4}{7}\pi) \cup (\tfrac{4}{7}\pi, \pi] \cup (4\pi, \tfrac{32}{7}\pi].$$

We will develop conditions from which it follows readily that this ψ is an orthonormal wavelet (see Example A of Chapter 7). Accepting this, for the moment, ψ cannot arise from an MRA. If this were the case, the scaling function must satisfy (2.16); from this equation we obtain

$$|\hat{\varphi}(\xi)| = \begin{cases} 1 & \text{if } 0 < |\xi| \leq \frac{4}{7}\pi \text{ or } \pi \leq |\xi| \leq \frac{8}{7}\pi \text{ or } 2\pi \leq |\xi| \leq \frac{16}{7}\pi, \\ 0 & \text{otherwise.} \end{cases}$$

If φ were the scaling function of an MRA, it would satisfy Proposition 1.11; that is, the squares of its translates by $2k\pi$, $k \in \mathbb{Z}$, should add to 1 almost everywhere. If $I = (0, \frac{2}{7}\pi)$, then $|\hat{\varphi}(\xi)| = 1$ for all $\xi \in I$; moreover, for $\xi \in I$, $2\pi < \xi + 2\pi < \frac{16}{7}\pi$ so that $|\hat{\varphi}(\xi + 2\pi)| = 1$. But this shows that

$$\sum_{k \in \mathbb{Z}} |\hat{\varphi}(\xi + 2k\pi)|^2 \geq |\hat{\varphi}(\xi)|^2 + |\hat{\varphi}(\xi + 2\pi)|^2 = 2 \qquad \text{for } \xi \in I,$$

which proves that the equality in Proposition 1.11 cannot hold.

It turns out that, in a real sense, those wavelets that do not arise from an MRA are "rare." In fact, it is known (and we will present this in Chapter 7) that, if ψ is an orthonormal wavelet such that $\hat{\psi}$ has some regularity and a mild decay at ∞, then ψ must be associated with an MRA. A particularly simple condition that characterizes those wavelets ψ that arise from an MRA is

$$\sum_{k \in \mathbb{Z}} \sum_{j=1}^{\infty} |\hat{\psi}(2^j(\xi + 2k\pi))|^2 = 1 \qquad \text{for } a.e. \ \xi \in \mathbb{R}. \qquad (2.18)$$

It follows immediately from (2.16) and Proposition 1.11 that this equality is necessary. That (2.18) is sufficient for a wavelet ψ to be associated with an MRA is much more difficult to prove and we postpone the argument until Chapter 7, where we shall present all the precise results concerning wavelets that must be associated with an MRA.

We know how to obtain the low pass filter m_0 from the scaling function φ (see (2.2) and (2.3)); in some cases it is possible to find the scaling function if we know the low pass filter. We shall consider this question in detail in Chapter 7; at this point we present a special case that will suffice for some of the constructions we carry out in the next section.

Given a function φ we define its **radial (or even) majorant** R_φ by

$$R_\varphi(x) = \operatorname*{ess.sup}_{|t| \geq |x|} |\varphi(t)|. \qquad (2.19)$$

Assume that the radial majorant belongs to $L^1(\mathbb{R})$. This is certainly true for the Haar wavelet and for each of the Lemarié-Meyer wavelets described in Example D. It is also true for all compactly supported wavelets (which we shall develop in the next section) and for spline wavelets (that will be described in Chapter 4). Notice that this is not true for the scaling function of the Shannon wavelet (presented in Example C), since $|\varphi(x)| \sim \frac{1}{|x|}$ (at $\pm\infty$) for this wavelet. For simplicity we assume $\varphi \in L^\infty(\mathbb{R})$.

Observe that R_φ is radial, decreasing on $[0, \infty)$, and $|\varphi(x)| \leq R_\varphi(x)$ for almost every $x \in \mathbb{R}$. We have, for $k \neq 0$,

$$|\alpha_k| = \tfrac{1}{2}\left| \int_{\mathbb{R}} \varphi(\tfrac{1}{2}x)\, \overline{\varphi(x+k)}\, dx \right|$$

$$= \tfrac{1}{2}\left| \int_{|x| < \frac{|k|}{2}} \varphi(\tfrac{1}{2}x)\, \overline{\varphi(x+k)}\, dx + \int_{|x| \geq \frac{|k|}{2}} \varphi(\tfrac{1}{2}x)\, \overline{\varphi(x+k)}\, dx \right|.$$

If $|x| < \tfrac{1}{2}|k|$, then $|x+k| > \tfrac{1}{2}|k|$ and, hence, $|\varphi(x+k)| \leq R_\varphi(\tfrac{1}{2}|k|)$. For $|x| \geq \tfrac{1}{2}|k|$, $|\varphi(\tfrac{1}{2}x)| \leq R_\varphi(\tfrac{1}{4}|k|)$. Therefore, since $\varphi \in L^1(\mathbb{R})$ (which follows from the assumption $R_\varphi \in L^1(\mathbb{R})$),

$$|\alpha_k| \leq \tfrac{1}{2} R_\varphi(\tfrac{1}{2}|k|) \int_{|x| < \frac{|k|}{2}} |\varphi(\tfrac{1}{2}x)|\, dx + \tfrac{1}{2} R_\varphi(\tfrac{1}{4}|k|) \int_{|x| \geq \frac{|k|}{2}} |\varphi(x+k)|\, dx$$

$$\leq c(\varphi) R_\varphi(\tfrac{1}{4}|k|).$$

Consequently, we have

$$\sum_{k \in \mathbb{Z}} |\alpha_k| \leq c(\varphi) \sum_{k \in \mathbb{Z}} R_\varphi(\tfrac{1}{4}|k|) \leq 2c(\varphi) \sum_{k=1}^{\infty} \int_{k-1}^{k} R_\varphi(\tfrac{1}{4}x)\, dx < \infty.$$

Since $m_0(\xi) = \sum_{k \in \mathbb{Z}} \alpha_k e^{ik\xi}$, this shows the following result:

PROPOSITION 2.20 *If φ is a scaling function for an MRA which is bounded and its radial majorant, defined by (2.19), belongs to $L^1(\mathbb{R})$, then the low-pass filter $m_0 \in C(\mathbb{T})$.*

Iterating $\hat{\varphi}(\xi) = m_0(\tfrac{1}{2}\xi)\hat{\varphi}(\tfrac{1}{2}\xi)$ (see (2.3)), we obtain

$$\hat{\varphi}(\xi) = \hat{\varphi}(2^{-N}\xi) \prod_{j=1}^{N} m_0(2^{-j}\xi) \qquad \text{for all integers } N \geq 1.$$

Let us assume that φ is integrable, then $\hat{\varphi}$ is continuous and, by Theorem 1.7, $|\hat{\varphi}(0)| = 1$. Hence,

$$m_0(0) = \lim_{\xi \to 0} \frac{\hat{\varphi}(2\xi)}{\hat{\varphi}(\xi)} = 1.$$

From this it follows that

$$\lim_{N \to \infty} \prod_{j=1}^{N} m_0(2^{-j}\xi) = \lim_{N \to \infty} \frac{\hat{\varphi}(\xi)}{\hat{\varphi}(2^{-N}\xi)} = \frac{\hat{\varphi}(\xi)}{\hat{\varphi}(0)}.$$

We can always assume that $\hat{\varphi}(0) = 1$, and we will, indeed, do so. We then have the following result.

PROPOSITION 2.21 *If φ is a bounded scaling function for an MRA, its radial majorant R_φ belongs to $L^1(\mathbb{R})$, and $\hat{\varphi}(0) = 1$, then we have*

$$\hat{\varphi}(\xi) = \prod_{j=1}^{\infty} m_0(2^{-j}\xi).$$

More regular low-pass filters m_0 can be obtained if we assume stronger conditions on φ. One of these conditions is to assume that there exists a positive number m such that

$$\int_{-\infty}^{\infty} (1 + |x|^2)^m |\varphi(x)|^2 \, dx = c_m < \infty \tag{2.22}$$

This condition implies

$$\int_{|x| \geq A} |\varphi(x)|^2 \, dx \leq \frac{c_m}{(1 + A^2)^m} \tag{2.23}$$

for all $A > 0$. Then, using Schwarz's inequality and the fact that $\|\varphi\|_2 = 1$, we obtain

$$|\alpha_k| = \frac{1}{2} \left| \int_{-\infty}^{\infty} \varphi(\tfrac{1}{2}x) \overline{\varphi(x+k)} \, dx \right|$$

$$= \frac{1}{2} \left| \int_{|x| < \frac{|k|}{2}} \varphi(\tfrac{1}{2}x) \overline{\varphi(x+k)} \, dx + \int_{|x| \geq \frac{|k|}{2}} \varphi(\tfrac{1}{2}x) \overline{\varphi(x+k)} \, dx \right|$$

$$\leq \frac{1}{2}\sqrt{2} \left(\int_{|x| < \frac{|k|}{2}} |\varphi(x+k)|^2 \, dx \right)^{\frac{1}{2}} + \frac{1}{2} 1 \left(\int_{|x| \geq \frac{|k|}{2}} |\varphi(\tfrac{1}{2}x)|^2 \, dx \right)^{\frac{1}{2}}.$$

With the change of variables $x + k = y$ in the first integral and $y = \frac{1}{2}x$ in the second, we can easily show that

$$|\alpha_k| \leq \frac{c'_m}{(1 + |k|^2)^{\frac{m}{2}}}.$$

If m is an integer, $m > 2$, this inequality implies that the low-pass filter $m_0(\xi) = \sum_{k \in \mathbb{Z}} \alpha_k e^{ik\xi}$ belongs to $C^{m-2}(\mathbb{T})$.

We say that a function φ is of **polynomial decay** in $L^2(\mathbb{R})$ if (2.22) is satisfied for every non-negative integer m. (Some authors use the term "localized" for this property. The reader will clearly observe that this property can also be defined in terms of Sobolev spaces, in the frequency variable). The Haar scaling function and, in fact, any compactly supported scaling function is of polynomial decay. Also the Lemarié-Meyer wavelets presented in Example D have polynomial decay when $b \in C^\infty(\mathbb{R})$, since, then, ψ belongs to the Schwartz class.

For scaling functions with polynomial decay we have the following result:

PROPOSITION 2.24 *If a scaling function φ for an MRA has polynomial decay, the low-pass filter m_0 belongs to $C^\infty(\mathbb{T})$. If, in addition, $\hat{\varphi}(0) = 1$, we have*

$$\hat{\varphi}(\xi) = \prod_{j=1}^{\infty} m_0(2^{-j}\xi).$$

These results lead to the problem of finding properties satisfied by a 2π-periodic function m_0 that guarantee that m_0 is a low-pass filter associated with an MRA. In fact, this point of view is fundamental for the material we develop in the following section.

2.3   The construction of compactly supported wavelets

In this section we will show that for any non-negative integer n there exists an orthonormal wavelet ψ with compact support such that all the derivatives of ψ up to order n exist and are bounded. The existence of this orthonormal wavelet will be shown using the concept of a multiresolution analysis developed in the above two sections of this chapter.

We have seen several examples of orthonormal wavelets in section 2.2. But only one of them had compact support: this is the Haar wavelet given in Example B; in this case $\psi(x) = \chi_{[-1,-\frac{1}{2})} - \chi_{[-\frac{1}{2},0)}$, which satisfies our requirements with $n = 0$.

We shall try to construct a scaling function φ from a "filter" m_0. We assume that m_0 satisfies

$$\begin{cases} m_0 \in C^1 \text{ and is } 2\pi\text{-periodic,} \\ |m_0(\xi)|^2 + |m_0(\xi + \pi)|^2 = 1, \\ m_0(0) = 1. \end{cases} \tag{3.1}$$

The two equalities in (3.1) are, of course, necessary (see (2.5) and the argument used for establishing Proposition 2.21).

Because of (2.3), that is $\hat{\varphi}(\xi) = m_0(\frac{1}{2}\xi)\hat{\varphi}(\frac{1}{2}\xi)$, if such a φ existed, we must have

$$\hat{\varphi}(\xi) = \hat{\varphi}(2^{-N}\xi) \prod_{j=1}^{N} m_0(2^{-j}\xi) \equiv \hat{\varphi}(2^{-N}\xi)\Pi_N(\xi) \tag{3.2}$$

for each $N \in \mathbb{N}$. Theorem 1.7 leads us to expect $\hat{\varphi}$ to be continuous at 0 and, moreover, $\hat{\varphi}(0) = 1$ (since we concluded that $|\hat{\varphi}(0)| = 1$, as we remarked above, there is no loss of generality in assuming $\hat{\varphi}(0) = 1$). The infinite product

$$\prod_{j=1}^{\infty} m_0(2^{-j}\xi)$$

clearly converges for each $\xi \in \mathbb{R}$. We also have, by (3.1), $|m_0(\xi)| \leq 1$ for all ξ and, thus, $|\Pi_N(\xi)| \leq 1$. Consequently,

$$\Pi_{N+1}(\xi) - \Pi_N(\xi) = \Pi_N(\xi)(m_0(2^{-N-1}\xi) - 1),$$

and, therefore,

$$|\Pi_{N+1}(\xi) - \Pi_N(\xi)| \leq |m_0(2^{-N-1}\xi) - m_0(0)|$$

$$\leq \|m_0'\|_\infty 2^{-(N+1)}|\xi|.$$

Hence,

$$|\Pi_{N+M}(\xi) - \Pi_N(\xi)| \leq \|m_0'\|_\infty |\xi|\{2^{-(N+1)} + \cdots + 2^{-(N+M)}\}$$

$$\leq \|m_0'\|_\infty |\xi|2^{-N}$$

for all $N, M \in \mathbb{N}$. This shows that the sequence $\{\Pi_N(\xi) : N \in \mathbb{N}\}$ not only converges but it does so uniformly on bounded sets. Observe that we only need m_0 to be in the class C^1 for this argument to be valid. Also observe that $m_0(2^{-j}\xi) = 1 + O(2^{-j})$ for j large, so that the convergence of $\prod_{j=1}^{\infty} m_0(2^{-j}\xi)$ is equivalent to the convergence of $\sum_{j=1}^{\infty} 2^{-j}$. This is, in fact, another, equivalent, way to show the result just proved. Moreover, it is an easy consequence of the fact that Π_N is bounded by 1, and converges uniformly on bounded sets towards φ, that the convergence holds in the sense of tempered distributions. Thus, we shall define φ by letting

$$\hat{\varphi}(\xi) = \prod_{j=1}^{\infty} m_0(2^{-j}\xi) \qquad \text{for } \xi \in \mathbb{R}. \tag{3.3}$$

We then look for other properties of m_0 that will assure us that φ is a scaling function of a compactly supported orthonormal wavelet ψ. For example, if φ is the desired scaling function, then, from (2.3) we would have

$$m_0(\xi) = \sum_{k \in \mathbb{Z}} \alpha_k e^{ik\xi}$$

with

$$\alpha_k = \frac{1}{2} \int_{-\infty}^{\infty} \varphi(\tfrac{1}{2}x) \overline{\varphi(x+k)} \, dx.$$

That is, the α_k's are the coefficients of the expansion of $\frac{1}{2}\varphi(\frac{\cdot}{2}) \in V_{-1}$ in terms of the orthonormal basis $\{\varphi(\cdot + k) : k \in \mathbb{Z}\}$ of V_0. If the scaling function φ is to be compactly supported, only a finite number of these α_k can be non-zero. Thus, $m_0(\xi)$ must be a trigonometric polynomial.

The "simplest" case is the Haar wavelet; with the notation of Example B, $\varphi(x) = \chi_{[-1,0)}(x)$ so that $\frac{1}{2}\varphi(\frac{1}{2}x) = \frac{1}{2}\varphi(x) + \frac{1}{2}\varphi(x+1)$ and, thus,

$$m_0(\xi) = \tfrac{1}{2}(1 + e^{i\xi}).$$

From (2.12) we then obtain $\hat{\psi}(2\xi) = e^{i\xi}\frac{1}{2}(1 + e^{-i\xi}e^{-i\pi})\hat{\varphi}(\xi)$, which gives us $\psi(x) = \varphi(2x+1) - \varphi(2x)$; hence

$$\psi(x) = \begin{cases} 1 & \text{if } -1 \le x < -\frac{1}{2}, \\ -1 & \text{if } -\frac{1}{2} \le x < 0, \\ 0 & \text{otherwise.} \end{cases}$$

We now make another observation that shows that we cannot expect a compactly supported orthonormal wavelet to belong to $C^\infty(\mathbb{R})$. This will be obtained as a consequence of the following result:

Theorem 3.4 *Let r be a non-negative integer. Let ψ be a function in $C^r(\mathbb{R})$ such that*

$$|\psi(x)| \leq \frac{C}{(1+|x|)^{r+1+\varepsilon}} \qquad \text{for some } \varepsilon > 0, \qquad (3.5)$$

and that $\psi^{(m)} \in L^\infty(\mathbb{R})$ for $m = 1, 2, \cdots, r$. If $\{\psi_{j,k} : j, k \in \mathbb{Z}\}$ is an orthonormal system in $L^2(\mathbb{R})$, then all moments of ψ up to order r are zero; that is,

$$\int_\mathbb{R} x^m \psi(x)\, dx = 0 \qquad \text{for all } m = 0, 1, 2, \cdots, r.$$

Proof : We proceed by induction. Let us first assume that $r = 0$. Let a be a dyadic number, $a = 2^{-j_o} k_o$ for some $j_o, k_o \in \mathbb{Z}$, such that $\psi(a) \neq 0$. Since $\|\psi\|_2 = 1$, and ψ is continuous, such an a exists. We know that

$$\int_\mathbb{R} \overline{\psi(x)}\, \psi(2^j x - k)\, dx = 0 \qquad \text{whenever } (j, k) \neq (0, 0).$$

Taking $k = 2^{j-j_o} k_o$, with $j > \max\{j_o, 0\}$, this equality becomes

$$\int_\mathbb{R} \overline{\psi(x)}\, \psi(2^j(x - a))\, dx = 0.$$

Let $y = 2^j(x - a)$. Then

$$\int_\mathbb{R} \overline{\psi(a + 2^{-j} y)}\, \psi(y)\, dy = 0.$$

Since the left-hand side tends to $\overline{\psi(a)} \int_\mathbb{R} \psi(y)\, dy$, when $j \to \infty$ (by the Lebesgue dominated convergence theorem), and $\psi(a) \neq 0$, we have

$$\int_\mathbb{R} \psi(y)\, dy = 0.$$

Before considering the general case, let us prove the theorem when $r = 1$ in order to explain the method in a simple context. Since $\int_\mathbb{R} \psi(y)\, dy = 0$,

$$\vartheta(x) \equiv \int_{-\infty}^x \psi(y)\, dy$$

tends to 0 at ∞. Moreover,

$$\vartheta(x) = \int_{-\infty}^{x} \psi(y) \, dy = -\int_{x}^{\infty} \psi(y) \, dy.$$

Since we assume that

$$|\psi(x)| \leq \frac{C}{(1+|x|)^{2+\varepsilon}} \qquad \text{for an } \varepsilon > 0,$$

then

$$|\vartheta(x)| \leq \frac{C_1}{(1+|x|)^{1+\varepsilon}} \, .$$

Integrating by parts,

$$\int_{-\infty}^{\infty} \vartheta(x) \, dx = -\int_{-\infty}^{\infty} x\psi(x) \, dx.$$

It suffices to prove that

$$\int_{-\infty}^{\infty} \vartheta(x) \, dx = 0.$$

We can adapt the argument we just used for ψ. Since ψ is not constant and ψ' is continuous, there exists an $a = 2^{-j_o} k_o$ such that $\psi'(a) \neq 0$. Let us write again

$$\int_{\mathbb{R}} \overline{\psi(x)} \, \psi(2^j(x-a)) \, dx = 0.$$

Then, integrating by parts, it follows that

$$\int_{\mathbb{R}} \overline{\psi'(x)} \, \vartheta(2^j(x-a)) \, dx = 0,$$

which, by a change of variables, gives us

$$\int_{\mathbb{R}} \overline{\psi'(a+2^{-j}x)} \, \vartheta(x) \, dx = 0.$$

We conclude, letting j tend to ∞, that

$$\int_{-\infty}^{\infty} x\psi(x) \, dx = -\int_{-\infty}^{\infty} \vartheta(x) \, dx = 0.$$

Let us now proceed by induction from $r - 1$ to r. Since all moments up to order $r - 1$ are 0, we can integrate ψ r-times and obtain functions $\vartheta_1 = \vartheta, \vartheta_2, \cdots, \vartheta_r$ such that $\vartheta'_\ell = \vartheta_{\ell-1}$ and $\vartheta_1, \vartheta_2, \cdots, \vartheta_r$ all tend to 0 at $\pm\infty$. Moreover, there exist constants C_ℓ such that

$$|\vartheta_\ell(x)| \leq \frac{C_\ell}{(1 + |x|)^{r-\ell+1+\varepsilon}}, \qquad \ell = 1, 2, \cdots, r.$$

Integrating by parts r-times, it is easy to see that

$$\int_{\mathbb{R}} x^r \psi(x)\, dx = 0 \qquad \text{is equivalent to} \qquad \int_{\mathbb{R}} \vartheta_r(x)\, dx = 0.$$

Again, since ψ is not a polynomial and $\psi^{(r)}$ is continuous, there exists an $a = 2^{-j_o} k_o$ with $\psi^{(r)}(a) \neq 0$. From

$$\int_{\mathbb{R}} \overline{\psi(x)}\, \psi(2^j(x-a))\, dx = 0$$

we obtain

$$\int_{\mathbb{R}} \overline{\psi^{(r)}(x)}\, \vartheta_r(2^j(x-a))\, dx = 0$$

by integration by parts r-times. We obtain the desired result by changing variables and letting j tend to ∞.

∎

REMARK 1: It is sufficient to assume that the last derivative be in $L^\infty(\mathbb{R})$. Let us prove it for $r = 0$. The same modification of the proof we presented will give us the general case.

PROPOSITION 3.6 Let ψ be an $L^\infty(\mathbb{R})$ function such that

$$|\psi(x)| \leq \frac{C}{(1 + |x|)^{1+\varepsilon}} \qquad a.e., \qquad \text{for some } \varepsilon > 0.$$

If $\{\psi_{j,k} : j, k \in \mathbb{Z}\}$ is an orthonormal system in $L^2(\mathbb{R})$, then

$$\int_{\mathbb{R}} \psi(x)\, dx = 0.$$

PROOF: Let a be a Lebesgue point for ψ such that $\psi(a) \neq 0$. Let a_j be a sequence of dyadic points $2^{-j} k_j$ such that $|a_j - a| < 2^{-j}$.

Then, as before, we have

$$\int_{\mathbb{R}} \overline{\psi(a_j + 2^{-j}x)}\, \psi(x)\, dx = 0.$$

To prove that $\int_{\mathbb{R}} \psi(x)\, dx = 0$, it is sufficient to show that

$$\int_{\mathbb{R}} \overline{\psi(a_j + 2^{-j}x)}\, \psi(x)\, dx \longrightarrow \overline{\psi(a)} \int_{\mathbb{R}} \psi(x)\, dx \qquad \text{as } j \to \infty.$$

Let

$$I_j = \int_{\mathbb{R}} \big|\psi(a_j + 2^{-j}x) - \psi(a)\big| \cdot |\psi(x)|\, dx.$$

Then

$$I_j \le \|\psi\|_{\infty} \int_{|x|<1} \big|\psi(a_j + 2^{-j}x) - \psi(a)\big|\, dx$$

$$+\, C \sum_{\ell \ge 1} 2^{-\ell-\varepsilon\ell} \int_{2^{\ell-1} \le |x| < 2^{\ell}} \big|\psi(a_j + 2^{-j}x) - \psi(a)\big|\, dx$$

$$\le C \sum_{\ell \ge 0} 2^{-\ell-\varepsilon\ell} \int_{|x|<2^{\ell}} \big|\psi(a_j + 2^{-j}x) - \psi(a)\big|\, dx.$$

But

$$u(\ell, j) \equiv 2^{-\ell} \int_{|x|<2^{\ell}} \big|\psi(a_j + 2^{-j}x) - \psi(a)\big|\, dx$$

is bounded by $4\|\psi\|_{\infty}$. Moreover, using the fact that $|a_j - a| < 2^{-j} \le 2^{\ell-j}$ for $\ell \ge 0$,

$$u(\ell, j) \le 2^{-\ell+j} \int_{a-2^{\ell-j+1}}^{a+2^{\ell-j+1}} \big|\psi(y) - \psi(a)\big|\, dy.$$

Since a is a Lebesgue point, for every ℓ, the sequence $\{u(\ell, j)\}_{j=1}^{\infty}$ tends to 0 as j tends to ∞. Therefore,

$$I_j \le C \sum_{\ell \ge 0} 2^{-\varepsilon\ell} u(\ell, j)$$

tends to 0 when j tends to ∞. This finishes the proof.   ∎

REMARK 2: Since

$$\frac{d^m\hat{\psi}}{d\xi^m}(0) = (-i)^m \int_{\mathbb{R}} x^m \psi(x)\, dx,$$

under the hypotheses of Theorem 3.4 we have

$$\frac{d^m\hat{\psi}}{d\xi^m}(0) = 0 \qquad \text{for}\;\; m = 0, 1, 2, \cdots, r.$$

COROLLARY 3.7 *Let $\psi \in L^2(\mathbb{R})$ be a function in the Schwartz class S such that $\{\psi_{j,k} : j, k \in \mathbb{Z}\}$ is an orthonormal system in $L^2(\mathbb{R})$. Then,*

$$\frac{d^m\hat{\psi}}{d\xi^m}(0) = 0 \qquad \text{for all}\;\; m = 0, 1, 2, \cdots;$$

or, equivalently,

$$\int_{\mathbb{R}} x^m \psi(x)\, dx = 0 \qquad \text{for all}\;\; m = 0, 1, 2, \cdots.$$

The following theorem is also a corollary of Theorem 3.4.

THEOREM 3.8 *Let $\psi \in L^2(\mathbb{R})$ be a compactly supported function such that $\psi \in C^\infty$; then $\{\psi_{j,k} : j, k \in \mathbb{Z}\}$ **cannot** be an orthonormal system in $L^2(\mathbb{R})$, where $\psi_{j,k}(x) = 2^{\frac{j}{2}}\psi(2^j x - k)$.*

PROOF: If $\{\psi_{j,k} : j, k \in \mathbb{Z}\}$ is an orthonormal system in $L^2(\mathbb{R})$, then Theorem 3.4, which can be applied since the compactly supported function ψ satisfies (3.5), implies that all the moments of ψ are zero.

It follows that for all polynomials p in the variable x, $\int_{\mathbb{R}} p(x)\overline{\psi(x)}\, dx = 0$. Since ψ has compact support, given $\varepsilon > 0$ we can find a polynomial $p(x)$ such that $\sup_{x \in K}|\psi(x) - p(x)| < \varepsilon$, where K is the support of ψ (by the Weierstrass approximation theorem). Hence,

$$\|\psi\|_2^2 = \int_{\mathbb{R}} \psi(x)\,\overline{\psi(x)}\, dx = \int_K [\psi(x) - p(x)]\,\overline{\psi(x)}\, dx$$

$$\leq \varepsilon \int_K |\psi(x)|\, dx = \varepsilon \|\psi\|_1.$$

Since $\|\psi\|_1 \le \infty$ and $\varepsilon > 0$ is arbitrary, we must have $\|\psi\|_2^2 = 0$, which contradicts the normality of the system.

∎

This theorem shows that we can only expect a compactly supported orthonormal wavelet to be, at most, in the class $C^n(\mathbb{R})$ for some finite non-negative integer n.

Observe that Proposition 3.6 shows that if ψ is a bounded orthonormal wavelet with compact support we must have $\int_{-\infty}^{\infty} \psi(x)\,dx = 0$, or equivalently, $\hat{\psi}(0) = 0$. Similarly, if we want a compactly supported orthonormal wavelet ψ such that $\psi \in C^n(\mathbb{R})$, we must have

$$\int_{-\infty}^{\infty} x^\ell \psi(x)\,dx = 0 \qquad \text{for } \ell = 0, 1, 2, \cdots, n.$$

Let us now show that the proposed scaling function φ given by (3.3) belongs to $L^2(\mathbb{R})$.

PROPOSITION 3.9 *Let m_0 be a function defined on $\mathbb{R}$ satisfying (3.1); then, the function φ defined by (3.3) is in $L^2(\mathbb{R})$ and $\|\varphi\|_2 \le 1$.*

PROOF : Let $\Pi_N(\xi) = \prod_{j=1}^{N} m_0(2^{-j}\xi)$, which is a $2^{N+1}\pi$-periodic function. Hence, by (3.1) we obtain

$$I_N \equiv \int_{-2^N\pi}^{2^N\pi} |\Pi_N(\xi)|^2\,d\xi = \int_0^{2^{N+1}\pi} |\Pi_N(\xi)|^2\,d\xi$$

$$= \int_0^{2^N\pi} |\Pi_N(\xi)|^2\,d\xi + \int_{2^N\pi}^{2^{N+1}\pi} |\Pi_N(\xi)|^2\,d\xi$$

$$= \int_0^{2^N\pi} \left\{ |\Pi_N(\xi)|^2 + |\Pi_N(\xi + 2^N\pi)|^2 \right\} d\xi$$

$$= \int_0^{2^N\pi} |\Pi_{N-1}(\xi)|^2 \left\{ |m_0(2^{-N}\xi)|^2 + |m_0(2^{-N}\xi + \pi)|^2 \right\} d\xi$$

$$= \int_0^{2^N\pi} |\Pi_{N-1}(\xi)|^2\,d\xi = I_{N-1}.$$

Repeating this argument we obtain

$$I_N = I_1 = \int_0^{4\pi} |m_0(\tfrac{1}{2}\xi)|^2 \, d\xi = 2 \int_0^{2\pi} |m_0(\xi)|^2 \, d\xi$$

$$= 2 \int_0^{\pi} \left\{ |m_0(\xi)|^2 + |m_0(\xi + \pi)|^2 \right\} d\xi = 2\pi.$$

Since $|\Pi_N(\xi)|^2 \to |\hat{\varphi}(\xi)|^2$, by Fatou's lemma (applied to the sequence $\{\chi_{[-2^N\pi, 2^N\pi]} \Pi_N\}$) we have

$$\frac{1}{2\pi} \int_{\mathbb{R}} |\hat{\varphi}(\xi)|^2 \, d\xi \leq \frac{1}{2\pi} \lim_{N \to \infty} \int_{-2^N\pi}^{2^N\pi} |\Pi_N(\xi)|^2 \, d\xi = \frac{1}{2\pi} \lim_{N \to \infty} I_N = 1.$$

Thus, $\hat{\varphi} \in L^2(\mathbb{R})$ and φ, therefore, is well defined and $\|\varphi\|_2^2 = \frac{1}{2\pi} \|\hat{\varphi}\|_2^2 \leq 1.$ ∎

EXAMPLE E: Let us again examine the Haar wavelet given in Example B of section 2.2. We have $\varphi(x) = \chi_{[-1,0)}(x)$ so that

$$\hat{\varphi}(\xi) = \frac{e^{i\xi} - 1}{i\xi}, \qquad m_0(\xi) = \frac{1 + e^{i\xi}}{2}$$

and

$$\hat{\psi}(\xi) = i e^{i\frac{\xi}{2}} \frac{\sin^2(\xi/4)}{\xi/4}.$$

Let us show that, in this case, (3.3) is satisfied; that is, we have to prove

$$\prod_{j=1}^{\infty} \frac{(1 + e^{i2^{-j}\xi})}{2} = \frac{e^{i\xi} - 1}{i\xi} \tag{3.10}$$

or, equivalently,

$$\prod_{j=1}^{\infty} e^{i2^{-j-1}\xi} \cos(2^{-j-1}\xi) = e^{i\frac{\xi}{2}} \frac{\sin(\xi/2)}{\xi/2}.$$

Since the product of the exponentials on the left-hand side equals $e^{i\frac{\xi}{2}}$ we must show

$$\prod_{j=1}^{\infty} \cos(2^{-j-1}\xi) = \frac{\sin(\xi/2)}{\xi/2}.$$

Using the formula $\sin 2\alpha = 2\cos\alpha\sin\alpha$ we can write

$$\prod_{j=1}^{n}\cos(2^{-j-1}\xi) = \prod_{j=1}^{n}\frac{\sin(2^{-j}\xi)}{2\sin(2^{-j-1}\xi)} = \frac{\sin(\xi/2)}{2^n\sin(2^{-n-1}\xi)},$$

which tends to $\dfrac{\sin(\xi/2)}{\xi/2}$ when $n \to \infty$. This shows that, in this case, φ can indeed be defined by formula (3.3).

EXAMPLE F: Take $m_0(\xi) = \frac{1}{2}(1 + e^{i3\xi})$, which is a 2π-periodic function satisfying (3.1). If we define φ by (3.3) we deduce from (3.10) that

$$\hat{\varphi}(\xi) = \frac{e^{i3\xi} - 1}{i3\xi}.$$

It is easy to see that $\varphi(x) = \frac{1}{3}\chi_{[-3,0)}(x)$. Observe that then the integer translates of φ do not form an orthonormal system in $L^2(\mathbb{R})$, even if we normalize φ.

The above example shows that the conditions given in (3.1) are not sufficient to produce a function φ given by (3.3) such that the integer translates of φ form an orthonormal system. We will show that a sufficient condition on m_0 to have this property is

$$m_0(\xi) \neq 0 \qquad \text{for all } \xi \in [-\tfrac{1}{2}\pi, \tfrac{1}{2}\pi]. \tag{3.11}$$

We shall present in Chapter 7 necessary and sufficient conditions on a function m_0 satisfying (3.1) to produce, via (3.3), a φ whose integer translates form an orthonormal system. At present we only need the special case described by (3.11).

Recall that by Proposition 1.11, $\{\varphi(\cdot - k) : k \in \mathbb{Z}\}$ is an orthonormal system if and only if

$$\sum_{k\in\mathbb{Z}} |\hat{\varphi}(\xi + 2k\pi)|^2 = 1 \qquad \text{a.e. on } \mathbb{R},$$

so that if

$$G(\xi) = \sum_{k\in\mathbb{Z}} |\hat{\varphi}(\xi + 2k\pi)|^2$$

we need to show that (3.1) and (3.11) imply $G(\xi) \equiv 1$ almost everywhere.

From now on we shall assume also that m_0 is a **trigonometric poly-nomial** of the form

$$m_0(\xi) = \sum_{k=-M}^{M} \alpha_k e^{ik\xi}.$$

Then $\Pi_N(\xi)$ defined in (3.2) is a finite linear combination of terms of the form

$$e^{i\{\sum_{j=1}^{N} 2^{-j}\ell_j\}\xi}$$

where $-M \le \ell_j \le M$. Hence, the inverse Fourier transform of Π_N (as a distribution) is a finite linear combination of "Dirac δ-functions" of the form δ_a with $a = -\sum_{j=1}^{N} 2^{-j}\ell_j$ (notice that $(\delta_a)^\wedge(\xi) = e^{-ia\xi}$). Since $a \in [-M, M]$, the function φ defined by (3.3) is a function in $L^2(\mathbb{R})$ (Proposition 3.9) with support contained in $[-M, M]$, which is a limit, in the distribution sense, of measures supported on $[-M, M]$.

Since $\hat{\varphi} \in L^2(\mathbb{R})$,

$$G(\xi) = \sum_{k \in \mathbb{Z}} |\hat{\varphi}(\xi + 2k\pi)|^2$$

is a 2π-periodic function in $L^1([0, 2\pi))$. Moreover, by the Plancherel theorem,

$$c_\ell \equiv \frac{1}{2\pi} \int_0^{2\pi} G(\xi) e^{-i\ell\xi} d\xi = \frac{1}{2\pi} \sum_{k \in \mathbb{Z}} \int_{2k\pi}^{2(k+1)\pi} e^{-i\ell\xi} \hat{\varphi}(\xi) \overline{\hat{\varphi}(\xi)} \, d\xi$$

$$= \frac{1}{2\pi} \int_{-\infty}^{\infty} [e^{-i\ell\xi} \hat{\varphi}(\xi)] \overline{\hat{\varphi}(\xi)} \, d\xi = \int_{-\infty}^{\infty} \varphi(x - \ell) \overline{\varphi(x)} \, dx.$$

But, if φ is compactly supported, the last expression is 0 except for a finite number of integers ℓ. Therefore, we have proved

LEMMA 3.12 *If m_0 is a trigonometric polynomial satisfying (3.1), and φ is defined by (3.3), then φ is a compactly supported function in $L^2(\mathbb{R})$ and*

$$G(\xi) = \sum_{\ell \in \mathbb{Z}} |\hat{\varphi}(\xi + 2\ell\pi)|^2 = \sum_{\text{finite}} c_\ell e^{i\ell\xi}$$

is a trigonometric polynomial.

THEOREM 3.13 *Suppose that m_0 is a trigonometric polynomial which satisfies* (3.1) *and*

$$m_0(\xi) \neq 0 \qquad \text{for all } \xi \in [-\tfrac{1}{2}\pi, \tfrac{1}{2}\pi].$$

If φ is defined by (3.3), *then $\{\varphi(\cdot - k) : k \in \mathbb{Z}\}$ is an orthonormal system.*

PROOF : All we need to show is $G(\xi) \equiv 1$ on $[-\pi, \pi]$, where G is defined in Lemma 3.12 (see Proposition 1.11). We claim that if $\ell \in \mathbb{Z}$ and $\ell \neq 0$, $\hat{\varphi}(2\ell\pi) = 0$; in fact, $\ell = 2^p q$ with q odd, so that the product

$$\prod_{j=1}^{N} m_0(2^{-j}2^{p+1}q\pi)$$

contains the factor $m_0(q\pi)$ if $N \geq p+1$ and $m_0(q\pi) = m_0(\pi) = 0$ (because of (3.1)). Thus $G(0) = |\hat{\varphi}(0)|^2 = 1$.

We claim that

$$G(2\xi) = |m_0(\xi)|^2 G(\xi) + |m_0(\xi + \pi)|^2 G(\xi + \pi). \tag{3.14}$$

The idea is to separate the odd and even terms in the sum that defines $G(2\xi)$ as we did in the proof of (2.5):

$$G(2\xi) = \sum_{\ell \in \mathbb{Z}} |\hat{\varphi}(2(\xi + \ell\pi))|^2 = \sum_{\ell \in \mathbb{Z}} |m_0(\xi + \ell\pi)|^2 |\hat{\varphi}(\xi + \ell\pi)|^2$$

$$= \sum_{k \in \mathbb{Z}} |m_0(\xi)|^2 |\hat{\varphi}(\xi + 2k\pi)|^2 + \sum_{k \in \mathbb{Z}} |m_0(\xi + \pi)|^2 |\hat{\varphi}(\xi + \pi + 2k\pi)|^2$$

$$= |m_0(\xi)|^2 G(\xi) + |m_0(\xi + \pi)|^2 G(\xi + \pi).$$

Let $m = \min_{\xi \in [-\pi, \pi]} G(\xi)$ and choose $\xi_o \in [-\pi, \pi]$ such that $G(\xi_o) = m$. If $G(\tfrac{1}{2}\xi_o) > m$ we use (3.14) and the assumption $m_0(\tfrac{1}{2}\xi_o) \neq 0$ to obtain

$$m = G(\xi_o) = |m_0(\tfrac{1}{2}\xi_o)|^2 G(\tfrac{1}{2}\xi_o) + |m_0(\tfrac{1}{2}\xi_o + \pi)|^2 G(\tfrac{1}{2}\xi_o + \pi)$$

$$> m\,|m_0(\tfrac{1}{2}\xi_o)|^2 + m\,|m_0(\tfrac{1}{2}\xi_o + \pi)|^2 = m,$$

which is a contradition. Thus $G(\tfrac{1}{2}\xi_o) = m$; iterating, we obtain

$$G(2^{-j}\xi_o) = m \qquad \text{for all } j = 0, 1, 2, \cdots.$$

Hence, $1 = G(0) = m$.

Let $M = \sup_{\xi \in [-\pi, \pi]} G(\xi)$ and take $\xi_1 \in [-\pi, \pi]$ such that $G(\xi_1) = M$. If $G(\frac{1}{2}\xi_1) < M$, we have (since $m_0(\frac{1}{2}\xi_1) \neq 0$)

$$M = G(\xi_1) = |m_0(\tfrac{1}{2}\xi_1)|^2 G(\tfrac{1}{2}\xi_1) + |m_0(\tfrac{1}{2}\xi_1 + \pi)|^2 G(\tfrac{1}{2}\xi_1 + \pi)$$

$$< M|m_0(\tfrac{1}{2}\xi_1)|^2 + M|m_0(\tfrac{1}{2}\xi_1 + \pi)|^2 = M.$$

Thus $G(\frac{1}{2}\xi_1) = M$, and by iteration and taking limits, $G(0) = M$. Thus, $m = M = G(0)$, and, since we have already proved that $G(0) = 1$, this finishes the proof of Theorem 3.13.

∎

Observe that we could have finished the above proof by another argument once we established that $m = 1$: since, then, $G(\xi) \geq 1$, and by a periodization argument and Proposition 3.9, we have

$$\int_0^{2\pi} G(\xi)\, d\xi \leq 2\pi,$$

which implies $G(\xi) \equiv 1$.

We can now easily construct the multiresolution analysis associated with the trigonometric polynomial "filter" m_0, which satisfies (3.1) and (3.11). Let

$$V_0 = \overline{\operatorname{span}\{\varphi(\cdot - k) : k \in \mathbb{Z}\}}.$$

If $(\rho f)(x) = f(2x)$ and we define $V_j = \rho^j(V_0)$, $j \in \mathbb{Z} \setminus \{0\}$, we have the desired MRA. Property (1.1) is equivalent to showing $V_{-1} \subset V_0$; the definition of $\hat{\varphi}$ (see (3.3)) gives us $\hat{\varphi}(2\xi) = m_0(\xi)\hat{\varphi}(\xi)$, which, in turn, shows that $\frac{1}{2}\varphi(\frac{\cdot}{2}) \in V_0$; since V_0 is invariant under translations by integers, we have the desired result. Property (1.2) in the definition of an MRA follows from the definition of the spaces V_j, and property (1.5) is given in Theorem 3.13. Since $\hat{\varphi}$ is continuous at zero, Theorems 1.6 and Theorem 1.7 give us properties (1.3) and (1.4) of an MRA.

We have already observed that φ has a compact support contained in $[-M, M]$, where M is the degree of the trigonometric polynomial m_0. We can also see that the orthonormal wavelet ψ, given by (2.12), must have compact support. By (2.15)

$$\psi(x) = 2 \sum_{k \in \mathbb{Z}} (-1)^k \,\overline{\alpha_k}\, \varphi(2x - (k-1)).$$

Since $-M \le k \le M$ (because we are assuming that m_0 is a trigonometric polynomial of degree M) and $-M \le 2x - (k-1) \le M$ we see that ψ has compact support within $[-M - \frac{1}{2}, M + \frac{1}{2}]$.

We still have to construct a low-pass filter m_0 satisfying the above properties. Let us first construct a function that equals the square of the modulus of m_0. We are interested in finding a non-negative trigonometric polynomial $g(\xi)$ that satisfies

$$
\begin{cases}
\text{(i)} & g(\xi) + g(\xi + \pi) = 1 \ \text{ for all } \ \xi \in \mathbb{T}, \\
\text{(ii)} & g(0) = 1, \\
\text{(iii)} & g(\xi) > 0 \ \text{ for } \ \xi \in [-\frac{1}{2}\pi, \frac{1}{2}\pi].
\end{cases}
\tag{3.15}
$$

After we establish the existence of such a g we shall show how to obtain the trigonometric polynomial m_0 such that $|m_0(\xi)|^2 = g(\xi)$ for all $\xi \in \mathbb{R}$. Observe that (i), (ii) and (iii) give us (3.1) and (3.11).

If we want to find a function g, not necessarily a trigonometric polynomial, such that (3.15) holds, it is enough to take

$$
g(\xi) = \sum_{k \in \mathbb{Z}} \chi_{[-\frac{1}{2}\pi, \frac{1}{2}\pi)}(\xi - 2k\pi),
$$

which is a 2π-periodic function. The idea is to find a smooth version of this example. Thus, we want a trigonometric polynomial satisfying (3.15) which is 1 near the origin and 0 around π.

We now construct the trigonometric polynomial $g(\xi)$ satisfying (3.15). For $k = 0, 1, 2, \cdots$ define c_k by

$$
c_k^{-1} = \int_0^{\pi} (\sin t)^{2k+1} dt,
$$

which is clearly a positive number. Then, consider

$$
g_k(\xi) = 1 - c_k \int_0^{\xi} (\sin t)^{2k+1} dt.
$$

We have $g_k(0) = 1$, $g_k(\pi) = 0$ and

$$
g_k(\xi) + g_k(\xi + \pi) = 2 - c_k \int_0^{\xi} (\sin t)^{2k+1} dt - c_k \int_0^{\xi+\pi} (\sin t)^{2k+1} dt
$$

$$
= 1 - c_k \int_0^{\xi} (\sin t)^{2k+1} dt - c_k \int_{\pi}^{\xi+\pi} (\sin t)^{2k+1} dt = 1,
$$

since $\sin(t+\pi) = -\sin t$. This shows that (i) and (ii) of (3.15) are satisfied. Also $0 \le g_k(\xi) \le 1$ and $g_k(\xi) > 0$ on $(-\pi, \pi)$, which gives us (iii) of (3.15). The graphs of $g_k(\xi)$ for $k = 0$ and $k = 5$ are given in Figure 2.7.

EXAMPLE G: For $k = 0$ we have $g_0(\xi) = \frac{1}{2}(1 + \cos \xi)$, which satisfies (3.15); to find the filter m_0 such that $|m_0(\xi)|^2 = g_0(\xi)$ observe that

$$(1 + e^{i\xi})(1 + e^{-i\xi}) = 2 + 2\cos \xi;$$

thus, $m_0(\xi) = \frac{1}{2}(1 + e^{i\xi})$, which is the low-pass filter for the Haar wavelet (see Example B).

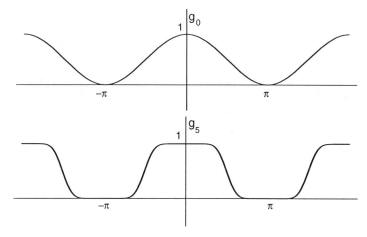

Figure 2.7: Computer graphs of $g_k(\xi)$ for $k = 0$ and $k = 5$.

Writing $\sin(\xi) = \frac{1}{2i}(e^{i\xi} - e^{-i\xi})$ we see that $g_k(\xi)$ is a trigonometric polynomial of degree $2k + 1$. The existence of the filter $m_0(\xi)$ such that $|m_0(\xi)|^2 = g_k(\xi)$, which was easily obtained when $k = 0$ in Example G, follows from the following result of L. Fejér and F. Riesz:

LEMMA 3.16 Let $g(\xi) = \sum_{k=-M}^{M} \gamma_k e^{ik\xi}$ be a trigonometric polynomial such that $g(\xi) \ge 0$ for all $\xi \in \mathbb{R}$. There exists a trigonometric polynomial m_0 of the form $m_0(\xi) = \sum_{k=0}^{M} \alpha_k e^{ik\xi}$ such that $|m_0(\xi)|^2 = g(\xi)$.

PROOF: We can assume $g(\xi) > 0$; otherwise we can take the polynomial $g(\xi) + \varepsilon$ and let $\varepsilon \to 0$. Since $g(\xi)$ is positive, $g(\xi)$ coincides with its conjugate, and, therefore, we must have $\gamma_k = \overline{\gamma_{-k}}$. If we consider the polynomial

$$P(z) = \sum_{k=-M}^{M} \gamma_k z^{k+M}, \qquad z \in \mathbb{C},$$

the above observation allows us to prove the relation

$$P(z) = z^{2M} \overline{P(1/\overline{z})}. \qquad (3.17)$$

The zeroes of P cannot be on the boundary of the unit circle since

$$e^{-iM\xi} P(e^{i\xi}) = g(\xi) > 0.$$

We can also assume that $P(0) \neq 0$. For, if $P(0) = 0$, then $\gamma_{-M} = 0$; therefore, $\gamma_M = \overline{\gamma_{-M}} = 0$. This implies that $g(\xi)$ is of order not exceeding $M - 1$, but it suffices to consider the case where g is precisely of degree M. Thus, if a is a zero of $P(z)$, (3.17) implies that $1/\overline{a}$ is also a zero of $P(z)$. This shows that if $a_1, \cdots, a_n$ are the zeroes of $P(z)$ in the interior of the unit circle, $1/\overline{a_1}, \cdots, 1/\overline{a_n}$ are the zeroes of $P(z)$ in the exterior of the unit circle. If r_k is the multiplicity of a_k we have

$$P(z) = \gamma_M \prod_{k=1}^{n} (z - a_k)^{r_k} (z - 1/\overline{a_k})^{r_k},$$

where $2\sum_{k=1}^{n} r_k = 2M$. Let us assume, for simplicity, that $\gamma_M = 1$. Since $z - 1/\overline{a} = z(\overline{a} - 1/z)/\overline{a}$ we have

$$P(e^{i\xi}) = C e^{iM\xi} \prod_{k=1}^{n} (e^{i\xi} - a_k)^{r_k} (e^{-i\xi} - \overline{a_k})^{r_k}.$$

Hence,

$$g(\xi) = e^{-iM\xi} P(e^{i\xi}) = C \left[\prod_{k=1}^{n} |e^{i\xi} - a_k|^{r_k} \right]^2.$$

Since $g(\xi) > 0$ we must have $C > 0$, and, thus,

$$m_0(\xi) = \sqrt{C} \prod_{k=1}^{n} (e^{i\xi} - a_k)^{r_k} \qquad (3.18)$$

is the desired trigonometric polynomial. Observe that the degree of $m_0(\xi)$ is M.

Notice that if the coefficients, γ_k, of g are real, then the coefficients, α_k, of m_0 can be chosen to be real. To see this observe that $P(z)$ is a polynomial in z with real coefficients; therefore, if a is a root of $P(z)$, $\overline{a}$ is also a root of $P(z)$. Hence, if a_k is a non-real root of P, $\overline{a_k}$ is also a root of P, and, then,

$$(e^{i\xi} - a_k)(e^{i\xi} - \overline{a_k}) = e^{2i\xi} - e^{i\xi}2\mathrm{Re}\{a_k\} + |a_k|^2,$$

which implies that m_0 is a polynomial with real coefficients.

For the trigonometric polynomial $g_k(\xi)$ we constructed above, let $m_0^{(k)}(\xi)$ be the trigonometric polynomial whose existence is proved in Lemma 3.16. Observe that the coefficients of $g_k(\xi)$ are real, so that $m_0^{(k)}(\xi)$ can be written in the form

$$m_0^{(k)}(\xi) = \sum_{j=0}^{2k+1} \alpha_j e^{ij\xi}$$

with $\alpha_j \in \mathbb{R}$. We now obtain a more detailed description of $g_k(\xi)$ and $m_0^{(k)}(\xi)$. It is immediate from the definition that g_k is an even function and, thus, we can express it as a function of $w = \cos\xi$: $g_k(\xi) = G_k(\cos\xi)$. Observe that $g_k(\pi) = 0$; thus, $G_k(w)$ has value zero at $w = -1$. Moreover,

$$\frac{dG_k}{dw} = \frac{dg_k}{d\xi}\frac{d\xi}{dw} = -c_k(\sin\xi)^{2k+1}\left(-\frac{1}{\sin\xi}\right) = c_k(\sin\xi)^{2k}$$

$$= c_k(1 - \cos^2\xi)^k = c_k(1 - w)^k(1 + w)^k.$$

We claim that G_k has a zero of multiplicity $k + 1$ at $w = -1$. To see this we observe that G_k' can be written in the form

$$G_k'(w) = 2^k c_k(1 + w)^k + \text{terms of higher order in } (1 + w).$$

The result follows by integrating both terms of this equality and using $G_k(-1) = 0$. Hence,

$$g_k(\xi) = (1 + \cos\xi)^{k+1}\tilde{g}_k(\xi), \tag{3.19}$$

where $\tilde{g}_k(\xi)$ is a non-negative trigonometric polynomial of degree k. Since $1 + \cos\xi = \frac{1}{2}(1 + e^{i\xi})(1 + e^{-i\xi})$, we have

$$m_0^{(k)}(\xi) = \left(\frac{1 + e^{i\xi}}{2}\right)^{k+1} 2^{\frac{k+1}{2}}\tilde{m}^{(k)}(\xi), \tag{3.20}$$

where $\tilde{m}^{(k)}(\xi)$ is a trigonometric polynomial of degree k of the form

$$\sum_{j=0}^{k} \gamma_k e^{ij\xi},$$

such that $|\tilde{m}^{(k)}(\xi)|^2 = \tilde{g}_k(\xi)$, whose existence is given by Lemma 3.16.

By (3.3) and formula (3.10) we have

$$\hat{\varphi}_k(\xi) = \prod_{j=1}^{\infty} m_0^{(k)}(2^{-j}\xi)$$

$$= e^{i\frac{k+1}{2}\xi}\left(\frac{\sin(\xi/2)}{\xi/2}\right)^{k+1} \prod_{j=1}^{\infty} 2^{\frac{k+1}{2}} \tilde{m}^{(k)}(2^{-j}\xi). \qquad (3.21)$$

We shall deduce the smoothness of $\varphi \equiv \varphi_k$ from this relation. It is immediate from (3.21) that $|\hat{\varphi}_k(\xi)| \sim c|\xi|^{-(k+1)}|P(\xi)|$ (for large ξ), where $P(\xi)$ is the product on the right in the above formula. The smoothness will follow if we can appropriately control $P(\xi)$; that is, more precisely, if we can prove that $P(\xi) \sim |\xi|^{\alpha(k)}$ with $\alpha(k) < k + 1$. Since $\psi_k(\frac{1}{2}\cdot)$ is a finite linear combination of integer translates of φ_k (see (2.15)), the wavelet ψ_k has the same smoothness as φ_k. To do this we need a result that will be presented after we make the following definition.

For $\alpha = n + \beta$, $n \in \mathbb{N}$, $0 < \beta \leq 1$, we define $\Lambda_\alpha(\mathbb{R})$ to be the set of bounded functions f which are n times continuously differentiable and such that the nth derivative $f^{(n)}$ is Hölder continuous of order β; that is,

$$|f^{(n)}(x+h) - f^{(n)}(x)| \leq c|h|^\beta$$

for all $x, h \in \mathbb{R}$.

LEMMA 3.22 *Let f be a function in $L^2(\mathbb{R})$ and $\alpha > 0$; suppose*

$$\int_{2^j\pi \leq |\xi| \leq 2^{j+1}\pi} |\hat{f}(\xi)| \, d\xi \leq C2^{-\alpha j}, \qquad j = 0, 1, 2, \cdots. \qquad (3.23)$$

Then, if $\alpha \notin \mathbb{N}$, $f \in \Lambda_\alpha(\mathbb{R})$, and if $\alpha \in \mathbb{N}$, $f \in \Lambda_{\alpha-\varepsilon}(\mathbb{R})$ for all $0 < \varepsilon < \alpha$.

PROOF : Since $\alpha > 0$, (3.23) implies that $\hat{f} \in L^1(\mathbb{R})$. Hence, f is bounded and if $|h| \geq 1$,

$$|f(x+h) - f(x)| \leq 2\|f\|_\infty \leq 2\|f\|_\infty |h|^\alpha.$$

Thus, we only need to consider the case $|h| < 1$. We first prove the lemma in the case $0 < \alpha < 1$. Writing $f(x + h) - f(x)$ in terms of the inverse Fourier transform we obtain

$$f(x + h) - f(x) = \frac{1}{2\pi} \int_{\mathbb{R}} e^{ix\xi}(e^{ih\xi} - 1)\hat{f}(\xi)\,d\xi.$$

Given $|h| < 1$, choose $j \equiv j(h) \in \mathbb{N}$ such that $2^{-j-1} \le |h| < 2^{-j}$. Let

$$A_k = \{\xi \in \mathbb{R} : 2^k\pi < |\xi| \le 2^{k+1}\pi\} \qquad \text{for } k = 0, 1, 2, \cdots.$$

Using $|e^{ix} - 1| \le \min\{|x|, 2\}$ for all $x \in \mathbb{R}$ we deduce

$$\left|f(x + h) - f(x)\right| \le \frac{1}{2\pi}\Big\{\int_{|\xi|\le\pi} |h\xi|\cdot|\hat{f}(\xi)|\,d\xi$$

$$+ \int_{\pi<|\xi|\le 2^j\pi} |h\xi|\cdot|\hat{f}(\xi)|\,d\xi + \int_{|\xi|>2^j\pi} 2|\hat{f}(\xi)|\,d\xi\Big\}$$

$$= \frac{1}{2\pi}\{I + II + III\}. \tag{3.24}$$

To estimate III we have

$$III \le 2\sum_{k=j}^{\infty} \int_{A_k} |\hat{f}(\xi)|\,d\xi \le 2C\sum_{k=j}^{\infty} 2^{-k\alpha} = 2C(\alpha)2^{-j\alpha} \le 2C(\alpha)|h|^{\alpha}.$$

To estimate I observe that for $|h| < 1$ and $0 < \alpha < 1$ we have $|h| < |h|^{\alpha}$; hence, $I \le \pi|h|^{\alpha}\|\hat{f}\|_{L^1}$. Observe that the estimates for I and III hold even for $\alpha = 1$. To estimate II we write

$$II \le |h|\sum_{k=0}^{j-1} \int_{A_k} |\xi|\cdot|\hat{f}(\xi)|\,d\xi \le C|h|\sum_{k=0}^{j-1} 2^{k+1}2^{-\alpha k}$$

$$\le 2C|h|\sum_{k=0}^{j-1} 2^{k(1-\alpha)} = C(\alpha)|h|2^{j(1-\alpha)} \le C(\alpha)|h|^{\alpha},$$

where we used $\alpha < 1$ to obtain the precise value of the last sum. The result now follows from the estimates for I, II and III, and (3.24).

To prove the result when $\alpha = 1$ we recall that (3.24) and the estimates for I and III are still valid in this case. Thus, we need to find a bound for

II. If $|\xi| \leq 2^j \pi$, we have $2^{-j}|\xi|\frac{1}{\pi} \leq 1$, so that for any $\varepsilon \in (0,1)$,

$$2^{-j}|\xi|\frac{1}{\pi} \leq \left(2^{-j}|\xi|\frac{1}{\pi}\right)^{1-\varepsilon}.$$

Thus,

$$II \leq C(\varepsilon)2^{-j(1-\varepsilon)} \int_{\pi < |\xi| \leq 2^j \pi} |\xi|^{1-\varepsilon}|\hat{f}(\xi)|\, d\xi$$

$$\leq C(\varepsilon)2^{-j(1-\varepsilon)} \sum_{k=0}^{j-1} \int_{A_k} |\xi|^{1-\varepsilon}|\hat{f}(\xi)|\, d\xi$$

$$\leq C(\varepsilon)2^{-j(1-\varepsilon)} \sum_{k=0}^{j-1} 2^{(k+1)(1-\varepsilon)}2^{-k}$$

$$\leq C(\varepsilon)2^{-j(1-\varepsilon)} \sum_{k=0}^{\infty} 2^{-\varepsilon k} \leq C(\varepsilon)|h|^{1-\varepsilon}.$$

Hence,

$$\left|f(x+h) - f(x)\right| \leq C(\varepsilon)\{|h| + |h|^{1-\varepsilon}\} \leq C(\varepsilon)|h|^{1-\varepsilon}$$

when $|h| < 1$. This finishes the proof for $\alpha = 1$.

For a general α write $\alpha = n + \beta$ with $n \in \mathbb{N}$ and $0 < \beta \leq 1$, and apply the above argument to $g(x) = f^{(n)}(x)$. ∎

REMARK: The definition of the spaces $\Lambda_\alpha(\mathbb{R})$ we have given is simple; however, when α is a positive integer, we do not obtain the "correct" version of these spaces. A somewhat more technical definition is the "appropriate" one for α being a positive integer. The fact that in Lemma 3.22 condition (3.23) gives us a weaker result ($f \in \Lambda_{\alpha-\varepsilon}(\mathbb{R})$) is a manifestation of this situation. This phenomenon can be understood better if we replace these exceptional spaces by the "Zygmund class" we shall present in section 6.7.

What we now need to do is to apply Lemma 3.22 to the function φ_k defined by (3.21). Thus, we need to show that (3.23) is satisfied. To do this we write $A_j = \{\xi \in \mathbb{R} : 2^j \pi \leq |\xi| \leq 2^{j+1}\pi\}$ and

$$I_j = \int_{A_j} |\hat{\varphi}_k(\xi)|\, d\xi,$$

for $j = 0, 1, 2, \cdots$. Using the Cauchy-Schwarz inequality we obtain

$$I_j \leq (2 \cdot 2^j \pi)^{\frac{1}{2}} \left(\int_{A_j} |\hat{\varphi}_k(\xi)|^2 \, d\xi \right)^{\frac{1}{2}}.$$

Since $|m_0^{(k)}(\xi)| \leq 1$, using (3.19) we have

$$|\hat{\varphi}_k(\xi)|^2 \leq \prod_{\ell=1}^{j+1} |m_0^{(k)}(2^{-\ell}\xi)|^2 = \prod_{\ell=1}^{j+1} g_k(2^{-\ell}\xi)$$

$$= \prod_{\ell=1}^{j+1} \left(\frac{1 + \cos(2^{-\ell}\xi)}{2} \right)^{k+1} 2^{k+1} \tilde{g}_k(2^{-\ell}\xi).$$

As in Example E, or using directly the formula $\sin 2\alpha = 2 \sin \alpha \cos \alpha$, we deduce that the product

$$\prod_{\ell=1}^{j+1} \frac{1 + \cos(2^{-\ell}\xi)}{2} = \prod_{\ell=1}^{j+1} \cos^2(2^{-\ell-1}\xi)$$

coincides with

$$\left| \frac{\sin(\xi/2)}{2^{j+1} \sin(2^{-j-2}\xi)} \right|^2.$$

When $\xi \in A_j$ the last expression is bounded by $(\sqrt{2}\, 2^{-(j+1)})^2$. Using this we obtain

$$I_j \leq (2 \cdot 2^j \pi)^{\frac{1}{2}} (\sqrt{2}\, 2^{-(j+1)})^{k+1} \left(\int_{A_j} \left\{ \prod_{\ell=1}^{j+1} 2^{k+1} \tilde{g}_k(2^{-\ell}\xi) \right\} d\xi \right)^{\frac{1}{2}}.$$

Collecting together the various powers of 2, we conclude

$$I_j \leq \sqrt{2\pi}\, 2^{-\frac{jk}{2}} \mathcal{G}_j^{\frac{1}{2}}, \tag{3.25}$$

where

$$\mathcal{G}_j = \int_{A_j} \left\{ \prod_{\ell=1}^{j+1} \tilde{g}_k(2^{-\ell}\xi) \right\} d\xi. \tag{3.26}$$

We now develop some precise formulae and estimates for the function $\tilde{g}_k$ so that we can estimate $\mathcal{G}_j$. By repeated integration by parts we have

$$\frac{1}{c_k} g_k(\xi) = \int_{\xi}^{\pi} (1 + \cos t)^k (1 - \cos t)^k \sin t \, dt$$

$$= (1 - \cos \xi)^k \frac{1}{k+1} (1 + \cos \xi)^{k+1}$$

$$+ \int_\xi^\pi \frac{k}{k+1} (1 - \cos t)^{k-1} (1 + \cos t)^{k+1} \sin t \, dt$$

$$= (1 + \cos \xi)^{k+1} \left\{ \frac{1}{k+1} (1 - \cos \xi)^k \right.$$

$$+ \frac{k}{(k+1)(k+2)} (1 - \cos \xi)^{k-1} (1 + \cos \xi)$$

$$+ \frac{k(k-1)}{(k+1)(k+2)(k+3)} (1 - \cos \xi)^{k-2} (1 + \cos \xi)^2$$

$$\left. + \cdots + \frac{k! \, k!}{(2k+1)!} (1 + \cos \xi)^k \right\}$$

$$= (1 + \cos \xi)^{k+1} \sum_{\ell=0}^k \frac{k! \, k!}{(k-\ell)!(k+\ell+1)!} (1 - \cos \xi)^{k-\ell} (1 + \cos \xi)^\ell.$$

Thus, using the notation of (3.19), $g_k(\xi) = (1 + \cos \xi)^{k+1} \tilde{g}_k(\xi)$, where

$$\tilde{g}_k(\xi) = c_k \sum_{\ell=0}^k \frac{(k!)^2}{(k-\ell)!(k+\ell+1)!} (1 - \cos \xi)^{k-\ell} (1 + \cos \xi)^\ell.$$

When $\xi = 0$ all the terms in this sum are zero except for the term in which $\ell = k$. Thus,

$$1 = g_k(0) = c_k 2^{k+1} \frac{(k!)^2}{(2k+1)!} 2^k.$$

We therefore obtain the following value for c_k:

$$c_k = \frac{k+1}{2^{2k+1}} \binom{2k+1}{k}.$$

With this value of c_k the formula for $\tilde{g}_k(\xi)$ is

$$\tilde{g}_k(\xi) = \frac{1}{2^{2k+1}} \sum_{\ell=0}^k \binom{2k+1}{k-\ell} (1 - \cos \xi)^{k-\ell} (1 + \cos \xi)^\ell.$$

As we did immediately before (3.19), we express $g_k(\xi)$ and, also, $\tilde{g}_k(\xi)$, as functions $G_k(w)$ and $\widetilde{G}_k(w)$ of $w = \cos \xi$. Since $G_k(w) = (1+w)^{k+1} \widetilde{G}_k(w)$

and $G'_k(w) = c_k(1-w)^k(1+w)^k$ we have

$$\widetilde{G}'_k(w) = \frac{c_k(1-w)^k - (k+1)\widetilde{G}_k(w)}{1+w}.$$

Since

$$\widetilde{G}_k(w) = \frac{1}{2^{2k+1}}\binom{2k+1}{k}(1-w)^k + \frac{1}{2^{2k+1}}\sum_{\ell=1}^{k}\binom{2k+1}{k-\ell}(1-w)^{k-\ell}(1+w)^{\ell},$$

we have

$$(1+w)\widetilde{G}'_k(w) = -(k+1)\frac{1}{2^{2k+1}}\sum_{\ell=1}^{k}\binom{2k+1}{k-\ell}(1-w)^{k-\ell}(1+w)^{\ell}.$$

In particular, we see that $\widetilde{G}'_k(w) \leq 0$ for $w \in [-1,1]$. It follows that the maximal value of $\widetilde{G}_k(w)$ on $[-1,1]$ is attained at $w = -1$. At this point the value of $\widetilde{G}_k$ is

$$\frac{1}{2^{2k+1}}\binom{2k+1}{k}2^k = \frac{1}{2^{k+1}}\binom{2k+1}{k}.$$

We summarize these results in the following lemma:

LEMMA 3.27 *If we define the function* $g_k(\xi) = 1 - c_k \int_0^{\xi}(\sin t)^{2k+1}\,dt$,
where $c_k^{-1} = \int_0^{\pi}(\sin t)^{2k+1}\,dt$, *then*

(1) $g_k(\xi) = (1 + \cos\xi)^{k+1}\tilde{g}_k(\xi),$

   where $\tilde{g}_k(\xi) = \dfrac{1}{2^{2k+1}}\displaystyle\sum_{\ell=0}^{k}\binom{2k+1}{k-\ell}(1-\cos\xi)^{k-\ell}(1+\cos\xi)^{\ell};$

(2) *The maximal value of* $\tilde{g}_k$ *is* $\dfrac{1}{2^{k+1}}\dbinom{2k+1}{k}$; *thus,*

(3) $\displaystyle\prod_{\ell=1}^{j+1}\tilde{g}_k(2^{-\ell}\xi) \leq 2^{-(j+1)(k+1)}\binom{2k+1}{k}^{j+1}.$

The estimate in (3) is obtained immediately from (2) applied to each factor.

We are now ready to estimate $\mathcal{G}_j$ in formula (3.26), so that we can apply Lemma 3.22 in order to obtain the corresponding smoothness of the wavelet we are constructing. Using (3) of Lemma 3.27 we see from (3.26) that

$$\mathcal{G}_j \leq 2 \cdot 2^j \pi \cdot 2^{-(j+1)(k+1)} \binom{2k+1}{k}^{j+1} = \pi 2^{-k} 2^{-jk} \binom{2k+1}{k}^{j+1}.$$

Thus, from (3.25), we obtain

$$I_j \leq \sqrt{2}\,\pi\, 2^{-\frac{k}{2}} 2^{-jk} \binom{2k+1}{k}^{\frac{j+1}{2}} = C(k) 2^{-j\alpha(k)}, \qquad (3.28)$$

where $C(k) = \sqrt{2}\,\pi\, 2^{-\frac{k}{2}} \binom{2k+1}{k}^{\frac{1}{2}}$ and $\alpha(k) = k - \frac{1}{2}\log_2 \binom{2k+1}{k}$.

Observe that $\alpha(1) = 1 - \frac{1}{2}\log_2 3 \cong 0.207518 > 0$; also $\{\alpha(k)\}_{k \geq 0}$ is an increasing sequence since

$$\alpha(k+1) - \alpha(k) = 1 + \frac{1}{2}\log_2\left[\binom{2k+1}{k} \Big/ \binom{2k+3}{k+1}\right] = \frac{1}{2}\log_2\left(1 + \frac{1}{2k+3}\right)$$

is obviously greater than zero. Also

$$\alpha(k+1) - \alpha(k) = \frac{1}{2\ln 2}\left(\frac{1}{2k+3} + O(k^{-2})\right) \geq C\frac{1}{k};$$

hence $\alpha(k) \geq c \log k$ and, asymptotically, $\alpha(k) \sim \frac{1}{4}\log_2 k$. This result can be obtained by using Stirling's formula

$$n! \sim \sqrt{2\pi}\, n^{n+\frac{1}{2}} e^{-n}.$$

In fact, we obtain the precise statement

$$\lim_{k \to \infty} \frac{\alpha(k)}{\frac{1}{4}\log_2 \frac{\pi k}{4}} = 1.$$

Our construction, therefore, shows the existence of compactly supported orthonormal wavelets belonging to each class $C^r(\mathbb{R})$, $r = 0, 1, 2, \cdots$. This is the content of the following result.

THEOREM 3.29 *For any integer $r = 0, 1, 2, \cdots$ there exists an orthonormal wavelet ψ with compact support such that ψ has bounded derivatives up to order r. Moreover, ψ can be obtained from an MRA whose scaling function φ also has compact support and the same smoothness as ψ.*

For future reference we present below the approximate value of $\alpha(k)$ for some k:

k	$\alpha(k)$
1	0.207518
2	0.339035
3	0.435358
4	0.511359
5	0.574125

k	$\alpha(k)$
10	0.78593
50	1.33274
100	1.57832
200	1.82608
400	2.07494

2.4 Better estimates for the smoothness of compactly supported wavelets

The degree of smoothness of the compactly supported wavelet ψ_k that we obtained was shown to depend on $\alpha(k) = k - \frac{1}{2}\log_2\binom{2k+1}{k}$; however, estimate (3) of Lemma 3.27 is not best possible and does not give us sharp results. In this section we will see that this is true even for the wavelet ψ_1. More generally, we will show how to improve the smoothness estimates for the compactly supported wavelets constructed in section 2.3.

For $k = 1$ a simple computation using the definition of $g_1(\xi)$ (or, more simply, using the formula given in (1) of Lemma 3.27) shows that

$$g_1(\xi) = \left(\frac{1 + \cos\xi}{2}\right)^2 (2 - \cos\xi). \tag{4.1}$$

We know that a "square root" of $\frac{1}{2}(1 + \cos\xi)$ in the sense of Fejér-Riesz is $\frac{1}{2}(1 + e^{i\xi})$; for the polynomial $2 - \cos\xi$ a "square root" of the form $a + be^{i\xi}$ can be easily computed, and we see that the filter associated with $g_1(\xi)$ is

$$m_0(\xi) = \left(\frac{1 + e^{i\xi}}{2}\right)^2 \frac{1}{\alpha}[(1 + \alpha)e^{i\xi} - 1],$$

where $\alpha = 1 + \sqrt{3}$. A formula for the Fourier transform of the scaling function can then be found with the recipe given by (3.3):

$$\hat{\varphi}_1(\xi) = e^{i\xi} \frac{\sin^2(\xi/2)}{(\xi/2)^2} \prod_{j=1}^{\infty} \frac{1}{\alpha}[(1 + \alpha)e^{i2^{-j}\xi} - 1].$$

There is also a similar formula for $\hat{\psi}_1$ that can be obtained from the identity $\hat{\psi}_1(\xi) = e^{i\frac{\xi}{2}} \overline{m_0(\frac{1}{2}\xi + \pi)} \, \hat{\varphi}(\frac{1}{2}\xi)$ (see (2.12)).

We shall show, in this case, how to obtain a more "precise" estimate for $\mathcal{G}_j$ (see (3.26)) using the explicit formula for g_1 given by (4.1). This will lead us to a better estimate for the smoothness of the wavelet ψ_1 than we previously obtained ($\alpha(1) \sim 0.207518$). In fact, we shall show that φ_1 and ψ_1 satisfy the Lipschitz condition $\Lambda_{\frac{1}{2}}(\mathbb{R})$.

Since $\tilde{g}_1(\xi) = \frac{1}{4}(2 - \cos \xi)$, enlarging the set of integration and changing variables ($\xi = 2^{j+1}\mu$), we obtain

$$\mathcal{G}_j \leq 2^{j+1} \int_{|\mu| \leq \pi} \left\{ \prod_{\ell=1}^{j+1} \tilde{g}_1(2^{j+1-\ell}\mu) \right\} d\mu$$

$$= 2^{j+1} (\tfrac{1}{4})^{j+1} \int_{|\mu| \leq \pi} \left\{ \prod_{m=0}^{j} (2 - \cos 2^m \mu) \right\} d\mu$$

$$= \frac{1}{2^{(j+1)}} \int_{-\pi}^{\pi} (2 - \cos \mu)(2 - \cos 2\mu) \cdots (2 - \cos 2^j \mu) \, d\mu.$$

When we multiply out the integrand we observe that all the integrals in which the product of one or more terms $\cos 2^m \mu$ appear are zero (since $\cos 2^m \mu$ is orthogonal to constant functions in $[-\pi, \pi]$); thus, $\mathcal{G}_j \leq 2\pi$. We can use this estimate in (3.25), which, for $k = 1$, gives us $I_j \leq 2\pi 2^{-\frac{j}{2}}$. Lemma 3.22 then shows that ψ_1 and φ_1 belong to the Lipschitz class $\Lambda_{\frac{1}{2}}(\mathbb{R})$.

The better estimate for the smoothness of ψ_1 we just obtained is due to a precise calculation of the last integral. Such a precise calculation is not possible for the general compactly supported wavelet ψ_k that we have constructed in section 2.3. Nevertheless, one reason why $\alpha(k) = k - \log_2 \binom{2k+1}{k}$ is not sharp is due to the "gross" estimate (3) in Lemma 3.27, where we have estimated the maximum of a product by the product of the maxima. Since the $\tilde{g}_k$ are trigonometric polynomials, it is not reasonable to expect that the maximum values of $\tilde{g}_k(\frac{1}{2}\xi)$ and $\tilde{g}_k(\frac{1}{4}\xi)$ are attained at the same point. We can observe, in fact, that this is not true in the case $k = 1$; below is a table that compares the approximate maximum of $\prod_{\ell=1}^{j+1} \tilde{g}_1(2^{-\ell}\xi)$ (obtained by a computer calculation) with the estimate $2^{-2(j+1)}3^{j+1}$, which is obtained from (3) of Lemma 3.27, for $j = 0, 1, 2$.

j	(3) of Lemma 3.27	Approximate maximum
0	3/4	3/4
1	9/16	6.48/16
2	27/32	16.88/32

These data indicate that better estimates can be obtained by some appropriate grouping of the factors $\tilde{g}_k$. More precisely, we have the following result:

LEMMA 4.2 *For* $m \in \mathbb{N}$ *fixed, define*

$$B_m^{(k)} = \sup_{\xi \in \mathbb{R}} \prod_{j=1}^{m} \tilde{g}_k(2^{-j}\xi).$$

If $A_j^{(m)} = \{\xi \in \mathbb{R} : 2^{mj}\pi \leq |\xi| \leq 2^{m(j+1)}\pi\}$ *and*

$$\mathcal{G}_j^{(m)} = \int_{A_j^{(m)}} \Big\{ \prod_{\ell=1}^{m(j+1)} \tilde{g}_k(2^{-\ell}\xi) \Big\} d\xi,$$

we have $\mathcal{G}_j^{(m)} \leq 2^{mj}(2^m - 1)2\pi \big(B_m^{(k)}\big)^{j+1}$.

PROOF : There are $m(j+1)$ factors in the product that appears inside the integral that defines $\mathcal{G}_j^{(m)}$; group the first m factors together, then the following m factors, and so on, and estimate each one of the groups by $B_m^{(k)}$. ∎

We now need a version of Lemma 3.22 adapted to our new situation.

LEMMA 4.3 *Let* f *be a function in* $L^2(\mathbb{R})$ *and* $\alpha > 0$; *if* $m \in \mathbb{N}$ *and*

$$\int_{A_j^{(m)}} |\hat{f}(\xi)| \, d\xi \leq C2^{-\alpha mj}, \qquad j = 0, 1, 2, \cdots,$$

where $A_j^{(m)} = \{\xi \in \mathbb{R} : 2^{jm}\pi \leq |\xi| \leq 2^{(j+1)m}\pi\}$, *then* $f \in \Lambda_\alpha(\mathbb{R})$ *when* $\alpha \notin \mathbb{Z}$, *and* $f \in \Lambda_{\alpha-\varepsilon}(\mathbb{R})$ *for any* $\varepsilon > 0$ *when* $\alpha \in \mathbb{Z}$.

PROOF : The proof is similar to the one that established Lemma 3.22, and we leave the details for the reader. ∎

For k and m fixed natural numbers, define

$$I_j^{(m)} = \int_{A_j^{(m)}} |\hat{\varphi}_k(\xi)| \, d\xi, \qquad j \in \mathbb{N}.$$

Using the Cauchy-Schwarz inequality we obtain

$$I_j^{(m)} \le \left(2^{jm}(2^m - 1)2\pi\right)^{\frac{1}{2}} \left(\int_{A_j^{(m)}} |\hat{\varphi}_k(\xi)|^2 \, d\xi\right)^{\frac{1}{2}}.$$

Now

$$|\hat{\varphi}_k(\xi)|^2 \le \prod_{\ell=1}^{m(j+1)} |m_k(2^{-\ell}\xi)|^2 = \prod_{\ell=1}^{m(j+1)} \left(\frac{1 + \cos(2^{-\ell}\xi)}{2}\right)^{k+1} 2^{k+1} \tilde{g}_k(2^{-\ell}\xi).$$

Since

$$\prod_{\ell=1}^{m(j+1)} \left(\frac{1 + \cos(2^{-\ell}\xi)}{2}\right) = \prod_{\ell=1}^{m(j+1)} \cos^2(2^{-\ell-1}\xi)$$

$$= \left|\frac{\sin(\xi/2)}{2^{m(j+1)} \sin(2^{-m(j+1)-1}\xi)}\right|^2 \le 2^{-2m(j+1)}|\sin(2^{-m-1}\pi)|^{-2}$$

when $\xi \in A_j^{(m)}$, we obtain

$$I_j^{(m)} \le C(k,m)2^{-\frac{1}{2}mjk}\left(\mathcal{G}_j^{(m)}\right)^{\frac{1}{2}}. \qquad (4.4)$$

From Lemma 4.2 we deduce

$$I_j^{(m)} \le C(k,m)2^{-\frac{mjk}{2}}2^{\frac{mj}{2}}\left(B_m^{(k)}\right)^{\frac{j+1}{2}}$$

$$= C(k,m)2^{-mj[\frac{k}{2}-\frac{1}{2}-\frac{1}{2m}\log_2 B_m^{(k)}]}$$

$$= C(k,m)2^{-mj[k-\frac{1}{2m}\log_2(2^{m(k+1)}B_m^{(k)})]}.$$

Combining this with Lemma 4.3 we deduce that φ_k, and also the wavelet ψ_k, belong to $\Lambda_{\alpha_m(k)}(\mathbb{R})$ if $\alpha_m(k) \notin \mathbb{Z}$, and to $\Lambda_{\alpha_m(k)-\varepsilon}(\mathbb{R})$ for any $\varepsilon > 0$ if $\alpha_m(k) \in \mathbb{Z}$, where

$$\alpha_m(k) = k - \frac{1}{2m}\log_2\left(2^{m(k+1)}B_m^{(k)}\right).$$

With $k = 1$ and $m = 3$, $\alpha_3(1) \sim 0.320459$, which is an improvement over the value $\alpha(1) \sim 0.207518$ obtained in section 2.3.

But this is not the best we can do. The best will be produced if we can find the supremum of $\{\alpha_m(k) : m = 1, 2, \cdots\}$, or, equivalently, the infimum of $\{b_m^{(k)} : m = 1, 2, \cdots\}$, where

$$b_m^{(k)} = \frac{1}{m} \log_2 \left(2^{m(k+1)} B_m^{(k)}\right).$$

We shall show, following [Co2], that

$$b^{(k)} \equiv \inf\{b_m^{(k)} : m = 1, 2, \cdots\} = \lim_{m \to \infty} b_m^{(k)}. \tag{4.5}$$

To see this, let $\varepsilon > 0$ and choose $m_\varepsilon \in \mathbb{N}$ such that $|b_{m_\varepsilon}^{(k)} - b^{(k)}| \leq \frac{1}{2}\varepsilon$; by the Euclidean algorithm we can write $m = c_m m_\varepsilon + r_m$ with $c_m, r_m \in \mathbb{N}$ and $0 \leq r_m < m_\varepsilon$. Hence,

$$2^{m(k+1)} B_m^{(k)} \leq 2^{(c_m m_\varepsilon + r_m)(k+1)} \left(B_{m_\varepsilon}^{(k)}\right)^{c_m} \left(B_1^{(k)}\right)^{r_m}$$

$$= \left(2^{m_\varepsilon(k+1)} B_{m_\varepsilon}^{(k)}\right)^{c_m} \left(2^{(k+1)} B_1^{(k)}\right)^{r_m}.$$

Taking logarithms we obtain

$$b_m^{(k)} \leq b_{m_\varepsilon}^{(k)} + \frac{b_1^{(k)}}{c_m}.$$

For m large enough, $\frac{1}{c_m} b_1^{(k)} < \frac{1}{2}\varepsilon$ and, hence,

$$b^{(k)} \leq b_m^{(k)} \leq b_{m_\varepsilon}^{(k)} + \frac{1}{2}\varepsilon \leq b^{(k)} + \varepsilon,$$

which proves the desired result.

The number $\gamma(k) = k - \frac{1}{2}b^{(k)}$ is called the **critical exponent** of the wavelet ψ_k and measures the maximum smoothness that this wavelet might have. To get an upper bound of this critical exponent we can use the ergodic theorem for the ergodic transformation $\xi \mapsto 2\xi \pmod{2\pi}$. In fact,

$$\lim_{m \to \infty} \frac{1}{m} \log_2 2^{m(k+1)} \prod_{\ell=1}^{m} \tilde{g}_k(2^\ell \xi)$$

$$= \lim_{m \to \infty} \frac{1}{m} \sum_{\ell=1}^{m} \log_2 \left(2^{(k+1)} \tilde{g}_k(2^\ell \xi)\right)$$

$$= \frac{1}{2\pi} \int_0^{2\pi} \log_2 \left(2^{k+1} \tilde{g}_k(\xi) \right) d\xi \equiv A^{(k)}$$

by the ergodic theorem; thus, $b^{(k)} \geq A^{(k)}$, so that the critical exponent cannot be bigger than $k - \frac{1}{2} A^{(k)}$. The following table shows the approximate values of this bound for the critical exponent for $k = 1, 2, 3$ obtained with a computer.

k	$k - \frac{1}{2} A^{(k)}$
1	0.55001
2	1.08783
3	1.61792

2.5 Notes and references

1. The idea of an MRA was introduced by S. Mallat (see [Mal1]) where the general theory of finding a wavelet starting from an MRA scheme was developed. That the five conditions that define an MRA are not independent appears not to have been noticed during the first stages of the development of this theory. An indication that this is the case can be found in section 5.3.2 of [Da1]; a precise statement can be found in [Mad]. As far as we know the proofs of Theorems 1.6 and 1.7 have not appeared in this form in the literature. We have treated the concept of MRA in one dimension. For the extension to n-dimensions the best reference is [Me1].

2. Multiresolution analyses with dilation factors other than 2 can be considered. When $\lambda > 1$ is rational, say $\lambda = p/q$ with p and q being relatively prime natural numbers, the examples are more varied. Although we start with only one scaling function in the definition of this generalized MRA, the number of corresponding functions ψ needed to generate an orthonormal basis for $L^2(\mathbb{R})$ could be more than one; in fact, it is $p - q$. The case $\lambda = 1 + \frac{1}{m} = (m+1)/m$ was mentioned in note 3 of section 1.6, and only one function ψ is needed. It can be shown that if $p > q > 1$, the scaling function φ, and consequently the wavelets associated with it, cannot have compact support. Thus, the behavior is different from the case $p = 2$, $q = 1$, where we have constructed the compactly supported wavelets of Daubechies (see sections 2.3 and 2.4). For details about these wavelets we refer the reader to [Au3].

3. I. Daubechies was the first to construct compactly supported orthonormal wavelets (see [Da2]) with a fixed degree of smoothness. She observed that if φ and ψ belong to $C^r(\mathbb{R})$ then the low-pass filter m_0 must be of the form

$$m_0(\xi) = \left(\frac{1 + e^{-i\xi}}{2}\right)^{r+1} \mathcal{F}(\xi) \qquad (5.1)$$

with $\mathcal{F}$ being 2π-periodic and $\mathcal{F} \in C^r(\mathbb{R})$. This can be deduced from Theorem 3.4. Writing $\mathcal{M}(\xi) = |m_0(\xi)|^2$, one can see from (5.1) that

$$\mathcal{M}(\xi) = \left(\cos^2(\tfrac{1}{2}\xi)\right)^{r+1} \mathcal{G}(\xi),$$

where $\mathcal{G}(\xi)$ is a polynomial in $\cos(\xi)$. Since $1 - \cos(\xi) = 2\sin^2(\tfrac{1}{2}\xi)$, we can write

$$\mathcal{M}(\xi) = \left(\cos^2(\tfrac{1}{2}\xi)\right)^{r+1} \mathcal{P}(\sin^2(\tfrac{1}{2}\xi)), \qquad (5.2)$$

where $\mathcal{P}$ is a polynomial. This equality, together with $\mathcal{M}(\xi) + \mathcal{M}(\xi+\pi) = 1$ (which is (2.5)) imply that $\mathcal{P}$ must satisfy the equality

$$(1 - y)^{r+1} \mathcal{P}(y) + y^{r+1} \mathcal{P}(1 - y) = 1 \qquad (y = \sin^2(\tfrac{1}{2}\xi)). \qquad (5.3)$$

Some "involved" combinatorial lemmas are used in [Da2] to find $\mathcal{P}$ from (5.3). A more elegant solution is presented in section 6.1 of [Da1]. Observe also that the solution of (5.3) can be found using the Euclidean algorithm since the polynomials $(1 - y)^{r+1}$ and y^{r+1} are relatively prime. This approach was pointed out by Y. Meyer. The solution to (5.3) is

$$\mathcal{P}_r(y) \equiv \mathcal{P}(\sin^2(\tfrac{1}{2}\xi)) = \sum_{j=0}^{r} \binom{r + j}{j} \left(\sin^2(\tfrac{1}{2}\xi)\right)^j.$$

A "square root" of $\mathcal{M}$ is then found using Lemma 3.16. Lemma 3.16 is often referred to as the "Factorization Theorem of Fejér-Riesz"; our proof is borrowed from [RS].

4. More on the regularity of compactly supported wavelets (which we have treated in sections 2.3 and 2.4) can be found in [Da1], [Da2], [Co2], [Co3], [CC], [DL1] and [DL2]. We mention that, asymptotically, the critical exponent of regularity $\gamma(k)$ of the compactly supported wavelet ψ_k is

$$(1 - \tfrac{1}{2}\log_2 3)k \cong 0.2075k;$$

hence, asymptotically, a C^k compactly supported wavelet has a support whose measure is, roughly, $5k$. A more detailed study of the differentiability properties of φ_1 and ψ_1, that goes beyond what is presented at the beginning of section 2.4, can be found in [Pol].

3

Band-limited wavelets

A function $f \in L^2(\mathbb{R})$ is said to be **band-limited** if the support of $\hat{f}$ is contained in a finite interval. Our purpose in this chapter is to study orthonormal wavelets that are band-limited. In this setting some of the convergence problems that we shall encounter are easier to treat than they are in the general (not necessarily band-limited) setting. Nevertheless, many of the results we prove, as we shall see later, are valid for general wavelets. These more restricted results provide a good motivation for their extensions which will be presented later (mostly in Chapter 7).

Two simple equations that characterize the orthonormality of a system of the form

$$\left\{ \psi_{j,k}(x) = 2^{\frac{j}{2}} \psi(2^j x - k) : j, k \in \mathbb{Z} \right\}$$

are presented in section 3.1. Two more equations, (2.1) and (2.2), that characterize the completeness of the above system in $L^2(\mathbb{R})$ when ψ is band-limited (this restriction will be removed in Chapter 7) are derived in the second section.

Armed with this quantitative description, we characterize all orthonormal wavelets whose Fourier transform has support contained in $[-\frac{8}{3}\pi, \frac{8}{3}\pi]$. This includes all the Lemarié-Meyer wavelets already presented in the first two chapters.

3.1 Orthonormality

In this section the assumption that ψ is band-limited is not needed. Suppose that ψ is an orthonormal wavelet; then, in particular, the system

$\{\psi_{j,k} : j, k \in \mathbb{Z}\}$ is orthonormal, where $\psi_{j,k}(x) = 2^{\frac{j}{2}}\psi(2^j x - k)$. In this section we exploit these orthonormality relations to find two equations that characterize them.

The fact that the system $\{\psi_{0,k} : k \in \mathbb{Z}\}$ is orthonormal is, as pointed out in Proposition 1.11 of Chapter 2, equivalent to

$$\sum_{k \in \mathbb{Z}} |\hat{\psi}(\xi + 2k\pi)|^2 = 1 \qquad \text{for } a.e. \ \xi \in \mathbb{R}. \tag{1.1}$$

Performing a change of variables, we see that $<\psi_{j,k}, \psi_{j,\ell}> = <\psi_{0,k}, \psi_{0,\ell}>$; this tells us that the system $\{\psi_{j,k} : k \in \mathbb{Z}\}$ is orthonormal for each fixed j when (1.1) is satisfied.

If, say, $j > n$, then the change of variables $x = 2^{-n}(y + m)$ shows that $<\psi_{j,k}, \psi_{n,m}> = <\psi_{\ell,p}, \psi_{0,0}>$, where $\ell = j - n$ and $p = k - 2^{j-n}m$. Relabeling this we see that the orthogonality between $\psi_{j,k}$ and $\psi_{n,m}$, for $j > n$ and $k, m \in \mathbb{Z}$, can be reduced to the orthogonality between $\psi_{j,k}$ and ψ, when $j > 0$ and $k \in \mathbb{Z}$. By the Plancherel theorem this means that, for $j \geq 1$ and $k \in \mathbb{Z}$,

$$0 = <\psi, \psi_{j,k}> = \frac{1}{2\pi} \int_{\mathbb{R}} \hat{\psi}(\xi) 2^{-\frac{j}{2}} \overline{\hat{\psi}(2^{-j}\xi)} e^{i2^{-j}k\xi} d\xi$$

$$= \frac{1}{2\pi} \int_{\mathbb{R}} 2^{\frac{j}{2}} \hat{\psi}(2^j \mu) \overline{\hat{\psi}(\mu)} e^{ik\mu} d\mu.$$

Thus,

$$0 = \sum_{\ell \in \mathbb{Z}} \int_{2\ell\pi}^{2(\ell+1)\pi} \hat{\psi}(2^j \xi) \overline{\hat{\psi}(\xi)} e^{ik\xi} d\xi$$

$$= \int_0^{2\pi} \left\{ \sum_{\ell \in \mathbb{Z}} \hat{\psi}(2^j(\xi + 2\ell\pi)) \overline{\hat{\psi}(\xi + 2\ell\pi)} \right\} e^{ik\xi} d\xi$$

for all $k \in \mathbb{Z}$ when $j \geq 1$. This shows that

$$\sum_{k \in \mathbb{Z}} \hat{\psi}(2^j(\xi + 2k\pi)) \overline{\hat{\psi}(\xi + 2k\pi)} = 0 \qquad \text{a.e. on } \mathbb{R}, \quad j \geq 1. \tag{1.2}$$

Thus, (1.1) and (1.2) are **necessary and sufficient** conditions for the orthonormality of the system $\{\psi_{j,k} : j, k \in \mathbb{Z}\}$.

Let us observe that the two series (1.1) and (1.2) converge for a.e. $\xi \in \mathbb{R}$. In fact, if we define

$$\zeta_j(\xi) \equiv \sum_{k \in \mathbb{Z}} \hat{\psi}(2^j(\xi + 2k\pi)) \overline{\hat{\psi}(\xi + 2k\pi)}, \qquad j \in \mathbb{Z}, \qquad (1.3)$$

then a change of variables combined with the Schwarz inequality give us

$$\int_{\mathbb{T}} |\zeta_j(\xi)| \, d\xi \leq \int_{\mathbb{R}} |\hat{\psi}(2^j\xi) \, \hat{\psi}(\xi)| \, d\xi \leq \left(\int_{\mathbb{R}} |\hat{\psi}(2^j\xi)|^2 d\xi \right)^{\frac{1}{2}} \left(\int_{\mathbb{R}} |\hat{\psi}(\xi)|^2 d\xi \right)^{\frac{1}{2}}$$

$$= 2^{-\frac{j}{2}} \int_{\mathbb{R}} |\hat{\psi}(\xi)|^2 d\xi = 2^{-\frac{j}{2}} 2\pi \|\psi\|_2^2 < \infty.$$

This shows that $\zeta_j \in L^1(\mathbb{T})$ for all $j \in \mathbb{Z}$, and, hence, gives us the a.e. absolute convergence of the series in (1.1) and (1.2) when $j = 0$ and $j \neq 0$, respectively.

We now examine the orthogonal projection Q_j of $L^2(\mathbb{R})$ onto W_j, where

$$W_j = \overline{\text{span}} \{\psi_{j,k} : k \in \mathbb{Z}\}, \qquad j \in \mathbb{Z}.$$

We know that $\{\psi_{j,k} : k \in \mathbb{Z}\}$ is an orthonormal basis for W_j, therefore,

$$Q_j f = \sum_{k \in \mathbb{Z}} <f, \psi_{j,k}> \psi_{j,k} \qquad \text{for all } f \in L^2(\mathbb{R}). \qquad (1.4)$$

Since

$$(\psi_{j,k})^\wedge(\xi) = 2^{-\frac{j}{2}} \hat{\psi}(2^{-j}\xi) e^{-i2^{-j}k\xi} \qquad \text{for all } j, k \in \mathbb{Z},$$

we see that $\{\gamma_{j,k} : k \in \mathbb{Z}\}$ is an orthonormal basis for $(W_j)^\wedge$, where

$$\gamma_{j,k}(\xi) = \frac{2^{-\frac{j}{2}}}{\sqrt{2\pi}} e^{-i2^{-j}k\xi} \hat{\psi}(2^{-j}\xi), \qquad j, k \in \mathbb{Z}.$$

The factor $\frac{1}{\sqrt{2\pi}}$ comes from the Plancherel theorem and it normalizes the $\gamma_{j,k}$'s. Hence, we can write

$$(Q_j f)^\wedge = \sum_{k \in \mathbb{Z}} <\hat{f}, \gamma_{j,k}> \gamma_{j,k} \qquad \text{for } f \in L^2(\mathbb{R}).$$

Let $g_j(\mu) = \hat{f}(\mu) \overline{\hat{\psi}(2^{-j}\mu)}$ and

$$F_j(\xi) = \sum_{\ell \in \mathbb{Z}} g_j(\xi + 2^{j+1}\ell\pi).$$

The last function is $2^{j+1}\pi$-periodic and so are the functions

$$E_k^{(j)}(\xi) = \frac{2^{-\frac{j}{2}}}{\sqrt{2\pi}} e^{-i2^{-j}k\xi}, \qquad k \in \mathbb{Z}.$$

In fact, the system $\{E_k^{(j)} : k \in \mathbb{Z}\}$ is an orthonormal basis for $L^2([0, 2^{j+1}\pi))$. Using again a periodization argument, we obtain

$$
\begin{aligned}
<\hat{f}, \gamma_{j,k}> &= \int_{\mathbb{R}} \hat{f}(\xi) \frac{2^{-\frac{j}{2}}}{\sqrt{2\pi}} e^{i2^{-j}k\xi} \overline{\hat{\psi}(2^{-j}\xi)} \, d\xi \\
&= \int_0^{2^{j+1}\pi} \left(\sum_{\ell \in \mathbb{Z}} g_j(\xi + 2^{j+1}\ell\pi) \right) \frac{2^{-\frac{j}{2}}}{\sqrt{2\pi}} e^{i2^{-j}k\xi} \, d\xi \\
&= <F_j, E_k^{(j)}>,
\end{aligned}
\tag{1.5}
$$

where the last inner product is the one associated with $L^2([0, 2^{j+1}\pi))$. Hence,

$$F_j(\xi) = \sum_{k \in \mathbb{Z}} <\hat{f}, \gamma_{j,k}> E_k^{(j)}(\xi)$$

with convergence in $L^2([0, 2^{j+1}\pi))$. More explicitly, we have found

$$\sum_{\ell \in \mathbb{Z}} \hat{f}(\xi + 2^{j+1}\ell\pi) \overline{\hat{\psi}(2^{-j}\xi + 2\ell\pi)} = \sum_{k \in \mathbb{Z}} <\hat{f}, \gamma_{j,k}> E_k^{(j)}(\xi). \tag{1.6}$$

From this we obtain the following result:

THEOREM 1.7 *Suppose* $f \in L^2(\mathbb{R})$.

(A) *f is orthogonal to W_j if and only if*

$$\sum_{k \in \mathbb{Z}} \hat{f}(\xi + 2^{j+1}k\pi) \overline{\hat{\psi}(2^{-j}\xi + 2k\pi)} = 0 \qquad \text{for a.e. } \xi \in \mathbb{R};$$

(B) *For the projection operator Q_j we have*

$$(Q_j f)^{\wedge}(\xi) = \hat{\psi}(2^{-j}\xi) \sum_{k \in \mathbb{Z}} \hat{f}(\xi + 2^{j+1}k\pi) \overline{\hat{\psi}(2^{-j}\xi + 2k\pi)} \qquad \text{a.e. on } \mathbb{R}.$$

PROOF : (A) follows immediately from (1.6); notice that (A) is obtained by a periodization argument similar to the one that gave us (1.2).

To prove (B) multiply both sides of (1.6) by $\hat{\psi}(2^{-j}\xi)$ and observe that $\hat{\psi}(2^{-j}\xi)E_k^{(j)}(\xi) = \gamma_{j,k}(\xi)$.

3.2 Completeness

Conditions (1.1) and (1.2) of section 3.1 are not sufficient to assure us that $\{\psi_{j,k} : j, k \in \mathbb{Z}\}$ is complete. In fact, suppose that $\{\psi_{j,k} : j, k \in \mathbb{Z}\}$ is an orthonormal system; if we define $\alpha(x) = \sqrt{2}\,\psi(2x)$ we have

$$\alpha_{j,k}(x) = \psi_{j+1,2k}(x) \qquad \text{for all} \;\; j, k \in \mathbb{Z}.$$

Hence, $\{\alpha_{j,k} : j, k \in \mathbb{Z}\}$ is also an orthonormal system, but it **cannot** be complete. In this section we study conditions that will provide completeness of an orthonormal system of the type involved in wavelets.

All the results proved in the previous section hold for a general wavelet; in fact, the band-limited condition played no fundamental role in our proofs. The material that we shall develop in this section extends to the general wavelet; however, the band-limited assumption makes convergence questions, as well as others, much easier to treat.

The main result of this section is to prove that the conditions

$$\sum_{j \in \mathbb{Z}} |\hat{\psi}(2^j \xi)|^2 = 1 \qquad \text{a.e. on} \;\; \mathbb{R} - \{0\}, \tag{2.1}$$

and

$$\sum_{j=0}^{\infty} \hat{\psi}(2^j \xi)\, \overline{\hat{\psi}(2^j(\xi + 2k\pi))} = 0 \quad \text{a.e. on} \;\; \mathbb{R}, \quad k \in 2\mathbb{Z}+1, \tag{2.2}$$

characterize the completeness of the orthonormal system $\{\psi_{j,k} : j, k \in \mathbb{Z}\}$.

These two equations are **most** important: it turns out that they imply (1.1) and (1.2) and, consequently, they completely characterize orthonormal wavelets. This apparently simple and elementary fact is by no means easy to prove in the general (not necessarily band-limited) case. We shall present a proof of this and many of its consequences in Chapter 7.

We begin by proving some needed results. If ψ is a band-limited wavelet, we can always find an integer J such that supp $(\hat{\psi}) \subset (-2^J \pi, 2^J \pi)$. We will assume this inclusion relation throughout this section.

LEMMA 2.3 *Suppose that* $f \in L^2(\mathbb{R})$ *and* $\hat{f}$ *has a support contained in* $I = (a, b)$, *where* $b - a \leq 2^{-J} \pi$ *and* $I \cap [-\pi, \pi] = \emptyset$; *then, for all* $j \in \mathbb{Z}$,

$$(Q_j f)^\wedge(\xi) = \hat{f}(\xi)|\hat{\psi}(2^{-j}\xi)|^2 \qquad a.e. \ on \ I.$$

PROOF: If $-j \leq J$, $k \neq 0$ and $\xi \in I$, we have that $\xi + 2^{j+1} k\pi$ lies outside the support of $\hat{f}$. In fact, if $k > 0$,

$$b = a + (b - a) \leq a + 2^{-J}\pi \leq a + 2^{j+1} k\pi < \xi + 2^{j+1} k\pi$$

so that, in this case, $\xi + 2^{j+1} k\pi$ is to the right of I; if $k < 0$,

$$a = b + (a - b) \geq b - 2^{-J}\pi \geq b + 2^{-J} k\pi \geq b + 2^{j+1} k\pi > \xi + 2^{j+1} k\pi$$

which proves that $\xi + 2^{j+1} k\pi$ is to the left of I. Therefore, in the formula for $(Q_j f)^\wedge$ given in (B) of Theorem 1.7, all terms are zero except the one corresponding to $k = 0$ and this proves the lemma if $-j \leq J$. Notice that the condition $I \cap [-\pi, \pi] = \emptyset$ has not been used in this part.

If $-j > J$, then $|2^{-j}\xi| \geq 2^J |\xi| \geq 2^J \pi$ when $\xi \in I$ since $I \cap [-\pi, \pi] = \emptyset$; hence, $\hat{\psi}(2^{-j}\xi) = 0$, which proves that $\hat{f}(\xi)|\hat{\psi}(2^{-j}\xi)|^2 = 0$. On the other hand, if $\xi + 2^{j+1} k\pi \in I$,

$$|2^{-j}\xi + 2k\pi| \geq 2^J |\xi + 2^{j+1} k\pi| \geq 2^J \pi$$

and, hence, $\hat{\psi}(2^{-j}\xi + 2k\pi) = 0$; using (B) of Theorem 1.7, we deduce $(Q_j f)^\wedge(\xi) = 0$. Hence, $(Q_j f)^\wedge(\xi) = \hat{f}(\xi)|\psi(2^{-j}\xi)|^2$ when $-j > J$ and $\xi \in I$ (since both terms are 0), and the lemma is proved. ∎

We are now ready to establish condition (2.1).

THEOREM 2.4 *If* ψ *is a band-limited orthonormal wavelet, then*

$$\sum_{j \in \mathbb{Z}} |\hat{\psi}(2^j \xi)|^2 = 1 \qquad for \ a.e. \ \xi \in \mathbb{R} - \{0\}.$$

PROOF: Let us define

$$\omega(\xi) \equiv \sum_{j\in\mathbb{Z}} |\hat\psi(2^j\xi)|^2. \tag{2.5}$$

It is clear that $\omega(2^n\xi) = \omega(\xi)$ for all $n \in \mathbb{Z}$. Hence, it suffices to show that $\omega(\xi) = 1$ for a.e. $\xi \in (-2\pi, -\pi) \cup (\pi, 2\pi)$. Notice that the set $(\pi, 2\pi)$, except for a finite number of points, can be expressed as the union of 2^J intervals I with the properties described in Lemma 2.3; thus, it is enough to show the equality for almost every $\xi \in I$ where I is an interval described in the lemma.

On such an interval, by Lemma 2.3 we have

$$\int_I \left| \sum_{j=-M}^{M} (Q_j f)^\wedge(\xi) \right|^2 d\xi = \int_I |\hat f(\xi)|^2 \left(\sum_{j=-M}^{M} |\hat\psi(2^{-j}\xi)|^2 \right)^2 d\xi$$

for all positive integers M, when $\operatorname{supp}(\hat f) \subset I$. Since the Q_j's are mutually orthogonal projections, the left-hand side of the above formula is dominated by the square of the L^2-norm of $\hat f$. Hence,

$$\int_I |\hat f(\xi)|^2 \left(\sum_{j=-M}^{M} |\hat\psi(2^{-j}\xi)|^2 \right)^2 d\xi \le \int_I |\hat f(\xi)|^2 \, d\xi.$$

Since this inequality is true for all functions f in $L^2(\mathbb{R})$ with $\hat f$ supported in I, we must have

$$\sum_{j=-M}^{M} |\hat\psi(2^{-j}\xi)|^2 \le 1 \qquad \text{a.e. on } I,$$

for all positive integers M. This shows that the series $\sum_{j\in\mathbb{Z}} |\hat\psi(2^{-j}\xi)|^2$ converges. Moreover,

$$\int_I |\hat f(\xi)|^2 \left(1 - \sum_{j=-M}^{M} |\hat\psi(2^{-j}\xi)|^2 \right)^2 d\xi = \int_I \left| \hat f(\xi) - \sum_{j=-M}^{M} (Q_j f)^\wedge(\xi) \right|^2 d\xi$$

tends to zero as $M \to \infty$. An application of Fatou's lemma shows that the desired equality holds for almost every $\xi \in I$.

∎

Our next goal is to show that $\hat{\psi}$ has support away from the origin when ψ is a band-limited wavelet such that $|\hat{\psi}|$ is continuous at zero.

Suppose that ψ is band-limited and $|\hat{\psi}|$ is continuous at zero; assume also that $\{\psi_{j,k} : j, k \in \mathbb{Z}\}$ is an orthonormal system for $L^2(\mathbb{R})$. Since $\psi \perp W_j$ for all $j \neq 0$, part (A) of Theorem 1.7 shows that

$$\sum_{k \in \mathbb{Z}} \hat{\psi}(\xi + 2^{j+1}k\pi) \overline{\hat{\psi}(2^{-j}\xi + 2k\pi)} = 0 \qquad \text{a.e. on } \mathbb{R}, \quad \text{when } j \neq 0.$$

Recall that $\operatorname{supp}(\hat{\psi}) \subset (-2^J\pi, 2^J\pi)$; hence, the point $\xi + 2^{j+1}k\pi$ lies outside the support of $\hat{\psi}$ when $k \neq 0$, $\xi \in \operatorname{supp}(\hat{\psi})$ and $j \geq J$. Therefore, the above formula reduces to

$$\hat{\psi}(\xi) \overline{\hat{\psi}(2^{-j}\xi)} = 0 \qquad \text{a.e. on } \operatorname{supp}(\hat{\psi}), \quad \text{when } j \geq J.$$

But this equality is trivially true if $\xi \notin \operatorname{supp}(\hat{\psi})$, so that $|\hat{\psi}(\xi)\hat{\psi}(2^{-j}\xi)| = 0$ for a.e. $\xi \in \mathbb{R}$ when $j \geq J$. Letting $j \to \infty$ and using the continuity of $|\hat{\psi}|$ at zero, we obtain

$$|\hat{\psi}(\xi)\hat{\psi}(0)| = 0 \qquad \text{for a.e. } \xi \in \mathbb{R}.$$

Since $\hat{\psi}$ is not identically zero we deduce $\hat{\psi}(0) = 0$.

We announce this in the following proposition:

PROPOSITION 2.6 *If ψ is band-limited, $|\hat{\psi}|$ is continuous at zero and $\{\psi_{j,k} : j, k \in \mathbb{Z}\}$ is an orthonormal system, then $\hat{\psi}(0) = 0$.*

If, in addition, ψ is an orthonormal wavelet, we have a stronger result:

THEOREM 2.7 *If ψ is a band-limited orthonormal wavelet such that $|\hat{\psi}|$ is continuous at 0, then $\hat{\psi} = 0$ a.e. in an open neighborhood of the origin.*

PROOF: Again let us assume that $\operatorname{supp}(\hat{\psi}) \subset (-2^J\pi, 2^J\pi)$. As we showed in the proof of Proposition 2.6,

$$|\hat{\psi}(\xi)\hat{\psi}(2^{-j}\xi)| = 0 \qquad \text{a.e. on } \mathbb{R}, \quad \text{for } |j| \geq J.$$

This shows that, when $|j| \geq J$, $\hat{\psi}(2^j\xi) = 0$ for a.e. $\xi \in \operatorname{supp}(\hat{\psi})$ (when $j \leq -J$ we can put $\eta = 2^{-j}\xi$ and obtain $|\hat{\psi}(2^j\eta)\hat{\psi}(\eta)| = 0$ a.e. on $\mathbb{R}$). The equality in Theorem 2.4 thus becomes

$$\sum_{|j|<J} |\hat{\psi}(2^j\xi)|^2 = 1 \qquad \text{for} \;\; a.e. \;\; \xi \in \text{supp}\,(\hat{\psi}).$$

Since there are no more than $2J - 1$ terms in this sum, for each of them there must exist a $j_o \in (-J, J) \cap \mathbb{Z}$ such that

$$|\hat{\psi}(2^{j_o}\xi)| \geq \left(\frac{1}{2J-1}\right)^{\frac{1}{2}} \qquad \text{for} \;\; a.e. \;\; \xi \in \text{supp}\,(\hat{\psi}).$$

By Proposition 2.6 and the continuity of $|\hat{\psi}|$ at 0, there exists an $\varepsilon > 0$ such that

$$|\hat{\psi}(\mu)| \leq \frac{1}{2}\left(\frac{1}{2J-1}\right)^{\frac{1}{2}} \qquad \text{for} \;\; |\mu| < \varepsilon.$$

This shows that almost every $\xi \in \text{supp}\,(\hat{\psi})$ satisfies $\varepsilon \leq |2^{j_o}\xi| \leq 2^J|\xi|$, and, hence, $|\xi| \geq 2^{-J}\varepsilon$ (otherwise we could not have $\sum_{|j|<J} |\hat{\psi}(2^j\xi)|^2 = 1$ a.e. on supp $(\hat{\psi})$). Therefore, $\hat{\psi}(\xi) = 0$ for a.e. $\xi \in (-2^{-J}\varepsilon, 2^{-J}\varepsilon)$. ∎

We now show that any band-limited orthonormal wavelet such that $|\hat{\psi}|$ is continuous at 0 must satisfy equality (2.2).

THEOREM 2.8 *If ψ is a band-limited orthonormal wavelet such that $|\hat{\psi}|$ is continuous at 0, then for each odd integer q we have*

$$\sum_{j=0}^{\infty} \hat{\psi}(2^j\xi)\,\overline{\hat{\psi}(2^j(\xi + 2q\pi))} = 0 \qquad a.e. \;\; \text{on} \;\; \mathbb{R}.$$

PROOF : Isolating the term $k = 0$ in part (B) of Theorem 1.7, we have

$$(Q_j f)^{\wedge}(\xi) = |\hat{\psi}(2^{-j}\xi)|^2 \hat{f}(\xi) + \hat{\psi}(2^{-j}\xi)\sum_{k\neq 0} \hat{f}(\xi + 2^{j+1}k\pi)\,\overline{\hat{\psi}(2^{-j}\xi + 2k\pi)}.$$

Since $|\hat{\psi}|$ is continuous at 0, by Theorem 2.7 we can find $J \in \mathbb{N}$ such that supp $(\hat{\psi}) \subset [-2^J\pi, -2^{-J}\pi] \cup [2^{-J}\pi, 2^J\pi]$. If $0 < a < |\xi| < b$, only a finite number of j's (say, $|j| \leq M$) give us non-zero terms in the second summand of the above formula. Let us fix such a j; then there exists at most a finite number of k's such that $\hat{\psi}(2^{-j}\xi + 2k\pi) \neq 0$. Thus, there exists at most a finite number of values of the form $v = 2^j k$ that occur in the non-zero summands; when $k \neq 0$ they are of the (unique) form $v = 2^p q$ for a finite number of $p, q \in \mathbb{Z}$, $p \geq j$ and q odd. Thus,

$$(Q_j f)^\wedge(\xi) = |\hat{\psi}(2^{-j}\xi)|^2 \hat{f}(\xi)$$

$$+ \hat{\psi}(2^{-j}\xi) \sum_{p \geq j} \sum_{q \in 2\mathbb{Z}+1} \hat{f}(\xi + 2^{p+1}q\pi) \overline{\hat{\psi}(2^{-j}(\xi + 2^{p+1}q\pi))}. \qquad (2.9)$$

Summing over all $j \in \mathbb{Z}$ and using Theorem 2.4 we obtain

$$\hat{f}(\xi) = \sum_{j \in \mathbb{Z}} (Q_j f)^\wedge(\xi)$$

$$= \hat{f}(\xi) + \sum_{q \in 2\mathbb{Z}+1} \sum_{p \in \mathbb{Z}} \hat{f}(\xi + 2^{p+1}q\pi) \sum_{j \leq p} \hat{\psi}(2^{-j}\xi) \overline{\hat{\psi}(2^{-j}(\xi + 2^{p+1}q\pi))},$$

where we have changed the order of summation, which we can do since all the sums involved are finite. Thus, for a.e. $\xi \in (-b, -a) \cup (a, b)$,

$$\sum_{q \in 2\mathbb{Z}+1} \sum_{p \in \mathbb{Z}} \hat{f}(\xi + 2^{p+1}q\pi) \sum_{j \leq p} \hat{\psi}(2^{-j}\xi) \overline{\hat{\psi}(2^{-j}(\xi + 2^{p+1}q\pi))} = 0.$$

Let us define

$$h_k(\xi) \equiv \sum_{j=0}^{\infty} \hat{\psi}(2^j\xi) \overline{\hat{\psi}(2^j(\xi + 2k\pi))}, \qquad k \in 2\mathbb{Z}+1. \qquad (2.10)$$

Then the change of indices $p - j = \ell$ give us, for a.e. $\xi \in (-b, -a) \cup (a, b)$,

$$0 = \sum_{q \in 2\mathbb{Z}+1} \sum_{p \in \mathbb{Z}} \hat{f}(\xi + 2^{p+1}q\pi) \sum_{\ell \geq 0} \hat{\psi}(2^\ell 2^{-p}\xi) \overline{\hat{\psi}(2^\ell(2^{-p}\xi + 2q\pi))}$$

$$= \sum_{q \in 2\mathbb{Z}+1} \sum_{p \in \mathbb{Z}} \hat{f}(\xi + 2^{p+1}q\pi) h_q(2^{-p}\xi).$$

If we fix $q_o \in 2\mathbb{Z}+1$ and $p_o = 0$, then for each $\xi_o \in (-b, -a) \cup (a, b)$ there exists a $\delta > 0$ such that the interval

$$U = (\xi_o + 2q_o\pi - \delta, \xi_o + 2q_o\pi + \delta)$$

contains no point of the form $\xi_o + 2^{p+1}q\pi$ if $(p, q) \neq (0, q_o)$ (which we can do since only a finite number of p and q are involved). With $\hat{f} = \chi_U$ in the above equality, we obtain, for a.e. $\xi \in (\xi_o - \delta, \xi_o + \delta) \cap \{(-b, -a) \cup (a, b)\}$,

$$0 = \sum_{q \in 2\mathbb{Z}+1} \sum_{p \in \mathbb{Z}} \hat{f}(\xi + 2^{p+1} q\pi) h_q(2^{-p}\xi) = h_{q_o}(\xi).$$

Since ξ_o is an arbitrary point of $(-b, -a) \cup (a, b)$ the above equality is true for almost every $\xi \in (-b, -a) \cup (a, b)$. Letting $a \to 0$ and $b \to \infty$ we see that $h_{q_o}(\xi) = 0$ for almost every $\xi \in \mathbb{R} - \{0\}$. Also observe that $h_{q_o}(\xi) = 0$ is true if $\xi = 0$ since $\hat{\psi}(0) = 0$. The desired result follows since q_o is an arbitrary odd integer.

∎

THEOREM 2.11 *Suppose $\psi \in L^2(\mathbb{R})$ is a band-limited function such that $\hat{\psi}$ is zero on an open neighborhood of the origin and $\{\psi_{j,k} : j, k \in \mathbb{Z}\}$ is an orthonormal system that satisfies (2.1) and (2.2); then ψ is an orthonormal wavelet.*

PROOF : It is enough to show that $\sum_{j \in \mathbb{Z}}(Q_j f)^\wedge(\xi) = \hat{f}(\xi)$ a.e. on $\mathbb{R}$ for all $f \in L^2(\mathbb{R})$, where Q_j is the orthogonal projection onto W_j (see (1.4)). By part (B) of Theorem 1.7 we can write

$$(Q_j f)^\wedge(\xi) = \hat{f}(\xi)|\hat{\psi}(2^{-j}\xi)|^2 + \hat{\psi}(2^{-j}\xi)\sum_{k \neq 0}\hat{f}(\xi + 2^{j+1}k\pi)\,\overline{\hat{\psi}(2^{-j}\xi + 2k\pi)}.$$

For j fixed and $k \neq 0$, we can write $2^j k = 2^j 2^p q = 2^n q$ for $p \geq 0$ and $q \in 2\mathbb{Z} + 1$. Since ψ is band-limited and $\hat{\psi}$ is zero in an open neighborhood of the origin, when we sum the above expression over j we can interchange the order of summation; thus, using (2.1), we obtain

$$\sum_{j \in \mathbb{Z}}(Q_j f)^\wedge(\xi) = \hat{f}(\xi) \cdot 1$$

$$+ \sum_{q \in 2\mathbb{Z}+1} \sum_{n \in \mathbb{Z}} \hat{f}(\xi + 2^{n+1}q\pi) \sum_{j \leq n} \hat{\psi}(2^{-j}\xi)\,\overline{\hat{\psi}(2^{-j}(\xi + 2^{n+1}q\pi))}.$$

Writing $\xi = 2^n \mu$ and using the change of indices $k = n - j$, we see that the last sum coincides with the left-hand side of (2.2), which we assume to be zero. Hence, $\sum_{j \in \mathbb{Z}}(Q_j f)^\wedge(\xi) = \hat{f}(\xi)$ a.e. on $\mathbb{R}$, which proves the completeness of the system.

∎

We summarize the above results:

THEOREM 2.12 *If $\psi \in L^2(\mathbb{R})$ is a band-limited function such that $\hat{\psi}$ is zero in a neighborhood of the origin and $\{\psi_{j,k} : j, k \in \mathbb{Z}\}$ is an orthonormal system, then this system is complete if and only if*

$$\sum_{j \in \mathbb{Z}} |\hat{\psi}(2^j \xi)|^2 = 1 \qquad a.e. \ on \ \mathbb{R} - \{0\} \tag{2.13}$$

and

$$\sum_{j=0}^{\infty} \hat{\psi}(2^j \xi) \overline{\hat{\psi}(2^j(\xi + 2k\pi))} = 0 \quad a.e. \ on \ \mathbb{R}, \quad k \in 2\mathbb{Z}+1. \tag{2.14}$$

Observe that for the necessity it is enough to assume the continuity of $|\hat{\psi}|$ at the origin, but for the sufficiency we need to assume that $\hat{\psi}$ is almost everywhere zero in a neighborhood of the origin. These conditions arise because of the type of proof we have given; they will be removed in Chapter 7.

We have pointed out at the beginning of this section that orthonormality is a consequence of the characterization of completeness given in Theorem 2.12. We present an idea of why this is possible. This can be done in more general settings using "frames." A **frame** on a Hilbert space $\mathbb{H}$ is a family $\{\varphi_j : j \in \mathbb{J}\}$ of elements of $\mathbb{H}$ satisfying the following property: there exist constants A and B, $0 < A \leq B < \infty$ such that

$$A\|f\|_{\mathbb{H}}^2 \leq \sum_j |(f, \varphi_j)|^2 \leq B\|f\|_{\mathbb{H}}^2 \qquad \text{for all} \ f \in \mathbb{H},$$

where $(\cdot, \cdot)$ is the inner product in the Hilbert space $\mathbb{H}$.

Suppose the family $\{\gamma_{j,k}(x) = 2^{\frac{j}{2}} \gamma(2^j x - k) : j, k \in \mathbb{Z}\}$ is a frame in $L^2(\mathbb{R})$; we can then define the "frame operator" $\mathcal{F}$ by

$$\mathcal{F}f = \sum_{j \in \mathbb{Z}} \sum_{k \in \mathbb{Z}} <f, \gamma_{j,k}> \gamma_{j,k}.$$

Observe that if γ has norm 1 in $L^2(\mathbb{R})$, all the $\gamma_{j,k}$ also have norm 1 in $L^2(\mathbb{R})$. If we assume that the frame operator $\mathcal{F}$ is the identity (that is, $\mathcal{F}f = f$ in the L^2-norm, which is clearly a statement of the completeness of the system), then

$$1 = <\gamma_{m,n}, \gamma_{m,n}> = <\mathcal{F}\gamma_{m,n}, \gamma_{m,n}>$$

$$= \sum_{j,k} <\gamma_{m,n}, \gamma_{j,k}><\gamma_{j,k}, \gamma_{m,n}> = \sum_{j,k} |<\gamma_{m,n}, \gamma_{j,k}>|^2.$$

Since the last sum contains the term $|<\gamma_{m,n}, \gamma_{m,n}>|^2 = 1$, the remainder of the non-negative terms must be all zero; this shows that the system is orthonormal. We shall elaborate on this later (see Chapter 8).

REMARK 1: The length of the interval around the origin found in Theorem 2.7 can be as small as we wish. In fact, given $\varepsilon > 0$, one can construct a band-limited orthonormal wavelet ψ such that $\hat{\psi}$ is continuous at 0, $|\hat{\psi}|$ is even, and $\hat{\psi}$ is not identically 0 on $(-\varepsilon, \varepsilon)$ (of course $\hat{\psi}$ is identically 0 on a smaller interval by Theorem 2.7).

This wavelet ψ will be constructed by using the MRA method developed in Chapter 2. We observe that we can assume $\varepsilon \leq \pi$ since, otherwise, the Shannon wavelet (see Example C of Chapter 2) will satisfy the above requirements.

We begin by constructing a low-pass filter m_0. For $J \in \mathbb{N}$ such that $2^{-J}\pi < \frac{1}{2}\varepsilon$, let $I = [a, b]$ be an interval centered at $2^{-J}\pi$ of length 2ℓ, with

$$\ell < \frac{\pi}{2^J(2^J + 1)}.$$

Choose positive numbers α and β such that $\alpha^2 + \beta^2 = 1$. For $\xi \in [0, \pi]$ we define

$$m_0(\xi) = \chi_{[0,\frac{\pi}{2}]\setminus I}(\xi) + \alpha\chi_I(\xi) + \beta\chi_{\pi-I}(\xi), \tag{2.15}$$

where $\pi - I = [\pi - b, \pi - a]$; if $\xi \in [-\pi, 0)$ we let $m_0(\xi) = m_0(-\xi)$. Finally, m_0 is extended to $\mathbb{R}$ so that it is 2π-periodic. The graph of this low-pass filter is given in Figure 3.1.

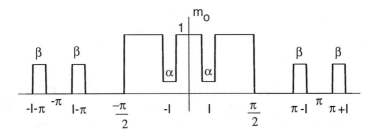

Figure 3.1: The low-pass filter defined by (2.15)

It is clear that

$$|m_0(\xi)|^2 + |m_0(\xi + \pi)|^2 = 1.$$

Since $m_0(\xi) \equiv 1$ on a neighborhood of the origin the infinite product $\prod_{j=1}^{\infty} m_0(2^{-j}\xi)$ converges for every $\xi \in \mathbb{R}$. Define φ by letting

$$\hat{\varphi}(\xi) = \prod_{j=1}^{\infty} m_0(2^{-j}\xi), \qquad \xi \in \mathbb{R}.$$

We now show that φ is band-limited; in fact, we shall show that $\operatorname{supp}(\hat{\varphi}) \subset [-2^J\pi, 2^J\pi]$.

Let $\xi \in [2^{j-1}\pi, 2^j\pi]$ for $j \geq J+1$ and let $\eta = 2^{-j}\xi$ so that $\eta \in [\frac{1}{2}\pi, \pi]$. From the definition of $\hat{\varphi}$ we have

$$\hat{\varphi}(\xi) = m_0(2^{j-1}\eta) \cdots m_0(2\eta)m_0(\eta)\hat{\varphi}(\eta).$$

If $\eta \notin \pi - I$, $m_0(\eta) = 0$ and, consequently, $\hat{\varphi}(\xi) = 0$. If $\eta \in \pi - I = [\pi - 2^{-J}\pi - \ell, \pi - 2^{-J}\pi + \ell]$ we have $2^J\eta + 2\pi - 2^J\pi \in [\pi - 2^J\ell, \pi + 2^J\ell]$. Since we have chosen $\ell < 2^{-J}\pi/(2^J + 1)$ the interval $[\pi - 2^J\ell, \pi + 2^J\ell]$ lies between $\pi - I$ and $\pi + I$; hence, the 2π-periodicity of m_0 implies

$$m_0(2^J\eta) = m_0(2^J\eta + 2\pi - 2^J\pi) = 0.$$

Since $j - 1 \geq J$, $m_0(2^J\eta)$ is one of the factors in the above product for $\hat{\varphi}$ and hence $\hat{\varphi}(\xi) = 0$. This shows that $\hat{\varphi}(\xi) \equiv 0$ on $[2^{j-1}\pi, 2^j\pi]$ if $j \geq J+1$. Since $\hat{\varphi}$ is even, this shows the desired result.

Observe that $\hat{\varphi}$ is a bounded, compactly supported function; thus, both $\hat{\varphi}$ and φ are in $L^2(\mathbb{R})$ (of course, this also follows as in the proof of Proposition 3.9 of Chapter 2).

Consider the expression, as in (1.10) of Chapter 2,

$$\sigma_\varphi^2(\xi) = \sum_{k \in \mathbb{Z}} |\hat{\varphi}(\xi + 2k\pi)|^2$$

(since φ is band-limited the sum contains only a finite number of terms for each $\xi \in \mathbb{R}$. The proof of Theorem 3.13 in Chapter 2 shows that $\sigma_\varphi(\xi) \equiv 1$ (observe that $m_0(\xi) \neq 0$ on $[-\frac{1}{2}\pi, \frac{1}{2}\pi]$). By Proposition 1.11 of Chapter 2, $\{\varphi(\cdot - k) : k \in \mathbb{Z}\}$ is an orthonormal system.

The argument we gave after the proof of Theorem 3.13 of Chapter 2 shows that φ is a scaling function for an MRA, and one of its associated wavelets ψ satisfies

$$\hat{\psi}(\xi) = e^{i\frac{\xi}{2}} \overline{m_0(\tfrac{1}{2}\xi + \pi)}\, \hat{\varphi}(\tfrac{1}{2}\xi) \qquad (2.16)$$

(see (2.12) of Chapter 2). Clearly, ψ is band-limited, since $\hat{\varphi}$ has a compact support. Since $\hat{\varphi}$ is even and m_0 is 2π-periodic and even, we can deduce that $|\hat{\psi}|$ is even:

$$|\hat{\psi}(-\xi)| = m_0(-\tfrac{1}{2}\xi + \pi)\hat{\varphi}(-\tfrac{1}{2}\xi) = m_0(-\tfrac{1}{2}\xi + \pi - 2\pi)\hat{\varphi}(\tfrac{1}{2}\xi)$$

$$= m_0(\tfrac{1}{2}\xi + \pi)\hat{\varphi}(\tfrac{1}{2}\xi) = |\hat{\psi}(\xi)|.$$

It is now easy to show that $\hat{\psi}$ cannot be identically zero on $(-\varepsilon, \varepsilon)$. Recall that we are assuming $\varepsilon \leq \pi$. If $\hat{\psi}(\xi) \equiv 0$ on $(-\varepsilon, \varepsilon)$, since $\hat{\varphi}(\tfrac{1}{2}\xi)$ is never zero on $[-\pi, \pi]$ we deduce from (2.16) that $m_0(\tfrac{1}{2}\xi + \pi) \equiv 0$ on $(-\varepsilon, \varepsilon)$. This implies that $m_0(\xi) \equiv 0$ on $(\pi - \tfrac{1}{2}\varepsilon, \pi + \tfrac{1}{2}\varepsilon)$; but this is not true since our choice of I implies $(\pi - I) \cap (\pi - \tfrac{1}{2}\varepsilon, \pi + \tfrac{1}{2}\varepsilon) \neq \emptyset$ (in fact, since $2^{-J}\pi < \tfrac{1}{2}\varepsilon$, $\pi - \tfrac{1}{2}\varepsilon < \pi - 2^{-J}\pi$ and $\pi - 2^{-J}\pi$ is the central point of $\pi - I$).

This example gives us a band-limited orthonormal wavelet ψ such that $\hat{\psi}$ is continuous at 0 and the length of the open interval about zero on which $\hat{\psi} \equiv 0$ is arbitrarily small.

The process we just described can be smoothed out by doing a little more work. This has been done in [BSW]: for every $\varepsilon > 0$, there exists a band-limited orthonormal wavelet ψ such that $|\hat{\psi}|$ is even, $\hat{\psi} \in C^\infty$ and $\hat{\psi}(\xi)$ is not identically zero on $(-\varepsilon, \varepsilon)$. More results concerning the support of band-limited orthonormal wavelets are given in section 3.4 (see Theorem 4.18 of this chapter).

3.3 The Lemarié-Meyer wavelets revisited

In section 1.4 we presented the Lemarié-Meyer wavelets and showed how they can be constructed from the local sine and cosine bases. In this section we give a complete characterization of them.

THEOREM 3.1 *Suppose* $\psi \in L^2(\mathbb{R})$ *and* $b = |\hat{\psi}|$ *is an even continuous function with support contained in* $[-2\pi - \varepsilon', -\pi + \varepsilon] \cup [\pi - \varepsilon, 2\pi + \varepsilon']$, *where* $0 < \varepsilon, \varepsilon', \varepsilon + \varepsilon' \leq \pi$ *and* $\varepsilon, \varepsilon'$ *are minimal for the support of* $\hat{\psi}$. *Then* ψ *is an orthonormal wavelet if and only if*

(i) $b^2(\xi) + b^2(4\pi - \xi) = 1$ *if* $\xi \in [2\pi - \varepsilon', 2\pi + \varepsilon']$,

(ii) $b^2(\xi) + b^2(2\pi - \xi) = 1$ *if* $\xi \in [\pi - \varepsilon, \pi + \varepsilon]$,

(iii) $b(\xi) = 1$ *if* $\xi \in [\pi + \varepsilon, 2\pi - \varepsilon']$,

(iv) $\varepsilon' = 2\varepsilon$,

(v) $b(\pi + \xi) = b(2(\pi - \xi))$ *if* $\xi \in [-\varepsilon, \varepsilon]$,

(vi) $\hat{\psi}(\xi) = e^{i\alpha(\xi)} b(\xi)$, *where* α *satisfies*

$$\alpha(\xi + \pi) - \alpha(2(\xi + \pi)) - \alpha(\xi - \pi) + \alpha(2(\xi - \pi)) = (2m(\xi) + 1)\pi$$

for some $m(\xi) \in \mathbb{Z}$, *for all* $\xi \in [-\varepsilon, \varepsilon]$.

REMARK 1:   Observe that (i), (ii), (iii) and (v) are conditions that a bell function for $[\pi, 2\pi]$ must satisfy if we extend this interval by ε to the left of π and $\varepsilon' = 2\varepsilon$ to the right of 2π. See section 1.3 for the definition and properties of bell functions. Also compare this theorem with the result given in Corollary 4.7 of Chapter 1. When constructing the wavelet basis from one of these wavelets, the associated bell function is dilated by 2^j, $j \in \mathbb{Z}$, forming adjacent ones that must be compatible. Because of this, the conditions (i), (ii), (iv) and (v) are not independent. The following simple argument shows that (ii), (iv) and (v) imply (i):

For $\eta \in [-\varepsilon, \varepsilon]$ (v) gives us the two equations

(1) $b(\pi + \eta) = b(2\pi - 2\eta)$, and

(2) $b(\pi - \eta) = b(2\pi + 2\eta)$.

From (ii) we also have

(3) $b^2(\pi + \eta) + b^2(\pi - \eta) = b^2(\pi + \eta) + b^2(2\pi - (\pi + \eta)) = 1$.

Substituting (1) and (2) in (3) we obtain

(4) $b^2(2\pi - 2\eta) + b^2(2\pi + 2\eta) = 1$ for $\eta \in [-\varepsilon, \varepsilon]$.

With $\xi = 2\pi + 2\eta \in [2\pi - 2\varepsilon, 2\pi + 2\varepsilon] = [2\pi - \varepsilon', 2\pi + \varepsilon']$ we can write condition (4) in the form $b^2(\xi) + b^2(4\pi - \xi) = 1$ if $\xi \in [2\pi - \varepsilon', 2\pi + \varepsilon']$, which is condition (i).

REMARK 2: The above theorem characterizes those wavelets for which $|\hat{\psi}|$ is even and continuous and with support that is close to the support of the Fourier transform of the Shannon wavelet. In Chapter 7 we shall show that these wavelets form a subset of a large class of related wavelets. In the next section of this chapter we also extend this class of wavelets.

REMARK 3: The most elementary solution of the functional equation satisfied by the phase function is $\alpha(\xi) = \frac{1}{2}\xi$, as can be easily checked. Other solutions are

$$\alpha(\xi) = \tfrac{1}{2}\xi + \gamma(\xi),$$

where γ is a 2π-periodic function. Another solution is

$$\alpha(\xi) = \tfrac{1}{2}\xi - \text{sign}(\xi)\tfrac{1}{2}\pi;$$

observe that if ψ is defined by $\hat{\psi}(\xi) = e^{i\frac{\xi}{2}}b(\xi)$, then the Hilbert transform of ψ has a Fourier transform equal to

$$-i(\text{sign}(\xi))e^{i\frac{\xi}{2}}b(\xi) = e^{i(\frac{\xi}{2}-\text{sign}(\xi)\frac{\pi}{2})}b(\xi).$$

PROOF: Let us first assume that ψ is an orthonormal wavelet. Equality (1.1) can be written in the form

$$\sigma_\psi^2(\xi) = \sum_{k=-\infty}^{\infty} b^2(\xi + 2k\pi) = 1. \tag{3.2}$$

When $\xi \in [2\pi - \varepsilon', 2\pi + \varepsilon']$ a simple calculation shows that $\xi + 2k\pi$ lies outside the support of b, except when $k = 0$ or $k = -2$ (use $\varepsilon + \varepsilon' \leq \pi$ to rule out the case $k = -1$). Since b is even, we see that (3.2) reduces to (i) for such ξ. When $\xi \in [\pi - \varepsilon, \pi + \varepsilon]$, $\xi + 2k\pi$ lies in the support of b only if $k = 0$ or $k = -1$; in this case we obtain (ii). If $\xi \in [\pi + \varepsilon, 2\pi - \varepsilon']$, then $\xi + 2k\pi$ lies outside the support of b when $k \neq 0$ and we obtain (iii).

Since ψ is a band-limited wavelet and $\hat{\psi} = 0$ in a neighborhood of the origin, we can apply Theorem 2.12. Condition (2.14) of this theorem when $k = -1$ is

$$h_{-1}(\xi) = \sum_{j=0}^{\infty} \hat{\psi}(2^j\xi)\,\overline{\hat{\psi}(2^j(\xi - 2\pi))} = 0, \qquad \xi \in \mathbb{R}. \tag{3.3}$$

If $2^j \xi$ belongs to the support of $\hat{\psi}$ and $j \geq 2$, the point $2^j(\xi - 2\pi)$ does not belong to the support of $\hat{\psi}$; therefore, (3.3) is equivalent to

$$\hat{\psi}(\xi)\overline{\hat{\psi}(\xi - 2\pi)} + \hat{\psi}(2\xi)\overline{\hat{\psi}(2(\xi - 2\pi))} = 0, \qquad \xi \in \mathbb{R}.$$

But this means that if $\xi \in \mathbb{R}$, then

$$b(\xi)b(2\pi - \xi) = b(2\xi)b(4\pi - 2\xi). \tag{3.4}$$

Let us assume $\varepsilon' > 2\varepsilon$; then the set $(2\pi - \varepsilon', 2\pi - 2\varepsilon) \subset [2\pi - \varepsilon', 2\pi + \varepsilon']$ is non-empty. It follows that equality (i) holds for all $\xi \in (2\pi - \varepsilon', 2\pi - 2\varepsilon)$ and, hence,

$$(*) \qquad b^2(\xi) + b^2(4\pi - \xi) = 1 \qquad \text{if } \xi \in (2\pi - \varepsilon', 2\pi - 2\varepsilon);$$

for such a ξ, $\frac{1}{2}\xi \in (\pi - \frac{1}{2}\varepsilon', \pi - \varepsilon)$ and (3.4) with ξ replaced by $\frac{1}{2}\xi$ gives us

$$(\star) \qquad b(\xi)b(4\pi - \xi) = 0 \qquad \text{if } \xi \in (2\pi - \varepsilon', 2\pi - 2\varepsilon).$$

By (iii) (that has just been proved), $b(\xi) = 1$ when $\xi = 2\pi - \varepsilon'$; thus, $b(\xi) > 0$ in a neighborhood of $2\pi - \varepsilon'$. It follows from $(*)$ and $(\star)$ that the continuous unit vector

$$(b(\xi), b(4\pi - \xi))$$

must coincide with $(1, 0)$ in this neighborhood. By continuity we must have $b(4\pi - \xi) = 0$ on $(2\pi - \varepsilon', 2\pi - 2\varepsilon)$. Letting $\mu = 4\pi - \xi$, we see that $b(\mu) = 0$ for $\mu \in (2\pi + 2\varepsilon, 2\pi + \varepsilon')$, contradicting the minimality of ε'. Thus $\varepsilon' \leq 2\varepsilon$.

Let us now assume $\varepsilon' < 2\varepsilon$; then the set $(\pi - \varepsilon, \pi - \frac{1}{2}\varepsilon') \subset [0, \pi - \frac{1}{2}\varepsilon']$ is non-empty. Since $4\pi - 2\xi > 2\pi + \varepsilon'$ if $\xi \in (\pi - \varepsilon, \pi - \frac{1}{2}\varepsilon')$, we have $b(4\pi - 2\xi) = 0$ so that by (3.4) we obtain

$$(\dagger) \qquad b(\xi)b(2\pi - \xi) = 0 \qquad \text{if } \xi \in (\pi - \varepsilon, \pi - \frac{1}{2}\varepsilon');$$

moreover, (ii) is also true on this interval, and we have

$$(\ddagger) \qquad b^2(\xi) + b^2(2\pi - \xi) = 1 \qquad \text{if } \xi \in (\pi - \varepsilon, \pi - \frac{1}{2}\varepsilon').$$

But $b(\pi - \varepsilon) = 0$ and, by continuity and $(\dagger)$, $b(2\pi - \xi) > 0$ on an interval of the form $(\pi - \varepsilon, \pi - \varepsilon + \delta) \subset (\pi - \varepsilon, \pi - \frac{1}{2}\varepsilon')$. By $(\ddagger)$, $(\dagger)$ and continuity, we have

$$b(\xi) = 0 \quad \text{and} \quad b(2\pi - \xi) = 1 \quad \text{for } \xi \in (\pi - \varepsilon, \pi - \varepsilon + \delta) \subset (\pi - \varepsilon, \pi - \frac{1}{2}\varepsilon').$$

As in the previous case, the continuous unit vector-valued function

$$(b(\xi), b(2\pi - \xi)) = (0, 1) \qquad \text{on } (\pi - \varepsilon, \pi - \varepsilon + \delta),$$

and it is also continuous on the larger interval $(\pi - \varepsilon, \pi - \frac{1}{2}\varepsilon')$. It follows that $b(\xi) = 0$ on this larger interval, contradicting the minimality of ε. Thus $\varepsilon' = 2\varepsilon$ and we obtain (iv).

We want to prove (v). By part (A) of Theorem 1.7 with $f = \psi \perp W_{-1}$ we have

$$\sum_{k \in \mathbb{Z}} \hat{\psi}(2(\xi + k\pi)) \overline{\hat{\psi}(\xi + k\pi)} = 0.$$

When $\xi \in [-\varepsilon, \varepsilon]$ only the terms corresponding to $k = -1$ and $k = 1$ can be non-zero (use $\varepsilon' = 2\varepsilon$ and $\varepsilon + \varepsilon' \leq \pi$). Hence,

$$\hat{\psi}(2(\xi - \pi)) \overline{\hat{\psi}(\xi - \pi)} + \hat{\psi}(2(\xi + \pi)) \overline{\hat{\psi}(\xi + \pi)} = 0 \qquad (3.5)$$

if $\xi \in [-\varepsilon, \varepsilon]$. This implies

$$b(\xi - \pi)b(2(\xi - \pi)) = b(\xi + \pi)b(2(\xi + \pi)) \qquad \text{for } \xi \in [-\varepsilon, \varepsilon].$$

Having shown (i) and (ii) and $\varepsilon' = 2\varepsilon$, we have, for $\xi \in [-\varepsilon, \varepsilon]$,

$$\begin{aligned}
b^2(\xi + \pi) &= b^2(\xi + \pi)\left[b^2(2(\xi + \pi)) + b^2(2(\xi - \pi))\right] \\
&= b^2(\xi - \pi)b^2(2(\xi - \pi)) + b^2(\xi + \pi)b^2(2(\xi - \pi)) \\
&= b^2(2(\xi - \pi))\left[b^2(\pi - \xi) + b^2(\pi + \xi)\right] \\
&= b^2(2(\xi - \pi)),
\end{aligned}$$

which is property (v).

It remains to show (vi). Equation (3.5) tells us that the vectors

$$(\hat{\psi}(\xi + \pi), \hat{\psi}(\xi - \pi)) \qquad \text{and} \qquad (\hat{\psi}(2(\xi + \pi)), \hat{\psi}(2(\xi - \pi)))$$

are orthogonal; (i) and (ii) imply that they are unit vectors on $[-\varepsilon, \varepsilon]$. Thus, there exists $\delta(\xi)$, $\xi \in [-\varepsilon, \varepsilon]$, such that

$$e^{i\delta(\xi)}\left(b(\xi + \pi)e^{i\alpha(\xi + \pi)}, b(\xi - \pi)e^{i\alpha(\xi - \pi)}\right)$$

$$= \left(-b(2(\xi - \pi))e^{-i\alpha(2(\xi - \pi))}, b(2(\xi + \pi))e^{-i\alpha(2(\xi + \pi))}\right).$$

Now $b(\xi + \pi) = b(2(\xi - \pi))$ and $b(\xi - \pi) = b(2(\xi + \pi))$ by (v) and the evenness of b. Thus

$$b(\xi + \pi)e^{i\{\alpha(\xi+\pi)+\alpha(2(\xi-\pi))-\pi\}} = e^{-i\delta(\xi)}b(\xi + \pi)$$

and

$$b(\xi - \pi)e^{i\{\alpha(\xi-\pi)+\alpha(2(\xi+\pi))\}} = e^{-i\delta(\xi)}b(\xi - \pi)$$

if $\xi \in [-\varepsilon, \varepsilon]$. When $b(\xi + \pi)$ and $b(\xi - \pi)$ are not zero we can divide by these values to obtain the functional equation for α given in (vi); if these values are zero, we can choose α so that the equation holds.

We will now show that conditions (i) – (vi) imply orthonormality, and then we will show completeness.

To prove orthonormality we need to show (1.1) and (1.2). In our case (1.1) is equivalent to (3.2) and the argument given to show (i), (ii) and (iii) from (3.2) can be reversed to show that (i), (ii) and (iii) is all we need to prove (3.2) and, equivalently, (1.1).

To prove (1.2) we need to show that $\zeta_j(\xi) = 0$ for almost every $\xi \in \mathbb{R}$ and $j \geq 1$, where

$$\zeta_j(\xi) = \sum_{k \in \mathbb{Z}} \hat{\psi}(2^j(\xi + 2k\pi))\,\overline{\hat{\psi}(\xi + 2k\pi)},$$

as defined in (1.3). Since $\zeta_j(\xi)$ is 2π-periodic we only need to check this result on an interval of length 2π. If $\xi \in [\varepsilon, 2\pi + \varepsilon]$ and $j \geq 2$, then either $\xi + 2k\pi$ or $2^j(\xi + 2k\pi)$ lie outside the support of $\hat{\psi}$ and, thus, $\zeta_j(\xi) = 0$. For the remaining case $j = 1$ we observe that (1.2) is equivalent to

$$\hat{\psi}(2\xi)\,\overline{\hat{\psi}(\xi)} + \hat{\psi}(2(\xi - 2\pi))\,\overline{\hat{\psi}(\xi - 2\pi)} = 0 \qquad \text{for } \xi \in [0, 2\pi];$$

this equality, in turn, is equivalent to

$$\hat{\psi}(2(\xi + \pi))\,\overline{\hat{\psi}(\xi + \pi)} + \hat{\psi}(2(\xi - \pi))\,\overline{\hat{\psi}(\xi - \pi)} = 0 \qquad \text{for } \xi \in [-\pi, \pi].$$

When $\xi \in [-\pi, -\varepsilon] \cup [\varepsilon, \pi]$ one of the factors in each term is zero. The case $\xi \in [-\varepsilon, \varepsilon]$ follows by "reversing" the argument that established (v) from equality (3.5).

This shows that $\{\psi_{j,k} : j, k \in \mathbb{Z}\}$ is an orthonormal system. Observe that in this argument we have not used the minimality of ε and $\varepsilon' = 2\varepsilon$.

Finally we need to show that the system $\{\psi_{j,k} : j, k \in \mathbb{Z}\}$ is complete. One way to do this is to prove that it satisfies (2.13) and (2.14) of Theorem 2.12. We shall give a different proof that will allow us to make some observations concerning the hypothesis necessary to prove completeness.

Suppose that $f \in L^2(\mathbb{R})$ is orthogonal to all the elements in the system $\{\psi_{j,k} : j, k \in \mathbb{Z}\}$; we shall show $f = 0$ in $L^2(\mathbb{R})$ by showing that $\hat{f}(\xi) = 0$ for almost every $\xi \in \mathbb{R}$. If for any g,

$$\tilde{g}(\xi) \equiv g(-\xi),$$

then it is easily seen that $\tilde{\hat{g}} = \hat{\tilde{g}}$. Therefore, $f \perp \hat{\psi}_{j,k}$ (that is, $\hat{f} \perp (\psi_{j,k})^\wedge$) is equivalent to $\tilde{\hat{f}} \perp (\widetilde{\psi_{j,k}})^\wedge = (\tilde{\psi}_{j,-k})^\wedge$. Since $\{\psi_{j,k} : j, k \in \mathbb{Z}\}$ is an orthonormal basis if and only if $\{\tilde{\psi}_{j,k} : j, k \in \mathbb{Z}\}$ is an orthonormal basis, it suffices to show $\hat{f}(\xi) = 0$ for almost every $\xi \in (0, \infty)$. Moreover, it suffices to show that $\hat{f}(\xi) = 0$ for almost every $\xi \in [\pi - \varepsilon, 2(\pi - \varepsilon)]$ (or any interval of the form $(a, 2a)$, $a > 0$), since for every $m \in \mathbb{Z}$, $g = f(2^m \cdot) \perp \psi_{j,k}$ for all $j, k \in \mathbb{Z}$ is equivalent to $f \perp \psi_{j,k}$ for all $j, k \in \mathbb{Z}$ (observe that $<g, \psi_{j,k}> = 2^{-\frac{m}{2}} <f, \psi_{j-m,k}>$).

We now use part (A) of Theorem 1.7 with $j = 0$ (that is, $f \perp W_0$) to obtain

$$\sum_{k \in \mathbb{Z}} \hat{f}(\xi + 2k\pi)\overline{\hat{\psi}(\xi + 2k\pi)} = 0 \qquad \text{for a.e. } \xi \in \mathbb{R}.$$

If $\xi \in [\pi + \varepsilon, 2(\pi - \varepsilon)]$ all the terms in the above sum are zero except the one corresponding to $k = 0$. Hence, $\hat{f}(\xi)\overline{\hat{\psi}(\xi)} = 0$ for $\xi \in [\pi + \varepsilon, 2(\pi - \varepsilon)]$; since $|\hat{\psi}(\xi)| = 1$, by (iii) and (iv) we deduce that $\hat{f}(\xi) = 0$ for $\xi \in [\pi + \varepsilon, 2(\pi - \varepsilon)]$.

It remains to show that $\hat{f} \equiv 0$ on $[\pi - \varepsilon, \pi + \varepsilon]$. On this interval the above equality is equivalent to

(I) $\qquad \hat{f}(\xi - \pi)\overline{\hat{\psi}(\xi - \pi)} + \hat{f}(\xi + \pi)\overline{\hat{\psi}(\xi + \pi)} = 0 \qquad$ on $[-\varepsilon, \varepsilon]$.

If we use part (A) of Theorem 1.7 with $j = -1$ (that is, $f \perp W_{-1}$), we obtain

(II) $\qquad \hat{f}(\xi - \pi)\overline{\hat{\psi}(2(\xi - \pi))} + \hat{f}(\xi + \pi)\overline{\hat{\psi}(2(\xi + \pi))} = 0 \quad$ on $[-\varepsilon, \varepsilon]$.

But (I) and (II) form a system of two-linear equations with $\hat{f}(\xi - \pi)$ and $\hat{f}(\xi + \pi)$ as unknowns whose determinant is

$$\Delta(\xi) = \overline{\hat{\psi}(\xi - \pi)}\,\overline{\hat{\psi}(2(\xi + \pi))} - \overline{\hat{\psi}(\xi + \pi)}\,\overline{\hat{\psi}(2(\xi - \pi))}. \qquad (3.6)$$

Using that $|\hat{\psi}|$ is even, (v), (vi) and (ii) we deduce

$$\Delta(\xi) = \left[b^2(\xi - \pi) + b^2(\xi + \pi) \right] e^{-i\{\alpha(\xi-\pi)+2(2(\xi+\pi))\}}$$

$$= e^{-i\{\alpha(\xi-\pi)+\alpha(2(\xi+\pi))\}} \neq 0.$$

Hence $\hat{f}(\xi) = 0$ on $[\pi - \varepsilon, \pi + \varepsilon]$, which completes the proof. $\blacksquare$

If one looks carefully at the argument that proves completeness in the above theorem, it can be seen that we have proved the following result:

> Suppose that $\{\psi_{j,k} : j, k \in \mathbb{Z}\}$ is an orthonormal system with supp $(\hat{\psi}) = [-2\pi - 2\varepsilon, -\pi + \varepsilon] \cup [\pi - \varepsilon, 2\pi + 2\varepsilon]$ and $|\hat{\psi}|$ is even; then the system $\{\psi_{j,k} : j, k \in \mathbb{Z}\}$ is complete.

Notice that continuity is only used to prove (iv), which we are assuming in this statement.

If $\Delta(\xi) = 0$ we use formula (3.6) to deduce that

$$b(\xi - \pi)b(2(\xi + \pi)) = b(\xi + \pi)b(2(\xi - \pi));$$

by (v) and the evenness of b we deduce $b(\xi - \pi) = b(\xi + \pi)$ on $[-\varepsilon, \varepsilon]$. But then, (ii) implies $b(\xi - \pi) = b(\xi + \pi) = \frac{1}{\sqrt{2}}$ for almost every $\xi \in [-\varepsilon, \varepsilon]$. We have shown the following result:

> Let $\psi \in L^2(\mathbb{R})$ be such that
>
> $$\text{supp}\,(\hat{\psi}) = [-2\pi - \varepsilon, -\pi + \varepsilon] \cup [\pi - \varepsilon, 2\pi + 2\varepsilon]$$
>
> and $b = |\hat{\psi}|$ is even. If b satisfies (i), (ii), (iii) and (v) of Theorem 3.1 and the set
>
> $$\{\xi \in [-\varepsilon, \varepsilon] : b(\xi - \pi) = b(\xi + \pi) = \frac{1}{\sqrt{2}}\}$$
>
> has measure zero, then $\{\tilde{\psi}_{j,k} : j, k \in \mathbb{Z}\}$ is complete in $L^2(\mathbb{R})$, where $(\tilde{\psi})^\wedge(\xi) = e^{i\beta(\xi)}b(\xi)$ and $\beta(\xi)$ is any measurable function on $[-\varepsilon, \varepsilon]$.

Observe that $\beta(\xi)$ does not necessarily satisfy the phase equation given in (vi) of Theorem 3.1.

3.4   Characterization of some band-limited wavelets

We shall now state a result that characterizes **all** wavelets ψ with supp $(\hat{\psi})$ contained in

$$K = [-\tfrac{8}{3}\pi, -\tfrac{2}{3}\pi] \cup [\tfrac{2}{3}\pi, \tfrac{8}{3}\pi].$$

We shall see that on $[\tfrac{2}{3}\pi, \tfrac{4}{3}\pi]$ we can choose $b(\xi) = |\hat{\psi}(\xi)|$ to be an arbitrary measurable function such that $0 \leq b(\xi) \leq 1$ almost everywhere; for the other points of K, b is determined by its values on $[\tfrac{2}{3}\pi, \tfrac{4}{3}\pi]$. We will show that all these wavelets do indeed arise from multiresolution analyses. Finally, we will show that any wavelet ψ with supp $(\hat{\psi})$ contained in $[-\tfrac{8}{3}\pi, \tfrac{8}{3}\pi]$ must satisfy

$$\hat{\psi}(\xi) = 0 \qquad \text{a.e. on } [-\tfrac{2}{3}\pi, \tfrac{2}{3}\pi].$$

This section provides us with examples of applications of some of the results established in sections 3.1 and 3.2.

THEOREM 4.1 *Suppose $\psi \in L^2(\mathbb{R})$ and $b = |\hat{\psi}|$ has support contained in $[-\tfrac{8}{3}\pi, -\tfrac{2}{3}\pi] \cup [\tfrac{2}{3}\pi, \tfrac{8}{3}\pi]$. Then ψ is an orthonormal wavelet if and only if*

(i) $\quad b^2(\xi) + b^2(\tfrac{1}{2}\xi) = 1 \quad$ *for a.e. $\xi \in [\tfrac{4}{3}\pi, \tfrac{8}{3}\pi]$,*

(ii) $\quad b^2(\xi) + b^2(\xi + 2\pi) = 1 \quad$ *for a.e. $\xi \in [-\tfrac{4}{3}\pi, -\tfrac{2}{3}\pi]$,*

(iii) $\quad b(\xi) = b(\tfrac{1}{2}\xi + 2\pi) \quad$ *for a.e. $\xi \in [-\tfrac{8}{3}\pi, -\tfrac{4}{3}\pi]$,*

(iv) $\quad \hat{\psi}(\xi) = e^{i\alpha(\xi)}b(\xi),$ *where α satisfies*

$$\alpha(\xi) + \alpha(2(\xi - 2\pi)) - \alpha(2\xi) - \alpha(\xi - 2\pi) = (2m(\xi) + 1)\pi$$

for some $m(\xi) \in \mathbb{Z}$, for a.e. $\xi \in [\tfrac{2}{3}\pi, \tfrac{4}{3}\pi] \cap \text{supp}(b) \cap (\tfrac{1}{2}\text{supp}(b))$.

PROOF : The orthonormality of the system $\{\psi_{j,k} : j, k \in \mathbb{Z}\}$ implies

$$\sigma_\psi^2(\xi) = \sum_{k \in \mathbb{Z}} b^2(\xi + 2k\pi) = 1 \qquad \text{for a.e. } \xi \in \mathbb{R}$$

(see (1.1)). It follows from the 2π-periodicity of $\sigma_\psi(\xi)$ that it is enough to examine it on the interval $[\frac{2}{3}\pi, \frac{8}{3}\pi]$; if ξ is in this interval, $\xi + 2k\pi$ lies outside the support of b if $k \geq 1$ or $k \leq -3$. Hence, the condition $\sigma_\psi^2(\xi) = 1$ for almost every $\xi \in \mathbb{R}$ is equivalent to

$$b^2(\xi - 4\pi) + b^2(\xi - 2\pi) + b^2(\xi) = 1 \qquad \text{for a.e. } \xi \in [\tfrac{2}{3}\pi, \tfrac{8}{3}\pi]. \qquad (4.2)$$

On $[\frac{2}{3}\pi, \frac{4}{3}\pi]$, (4.2) implies $b^2(\xi - 2\pi) + b^2(\xi) = 1$, which is condition (ii). On $[\frac{4}{3}\pi, \frac{8}{3}\pi]$, (4.2) implies

$$b^2(\xi - 4\pi) + b^2(\xi) = 1 \qquad \text{for a.e. } \xi \in [\tfrac{4}{3}\pi, \tfrac{8}{3}\pi]. \qquad (4.3)$$

Observe that ψ is band-limited, so that we can use Theorem 2.4 to write

$$\sum_{j \in \mathbb{Z}} b^2(2^j \xi) = 1 \qquad \text{for a.e. } \xi \in \mathbb{R}.$$

If $\xi \in [\frac{2}{3}\pi, \frac{4}{3}\pi]$, $2^j\xi$ lies outside the support of b if $j \neq 0$ and $j \neq 1$; hence,

$$b^2(\xi) + b^2(2\xi) = 1 \qquad \text{for a.e. } \xi \in [\tfrac{2}{3}\pi, \tfrac{4}{3}\pi],$$

which proves (i). Condition (iii) can now be proved using (i) and (4.3); in fact, these two conditions imply

$$b^2(\xi - 4\pi) = b^2(\tfrac{1}{2}\xi) \qquad \text{for a.e. } \xi \in [\tfrac{4}{3}\pi, \tfrac{8}{3}\pi],$$

which is (iii).

It remains to prove (iv). The orthonormality of the system also implies (1.2), which, for $j = 1$, is

$$\zeta_1(\xi) = \sum_{k \in \mathbb{Z}} \hat\psi(2(\xi + 2k\pi)) \overline{\hat\psi(\xi + 2k\pi)} = 0 \qquad \text{for a.e. } \xi \in \mathbb{R}.$$

If $\xi \in [\frac{2}{3}\pi, \frac{8}{3}\pi]$, $2(\xi + 2k\pi)$ lies outside the support of $\hat\psi$ when $k < -1$ and $k > 0$. Thus,

$$\hat\psi(\xi) \overline{\hat\psi(2\xi)} + \hat\psi(\xi - 2\pi) \overline{\hat\psi(2(\xi - 2\pi))} = 0 \qquad \text{for a.e. } \xi \in [\tfrac{2}{3}\pi, \tfrac{8}{3}\pi]. \qquad (4.4)$$

The system of two vectors

$$(\hat{\psi}(\xi), \hat{\psi}(\xi - 2\pi)) \qquad \text{and} \qquad (\hat{\psi}(2\xi), \hat{\psi}(2(\xi - 2\pi)))$$

is an orthonormal system for a.e. $\xi \in [\frac{2}{3}\pi, \frac{4}{3}\pi]$. The orthogonality follows from (4.4); the normality follows from (ii) and (4.3). Hence, there exists $\delta(\xi)$ such that, for a.e. $\xi \in [\frac{2}{3}\pi, \frac{4}{3}\pi]$,

$$e^{i\delta(\xi)} \left(e^{i\alpha(\xi)} b(\xi), \; e^{i\alpha(\xi - 2\pi)} b(\xi - 2\pi) \right)$$

$$= \left(-b(2(\xi - 2\pi)) e^{-i\alpha(2(\xi - 2\pi))}, \; b(2\xi) e^{-i\alpha(2\xi)} \right).$$

For $\xi \in [\frac{2}{3}\pi, \frac{4}{3}\pi]$, $2(\xi - 2\pi) \in [-\frac{8}{3}\pi, -\frac{4}{3}\pi]$; moreover, (iii) implies

$$b(2(\xi - 2\pi)) = b(\xi) \qquad \text{for a.e. } \xi \in [\tfrac{2}{3}\pi, \tfrac{4}{3}\pi]. \tag{4.5}$$

Also, for $\xi \in [\frac{2}{3}\pi, \frac{4}{3}\pi]$, we have $\xi - 2\pi \in [-\frac{4}{3}\pi, -\frac{2}{3}\pi]$; so that (ii) implies $b^2(\xi - 2\pi) + b^2(\xi) = 1$. Applying equality (4.3) to 2ξ we deduce that $b^2(2\xi - 4\pi) + b^2(2\xi) = 1$; these two equalities and (4.5) imply

$$b(\xi - 2\pi) = b(2\xi) \qquad \text{for a.e. } \xi \in [\tfrac{2}{3}\pi, \tfrac{4}{3}\pi]. \tag{4.6}$$

Hence, we have

$$e^{i\delta(\xi)} e^{i\alpha(\xi)} = e^{-i\alpha(2(\xi - 2\pi)) + i\pi} \qquad \text{for a.e. } \xi \in [\tfrac{2}{3}\pi, \tfrac{4}{3}\pi] \cap \text{supp}(b),$$

and

$$e^{i\delta(\xi)} e^{i\alpha(\xi - 2\pi)} = e^{-i\alpha(2\xi)} \qquad \text{for a.e. } \xi \in [\tfrac{2}{3}\pi, \tfrac{4}{3}\pi] \cap (\tfrac{1}{2}\text{supp}(b)).$$

This implies

$$\alpha(\xi) + \alpha(2(\xi - 2\pi)) - \alpha(\xi - 2\pi) - \alpha(2\xi) = (2m(\xi) + 1)\pi$$

for some $m(\xi) \in \mathbb{Z}$, for a.e. $\xi \in [\frac{2}{3}\pi, \frac{4}{3}\pi] \cap \text{supp}(b) \cap (\frac{1}{2}\text{supp}(b))$.

Assume now that (i), (ii), (iii) and (iv) hold and we want to prove that ψ is an orthonormal wavelet. Since ψ is band-limited and $\hat{\psi} = 0$ in a neighborhood of zero, this will be proved if we verify (1.1), (1.2) (orthonormality), (2.13) and (2.14) (completeness).

Easy consequences of (i), (ii) and (iii) are:

(a) $b(2(\xi - 2\pi)) = b(\xi)$ a.e. on $[\frac{2}{3}\pi, \frac{4}{3}\pi]$,

(b) $b(\xi - 2\pi) = b(2\xi)$ a.e. on $[\frac{2}{3}\pi, \frac{4}{3}\pi]$.

(a) is equivalent to (iii) (see (4.5)) while (b) follows from (i) and (ii).

Let us now prove (1.1); that is, $\sigma_\psi^2(\xi) = 1$ a.e. on an interval of length 2π, say, $[\frac{2}{3}\pi, \frac{8}{3}\pi]$. Recall that

$$\sigma_\psi^2(\xi) = \sum_{k \in \mathbb{Z}} |\hat\psi(\xi + 2k\pi)|^2 = \sum_{k \in \mathbb{Z}} b^2(\xi + 2k\pi).$$

If $\xi \in [\frac{2}{3}\pi, \frac{4}{3}\pi]$ all the terms in the above series are zero except the ones corresponding to $k = -1$ and $k = 0$. Hence, by (ii), $\sigma_\psi^2(\xi) = 1$ a.e. on this interval. If $\xi \in [\frac{4}{3}\pi, \frac{8}{3}\pi]$, only the terms involving $k = -2$ and $k = 0$ can be non-zero; hence, using (iii) and (i), we obtain $\sigma_\psi^2(\xi) = 1$ a.e. on this interval. Combining these two cases together we obtain (1.1).

To prove (1.2) we recall, as in (1.3), that

$$\zeta_j(\xi) = \sum_{k \in \mathbb{Z}} \hat\psi(2^j(\xi + 2k\pi)) \overline{\hat\psi(\xi + 2k\pi)}.$$

If $j \geq 2$, either $2^j(\xi + 2k\pi)$ or $\xi + 2k\pi$ lie outside the support of $\hat\psi$; hence, we only have to consider $j = 1$. But ζ_1 is 2π-periodic, so that we only need to examine it on $[\frac{2}{3}\pi, \frac{8}{3}\pi]$. In this case $2(\xi + 2k\pi)$ is outside the support of $\hat\psi$ when $k < -1$ or $k > 0$; hence,

$$\zeta_1(\xi) = \hat\psi(2\xi)\overline{\hat\psi(\xi)} + \hat\psi(2(\xi - 2\pi))\overline{\hat\psi(\xi - 2\pi)}$$

$$= e^{i\alpha(2\xi)}b(2\xi)e^{-i\alpha(\xi)}b(\xi) + e^{i\alpha(2(\xi-2\pi))}b(2(\xi - 2\pi))e^{-i\alpha(\xi-2\pi)}b(\xi - 2\pi).$$

Using (a) and (b) we obtain

$$\zeta_1(\xi) = b(\xi)b(2\xi)\left[e^{i\{\alpha(2\xi)-\alpha(\xi)\}} + e^{i\{\alpha(2(\xi-2\pi))-\alpha(\xi-2\pi)\}}\right].$$

If $\xi \notin [\frac{2}{3}\pi, \frac{4}{3}\pi] \cap \mathrm{supp}\,(b) \cap (\frac{1}{2}\mathrm{supp}\,(b))$, $b(\xi) = 0$ or $b(2\xi) = 0$ so that $\zeta_1(\xi) = 0$ in this case. If $\xi \in [\frac{2}{3}\pi, \frac{4}{3}\pi] \cap \mathrm{supp}\,(b) \cap (\frac{1}{2}\mathrm{supp}\,(b))$ we use (iv) to obtain

$$\zeta_1(\xi) = b(\xi)b(2\xi)e^{i\{\alpha(2\xi)-\alpha(\xi)\}}\left[1 + e^{i(2m(\xi)+1)\pi}\right] = 0.$$

This completes the proof of the orthonormality.

To show (2.13) let us express ω (see (2.5)) in terms of b:

$$\omega(\xi) = \sum_{j \in \mathbb{Z}} |\hat{\psi}(2^j \xi)|^2 = \sum_{j \in \mathbb{Z}} b^2(2^j \xi).$$

Observe that $(0, \infty)$ is the disjoint union of the intervals $2^\ell [\frac{2}{3}\pi, \frac{4}{3}\pi)$, $\ell \in \mathbb{Z}$. If $\xi \in 2^\ell [\frac{2}{3}\pi, \frac{4}{3}\pi)$, then $2^{-\ell}\xi \in [\frac{2}{3}\pi, \frac{4}{3}\pi)$ and $2^{1-\ell}\xi \in [\frac{4}{3}\pi, \frac{8}{3}\pi)$; using (i), we obtain

$$\omega(\xi) = b^2(2^{-\ell}\xi) + b^2(2^{1-\ell}\xi) = 1 \qquad \text{for a.e. } \xi \in 2^\ell [\tfrac{2}{3}\pi, \tfrac{4}{3}\pi).$$

This shows that $\omega(\xi) = 1$ a.e. on $(0, \infty)$. A similar argument shows that $\omega(\xi) = 1$ a.e. on $(-\infty, 0)$. This establishes (2.13).

Finally we need to show (2.14); that is, $h_k(\xi) = 0$ a.e. on $\mathbb{R}$ for all odd integers k, where

$$h_k(\xi) = \sum_{j=0}^{\infty} \hat{\psi}(2^j \xi)\, \overline{\hat{\psi}(2^j(\xi + 2k\pi))},$$

as defined in (2.10). If $2^j \xi \in \text{supp}(\hat{\psi})$, the point $2^j(\xi + 2k\pi)$ lies outside the support of $\hat{\psi}$ if $k \geq 3$ or $k \leq -3$; hence, we only have to show $h_1(\xi) = 0$ and $h_{-1}(\xi) = 0$ for a.e. $\xi \in \mathbb{R}$.

For $k = -1$ and $\xi \in [\frac{2}{3}\pi, \frac{4}{3}\pi]$ we use (a) and (b) to obtain

$$h_{-1}(\xi) = \hat{\psi}(\xi)\, \overline{\hat{\psi}(\xi - 2\pi)} + \hat{\psi}(2\xi)\, \overline{\hat{\psi}(2(\xi - 2\pi))}$$

$$= b(\xi) b(2\xi) \big[e^{i\{\alpha(\xi) - \alpha(\xi - 2\pi)\}} + e^{i\{\alpha(2\xi) - \alpha(2(\xi - 2\pi))\}} \big].$$

Using (iv) it is easy to see that $h_{-1}(\xi) = 0$ for almost every $\xi \in [\frac{2}{3}\pi, \frac{4}{3}\pi]$. For ξ outside $[\frac{2}{3}\pi, \frac{4}{3}\pi]$ it is easy to see that $h_{-1}(\xi) = 0$ using the support condition on b.

Similarly, we can show $h_1(\xi) = 0$ a.e. on $\mathbb{R}$. Another way to prove this result is to observe that

$$h_1(\xi) = \sum_{j=0}^{\infty} \hat{\psi}(2^j \xi)\, \overline{\hat{\psi}(2^j(\xi + 2\pi))}$$

$$= \sum_{j=0}^{\infty} \hat{\psi}(2^j(\xi + 2\pi - 2\pi))\, \overline{\hat{\psi}(2^j(\xi + 2\pi))} = \overline{h_{-1}(\xi + 2\pi)} = 0$$

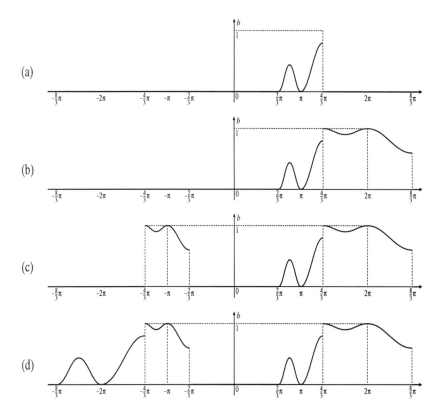

Figure 3.2: How to obtain the Fourier transform of wavelets from Theorem 4.1.

for almost every $\xi \in \mathbb{R}$.

∎

REMARK 1:   Conditions (i), (ii) and (iii) imply that b is completely determined by its values on the interval $[\frac{2}{3}\pi, \frac{4}{3}\pi]$. Choose a measurable function b defined on $[\frac{2}{3}\pi, \frac{4}{3}\pi]$ such that $0 \le b \le 1$ (see Figure 3.2(a)). Condition (i) extends b to $[\frac{4}{3}\pi, \frac{8}{3}\pi]$ as shown in Figure 3.2(b). Condition (ii) extends b to $[-\frac{4}{3}\pi, -\frac{2}{3}\pi]$ as shown in Figure 3.2(c). Finally, condition (iii) determines b on $[-\frac{8}{3}\pi, -\frac{4}{3}\pi]$ (see Figure 3.2(d)).

REMARK 2: $\alpha(\xi) = \frac{1}{2}\xi$ is a solution to the equation that the phase α satisfies. If $\text{supp}(b) \cap (\frac{1}{2}\text{supp}(b))$ has an empty interior on $[\frac{2}{3}\pi, \frac{4}{3}\pi]$ we can choose α to be any measurable function; in particular we can simply take $\alpha(\xi) = 0$.

EXAMPLE A: Starting with $b(\xi) = \chi_{[\pi,\frac{4}{3}\pi]}(\xi)$ on $[\frac{2}{3}\pi, \frac{4}{3}\pi]$, Remark 1 shows that

$$b(\xi) = \chi_{[-2\pi,-\pi]\cup[\pi,2\pi]}(\xi) \qquad \text{on } \mathbb{R}.$$

Since $[\frac{2}{3}\pi, \frac{4}{3}\pi] \cap \mathrm{supp}\,(b) \cap (\frac{1}{2}\mathrm{supp}\,(b)) = \{\pi\}$ we can choose $\alpha(\xi)$ to be any measurable function. In particular

$$\hat{\psi}(\xi) = b(\xi) \quad \text{or} \quad \hat{\psi}(\xi) = e^{i\frac{\xi}{2}}b(\xi)$$

make ψ an orthonormal wavelet. The graph of the wavelet corresponding to the second case can be seen in Figure 2.4 of Chapter 2.

EXAMPLE B: On $\mathbb{R}^+$ choose b to be a bell function for $[\pi, 2\pi]$ with associated extension intervals of length ε at π and of length 2ε at 2π; b satisfies (i) due to the properties of the bell functions. Extend b to $\mathbb{R}$ evenly; then (ii) and (iii) are satisfied. Here we have taken $0 < \varepsilon \le \frac{1}{3}\pi$. If $\hat{\psi}(\xi) = e^{i\frac{\xi}{2}}b(\xi)$, ψ is one of the Lemarié-Meyer wavelets. The graphs of $|\hat{\psi}|$ and ψ, for $\varepsilon = \frac{1}{3}\pi$ and $b = |\hat{\psi}|$ continuous, are given in Figure 1.5 and Figure 1.6 of Chapter 1.

EXAMPLE C: In Examples A and B, $b = |\hat{\psi}|$ is an even function; the theorem provides many examples of wavelets for which $b = |\hat{\psi}|$ is not even. Here is one of them. Take $b(\xi) = 0$ on $[\frac{2}{3}\pi, \frac{4}{3}\pi]$ and extend b to $\mathbb{R}$ according to the conditions of Theorem 4.1. Then

$$b(\xi) = \chi_{[-\frac{4}{3}\pi,-\frac{2}{3}\pi]\cup[\frac{4}{3}\pi,\frac{8}{3}\pi]}(\xi)$$

is not even. We can take $\alpha(\xi) = 0$ in this case since

$$[\tfrac{2}{3}\pi, \tfrac{4}{3}\pi] \cap \mathrm{supp}\,(b) = \emptyset,$$

If $\hat{\psi}(\xi) = b(\xi)$, ψ is an orthonormal wavelet (see Figure 3.3).

EXAMPLE D: We can easily construct a wavelet ψ such that $\mathrm{supp}\,(\hat{\psi})$ is disjoint from the support of the Fourier transform of the Shannon wavelet. Take $b(\xi) = \chi_{[\frac{2}{3}\pi,\pi]}(\xi)$ on $[\frac{2}{3}\pi, \frac{4}{3}\pi]$. In order to satisfy the conditions in Theorem 4.1 we must take (see Figure 3.4)

$$b(\xi) = \chi_{[-\frac{8}{3}\pi,-2\pi]\cup[-\pi,-\frac{2}{3}\pi]\cup[\frac{2}{3}\pi,\pi]\cup[2\pi,\frac{8}{3}\pi]}(\xi) \qquad \text{on } \mathbb{R}.$$

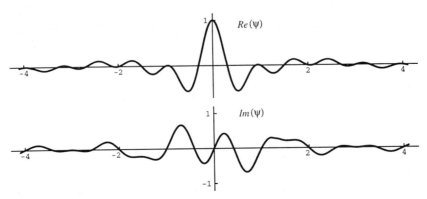

Figure 3.3: Graphs of $Re(\psi)$ and $Im(\psi)$ for $\hat{\psi}(\xi) = \chi_{[-\frac{4}{3}\pi, -\frac{2}{3}\pi] \cup [\frac{4}{3}\pi, \frac{8}{3}\pi]}(\xi)$.

Hence, if $\hat{\psi}(\xi) = e^{i\alpha(\xi)} b(\xi)$ for any choice of measurable α, ψ is an orthonormal wavelet. Observe that on the unit sphere of $L^2(\mathbb{R})$ this wavelet ψ is "far away" from the Shannon wavelet ψ_s since

$$\left\| \psi - \psi_s \right\|_2^2 = \frac{1}{2\pi} \left\| \hat{\psi} - \hat{\psi}_s \right\|_2^2 = \frac{4\pi}{2\pi} = 2$$

so that $\left\| \psi - \psi_s \right\|_2 = \sqrt{2}$.

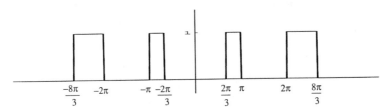

Figure 3.4: $|\hat{\psi}|$ for $b(\xi) = \chi_{[\frac{2}{3}\pi, \pi]}(\xi)$ on $[\frac{2}{3}\pi, \frac{4}{3}\pi]$.

EXAMPLE E: If $b(\xi) = \frac{1}{\sqrt{2}} \chi_{[\frac{2}{3}\pi, \frac{4}{3}\pi]}(\xi)$ on $[\frac{2}{3}\pi, \frac{4}{3}\pi]$, then

$$b(\xi) = \frac{1}{\sqrt{2}} \chi_{[-\frac{8}{3}\pi, -\frac{2}{3}\pi] \cup [\frac{2}{3}\pi, \frac{8}{3}\pi]}(\xi) \qquad \text{on } \mathbb{R},$$

which is constant on the interval $B = [-\frac{8}{3}\pi, -\frac{2}{3}\pi] \cup [\frac{2}{3}\pi, \frac{8}{3}\pi]$. Then, if $\hat{\psi}(\xi) = e^{i\frac{\xi}{2}} b(\xi)$, ψ is an orthonormal wavelet. This is the only wavelet (up to a phase) whose Fourier transform $\hat{\psi}$ is supported on B and $|\hat{\psi}|$ is constant on this interval B (see Figure 3.5).

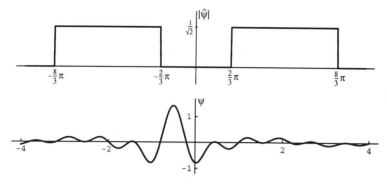

Figure 3.5: Graphs of $|\hat{\psi}|$ and ψ for which $|\hat{\psi}| = \frac{1}{\sqrt{2}}\chi_{[-\frac{8}{3}\pi, -\frac{2}{3}\pi] \cup [\frac{2}{3}\pi, \frac{8}{3}\pi]}$.

EXAMPLE F : Let C be a Cantor type set contained in $[\frac{2}{3}\pi, \frac{4}{3}\pi]$ having positive measure; let $b(\xi) = \chi_C(\xi)$ on $[\frac{2}{3}\pi, \frac{4}{3}\pi]$. This function can then be extended to $[-\frac{8}{3}\pi, -\frac{2}{3}\pi] \cup [\frac{2}{3}\pi, \frac{8}{3}\pi]$ using (i), (ii) and (iii) of Theorem 4.1. Hence, we have a wavelet ψ such that $\hat{\psi}(\xi) = e^{i\frac{\xi}{2}}b(\xi)$ and supp $(\hat{\psi})$ is a Cantor type set.

One more condition characterizes the wavelets ψ with supp $(\hat{\psi})$ contained in the interval $[-\frac{8}{3}\pi, -\frac{2}{3}\pi] \cup [\frac{2}{3}\pi, \frac{8}{3}\pi]$ for which $b = |\hat{\psi}|$ is even.

PROPOSITION 4.7 *Suppose $\psi \in L^2(\mathbb{R})$, $b = |\hat{\psi}|$ has support contained in $[-\frac{8}{3}\pi, -\frac{2}{3}\pi] \cup [\frac{2}{3}\pi, \frac{8}{3}\pi]$ and ψ is an orthonormal wavelet. Then b is almost everywhere even if and only if*

$$b^2(\xi) + b^2(2\pi - \xi) = 1 \qquad for \ \ a.e. \ \xi \in [\tfrac{2}{3}\pi, \tfrac{4}{3}\pi]. \qquad (4.8)$$

PROOF : If b is even and $\xi \in [\frac{2}{3}\pi, \frac{4}{3}\pi]$, from (ii) of Theorem 4.1 we deduce

$$1 = b^2(-\xi) + b^2(-\xi + 2\pi) = b^2(\xi) + b^2(2\pi - \xi),$$

which proves (4.8).

Suppose that (4.8) holds. Then we have $b^2(-\xi) + b^2(2\pi + \xi) = 1$ for $\xi \in [-\frac{4}{3}\pi, -\frac{2}{3}\pi]$; comparing this with (ii), we obtain

$$b(\xi) = b(-\xi) \qquad for \ \ a.e. \ \xi \in [-\tfrac{4}{3}\pi, -\tfrac{2}{3}\pi]. \qquad (4.9)$$

For $\xi \in [\frac{4}{3}\pi, \frac{8}{3}\pi]$, we apply (ii) to $-\frac{1}{2}\xi$ to obtain $b^2(-\frac{1}{2}\xi) + b^2(-\frac{1}{2}\xi + 2\pi) = 1$. By (4.9), $b(-\frac{1}{2}\xi) = b(\frac{1}{2}\xi)$; since $-\xi \in [-\frac{8}{3}\pi, -\frac{4}{3}\pi]$, we use (iii) to obtain

$b(-\xi) = b(-\frac{1}{2}\xi + 2\pi)$. Hence,

$$b^2(\tfrac{1}{2}\xi) + b^2(-\xi) = 1 \qquad \text{for} \quad a.e. \ \xi \in [\tfrac{4}{3}\pi, \tfrac{8}{3}\pi].$$

Comparing with (i), we obtain $b(\xi) = b(-\xi)$ a.e. on $[\tfrac{4}{3}\pi, \tfrac{8}{3}\pi]$.   ∎

We now show that all the wavelets characterized in Theorem 4.1 arise from multiresolution analyses. To do this we have to find a scaling function φ whose translates form an orthonormal system; this scaling function must satisfy (2.16) of Chapter 2, that is

$$|\hat{\varphi}(\xi)|^2 = \sum_{j=1}^{\infty} |\hat{\psi}(2^j \xi)|^2.$$

Since $\{\varphi(\cdot - k) : k \in \mathbb{Z}\}$ must be an orthonormal system, by Proposition 1.11 of Chapter 2 we must have

$$D(\xi) \equiv \sum_{k \in \mathbb{Z}} |\hat{\varphi}(\xi + 2k\pi)|^2 = \sum_{k \in \mathbb{Z}} \sum_{j=1}^{\infty} |\hat{\psi}(2^j(\xi + 2k\pi))|^2 = 1 \qquad (4.10)$$

for almost everywhere $\xi \in \mathbb{R}$. For the wavelets characterized in Theorem 4.1 an easy computation shows that

$$|\hat{\varphi}(\xi)|^2 = \begin{cases} 1 & \text{if} \ |\xi| \leq \tfrac{2}{3}\pi, \\ b^2(2\xi) & \text{if} \ \tfrac{2}{3}\pi < |\xi| \leq \tfrac{4}{3}\pi, \\ 0 & \text{if} \ |\xi| > \tfrac{4}{3}\pi. \end{cases} \qquad (4.11)$$

To show that $D(\xi) = 1$ a.e. on $\mathbb{R}$ it is enough to show $D(\xi) = 1$ a.e. on $[-\tfrac{2}{3}\pi, \tfrac{4}{3}\pi]$ since $D(\xi)$ is 2π-periodic. If $\xi \in [-\tfrac{2}{3}\pi, \tfrac{2}{3}\pi]$, $\xi + 2k\pi$ is outside the support of $\hat{\varphi}$ if $k \neq 0$; thus $D(\xi) = |\hat{\varphi}(\xi)|^2 = 1$. On the interval $[\tfrac{2}{3}\pi, \tfrac{4}{3}\pi]$, $D(\xi) = |\hat{\varphi}(\xi)|^2 + |\hat{\varphi}(\xi - 2\pi)|^2$, and using (i) and (iii) of Theorem 4.1, we obtain

$$D(\xi) = b^2(2\xi) + b^2(2(\xi - 2\pi)) = 1 - b^2(\xi) + b^2(2(\xi - 2\pi)) = 1.$$

We shall show in a later chapter that if ψ is an orthonormal wavelet for $L^2(\mathbb{R})$, (4.10) is a necessary and sufficient condition for ψ to arise from an MRA. Assuming this, we have just shown that all the wavelets characterized in Theorem 4.1 arise from an MRA. But, in this case, we can give a proof based on results already proved in Chapter 2.

THEOREM 4.12 *Suppose* $\psi \in L^2(\mathbb{R})$, $b = |\hat{\psi}|$ *has support contained in* $[-\frac{8}{3}\pi, -\frac{2}{3}\pi] \cup [\frac{2}{3}\pi, \frac{8}{3}\pi]$ *and* ψ *is an orthonormal wavelet. Then* ψ *arises from a multiresolution analysis.*

PROOF : We write $\hat{\psi}(\xi) = e^{i\alpha(\xi)}b(\xi)$, where α satisfies the functional equation of part (iv) of Theorem 4.1. According to (2.16) of Chapter 2, if ψ arises from an MRA, there exists a scaling function φ such that

$$|\hat{\varphi}(\xi)|^2 = \sum_{j=1}^{\infty} |\hat{\psi}(2^j\xi)|^2.$$

In our case, as in (4.11),

$$|\hat{\varphi}(\xi)|^2 = \begin{cases} 1 & \text{if } |\xi| \leq \frac{2}{3}\pi, \\ b^2(2\xi) & \text{if } \frac{2}{3}\pi \leq |\xi| \leq \frac{4}{3}\pi, \\ 0 & \text{if } |\xi| \geq \frac{4}{3}\pi. \end{cases}$$

There are many choices for φ, but, as we shall see in the proof of the theorem, $\hat{\varphi}(\xi) = e^{i\alpha(2\xi)}|\hat{\varphi}(\xi)|$ is the appropriate one to obtain ψ.

For $j \in \mathbb{Z}$, let V_j be the closed span of $\{\varphi_{j,k} : k \in \mathbb{Z}\}$ in $L^2(\mathbb{R})$, where $\varphi_{j,k}(x) = 2^{\frac{j}{2}}\varphi(2^j x - k)$. We immediately have (1.2) of the definition of an MRA (see section 2.1); property (1.5) follows from the description of $|\hat{\varphi}(\xi)|^2$ given above and the properties of b given in Theorem 4.1, as we have already seen. Observe that $|\hat{\varphi}(0)| = 1$, so that by Theorem 1.6 and Theorem 1.7 of section 2.1, all we need to show is that (1.1) is satisfied.

To show (1.1) of the definition of an MRA, it is enough to show the inclusion $V_{-1} \subset V_0$; this is accomplished if we show that $\frac{1}{2}\varphi(\frac{1}{2}x)$ belongs to V_0, since this function generates V_{-1}; by taking Fourier transforms, we have already seen that this is equivalent to the existence of a 2π-periodic function $m_0(\xi)$, called the low-pass filter, such that $m_0 \in L^2(\mathbb{T})$ and $\hat{\varphi}(2\xi) = m_0(\xi)\hat{\varphi}(\xi)$ (see (2.3) of Chapter 2). This equation and the definition of $\hat{\varphi}$ lead us to define m_0 by letting

$$m_0(\xi) = e^{i\alpha(4\xi)}|\hat{\varphi}(2\xi)|e^{-i\alpha(2\xi)}\chi_{[-\frac{2}{3}\pi,\frac{2}{3}\pi]}(\xi) \qquad \text{if } \xi \in [-\pi, \pi],$$

and, then, to extend m_0 to $\mathbb{R}$ as a 2π-periodic function; since $m_0 \in L^2(\mathbb{T})$, we have shown that $V_{-1} \subset V_0$. This proves that the collection $\{V_j : j \in \mathbb{Z}\}$ is an MRA.

Observe that, since $|\hat{\varphi}(\xi)| = 1$ for $|\xi| \leq \frac{2}{3}\pi$, we have

$$|m_0(\xi)| = |\hat{\varphi}(2\xi)| = \begin{cases} 1 & \text{if } |\xi| \leq \frac{1}{3}\pi, \\ b(4\xi) & \text{if } \frac{1}{3}\pi \leq |\xi| \leq \frac{2}{3}\pi, \\ 0 & \text{if } \frac{2}{3}\pi \leq |\xi| \leq \pi. \end{cases}$$

and $|m_0(\xi + 2k\pi)| = |m_0(\xi)|$ for all $k \in \mathbb{Z}$, $\xi \in [-\pi, \pi]$.

We need to show that this MRA gives rise to the wavelet ψ. For $j \in \mathbb{Z}$, let W_j be the closed span in $L^2(\mathbb{R})$ of $\{\psi_{j,k} : k \in \mathbb{Z}\}$. It is enough to show that $W_0 \oplus V_0 = V_1$ or, equivalently, $W_{-1} \oplus V_{-1} = V_0$.

Observe that $W_{-1} \subset V_0$; this is equivalent to exhibiting the existence of a 2π-periodic function $m_1 \in L^2(\mathbb{T})$, such that $\hat{\psi}(2\xi) = m_1(\xi)\hat{\varphi}(\xi)$. The graphs of $\hat{\psi}(2\xi)$ and $\hat{\varphi}(\xi)$ lead us to define $m_1(\xi)$ on $[-\pi, \pi]$ by

$$m_1(\xi) = \begin{cases} 0 & \text{if } |\xi| \leq \frac{1}{3}\pi, \\ b(2\xi) & \text{if } \frac{1}{3}\pi \leq |\xi| \leq \frac{2}{3}\pi, \\ 1 & \text{if } \frac{2}{3}\pi \leq |\xi| \leq \pi. \end{cases}$$

and extended 2π-periodically to $\mathbb{R}$. Observe that $m_1(\xi) = |m_0(\xi+\pi)|$. This shows that $\psi \in W_0$ and $W_0 \subset V_1$.

A simple computation shows (look at the graphs in Figure 3.6) that

$$|m_0(\xi)|^2 + |m_1(\xi)|^2 = 1 \qquad \text{for} \quad a.e. \ \xi \in \mathbb{R},$$

due to the properties of $|\hat{\psi}| = b$ proved in Theorem 4.1. Hence,

$$\overline{m_1(\xi)}\,\hat{\psi}(2\xi) + \overline{m_0(\xi)}\,\hat{\varphi}(2\xi) = \{|m_1(\xi)|^2 + |m_0(\xi)|^2\}\hat{\varphi}(\xi) = \hat{\varphi}(\xi),$$

which shows that $V_0 = W_{-1} + V_{-1}$. To prove that this is an orthogonal direct sum, it is enough to show that V_0 is perpendicular to W_0. The characterization of V_0 given in Lemma 2.6 of Chapter 2 allows us to conclude that $V_0 \perp W_0$ if and only if $< m\hat{\varphi}, \hat{\psi} > = 0$ for all 2π-periodic functions $m \in L^2(\mathbb{T})$. But, since m is 2π-periodic, a periodization argument shows

$$0 = < m\hat{\varphi}, \hat{\psi} > = \int_{\mathbb{R}} m(\xi)\hat{\varphi}(\xi)\,\overline{\hat{\psi}(\xi)}\,d\xi$$

$$= \sum_{k \in \mathbb{Z}} \int_{2k\pi}^{2(k+1)\pi} m(\xi)\hat{\varphi}(\xi)\,\overline{\hat{\psi}(\xi)}\,d\xi$$

$$= \int_0^{2\pi} m(\xi) \Big\{ \sum_{k \in \mathbb{Z}} \hat{\varphi}(\xi + 2k\pi) \, \overline{\hat{\psi}(\xi + 2k\pi)} \Big\} \, d\xi,$$

which proves that $V_0 \perp W_0$ if and only if

$$\Upsilon(\xi) \equiv \sum_{k \in \mathbb{Z}} \hat{\varphi}(\xi + 2k\pi) \, \overline{\hat{\psi}(\xi + 2k\pi)} = 0 \qquad \text{for} \quad a.e. \ \xi \in \mathbb{R}. \tag{4.13}$$

In our case (4.13) is easy to verify; it is enough to examine it on an interval of length 2π since $\Upsilon(\xi)$ is 2π-periodic.

If $\xi \in [-\frac{2}{3}\pi, \frac{2}{3}\pi]$, either $\xi + 2k\pi \notin \text{supp}\,(\hat{\varphi})$ or $\xi + 2k\pi \notin \text{supp}\,(\hat{\psi})$, so that $\Upsilon(\xi) = 0$ on $[-\frac{2}{3}\pi, \frac{2}{3}\pi]$. If $\xi \in [\frac{2}{3}\pi, \frac{4}{3}\pi]$,

$$\Upsilon(\xi) = \hat{\varphi}(\xi) \, \overline{\hat{\psi}(\xi)} + \hat{\varphi}(\xi - 2\pi) \, \overline{\hat{\psi}(\xi - 2\pi)}$$

$$= e^{i\alpha(2\xi)} b(2\xi) e^{-i\alpha(\xi)} b(\xi) + e^{i\alpha(2(\xi - 2\pi))} b(2(\xi - 2\pi)) e^{-i\alpha(\xi - 2\pi)} b(\xi - 2\pi)$$

$$= b(\xi) b(2\xi) \big[e^{i\{\alpha(2\xi) - \alpha(\xi)\}} + e^{i\{\alpha(2(\xi - 2\pi)) - \alpha(\xi - 2\pi)\}} \big] = 0,$$

where we have used (a) and (b) in the proof of Theorem 4.1 and the functional equation that α satisfies stated in (iv) of the same theorem. This completes the proof. ∎

We have characterized all orthonormal wavelets ψ for which $\hat{\psi}$ has support contained in $[-\frac{8}{3}\pi, -\frac{2}{3}\pi] \cup [\frac{2}{3}\pi, \frac{8}{3}\pi]$, and proved that all these wavelets arise from multiresolution analyses. We shall prove that any orthonormal wavelet ψ for which $\hat{\psi}$ has support contained in $[-\frac{8}{3}\pi, \frac{8}{3}\pi]$ must satisfy $\hat{\psi}(\xi) = 0$ for almost every $\xi \in [-\frac{2}{3}\pi, \frac{2}{3}\pi]$; thus, Theorem 4.1 characterizes all orthonormal wavelets ψ for which $\text{supp}\,(\hat{\psi}) \subseteq [-\frac{8}{3}\pi, \frac{8}{3}\pi]$. We start by giving a simple argument which shows that this is true when ψ arises from an MRA.

If ψ arises from an MRA, by (4.10) we must have

$$D(\xi) = \sum_{k \in \mathbb{Z}} \sum_{j=1}^{\infty} |\hat{\psi}(2^j(\xi + 2k\pi))|^2 = 1 \qquad \text{for} \quad a.e. \ \xi \in \mathbb{R}.$$

If $\xi \in [-\frac{2}{3}\pi, \frac{2}{3}\pi]$, $2^j(\xi + 2k\pi) \notin \text{supp}\,(\hat{\psi})$ if $k \neq 0$; thus in our case

$$D(\xi) = \sum_{j=1}^{\infty} |\hat{\psi}(2^j \xi)|^2 = 1 \tag{4.14}$$

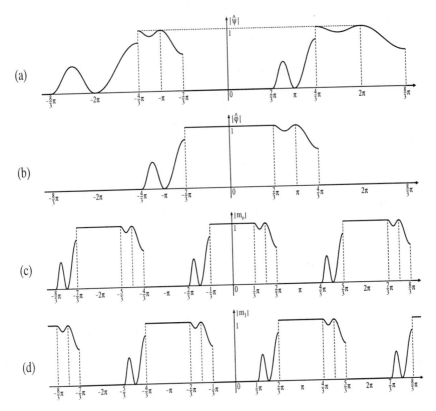

Figure 3.6: Illustration of Theorem 4.12 and its proof.

for almost every $\xi \in [-\frac{2}{3}\pi, \frac{2}{3}\pi]$. Since ψ is a band-limited orthonormal wavelet, by Theorem 2.4 we have

$$\sum_{j=-\infty}^{\infty} |\hat{\psi}(2^j \xi)|^2 = 1 \tag{4.15}$$

for almost every $\xi \in \mathbb{R}$. Comparing (4.14) and (4.15) on the interval $[-\frac{2}{3}\pi, \frac{2}{3}\pi]$ we obtain

$$\sum_{j=-\infty}^{0} |\hat{\psi}(2^j \xi)|^2 = 0$$

a.e. on this interval; from here it follows that $\hat{\psi}(\xi) = 0$ a.e. on $[-\frac{2}{3}\pi, \frac{2}{3}\pi]$.

In the next theorem we shall prove that this result, that $\hat{\psi}(\xi) = 0$ a.e. on $[-\frac{2}{3}\pi, \frac{2}{3}\pi]$, is true even when ψ **does not** arise from an MRA. We start with a simple lemma.

LEMMA 4.16 *For any orthonormal wavelet ψ,*

$$D(\xi) \equiv \sum_{k \in \mathbb{Z}} \sum_{j=1}^{\infty} |\hat{\psi}(2^j(\xi + 2k\pi))|^2 \tag{4.17}$$

is well defined and finite for almost every $\xi \in \mathbb{R}$. Moreover,

$$\int_I D(\xi)\, d\xi = 2\pi$$

for any interval I of length 2π in $\mathbb{R}$.

PROOF : $D(\xi)$ is well defined and cannot be infinite on a set of positive measure if the last part of the lemma is true. Since $D(\xi)$ is 2π-periodic, this second part follows if we prove the result for $I = [0, 2\pi]$. The proof is a simple periodization argument:

$$\int_0^{2\pi} D(\xi)\, d\xi = \sum_{j=1}^{\infty} \sum_{k \in \mathbb{Z}} \int_0^{2\pi} |\hat{\psi}(2^j(\xi + 2k\pi))|^2\, d\xi$$

$$= \sum_{j=1}^{\infty} \sum_{k \in \mathbb{Z}} \int_{2k\pi}^{2(k+1)\pi} |\hat{\psi}(2^j \xi)|^2\, d\xi$$

$$= \sum_{j=1}^{\infty} \int_{-\infty}^{\infty} |\hat{\psi}(2^j \xi)|^2\, d\xi = \sum_{j=1}^{\infty} 2^{-j} \int_{-\infty}^{\infty} |\hat{\psi}(\xi)|^2\, d\xi$$

$$= \|\hat{\psi}\|_2^2 \sum_{j=1}^{\infty} 2^{-j} = 2\pi \|\psi\|_2^2 = 2\pi.$$

∎

THEOREM 4.18 *Let ψ be an orthonormal wavelet such that $\hat{\psi}$ is supported in $[-\frac{8}{3}\pi, \frac{8}{3}\pi]$; then $\hat{\psi}(\xi) = 0$ for almost every $\xi \in [-\frac{2}{3}\pi, \frac{2}{3}\pi]$.*

PROOF : Let $D(\xi)$ be defined as in (4.17). The support condition on $\hat{\psi}$ and Theorem 2.4 give us

$$D(\xi) = \sum_{j=1}^{\infty} |\hat{\psi}(2^j \xi)|^2 = 1 - \sum_{j=-\infty}^{0} |\hat{\psi}(2^j \xi)|^2 \quad \text{a.e. on } [-\tfrac{2}{3}\pi, \tfrac{2}{3}\pi]. \tag{4.19}$$

If $\xi \in [\frac{2}{3}\pi, \frac{4}{3}\pi]$, $2^j(\xi + 2k\pi) \notin \text{supp}(\hat{\psi})$ if $j \geq 2$ and $2(\xi + 2k\pi) \notin \text{supp}(\hat{\psi})$ if $k > 0$ or $k < -1$; hence,

$$D(\xi) = |\hat{\psi}(2\xi)|^2 + |\hat{\psi}(2(\xi - 2\pi))|^2 \quad \text{for } a.e. \; \xi \in [\frac{2}{3}\pi, \frac{4}{3}\pi].$$

We now use (1.1) with ξ replaced by 2ξ to obtain

$$|\hat{\psi}(2\xi)|^2 + |\hat{\psi}(2\xi - 2\pi)|^2 + |\hat{\psi}(2\xi - 4\pi)|^2 = 1 \quad \text{a.e. on } [\frac{2}{3}\pi, \frac{4}{3}\pi].$$

Hence,

$$D(\xi) = 1 - |\hat{\psi}(2\xi - 2\pi)|^2 \quad \text{a.e. on } [\frac{2}{3}\pi, \frac{4}{3}\pi]. \tag{4.20}$$

Since $D(\xi)$ is non-negative (a.e.) and, by Lemma 4.16,

$$\int_{-\frac{2}{3}\pi}^{\frac{4}{3}\pi} D(\xi)\, d\xi = 2\pi,$$

(4.19) and (4.20) give us

$$\sum_{j=-\infty}^{0} |\hat{\psi}(2^j\xi)|^2 = 0 \quad \text{a.e. on } [-\frac{2}{3}\pi, \frac{2}{3}\pi],$$

and $|\hat{\psi}(2\xi - 2\pi)| = 0$ a.e. on $[\frac{2}{3}\pi, \frac{4}{3}\pi]$. Either of these conditions implies that $\hat{\psi}(\xi) = 0$ for almost every $\xi \in [-\frac{2}{3}\pi, \frac{2}{3}\pi]$. ∎

Observe that Theorem 4.1 can now be stated by only assuming that the support of b is contained in $[-\frac{8}{3}\pi, \frac{8}{3}\pi]$. Moreover, all the orthonormal wavelets ψ for which $\hat{\psi}$ have support contained in this interval must arise from multiresolution analyses because of Theorem 4.12.

3.5 Notes and references

1. Most of the results described in sections 3.1, 3.2 and 3.3 are contained in [BSW]. In particular, the proof of the fact that equations (1.1), (1.2), (2.1) and (2.2) characterize band-limited orthonormal wavelets can be found in the above mentioned paper. The example (2.15), showing that for any $\varepsilon > 0$ there exists a low-pass filter that produces an orthonormal wavelet ψ for which $\hat{\psi}$ is not identically zero on $(-\varepsilon, \varepsilon)$, is a modification of the example presented in [BSW].

2. Section 3.4 is new. Theorem 4.1, as well as other results in that section, can be generalized to give a characterization of all wavelets ψ for which $\hat{\psi}$ is supported in

$$S_\alpha = \left[-\tfrac{8}{3}\alpha, 4\pi - \tfrac{4}{3}\alpha\right], \qquad 0 < \alpha \leq \pi,$$

or

$$\widetilde{S}_\beta = \left[-4\pi + \tfrac{4}{3}\beta, \tfrac{8}{3}\beta\right], \qquad 0 < \beta \leq \pi.$$

Details of this result and several examples can be found in [HWW1]. In particular, for the case supp $(\hat{\psi}) \subseteq S_\alpha$, it is proved that $\hat{\psi}$ must be zero on $\left[-\tfrac{2}{3}\alpha, 2\pi - \tfrac{4}{3}\alpha\right]$ and that all the these wavelets arise from multiresolution analyses (similar statements hold for $\widetilde{S}_\beta$).

4

Other constructions of wavelets

We have constructed several types of wavelets in the previous chapters. Among them, the Haar wavelet, that appeared as an example of a multiresolution analysis (see Examples A and B of Chapter 2). This wavelet is obtained by taking V_0 to be the space of all functions in $L^2(\mathbb{R})$ which are constant on intervals of the form $[k, k + 1]$, $k \in \mathbb{Z}$. It is natural to generalize this construction by allowing a greater degree of smoothness for the functions of V_0. By doing so we obtain the Franklin and more general spline wavelets in the next two sections. It is important to indicate that these functions, that have been used in applications for more than 25 years, have found a natural and most useful place in the recent theory of wavelets.

The previous chapters treat the case of orthonormal wavelets in $L^2(\mathbb{R})$. There is a procedure to periodize the wavelets defined on the real line and obtain orthonormal, periodic wavelets for $L^2(\mathbb{T})$. This is presented successively in section 4.3 (for the Franklin wavelets), in section 4.4 (for the more general spline wavelets) and in section 4.5 (for very general wavelets arising from multiresolution analyses).

4.1   Franklin wavelets on the real line

We shall show how to construct a piecewise linear continuous function on $\mathbb{R}$ which is an orthonormal wavelet, and we will show that it is unique in a sense that is described below. This will be called a Franklin wavelet. We will construct this wavelet using the notion of a multiresolution analysis (MRA) introduced in Chapter 2.

Let V_0 be the space of all functions $f \in L^2(\mathbb{R})$ which are continuous on $\mathbb{R}$ and linear when restricted to each interval of the form $[k, k+1]$, $k \in \mathbb{Z}$. Define V_j, $j \in \mathbb{Z}$, as the space of all functions $f \in L^2(\mathbb{R})$ such that $f(2^{-j}\cdot) \in V_0$; the functions in V_j are continuous on $\mathbb{R}$ and linear on each interval of the form $[2^{-j}k, 2^{-j}(k+1)]$, $k \in \mathbb{Z}$. It is clear that properties (1.1), (1.2), (1.3) and (1.4) of the definition of a multiresolution analysis (see section 2.1) are satisfied by this sequence of subspaces. In order to show property (1.5) (on page 44) of an MRA we need to find a scaling function; that is, a function φ in V_0 whose integer translates form an orthonormal basis of V_0. We begin by studying the functions in V_0.

It is clear that the sequence of values $\{f(k)\}_{k \in \mathbb{Z}}$ completely determines the general function $f \in V_0$. In fact, this sequence $\{f(k)\}_{k \in \mathbb{Z}}$ must belong to $\ell^2(\mathbb{Z})$ and, conversely, any such sequence determines a unique function $f \in V_0$. We shall show, in fact,

$$\tfrac{1}{3}\big\|\{f(k)\}\big\|^2_{\ell^2(\mathbb{Z})} \leq \|f\|^2_{L^2(\mathbb{R})} \leq \big\|\{f(k)\}\big\|^2_{\ell^2(\mathbb{Z})} \qquad \text{for all } f \in V_0. \qquad (1.1)$$

The linearity of f on $[k, k+1]$ allows us to write

$$\int_k^{k+1} |f(x)|^2 \, dx = \int_0^1 \big[(1-t)f(k) + tf(k+1)\big]^2 dt$$

$$= \frac{[f(k)]^2 + [f(k+1)]^2}{3} + \frac{f(k)f(k+1)}{3}.$$

Now observe that

$$\frac{[f(k)]^2 + [f(k+1)]^2}{6} \leq \frac{[f(k)]^2 + [f(k+1)]^2}{3} + \frac{f(k)f(k+1)}{3}$$

$$\leq \frac{[f(k)]^2 + [f(k+1)]^2}{2}.$$

But

$$\|f\|^2_{L^2(\mathbb{R})} = \sum_{k=-\infty}^{\infty} \int_k^{k+1} |f(x)|^2 \, dx,$$

and, therefore, the above inequalities give us (1.1), since

$$\sum_{k \in \mathbb{Z}} [f(k)]^2 = \big\|\{f(k)\}\big\|^2_{\ell^2(\mathbb{Z})}.$$

Let

$$\Delta(x) = \begin{cases} x & \text{if } 0 \le x \le 1, \\ 2 - x & \text{if } 1 < x \le 2, \\ 0 & \text{otherwise}, \end{cases}$$

be the "triangle" function depicted in Figure 4.1. It is clear that every $f \in V_0$ has the representation

$$f(x) = \sum_{k \in \mathbb{Z}} f(k) \Delta(x - k + 1). \tag{1.2}$$

If $\chi = \chi_{[0,1]}$ we have $\hat{\chi}(\xi) = e^{-i\frac{\xi}{2}} \dfrac{\sin(\xi/2)}{\xi/2}$ and $\Delta = \chi * \chi$; hence

$$\hat{\Delta}(\xi) = [\hat{\chi}(\xi)]^2 = e^{-i\xi} \left(\frac{\sin(\xi/2)}{\xi/2} \right)^2.$$

From (1.2) we have

$$\hat{f}(\xi) = e^{-i\xi} \left(\frac{\sin(\xi/2)}{\xi/2} \right)^2 \sum_{k \in \mathbb{Z}} f(k) e^{-i(k-1)\xi}$$

$$= \left(\frac{\sin(\xi/2)}{\xi/2} \right)^2 \sum_{k \in \mathbb{Z}} f(k) e^{-ik\xi} \equiv \left(\frac{\sin(\xi/2)}{\xi/2} \right)^2 m_f(\xi),$$

where $m_f(\xi)$ is a 2π-periodic function on $\mathbb{R}$ belonging to $L^2(\mathbb{T})$. In fact,

$$\left\| m_f \right\|_{L^2(\mathbb{T})}^2 = 2\pi \left\| \{f(k)\} \right\|_{\ell^2(\mathbb{Z})}^2.$$

We have proved the following result (compare this result with the characterization of V_0 given in Lemma 2.6 of Chapter 2 for a general MRA).

Theorem 1.3 *A function f in $L^2(\mathbb{R})$ belongs to V_0 if and only if*

$$\hat{f}(\xi) = \left(\frac{\sin(\xi/2)}{\xi/2} \right)^2 m_f(\xi),$$

where m_f is a 2π-periodic function in $L^2(\mathbb{T})$. Moreover,

$$\| f \|_{L^2(\mathbb{R})} \approx \| m_f \|_{L^2(\mathbb{T})}.$$

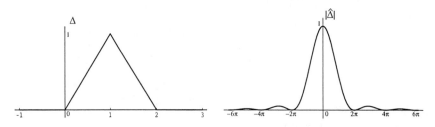

Figure 4.1: Graphs of Δ and $|\widehat{\Delta}|$.

There is another way of characterizing the elements of V_0:

THEOREM 1.4 *A function f in $L^2(\mathbb{R})$ belongs to V_0 if and only if $\xi^2 \hat{f}(\xi)$ is a 2π-periodic function on $\mathbb{R}$.*

PROOF: Theorem 1.3 immediately implies that, if $f \in V_0$, then

$$\xi^2 \hat{f}(\xi) = 4\sin^2(\tfrac{1}{2}\xi) m_f(\xi)$$

is a 2π-periodic function on $\mathbb{R}$.

Assume, now, that $\xi^2 \hat{f}(\xi)$ is a 2π-periodic function on $\mathbb{R}$. Define

$$\mu(\xi) = \frac{\xi^2 \hat{f}(\xi)}{4\sin^2(\tfrac{1}{2}\xi)}\,.$$

The 2π-periodicity of the functions involved and the boundedness of the function $\dfrac{\xi^4}{16\sin^4(\xi/2)}$ on $[-\pi, \pi]$ allow us to obtain

$$\int_0^{2\pi} |\mu(\xi)|^2 \, d\xi = \int_{-\pi}^{\pi} \frac{\xi^4 |\hat{f}(\xi)|^2}{16\sin^4(\xi/2)} \, d\xi \le C \int_{-\pi}^{\pi} |\hat{f}(\xi)|^2 \, d\xi \le C_o \|f\|_{L^2(\mathbb{R})}^2.$$

Thus $\mu \in L^2(\mathbb{T})$ and we can write its Fourier series in the form

$$\mu(\xi) \sim \sum_{k \in \mathbb{Z}} \hat{\mu}(k) e^{-ik\xi}$$

with $\sum_{k \in \mathbb{Z}} |\hat{\mu}(k)|^2 < \infty$. Since $\widehat{\Delta}(\xi) = e^{-i\xi} \dfrac{\sin^2(\xi/2)}{(\xi/2)^2}$ we obtain

$$\hat{f}(\xi) = \frac{\sin^2(\xi/2)}{(\xi/2)^2} \mu(\xi) = e^{-i\xi} \frac{\sin^2(\xi/2)}{(\xi/2)^2} \sum_{k \in \mathbb{Z}} \hat{\mu}(k) e^{-i(k-1)\xi}$$

$$= \sum_{k \in \mathbb{Z}} \hat{\mu}(k) \hat{\Delta}(\xi) e^{-i(k-1)\xi} = \sum_{k \in \mathbb{Z}} \hat{\mu}(k) (\tau_{k-1}\Delta)^{\wedge}(\xi),$$

where $\tau_k \Delta(x) = \Delta(x - k)$. By taking the inverse Fourier transform we deduce

$$f(x) = \sum_{k \in \mathbb{Z}} \hat{\mu}(k) \Delta(x - k + 1).$$

This shows that $f \in V_0$, since $\{\Delta(\cdot - k) : k \in \mathbb{Z}\}$ is a basis of V_0. ∎

Our aim is to prove that the sequence of subspaces $\{V_j : j \in \mathbb{Z}\}$ forms an MRA. The scaling function φ must belong to the space V_0 which we have characterized in Theorems 1.3 and 1.4. The wavelet ψ must belong to V_1. Two characterizations of this space can be easily obtained from the above theorems.

THEOREM 1.5 *Suppose $g \in L^2(\mathbb{R})$.*

(a) *$g \in V_1$ if and only if*

$$\hat{g}(\xi) = \frac{\sin^2(\xi/4)}{(\xi/4)^2} m_g(\xi),$$

where m_g is a 4π-periodic function in $L^2([0, 4\pi])$. Moreover,

$$\|g\|_{L^2(\mathbb{R})} \approx \|m_g\|_{L^2([0,4\pi])}.$$

(b) *$g \in V_1$ if and only if $\xi^2 \hat{g}(\xi)$ is a 4π-periodic function on $\mathbb{R}$.*

PROOF: For $g \in L^2(\mathbb{R})$ define $f(x) = g(\frac{1}{2}x)$. Then g belongs to V_1 if and only if f belongs to V_0. Moreover, $\hat{f}(\xi) = 2\hat{g}(2\xi)$. The result follows by using the characterizations of V_0 given in Theorem 1.3 and in Theorem 1.4. ∎

The system $\{\Delta(\cdot - k) : k \in \mathbb{Z}\}$ is a basis for V_0, but it is not an orthogonal system. As we stated above, we need to find a scaling function φ for which $\{\varphi(\cdot - k) : k \in \mathbb{Z}\}$ is an orthonormal basis of V_0. By Proposition 1.11 of Chapter 2, the orthonormality of the system $\{\varphi(\cdot - k) : k \in \mathbb{Z}\}$ is equivalent to

$$\sum_{k \in \mathbb{Z}} |\hat{\varphi}(\xi + 2k\pi)|^2 = 1 \qquad \text{for a.e. } \xi \in \mathbb{R}. \tag{1.6}$$

Theorem 1.3 gives us

$$\hat{\varphi}(\xi) = \frac{\sin^2(\xi/2)}{(\xi/2)^2} m_\varphi(\xi), \tag{1.7}$$

where m_φ is a 2π-periodic function belonging to $L^2(\mathbb{T})$.

From (1.6) we obtain

$$1 = \sum_{k \in \mathbb{Z}} \left(\frac{2 \sin(\frac{1}{2}\xi + k\pi)}{\xi + 2k\pi} \right)^4 |m_\varphi(\xi)|^2$$

$$= 16 \sin^4(\tfrac{1}{2}\xi) |m_\varphi(\xi)|^2 \sum_{k \in \mathbb{Z}} \frac{1}{(\xi + 2k\pi)^4} . \tag{1.8}$$

The following lemma gives us the value of the infinite sum on the right-hand side of the above equality.

LEMMA 1.9 *For every $\xi \in \mathbb{R}$ we have*

$$\sum_{k \in \mathbb{Z}} \frac{1}{(\xi + 2k\pi)^4} = \frac{1}{16 \sin^4(\frac{1}{2}\xi)} \left\{ 1 - \tfrac{2}{3} \sin^2(\tfrac{1}{2}\xi) \right\}.$$

PROOF : The lemma is proved by differentiating twice both sides of the equality

$$\sum_{k \in \mathbb{Z}} \frac{1}{(\xi + 2k\pi)^2} = \frac{1}{4 \sin^2(\frac{1}{2}\xi)} .$$

To prove this formula we consider the function $\chi = \chi_{[0,1]}$. Since

$$\hat{\chi}(\xi) = e^{-i\frac{\xi}{2}} \frac{\sin(\xi/2)}{\xi/2}$$

and $\{\chi(\cdot - k) : k \in \mathbb{Z}\}$ is clearly an orthonormal system in $L^2(\mathbb{R})$, we use Proposition 1.11 of Chapter 2 to obtain the result:

$$1 = \sum_{k \in \mathbb{Z}} |\hat{\chi}(\xi + 2k\pi)|^2 = \sum_{k \in \mathbb{Z}} \frac{4 \sin^2(\frac{1}{2}\xi + k\pi)}{(\xi + 2k\pi)^2} = 4 \sin^2(\tfrac{1}{2}\xi) \sum_{k \in \mathbb{Z}} \frac{1}{(\xi + 2k\pi)^2} .$$

∎

We now proceed with our search of the scaling function φ. Equality (1.8) and Lemma 1.9 give us

$$|m_\varphi(\xi)| = \left(1 - \tfrac{2}{3}\sin^2(\tfrac{1}{2}\xi)\right)^{-\frac{1}{2}}. \tag{1.10}$$

This shows that the orthonormality of the system $\{\varphi(\cdot - k) : k \in \mathbb{Z}\}$ completely determines the absolute values of m_φ and $\hat{\varphi}$ (see (1.7)). We choose

$$\hat{\varphi}(\xi) = \frac{\sin^2(\xi/2)}{(\xi/2)^2}\left(1 - \tfrac{2}{3}\sin^2(\tfrac{1}{2}\xi)\right)^{-\frac{1}{2}}. \tag{1.11}$$

Reversing the above steps we also see that if φ is given by (1.11), $\hat{\varphi}$ satisfies (1.6), showing, therefore, that $\{\varphi(\cdot - k) : k \in \mathbb{Z}\}$ is an orthonormal system in V_0. We shall show that this system is also complete in V_0. To see this consider the bijection from V_0 to $L^2(\mathbb{T})$ given by $f \mapsto m_f$ (see Theorem 1.3). The completeness of the translates of φ is seen to be equivalent to the completeness of the system $\{m_\varphi(\xi)e^{-ik\xi} : k \in \mathbb{Z}\}$ in $L^2(\mathbb{T})$. Since

$$\left(1 - \tfrac{2}{3}\sin^2(\tfrac{1}{2}\xi)\right)^{-\frac{1}{2}} = \left(\tfrac{2}{3} + \tfrac{1}{3}\cos\xi\right)^{-\frac{1}{2}}$$

is bounded above and below on $\mathbb{T}$, (1.10) shows that the completeness of the above system is equivalent to the well known completeness of the system $\{e^{-ik\xi} : k \in \mathbb{Z}\}$ in $L^2(\mathbb{T})$. This proves the desired result.

We have produced an MRA, $\{V_j : j \in \mathbb{Z}\}$, where V_j is the set of continuous functions belonging to $L^2(\mathbb{R})$ which are linear on each interval of the form $[2^{-j}k, 2^{-j}(k+1)]$, $k \in \mathbb{Z}$, whose scaling function φ can be chosen so that it satisfies (1.11). The low-pass filter m_0 of this MRA, associated with this scaling function φ, can be obtained from the equation $\hat{\varphi}(2\xi) = m_0(\xi)\hat{\varphi}(\xi)$ (see (2.3) of Chapter 2) and (1.11):

$$\begin{aligned}
m_0(\xi) &= \frac{(\sin\xi)^2\left(1 - \tfrac{2}{3}\sin^2(\tfrac{1}{2}\xi)\right)^{\frac{1}{2}}}{\left(2\sin(\tfrac{1}{2}\xi)\right)^2\left(1 - \tfrac{2}{3}\sin^2\xi\right)^{\frac{1}{2}}} \\
&= \left(\cos(\tfrac{1}{2}\xi)\right)^2\left(\frac{1 - \tfrac{2}{3}\sin^2(\tfrac{1}{2}\xi)}{1 - \tfrac{2}{3}\sin^2\xi}\right)^{\frac{1}{2}}.
\end{aligned} \tag{1.12}$$

An orthonormal wavelet ψ associated with the MRA we have just constructed can be given by (2.12) of Chapter 2; ψ satisfies

$$\hat{\psi}(2\xi) = e^{i\xi}\,\overline{m_0(\xi + \pi)}\,\hat{\varphi}(\xi)$$

$$= e^{i\xi}\sin^2(\tfrac{1}{2}\xi)\left(\frac{1-\tfrac{2}{3}\cos^2(\tfrac{1}{2}\xi)}{1-\tfrac{2}{3}\sin^2\xi}\right)^{\frac{1}{2}}\left(\frac{\sin(\xi/2)}{\xi/2}\right)^2\left(1-\tfrac{2}{3}\sin^2(\tfrac{1}{2}\xi)\right)^{-\frac{1}{2}}$$

$$= e^{i\xi}\frac{\sin^4(\tfrac{1}{2}\xi)}{(\tfrac{1}{2}\xi)^2}\left(\frac{1-\tfrac{2}{3}\cos^2(\tfrac{1}{2}\xi)}{(1-\tfrac{2}{3}\sin^2\xi)(1-\tfrac{2}{3}\sin^2(\tfrac{1}{2}\xi))}\right)^{\frac{1}{2}}.$$

For future reference the results we have obtained are collected in the following theorem:

THEOREM 1.13 *For each $j \in \mathbb{Z}$, let V_j be the subspace of $L^2(\mathbb{R})$ of all continuous functions on $\mathbb{R}$ which are linear on each interval of the form $[2^{-j}k, 2^{-j}(k+1)]$, $k \in \mathbb{Z}$. Then the sequence $\{V_j : j \in \mathbb{Z}\}$ forms an MRA for $L^2(\mathbb{R})$. An associated scaling function φ can be defined by (1.11). The orthonormal wavelet ψ associated with this scaling function via (2.12) of Chapter 2 satisfies*

$$\hat{\psi}(\xi) = e^{i\frac{\xi}{2}}\frac{\sin^4(\tfrac{1}{4}\xi)}{(\tfrac{1}{4}\xi)^2}\left(\frac{1-\tfrac{2}{3}\cos^2(\tfrac{1}{4}\xi)}{(1-\tfrac{2}{3}\sin^2(\tfrac{1}{2}\xi))(1-\tfrac{2}{3}\sin^2(\tfrac{1}{4}\xi))}\right)^{\frac{1}{2}}. \tag{1.14}$$

The wavelet we have constructed is often called the **Franklin wavelet**. (Though this name is also applied to "equivalent" wavelets in the sense of Proposition 2.13 of Chapter 2.) We shall obtain several properties of the Franklin wavelet ψ and the scaling function φ. The ones contained in the next result are a consequence of formula (1.14) that describes the Fourier transform of the Franklin wavelet ψ.

PROPOSITION 1.15 *The Franklin wavelet ψ, whose Fourier transform is given by (1.14), satisfies:*

    i) $\displaystyle\int_{\mathbb{R}}\psi(x)\,dx = 0 = \int_{\mathbb{R}}x\psi(x)\,dx;$

    ii) ψ *is symmetric with respect to* $x = -\tfrac{1}{2}$.

PROOF: Part i) follows from $\hat{\psi}(0) = 0$ and $\frac{d\hat{\psi}}{d\xi}(0) = 0$, which is an easy consequence of (1.14). Part ii) follows if we show $\psi(-1-x) = \psi(x)$. Write $\hat{\psi}(\xi) = e^{i\frac{\xi}{2}}\gamma(\xi)$; then

$$\psi(-1-x) = \frac{1}{2\pi}\int_{\mathbb{R}}\hat{\psi}(\xi)e^{i\xi(-1-x)}\,d\xi = \frac{1}{2\pi}\int_{\mathbb{R}}\hat{\psi}(\xi)e^{-i\xi}e^{-ix\xi}\,d\xi$$

$$= \frac{1}{2\pi}\int_{\mathbb{R}}\gamma(\xi)e^{-i\frac{\xi}{2}}e^{-ix\xi}\,d\xi.$$

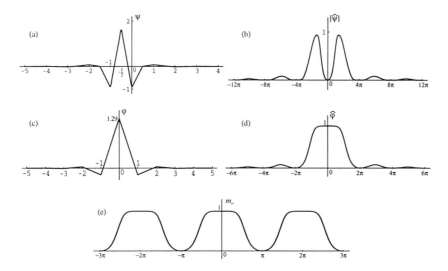

Figure 4.2: *Graphs of several functions related to the Franklin wavelet:*
  (a) *the Franklin wavelet ψ,*
  (b) *the modulus of the Fourier transform of the Franklin wavelet,*
  (c) *the scaling function φ given in (1.11),*
  (d) *the Fourier transform of the scaling function,*
  (e) *the corresponding low-pass filter m_0.*

By (1.14) $\gamma(\xi)$ is even and we obtain

$$\psi(-1-x) = \frac{1}{2\pi} \int_{\mathbb{R}} \gamma(\xi) e^{i\frac{\xi}{2}} e^{ix\xi} \, d\xi = \frac{1}{2\pi} \int_{\mathbb{R}} \hat{\psi}(\xi) e^{ix\xi} \, dx = \psi(x).$$

∎

The next property refers to the decay at infinity of the Franklin wavelet and the associated scaling function. We say that a function g has **exponential decay** if there exist constants $C > 0$ and $\alpha > 0$ such that

$$|g(x)| \leq Ce^{-\alpha|x|} \qquad \text{for all } x \in \mathbb{R}.$$

PROPOSITION 1.16 *The Franklin wavelet ψ defined by (1.14) and its associated scaling function φ defined by (1.11) have exponential decay.*

PROOF : We start by proving the result for the wavelet ψ. Write

$$\hat{\psi}(\xi) = e^{i\frac{\xi}{2}}\gamma(\xi)$$

as in the proof of Proposition 1.15. Consider the region $\Gamma(\xi_o, \eta_o)$ described in Figure 4.3 with η_o small enough so that

$$\gamma(z) = \frac{\sin^4(\frac{1}{4}z)}{(\frac{1}{4}z)^2}\left(\frac{1 - \frac{2}{3}\cos^2(\frac{1}{4}z)}{(1 - \frac{2}{3}\sin^2(\frac{1}{2}z))(1 - \frac{2}{3}\sin^2(\frac{1}{4}z))}\right)^{\frac{1}{2}}.$$

has no poles in $0 < |\Im m(z)| < \eta_o$. We then have

$$0 = \int_{\partial\Gamma(\xi_o, \eta_o)} \gamma(z)e^{ixz}\,dz = \int_{-\xi_o}^{\xi_o} \gamma(\xi)e^{ix\xi}\,d\xi + i\int_0^{\eta_o} \gamma(\xi_o + i\eta)e^{ix(\xi_o + i\eta)}\,d\eta$$

$$- \int_{-\xi_o}^{\xi_o} \gamma(\xi + i\eta_o)e^{ix(\xi + i\eta_o)}\,d\xi - i\int_0^{\eta_o} \gamma(-\xi_o + i\eta)e^{ix(-\xi_o + i\eta)}\,d\eta.$$

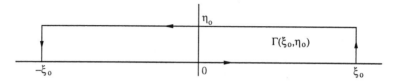

Figure 4.3: The region $\Gamma(\xi_o, \eta_o)$.

It is clear that there exists a constant C_o such that

$$\gamma(\xi + i\eta) \leq \frac{C_o}{\xi^2 + \eta^2} \qquad \text{for all } \xi, \eta \in \mathbb{R},$$

which shows that the second and the fourth terms in the above equality tend to 0 as $|\xi_o| \to \infty$. Therefore,

$$|\psi(x - \tfrac{1}{2})| = \frac{1}{2\pi}\left|\int_{\mathbb{R}} \gamma(\xi)e^{ix\xi}\,d\xi\right|$$

$$= \frac{1}{2\pi}\left|\lim_{|\xi_o| \to \infty} \int_{-\xi_o}^{\xi_o} \gamma(\xi + i\eta_o)e^{ix(\xi + i\eta_o)}\,d\xi\right| \leq Ce^{-x\eta_o}.$$

Since ψ is symmetric with respect to $-\frac{1}{2}$ we must have $|\psi(x-\frac{1}{2})| \leq Ce^{-|x|\eta_o}$, which shows the desired result.

The proof of the exponential decay of φ is similar since ξ^2 appears in the denominator of the formula for $\hat{\varphi}$ given in (1.11).　∎

We conclude this section by proving that the scaling function associated with the Franklin wavelet minimizes a certain functional.

THEOREM 1.17 *Among the functions $g \in V_0$ such that $\{g(\cdot - k) : k \in \mathbb{Z}\}$ is an orthonormal system, the scaling function φ given by (1.11) is the unique one, up to a constant factor of absolute value 1, that minimizes the functional*

$$\int_{\mathbb{R}} x^2 |g(x)|^2 \, dx.$$

PROOF : From the boundedness and the exponential decay of φ we deduce that

$$\int_{\mathbb{R}} x^2 |\varphi(x)|^2 \, dx < \infty.$$

Hence, it suffices to consider only those $g \in V_0$ such that $\{g(\cdot - k) : k \in \mathbb{Z}\}$ is an orthonormal system and $\int_{-\infty}^{\infty} x^2 |g(x)|^2 \, dx$ is finite. Thus, $\frac{d\hat{g}}{d\xi}$ exists almost everywhere and belongs to $L^2(\mathbb{R})$. As in the construction of the scaling function φ, the orthonormality of the system $\{g(\cdot - k) : k \in \mathbb{Z}\}$ completely determines $|\hat{g}(\xi)|$ (see (1.10) and (1.7)):

$$|\hat{g}(\xi)| = \left(\frac{\sin(\xi/2)}{\xi/2} \right)^2 \left(1 - \tfrac{2}{3} \sin^2(\tfrac{1}{2}\xi) \right)^{-\frac{1}{2}}.$$

Hence, by (1.11) we must have $\hat{g}(\xi) = e^{i\lambda(\xi)} \hat{\varphi}(\xi)$. Since both g and φ belong to V_0 we must have

$$e^{i\lambda(\xi)} = \frac{m_g(\xi)}{m_\varphi(\xi)},$$

where m_g and m_φ are the 2π-periodic functions associated with g and φ, respectively, given in Theorem 1.3. Moreover, $\lambda(\xi)$ can be chosen to be differentiable since $e^{i\lambda(\xi)}$ is the quotient of $\hat{g}(\xi)$ and $\hat{\varphi}(\xi)$ and this function never vanishes. Taking derivatives we obtain

$$\frac{d\hat{g}}{d\xi}(\xi) = i \frac{d\lambda}{d\xi}(\xi) e^{i\lambda(\xi)} \hat{\varphi}(\xi) + e^{i\lambda(\xi)} \frac{d\hat{\varphi}}{d\xi}(\xi).$$

Since $\frac{d\lambda}{d\xi}$, $\hat{\varphi}$, and $\frac{d\hat{\varphi}}{d\xi}$ are real valued we have

$$\int_{\mathbb{R}} |xg(x)|^2 \, dx = \int_{\mathbb{R}} \left| \frac{d\hat{g}}{d\xi}(\xi) \right|^2 d\xi = \int_{\mathbb{R}} \left| \frac{d\hat{\varphi}}{d\xi}(\xi) + i\frac{d\lambda}{d\xi}(\xi)\hat{\varphi}(\xi) \right|^2 d\xi$$

$$\geq \int_{\mathbb{R}} \left| \frac{d\hat{\varphi}}{d\xi}(\xi) \right|^2 d\xi = \int_{\mathbb{R}} x^2 |\varphi(x)|^2 \, dx.$$

This proves the desired minimality property. To see uniqueness, observe that, if we have only equalities in the above estimations, we must have

$$\int_{\mathbb{R}} \left| \frac{d\lambda}{d\xi}(\xi) \hat{\varphi}(\xi) \right|^2 d\xi = 0.$$

Since $\hat{\varphi}$ is never zero (see the definition of $\hat{\varphi}$ given in (1.11)), it follows that $\frac{d\lambda}{d\xi} = 0$ almost everywhere. Thus, $\lambda(\xi)$ is equal to a constant and, thus, $\hat{g}(\xi) = \tilde{k}\hat{\varphi}(\xi)$ with $|\tilde{k}| = 1$. This is, as promised, the uniqueness up to a unimodular multiplicative constant.

∎

4.2 Spline wavelets on the real line

In section 4.1 we have constructed an orthonormal basis of piecewise linear continuous functions on $\mathbb{R}$. The main purpose of this section is to show that a similar construction produces wavelets with greater degrees of smoothness.

For $n = 0, 1, 2, \cdots$, let $\mathcal{P}_n$ be the collection of all polynomials of degree at most n and $C^n \equiv C^n(\mathbb{R})$ be the set of all functions f defined on $\mathbb{R}$ such that all the derivatives of f up to order n exist and $f^{(n)}$ is continuous on $\mathbb{R}$. For $n \in \mathbb{N}$, the space $S^n \equiv S^n(\mathbb{R})$ of **splines of order** n is the set of all $f \in C^{n-1}$ such that the restrictions of f to any interval of the form $[k, k+1]$, $k \in \mathbb{Z}$, are in $\mathcal{P}_n$.

A basis for S^1 has been exhibited at the beginning of section 4.1. In fact, we observe that the representation (1.2) shows that $\{\Delta(\cdot - k) : k \in \mathbb{Z}\}$ is a basis of this space, with $\Delta = \chi * \chi$ and $\chi = \chi_{[0,1]}$. The linear independence follows immediately from the fact that $\Delta(\ell - k + 1) = \delta_{\ell,k}$ for all $\ell \in \mathbb{Z}$.

Let $\Delta^2(x) = (\Delta * \chi)(x)$; that is,

$$\Delta^2(x) = (\Delta * \chi)(x) = \int_{\mathbb{R}} \Delta(y)\chi(x - y)\, dy = \int_{x-1}^{x} \Delta(y)\, dy.$$

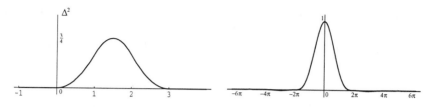

Figure 4.4: Graphs of Δ^2 (left) and $e^{i\frac{3}{2}\xi}\widehat{\Delta^2}(\xi)$ (right).

From this, it is easy to deduce

$$
\Delta^2(x) = \begin{cases}
\frac{1}{2}x^2 & \text{if } 0 \leq x \leq 1, \\
-(x - \frac{3}{2})^2 + \frac{3}{4} & \text{if } 1 \leq x \leq 2, \\
\frac{1}{2}(x - 3)^2 & \text{if } 2 \leq x \leq 3, \\
0 & \text{otherwise,}
\end{cases}
$$

and

$$
\widehat{\Delta^2}(\xi) = e^{-i\frac{3}{2}\xi}\left(\frac{\sin(\xi/2)}{\xi/2}\right)^3.
$$

It follows that $\Delta^2(\cdot - k) \in S^2$ for all $k \in \mathbb{Z}$. The functions $\Delta^2(\cdot - k)$ are called the **basic splines of order** 2. (See the graph of Δ^2 in Figure 4.4.)

The procedure just described can be extended to obtain splines of order n. For $n = 2, 3, 4, \cdots$, define

$$
\Delta^n = \Delta^{n-1} * \chi, \tag{2.1}
$$

where $\Delta^1 \equiv \Delta$. By the definition of convolution we have

$$
\Delta^n(x) = \int_{\mathbb{R}} \Delta^{n-1}(y)\chi(x - y)\, dy = \int_{x-1}^{x} \Delta^{n-1}(y)\, dy. \tag{2.2}
$$

We can now prove by induction that $\Delta^n \in S^n$, $n = 1, 2, 3, \cdots$. This is true for $n = 1$ (see Figure 4.1) and $n = 2$ (see Figure 4.4). If $x \in [k, k+1]$, (2.2) shows that

$$
\Delta^n(x) = \int_{x-1}^{k} \Delta^{n-1}(y)\, dy + \int_{k}^{x} \Delta^{n-1}(y)\, dy,
$$

where $\Delta^{n-1}\big|_{[x-1,k)}$ and $\Delta^{n-1}\big|_{[k,x)}$ are polynomials of degree at most $n - 1$; thus, $\Delta^n(x)$ is a polynomial of degree at most n when restricted to $[k, k+1)$. Since $\Delta^{n-1} \in C^{n-2}$, it is clear from (2.2) that $\Delta^n \in C^{n-1}$. This shows the desired result. It follows that $\Delta^n(\cdot - k) \in S^n$ for all $k \in \mathbb{Z}$. The functions $\Delta^n(\cdot - k)$, $k \in \mathbb{Z}$, are called the **basic splines of order** n. In the next lemma we collect some results about the basic splines.

LEMMA 2.3 *The basic spline of order* n, Δ^n, *satisfies the following properties:*

(i) $\qquad \widehat{\Delta^n}(\xi) = e^{-i\frac{n+1}{2}\xi} \left(\dfrac{\sin(\xi/2)}{\xi/2} \right)^{n+1}$;

(ii) $\qquad supp\,(\Delta^n) = [0, n+1]$ *and* $\Delta^n(x) > 0$ *for all* $x \in (0, n+1)$;

(iii) $\qquad \displaystyle\sum_{k=-\infty}^{\infty} \Delta^n(x-k) = 1$ *for all* $x \in \mathbb{R}$;

(iv) $\qquad \Delta^n$ *is symmetric with respect to* $\frac{1}{2}(n+1)$, *that is,*

$$\Delta^n\left(\frac{n+1}{2} + x\right) = \Delta^n\left(\frac{n+1}{2} - x\right) \qquad \textit{for all}\ \ x \in \mathbb{R}.$$

PROOF : (i) follows from $\hat{\chi}(\xi) = e^{-i\frac{\xi}{2}} \frac{\sin(\xi/2)}{\xi/2}$ and (2.1). Property (ii) is easily deduced by induction. Induction is again the main tool to prove (iii). The case $n = 1$ is easy, and left to the reader; by assuming the inductive hypothesis, we use (2.2) to obtain

$$\sum_{k \in \mathbb{Z}} \Delta^n(x-k) = \sum_{k \in \mathbb{Z}} \int_{x-k-1}^{x-k} \Delta^{n-1}(y)\,dy = \sum_{k \in \mathbb{Z}} \int_{x-1}^{x} \Delta^{n-1}(z-k)\,dz$$

$$= \int_{x-1}^{x} \left\{ \sum_{k \in \mathbb{Z}} \Delta^{n-1}(z-k) \right\} dz = 1.$$

Property (iv) follows, again, by induction and (2.2). Details are left to the reader. ∎

We shall now try to construct wavelets that belong to the class S^n, $n > 1$. The case $n = 1$ was developed in section 4.1 where we began the construction by introducing the space V_0 which was used to introduce an associated multiresolution analysis that produced the desired wavelet by the method developed in Chapter 2. We shall see that the cases $n > 1$ are somewhat more complicated and there are some differences between the cases when n is odd and when n is even. In particular, it is not possible to proceed in a way that is very close to that employed in the last section; however, this treatment does provide some good motivation for the following construction.

We shall try to find a function $\varphi \in S^n \cap L^2(\mathbb{R})$ that has the appropriate properties of a scaling function. Motivated by Theorems 1.3 and 1.4 we impose the condition that $\xi^{n+1}\hat{\varphi}(\xi) \equiv c_n(\xi)$ defines a 2π-periodic function in $L^2(\mathbb{T})$. It is also clear that we must impose the condition

$$\sum_{k \in \mathbb{Z}} |\hat{\varphi}(\xi + 2k\pi)|^2 = 1 \qquad \text{for } a.e. \ \xi \in \mathbb{R}, \tag{2.4}$$

which, by Proposition 1.11 of Chapter 2, guarantees the orthonormality of the system $\{\varphi(\cdot - k) : k \in \mathbb{Z}\}$ in $L^2(\mathbb{R})$.

REMARK: The condition $\varphi \in L^2(\mathbb{R}) \cap S^n$ requires $\xi^{n+1}\hat{\varphi}(\xi)$ to be a 2π-periodic function in $L^2(\mathbb{T})$. To see this, observe that on the interval $(k, k+1)$ the function φ agrees with a polynomial of order at most n; thus, $(\frac{d^{n+1}}{dx^{n+1}}\varphi)(x) = 0$ on $(k, k+1)$ and, therefore, $(\frac{d^n}{dx^n}\varphi)(x)$ equals a constant on this interval. In the sense of distributions this means that near k, the function $\frac{d^{n+1}}{dx^{n+1}}\varphi$ should coincide with a multiple of the Dirac "δ-function" at k; that is,

$$\left(\frac{d^{n+1}}{dx^{n+1}}\varphi\right)(x) = \sum_{k \in \mathbb{Z}} a_k \delta(x - k),$$

where a_k is the "jump" at k exhibited by the piecewise constant function $\frac{d^n}{dx^n}\varphi$ at the point k. Since the Fourier transform of $\delta(\cdot - k)$ is the function $e^{-ik\xi}$ we have, formally,

$$\xi^{n+1}\hat{\varphi}(\xi) = (-i)^{n+1} \sum_{k \in \mathbb{Z}} a_k e^{-ik\xi}.$$

We may consider the right-hand side of this equality to be the Fourier series of the function we denoted by $c_n(\xi)$.

If we replace $\hat{\varphi}(\xi)$ in (2.4) by $\xi^{-n-1} c_n(\xi)$ $(= \hat{\varphi}(\xi))$ and use the 2π-periodicity of $c_n(\xi)$ we obtain

$$1 = |c_n(\xi)|^2 \sum_{k \in \mathbb{Z}} \frac{1}{(\xi + 2k\pi)^{2n+2}}. \tag{2.5}$$

Recall that in the proof of Lemma 1.9 we established a precise formula for
the infinite sum in (2.5) when $n = 0$:

$$\sum_{k \in \mathbb{Z}} \frac{1}{(\xi + 2k\pi)^2} = \left(2\sin(\tfrac{1}{2}\xi)\right)^{-2}. \tag{2.6}$$

In fact, Lemma 1.9 gives us the value of the series in (2.5) when $n = 1$. For
all of the cases $N \geq 1$ we have

PROPOSITION 2.7 *Suppose $N = 1, 2, 3, \cdots$, then*

$$\sum_{k \in \mathbb{Z}} \frac{1}{(\xi + 2k\pi)^{N+1}} = \left(2\sin(\tfrac{1}{2}\xi)\right)^{-N-1} P_N(\tfrac{1}{2}\xi), \tag{2.8}$$

where P_N is a trigonometric polynomial satisfying:

(i) P_N *is an even function, and $P_N(k\pi) = (-1)^{k(N-1)}$ for all $k \in \mathbb{Z}$;*

(ii)   *when N is odd, P_N is π-periodic and $P_N(\xi) > 0$ for all $\xi \in \mathbb{R}$;*

(iii)   *when N is even, $P_N(\xi + \pi) = -P_N(\xi)$ for all $\xi \in \mathbb{R}$.*

PROOF : We prove the various assertions in this proposition by induction.
Equality (2.6) gives us the case $N = 1$, for which $P_1(\xi) \equiv 1$. In fact,
Lemma 1.9 shows that $P_3(\xi) = 1 - \tfrac{2}{3}\sin^2 \xi$. An even simpler calculation
leads us to $P_2(\xi) = \cos\xi$. Let us assume the validity of equality (2.8) for
the case $N - 1$:

$$\sum_{k \in \mathbb{Z}} \frac{1}{(\xi + 2k\pi)^N} = \frac{P_{N-1}(\tfrac{1}{2}\xi)}{(2\sin(\tfrac{1}{2}\xi))^N},$$

where $P_{N-1}(\xi)$ is a trigonometric polynomial. Differentiating both sides of
this equality we obtain (2.8) with

$$P_N(\xi) = (\cos\xi)P_{N-1}(\xi) - \frac{1}{N}(\sin\xi)P'_{N-1}(\xi), \tag{2.9}$$

which is clearly a trigonometric polynomial (that is, a polynomial in $\sin\xi$
and $\cos\xi$). From this and $P_1 \equiv 1$ we see that

$$P_N(k\pi) = (-1)^k P_{N-1}(k\pi) = (-1)^{k(N-1)} P_1(k\pi) = (-1)^{k(N-1)}$$

for all $k \in \mathbb{Z}$. If we assume the validity of (i), (ii) and (iii) for $N - 1$ we also obtain from (2.9) that

$$P_N(-\xi) = (\cos \xi) P_{N-1}(\xi) + \frac{1}{N}(\sin \xi) P'_{N-1}(-\xi)$$

$$= (\cos \xi) P_{N-1}(\xi) - \frac{1}{N}(\sin \xi) P'_{N-1}(\xi) = P_N(\xi),$$

since the evenness of P_{N-1} implies that P'_{N-1} is odd. This establishes (i).

We now prove (ii) and (iii) simultaneously. In particular, let us assume that

$$P_{N-1}(\xi + \pi) = (-1)^N P_{N-1}(\xi).$$

Then a direct computation involving (2.9) shows that this relation holds for $P_N(\xi)$ as follows:

$$P_N(\xi + \pi) = -(\cos \xi) P_{N-1}(\xi + \pi) + \frac{1}{N}(\sin \xi) P'_{N-1}(\xi + \pi)$$

$$= (-1)^N(\cos \xi) P_{N-1}(\xi) - (-1)^N \frac{1}{N}(\sin \xi) P'_{N-1}(\xi) = (-1)^{N+1} P_N(\xi).$$

The only thing remaining to be shown is that $P_N(\xi) > 0$ for all $\xi \in \mathbb{R}$ when N is odd. But from (2.8) we have

$$P_N(\xi) = \sum_{k \in \mathbb{Z}} \frac{(2 \sin \xi)^{N+1}}{(2\xi + 2k\pi)^{N+1}} = \sum_{k \in \mathbb{Z}} (-1)^{k(N+1)} \left(\frac{\sin(\xi + k\pi)}{\xi + k\pi} \right)^{N+1}.$$

When N is odd the last sum is clearly positive. ∎

We have already observed that $P_1(\xi) = 1$, $P_2(\xi) = \cos \xi$ and

$$P_3(\xi) = 1 - \tfrac{2}{3} \sin^2 \xi = \tfrac{2}{3} + \tfrac{1}{3} \cos(2\xi).$$

Simple calculations show that

$$\left. \begin{array}{l} P_4(\xi) = \tfrac{1}{3} \cos^3 \xi + \tfrac{2}{3} \cos \xi, \\ P_5(\xi) = \tfrac{1}{30} \cos^2(2\xi) + \tfrac{13}{30} \cos(2\xi) + \tfrac{8}{15}. \end{array} \right\} \tag{2.10}$$

The graphs of P_2, P_3, P_4 and P_5 are given in Figure 4.5. We see, in particular, that there is an important difference between the polynomials

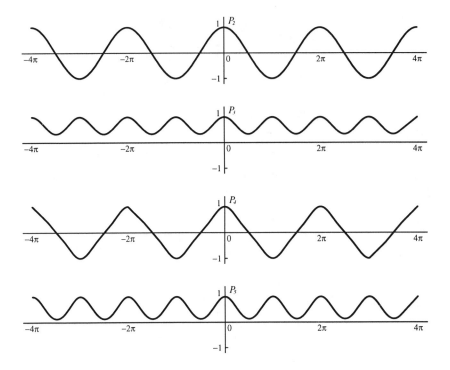

Figure 4.5: Graphs of P_2, P_3, P_4 and P_5.

P_N when N is even and N is odd: in the latter case $P_N(\xi)$ is always positive (see part (ii) of Proposition 2.7), while in the former case $P_N(\xi)$ assumes negative values (see part (iii) of Proposition 2.7). This phenomenon is consistent with equality (2.5), that must be satisfied by the scaling function φ we are seeking, since the sum on the right side has the form of the left side of (2.8) with $N = 2n + 1$. Since $\hat{\varphi}(\xi) = \xi^{-n-1}c_n(\xi)$, (2.5) and (2.8) completely determine $|\hat{\varphi}(\xi)|$. Indeed, we must have

$$|\hat{\varphi}(\xi)| = \left| \frac{\sin(\xi/2)}{\xi/2} \right|^{n+1} \frac{1}{\sqrt{P_{2n+1}(\xi/2)}} . \tag{2.11}$$

Thus, the scaling function we are seeking, φ^n, must have a Fourier transform $\widehat{\varphi^n}$ having absolute value given by the expression on the right in (2.11). For reasons that will become clear shortly, once again we find that the cases when n is odd or even must be treated differently. In fact, we choose φ^n so that

$$
\widehat{\varphi^n}(\xi) =
\begin{cases}
\left(\dfrac{\sin(\xi/2)}{\xi/2}\right)^{n+1} \dfrac{1}{\sqrt{P_{2n+1}(\xi/2)}} & \text{if } n \text{ is odd,} \\[4mm]
e^{-i\frac{\xi}{2}}\left(\dfrac{\sin(\xi/2)}{\xi/2}\right)^{n+1} \dfrac{1}{\sqrt{P_{2n+1}(\xi/2)}} & \text{if } n \text{ is even.}
\end{cases}
\tag{2.12}
$$

It then follows from Proposition 2.7 that, in either case, $\xi^{n+1}\widehat{\varphi^n}(\xi)$ is a 2π-periodic function in $L^\infty(\mathbb{T}) \subset L^2(\mathbb{T})$. Moreover, $\frac{d^j\varphi^n}{dx^j}$ exists and belongs to $L^2(\mathbb{R})$ for $j = 0, 1, \cdots, n$. In fact, $\frac{d^{n+1}\varphi^n}{dx^{n+1}}$ also exists as a tempered distribution. Let $\vartheta(\xi) = \xi^n\widehat{\varphi^n}(\xi)$; thus, since $\xi\vartheta(\xi)$ is 2π-periodic we can represent it in $L^2(\mathbb{T})$ by its Fourier series:

$$
\xi\vartheta(\xi) = \xi^{n+1}\widehat{\varphi^n}(\xi) = \sum_{\ell\in\mathbb{Z}} a_\ell e^{i\ell\xi}.
\tag{2.13}
$$

Then, if $t(x)$ is a tempered test function supported in the open interval $(\ell_o, \ell_o + 1)$,

$$
\int_{\ell_o}^{\ell_o+1} \left(\frac{d^n\varphi^n}{dx^n}\right)(x)\, \overline{t'(x)}\, dx = \int_{-\infty}^{\infty} \left(\frac{d^n\varphi^n}{dx^n}\right)(x)\, \overline{t'(x)}\, dx
$$

$$
= \frac{1}{2\pi i}\int_{-\infty}^{\infty} \vartheta(\xi)\left(\xi\,\overline{\hat{t}(\xi)}\right) d\xi = \frac{1}{2\pi i}\int_{-\infty}^{\infty} \overline{\hat{t}(\xi)} \sum_{\ell\in\mathbb{Z}} a_\ell e^{i\ell\xi}\, d\xi
$$

$$
= -i\sum_{\ell\in\mathbb{Z}} a_\ell\, \overline{t(\ell)} = 0,
$$

since ℓ lies outside the support of t for all $\ell \in \mathbb{Z}$. This shows that, for each $\ell_o \in \mathbb{Z}$,

$$
\int_{\ell_o}^{\ell_o+1} \left(\frac{d^n\varphi^n}{dx^n}\right)(x)\, s(x)\, dx = 0
$$

for any test function $s \in \mathcal{S}(\mathbb{R})$ supported in $(\ell_o, \ell_o + 1)$ with $\int_{\mathbb{R}} s(x)\, dx = 0$. But this means that $\frac{d^n\varphi^n}{dx^n}$ is constant in each interval of the form (ℓ_o, ℓ_o+1) and, consequently, $\varphi^n(x)$ restricted to (ℓ_o, ℓ_o+1) is a polynomial of order at most n. It is also evident that φ^n belongs to the class $C^{n-1}(\mathbb{R})$. (Observe that

$$
\left(\frac{d^{n-1}\varphi^n}{dx^{n-1}}\right)^\wedge(\xi) = \frac{b(\xi)}{1 + \xi^2},
$$

where $b \in L^\infty(\mathbb{R})$; thus, the Fourier transform of $\frac{d^{n-1}\varphi^n}{dx^{n-1}}$ is in $L^1(\mathbb{R})$, and this implies $\frac{d^{n-1}\varphi^n}{dx^{n-1}}$ is continuous on $\mathbb{R}$.) We collect these results in the following:

PROPOSITION 2.14 *The functions φ^n defined by the equality (2.12) belong to $S^n(\mathbb{R}) \cap L^2(\mathbb{R})$ and $\{\varphi^n(\cdot - k) : k \in \mathbb{Z}\}$ is an orthonormal system in $L^2(\mathbb{R})$.*

We now define $V_0 \subset L^2(\mathbb{R})$ to be the closure in the L^2-norm of the span of $\{\varphi^n(\cdot - k) : k \in \mathbb{Z}\}$, and for each $j \in \mathbb{Z}$ we let V_j be the closure of the span of $\{(\varphi^n)_{j,k} : k \in \mathbb{Z}\}$. Then conditions (1.2) and (1.5) (in Chapter 2) in the definition of an MRA are satisfied by the sequence $\{V_j\}_{j \in \mathbb{Z}}$. Suppose that condition (1.1) (also in Chapter 2) is satisfied; then the fact that $\widehat{\varphi^n}$ is continuous at 0 and $\widehat{\varphi^n}(0) = 1$, together with Theorems 1.6 and 1.7 in Chapter 2, imply the remaining two conditions in the definition of an MRA. To show (1.1) it suffices to show $V_{-1} \subset V_0$ and, by applying the inverse Fourier transform, this inclusion is true if there exists a 2π-periodic function $m_0^n \in L^2(\mathbb{T})$ such that $\widehat{\varphi^n}(2\xi) = \widehat{\varphi^n}(\xi) m_0^n(\xi)$ (see (2.3) of Chapter 2). But this is easily seen to be the case: we let

$$m_0^n(\xi) = \frac{\widehat{\varphi^n}(2\xi)}{\widehat{\varphi^n}(\xi)} = e^{i\alpha_n(\xi)} \left(\cos(\tfrac{1}{2}\xi)\right)^{n+1} \sqrt{\frac{P_{2n+1}(\tfrac{1}{2}\xi)}{P_{2n+1}(\xi)}}, \qquad (2.15)$$

where

$$\alpha_n(\xi) = \begin{cases} 0 & \text{if } n \text{ is odd,} \\ -\tfrac{1}{2}\xi & \text{if } n \text{ is even.} \end{cases}$$

From Proposition 2.7 we see that the square root factor on the right is 2π-periodic and, because of our choice of $\alpha_n(\xi)$, the same is true for

$$e^{i\alpha_n(\xi)} \left(\cos(\tfrac{1}{2}\xi)\right)^{n+1}.$$

It is also clear that $m_0^n \in L^2(\mathbb{T})$. We have constructed, therefore, an MRA with an associated scaling function that is a spline of order n. The corresponding low-pass filter is given by equality (2.15).

We can now use formula (2.12) in Chapter 2 to produce an orthonormal wavelet ψ^n from the scaling function φ^n and the low-pass filter m_0^n; that is, the Fourier transform of ψ^n is given by the formula

$$
\widehat{\psi^n}(2\xi) = \begin{cases} e^{i\xi} \dfrac{(\sin(\frac{1}{2}\xi))^{2n+2}}{(\frac{1}{2}\xi)^{n+1}} \sqrt{\dfrac{P_{2n+1}(\frac{1}{2}\xi + \frac{1}{2}\pi)}{P_{2n+1}(\xi)P_{2n+1}(\frac{1}{2}\xi)}} & \text{if } n \text{ is odd}, \\[4mm] -ie^{i\xi} \dfrac{(\sin(\frac{1}{2}\xi))^{2n+2}}{(\frac{1}{2}\xi)^{n+1}} \sqrt{\dfrac{P_{2n+1}(\frac{1}{2}\xi + \frac{1}{2}\pi)}{P_{2n+1}(\xi)P_{2n+1}(\frac{1}{2}\xi)}} & \text{if } n \text{ is even}. \end{cases} \tag{2.16}
$$

Let us examine these wavelets and their properties in more detail. In either case (when n is odd or n is even) formula (2.16) shows that

$$
\widehat{\psi^n}(\xi) = e^{i\frac{\xi}{2}} b_n(\xi) \xi^{-n-1},
$$

where b_n is an even function in $L^\infty(\mathbb{R})$; moreover, $\xi^{n+1} \widehat{\psi^n}(\xi) = e^{i\frac{\xi}{2}} b_n(\xi)$ is a 4π-periodic function. These properties are direct consequences of Proposition 2.7. For example, the fact that P_{2n+1} is even and π-periodic (part (i) and (ii) of Proposition 2.7) shows that

$$
P_{2n+1}\left(\frac{\xi + 2\pi}{4}\right) = P_{2n+1}\left(\frac{-\xi - 2\pi}{4}\right)
$$

$$
= P_{2n+1}\left(\frac{-\xi - 2\pi}{4} + \pi\right) = P_{2n+1}\left(\frac{-\xi + 2\pi}{4}\right);
$$

thus, $\xi \mapsto P_{2n+1}\left(\frac{\xi+2\pi}{4}\right)$ is an even function. Even simpler calculations show that the other factors making up b_n are even. The 4π-periodicity is precisely what we need to modify the argument we presented before Proposition 2.14 to show that, for the wavelet ψ^n, $\psi^n(\frac{1}{2}x)$ belongs to the class S^n of splines of order n. We collect these facts in the following:

THEOREM 2.17 *The wavelets ψ^n defined by equality (2.16) belong to the space V_1, and $\psi^n(\frac{1}{2}x)$ are splines of order n. When n is odd, ψ^n is even about $x = -\frac{1}{2}$, while, when n is even, ψ^n is odd about $x = -\frac{1}{2}$; that is,*

(i) $\psi^n(x) = \begin{cases} \psi^n(-1-x) & \text{if } n \text{ is odd}, \\ -\psi^n(-1-x) & \text{if } n \text{ is even}. \end{cases}$

Moreover,

(ii) $\displaystyle \int_{\mathbb{R}} x^k \psi^n(x)\, dx = 0 \qquad \text{for } k = 0, 1, 2, \cdots, n.$

PROOF: Since $\widehat{\psi^n}(\xi) = e^{i\frac{\xi}{2}} b_n(\xi) \xi^{-n-1}$ with $b_n(\xi)$ an even function, $b_n(\xi)\xi^{-n-1}$ is even when n is odd and, clearly, an odd function when n is even. The factor $e^{i\frac{\xi}{2}}$ represents a translation of ψ^n that shifts this parity property from 0 to $-\frac{1}{2}$.

The vanishing moment property (ii) follows from the fact that

$$\left(\frac{d^k \widehat{\psi^n}}{d\xi^k}\right)(0) = 0 \qquad \text{for } k = 0, 1, 2, \cdots, n,$$

which is a consequence of the identity (2.16). Observe that the power $2n+2$ of $\sin(\frac{1}{2}\xi)$ compensates for the (possible) singularity produced by ξ^{-n-1}. In the next proof we obtain an estimate that assures us of the integrability of the integrand in (ii). ∎

REMARK: The wavelet ψ^n discussed in Theorem 2.17 is often called a **spline wavelet of order n on the real line**. Notice that, although ψ^n is not in $S^n(\mathbb{R})$, its dilation $\psi^n(2^{-1}\cdot)$ does belong to $S^n(\mathbb{R})$.

In Proposition 1.16 we saw that the Franklin wavelet and its associated scaling function have exponential decays at ∞. This property, and more, is shared with the higher order splines:

THEOREM 2.18 *The spline wavelets ψ^n, $n = 1, 2, \cdots$, and their associated scaling functions φ^n, as well as their derivatives*

$$\frac{d^j \psi^n}{dx^j}, \ \frac{d^j \varphi^n}{dx^j}, \qquad j = 1, 2, \cdots, n-1,$$

have exponential decay at ∞.

PROOF: The proof of this theorem is done by using the same argument we employed for the proof of Proposition 1.16. It is easily seen that the same observation about the poles that was made about the analytic extensions of $\widehat{\psi^n}$ and $\widehat{\varphi^n}$ in the proof of Proposition 1.16 applies to the Fourier transforms of the derivatives of ψ^n and φ^n we are now considering. ∎

We end this section by establishing an identity that will prove to be most helpful for transferring the construction of spline wavelets from the real line to their periodic analogs on the torus $\mathbb{T}$.

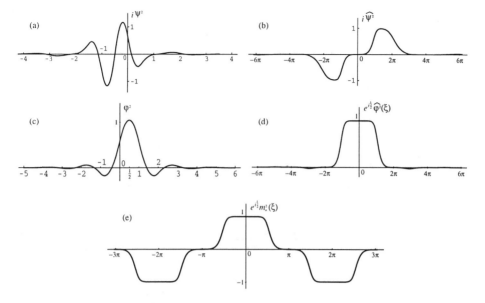

Figure 4.6: *Graphs of several functions related to the spline wavelet ψ^2:*
  (a) $i\psi^2$,
  (b) $ie^{-i\frac{\xi}{2}}\widehat{\psi^2}$,
  (c) *the scaling function φ^2 given in* (2.12),
  (d) $e^{i\frac{\xi}{2}}\widehat{\varphi^2}$,
  (e) $e^{i\frac{\xi}{2}}m_0^2(\xi)$, *where m_0^2 is the corresponding low-pass filter.*

THEOREM 2.19 *For any $n = 1, 2, 3, \cdots$, the spline wavelet ψ^n satisfies*

$$\sum_{k\in\mathbb{Z}} \frac{\widehat{\psi^n}(\xi + 2k\pi)}{(\xi + 2k\pi)^{n+1}} = 0 \qquad for\ \ a.e.\ \xi \in \mathbb{R}.$$

PROOF : Since $\widehat{\psi^n} \in L^\infty(\mathbb{R})$ it is clear that the above series converges for a.e. $\xi \in \mathbb{R}$. In order to obtain this identity we use an idea we exploited before in this book: we break up the sum over $\mathbb{Z}$ into the two sums obtained by considering the even and odd integers separately. Thus,

$$\sum_{k\in\mathbb{Z}} \frac{\widehat{\psi^n}(\xi + 2k\pi)}{(\xi + 2k\pi)^{n+1}} = \Sigma_e(\xi) + \Sigma_o(\xi),$$

where

$$\Sigma_e(\xi) = \sum_{\ell\in\mathbb{Z}} \frac{\widehat{\psi^n}(\xi + 4\ell\pi)}{(\xi + 4\ell\pi)^{n+1}}$$

and

$$\Sigma_o(\xi) = \sum_{\ell \in \mathbb{Z}} \frac{\widehat{\psi^n}(\xi + 2(2\ell + 1)\pi)}{(\xi + 2(2\ell + 1)\pi)^{n+1}}.$$

Let us first consider the case when n is odd. Using (2.16) and Proposition 2.7 we have

$$\widehat{\psi^n}(\xi + 4\ell\pi) = 4^{n+1} e^{i\frac{\xi}{2}} \frac{(\sin(\frac{1}{4}\xi))^{2n+2}}{(\xi + 4\ell\pi)^{n+1}} \left(\frac{P_{2n+1}(\frac{1}{4}(\xi + 2\pi))}{P_{2n+1}(\frac{1}{2}\xi)P_{2n+1}(\frac{1}{4}\xi)} \right)^{\frac{1}{2}}.$$

Substituting this value for $\widehat{\psi^n}(\xi + 4\ell\pi)$ in the expression for $\Sigma_e(\xi)$ and using (2.8) with $N = 2n + 1$ we then obtain

$$\Sigma_e(\xi) = 4^{n+1} e^{i\frac{\xi}{2}} \left(\frac{P_{2n+1}(\frac{1}{4}\xi + \frac{1}{2}\pi)}{P_{2n+1}(\frac{1}{2}\xi)P_{2n+1}(\frac{1}{4}\xi)} \right)^{\frac{1}{2}} \sum_{\ell \in \mathbb{Z}} \frac{(\sin(\frac{1}{4}\xi))^{2n+2}}{(\xi + 4\ell\pi)^{2n+2}}$$

$$= 4^{-n-1} e^{i\frac{\xi}{2}} \left(\frac{P_{2n+1}(\frac{1}{4}\xi + \frac{1}{2}\pi)P_{2n+1}(\frac{1}{4}\xi)}{P_{2n+1}(\frac{1}{2}\xi)} \right)^{\frac{1}{2}}.$$

In order to evaluate $\widehat{\psi^n}(\xi + 2(2\ell + 1)\pi)$ we use (2.16) and Proposition 2.7 again, then substitute this value in the expression for $\Sigma_o(\xi)$, use (2.8) and obtain

$$\Sigma_o(\xi) = -4^{-n-1} e^{i\frac{\xi}{2}} \left(\frac{P_{2n+1}(\frac{1}{4}\xi)P_{2n+1}(\frac{1}{4}\xi + \frac{1}{2}\pi)}{P_{2n+1}(\frac{1}{2}\xi)} \right)^{\frac{1}{2}}.$$

Hence, $\Sigma_e(\xi) + \Sigma_o(\xi) = 0$, and we obtain the desired equality.

When n is even we can repeat this argument by using the other case in (2.16). The only difference is the factor $-i$ that does not affect the conclusion $\Sigma_e(\xi) + \Sigma_o(\xi) = 0$.   ∎

4.3 Orthonormal bases of piecewise linear continuous functions for $L^2(\mathbb{T})$

The boundary of the unit disk $\mathbb{T} = \{e^{2\pi i x} : 0 \le x < 1\}$ can be identified with the interval $[0, 1)$ via the correspondence

$$e^{2\pi i x} \longleftrightarrow x.$$

With this identification, the space $L^2(\mathbb{T})$ of square integrable functions on $\mathbb{T}$ can be viewed as the space of 1-periodic functions defined on $\mathbb{R}$ such that

$$\|f\|^2_{L^2([0,1))} = \int_0^1 |f(x)|^2 \, dx < \infty.$$

The Haar function restricted to $[0, 1)$

$$h(x) = \begin{cases} 1 & \text{if } 0 \leq x < \frac{1}{2}, \\ -1 & \text{if } \frac{1}{2} \leq x < 1, \end{cases}$$

can be used to find an orthonormal basis for $L^2(\mathbb{T})$. Let us define, for $j = 0, 1, \cdots,$ and $k = 0, 1, \cdots, 2^j - 1,$

$$h_{j,k}(x) = 2^{\frac{j}{2}} h(2^j x - k), \qquad x \in [0, 1). \tag{3.1}$$

Extend each $h_{j,k}$ to a 1-periodic function on $\mathbb{R}$. With the identification described above, the system of 1-periodic functions

$$\{h_{j,k} : j = 0, 1, \cdots, \text{ and } k = 0, 1, \cdots, 2^j - 1\}$$

is easily seen to be an orthonormal basis for $L^2(\mathbb{T})$. As was the case for the Haar wavelet on $\mathbb{R}$, this periodization of it furnishes us with a simple example of the periodic wavelets we shall construct in this section.

In fact, we shall construct orthonormal bases for $L^2(\mathbb{T})$ whose elements are piecewise linear continuous functions. One of these bases will be obtained by using the Franklin wavelet constructed in section 4.1 (see Theorem 1.13 in that section).

Let $\mathcal{B}$ be the space of continuous 1-periodic functions on $\mathbb{R}$. For $j \geq 0$, let $\mathcal{B}_j$ be the subspace of $\mathcal{B}$ consisting of all piecewise linear functions with nodes at $2^{-j}k$, $k = 0, 1, \cdots, 2^j - 1$. In particular, the functions in $\mathcal{B}_j$ are continuous on $[0, 1)$ and linear on each interval of the form $[2^{-j}k, 2^{-j}(k+1)]$, $k = 0, 1, \cdots, 2^j - 1$. We shall write $N = N_j = 2^j - 1$ for convenience. The space $\mathcal{B}_0$ is the space of constant functions. Clearly,

$$\mathcal{B}_0 \subset \mathcal{B}_1 \subset \mathcal{B}_2 \subset \cdots \subset \mathcal{B}_j \subset \cdots \subset \mathcal{B} \subset L^2(\mathbb{T}) = L^2([0,1)),$$

and the dimension of $\mathcal{B}_j$ is $2^j = N + 1$.

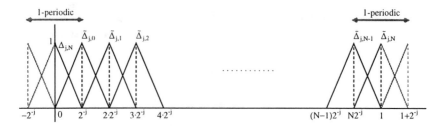

Figure 4.7: Graphs of $\tilde{\Delta}_{j,k}$, $k = 0, 1, \cdots, N$.

A basis for $\mathcal{B}_j$ can be obtained as follows. We first introduce the 1-periodic "triangle" functions defined by

$$\tilde{\Delta}_j(x) = \begin{cases} 2^j x & \text{if } 0 \le x < 2^{-j}, \\ 2 - 2^j x & \text{if } 2^{-j} \le x < 2^{-j+1}, \\ 0 & \text{if } 2^{-j+1} \le x < 1, \end{cases}$$

on $[0, 1)$, for $j = 1, 2, \cdots$. We then introduce the translates of these functions

$$\tilde{\Delta}_{j,k}(x) = \tilde{\Delta}_j(x - 2^{-j}k), \qquad k = 0, 1, 2, \cdots, N;$$

(see Figure 4.7). It is clear that

$$\{\tilde{\Delta}_{j,0}, \tilde{\Delta}_{j,1}, \cdots, \tilde{\Delta}_{j,N}\}$$

is a basis for $\mathcal{B}_j$. This is not, however, an orthonormal basis.

An orthonormal basis for $\mathcal{B}_j$ can be constructed with the aid of the group of $(2^j)^{\text{th}}$ roots of unity

$$\Gamma_j = \left\{1, e^{2\pi i 2^{-j}}, e^{2\pi i 2 \cdot 2^{-j}}, \cdots, e^{2\pi i N 2^{-j}}\right\}.$$

Let $U = U_j$ be the unitary operator in $L^2(\mathbb{T})$ defined by

$$(U_j f)(x) = f(x - 2^{-j}) \qquad \text{for all } f \in L^2(\mathbb{T}).$$

The powers of U give us an action of Γ_j on $\mathcal{B}_j$. Observe that

$$U^k(\tilde{\Delta}_j) = \tilde{\Delta}_{j,k} \qquad \text{for } k = 0, 1, \cdots, N,$$

and that $\tilde{\Delta}_j$ is a cyclic vector under this action: $U^{N+1}\tilde{\Delta}_j = \tilde{\Delta}_j$. The minimal polynomial of U is $x^{N+1} - 1$ (it coincides with its characteristic polynomial) and the (distinct) proper values of U are

$$\lambda_j = \lambda_j^1 = e^{2\pi i 2^{-j}}, \ \lambda_j^2, \ \cdots, \ \lambda_j^N, \ \lambda_j^{N+1} = 1.$$

Since U is unitary and the proper values are distinct, there exists an orthonormal set of associated proper vectors. This is, then, an orthonormal basis for $\mathcal{B}_j$.

The family of spaces $\mathcal{B}_j$, $j = 0, 1, \cdots$, can be used to obtain an orthonormal basis for $L^2(\mathbb{T})$. Let $\mathcal{C}_j$ be the orthogonal complement of $\mathcal{B}_j$ in $\mathcal{B}_{j+1}$. Then $\mathcal{C}_j \oplus \mathcal{B}_j = \mathcal{B}_{j+1}$ and $\dim(\mathcal{C}_j) = 2^j$. Since

$$\mathcal{B}_0 \oplus \left(\bigoplus_{j=0}^{\infty} \mathcal{C}_j \right) = \overline{\mathcal{B}},$$

and $\mathcal{B}$ is dense in $L^2(\mathbb{T})$, a basis is obtained if we find an orthonormal basis for each $\mathcal{C}_j$, $j = 0, 1, \cdots$.

To do this, observe that U is unitary on $\mathcal{C}_j$ and on this space it has minimal polynomial $x^{N+1} - 1$. Therefore, there exists an orthonormal basis of proper vectors $\{z_0, z_1, \cdots, z_N\} \subset \mathcal{C}_j$ such that $U_j(z_\ell) = \lambda_j^\ell z_\ell$. Let

$$g_j = \frac{1}{\sqrt{N+1}} \sum_{\ell=0}^{N} z_\ell.$$

Clearly,

$$U_j^{N+1} g_j = \frac{1}{\sqrt{N+1}} \sum_{\ell=0}^{N} \lambda_j^{\ell(N+1)} z_\ell = g_j,$$

and, therefore, g_j is cyclic for U_j. Moreover,

$$(U_j^m g_j, U_j^n g_j) = \frac{1}{N+1} \sum_{\ell=0}^{N} \lambda_j^{\ell(m-n)} = \delta_{m,n}$$

since, for $n \neq m$, $\lambda_j^{(m-n)}$ is a root of the polynomial

$$x^N + x^{N-1} + \cdots + x + 1 = \frac{x^{N+1} - 1}{x - 1}.$$

We have proved the following result that presents us with an orthonormal basis for $L^2(\mathbb{T})$:

THEOREM 3.2 *If C_j is the orthogonal complement of B_j in B_{j+1}, then there exists a $g_j \in C_j$ such that $\{g_j, Ug_j, \cdots, U^N g_j\}$ is an orthonormal basis for C_j, $j = 0, 1, 2, \cdots$.*

The orthonormal basis provided by Theorem 3.2 consists of piecewise linear continuous functions in $L^2(\mathbb{T})$. This was accomplished by finding a $g_j \in C_j$ that generates a basis for C_j, $j = 0, 1, 2, \cdots$. The functions g_j are chosen in each C_j without any concern for a relation connecting any two of them. What we want to do is to find a single function on $\mathbb{R}$ that will generate a basis for each C_j by appropriate dilations, periodizations and translations much in the spirit of (3.1) and the wavelet bases of previous chapters.

To do this we need to characterize the spaces C_j and B_j. We begin by computing the Fourier coefficients of $\tilde{\Delta}_j$. Since we have identified $\mathbb{T}$ with the interval $[0, 1)$, the Fourier coefficients of a 1-periodic function f with $\|f\|_{L^2([0,1))} < \infty$ are given by

$$\mathcal{F}[f](k) \equiv \int_0^1 f(x)e^{-2\pi ikx}\,dx.$$

Recall that throughout this book we use the traditional symbol $\hat{f}(\xi)$ to denote the Fourier transform of an $L^2(\mathbb{R})$ function f, that is,

$$\hat{f}(\xi) = \int_{\mathbb{R}} f(x)e^{-ix\xi}\,dx.$$

LEMMA 3.3 *The Fourier coefficients of $\tilde{\Delta}_j$ are*

$$\mathcal{F}[\tilde{\Delta}_j](k) = \begin{cases} 2^{-j} & \text{if } k = 0, \\ -\dfrac{2^j}{4\pi^2}\left(\dfrac{e^{-2\pi i2^{-j}k} - 1}{k}\right)^2 & \text{if } k \neq 0. \end{cases}$$

PROOF: The proof is a straightforward computation. It can also be proved by observing that $\tilde{\Delta}_j = \tilde{\chi}_j * \tilde{\chi}_j$, where $\tilde{\chi}_j$ is the 1-periodic function whose values on $[0, 1)$ are $2^{\frac{j}{2}}\chi_{[0,2^{-j})}$. Since

$$\mathcal{F}[\tilde{\chi}_j](k) = -\frac{2^{\frac{j}{2}}}{2\pi ik}(e^{-2\pi i2^{-j}k} - 1) \qquad \text{for } k \neq 0,$$

the result follows from the equalities

$$\mathcal{F}[\tilde{\Delta}_j](k) = \mathcal{F}[\tilde{\chi}_j * \tilde{\chi}_j](k) = \mathcal{F}[\tilde{\chi}_j](k)\,\mathcal{F}[\tilde{\chi}_j](k).$$

∎

If we define

$$b_j(y) = \begin{cases} 1 & \text{if } y = 0, \\ \left(\dfrac{e^{-2\pi i 2^{-j} y} - 1}{2\pi i 2^{-j} y}\right)^2 & \text{if } y \neq 0, \end{cases} \tag{3.4}$$

we obtain a continuous function on $\mathbb{R}$. Moreover, we have

$$\mathcal{F}[\tilde{\Delta}_j](k) = 2^{-j} b_j(k) = \frac{1}{N+1} b_j(k) \qquad \text{for } k \in \mathbb{Z}. \tag{3.5}$$

In order to characterize those f that belong to $\mathcal{C}_j$ we shall use the fact that we must have

$$\sum_{k \in \mathbb{Z}} \mathcal{F}[g](k)\,\overline{\mathcal{F}[f](k)} = 0 \qquad \text{for all } g \in \mathcal{B}_j.$$

Since each $g \in \mathcal{B}_j$ can be expressed as

$$g = \sum_{m=0}^{N} a_m \tilde{\Delta}_{j,m} = \sum_{m=0}^{N} a_m \tilde{\Delta}_j(\cdot - 2^{-j} m),$$

we have

$$\sum_{m=0}^{N} \sum_{k \in \mathbb{Z}} \overline{\mathcal{F}[f](k)}\, a_m \mathcal{F}[\tilde{\Delta}_j] e^{-2\pi i 2^{-j} k m} = 0.$$

For each $0 \leq m \leq N$, applying this to $g = \tilde{\Delta}_{j,m} \in \mathcal{B}_j$ and using (3.5), we obtain

$$0 = \frac{1}{N+1}\overline{\mathcal{F}[f](0)}\, b_j(0) + \frac{1}{N+1}\sum_{k \neq 0} \overline{\mathcal{F}[f](k)}\, b_j(k) e^{-2\pi i 2^{-j} k m}.$$

Write $\lambda = e^{-2\pi i 2^{-j}}$, the non-zero k as $n + 2^j \ell$ with $n = 1, \cdots, N$ and $\ell \in \mathbb{Z}$, and observe that $\lambda^{mk} = \lambda^{mn}$. Then, the last equality is equivalent to

$$0 = \overline{\mathcal{F}[f](0)}\, b_j(0) + \sum_{n=1}^{N} (\lambda^m)^n \sum_{\ell \in \mathbb{Z}} \overline{\mathcal{F}[f](n + 2^j \ell)}\, b_j(n + 2^j \ell)$$

for $m = 0, 1, 2, \cdots, N$. Let

$$A_n = \sum_{\ell \in \mathbb{Z}} \overline{\mathcal{F}[f](n + 2^j \ell)} \, b_j(n + 2^j \ell) \qquad \text{for } n = 0, 1, 2, \cdots, N.$$

Since $b_j(2^j \ell) = \delta_{\ell,0}$, we have $A_0 = \overline{\mathcal{F}[f](0)}$. Thus, we have shown

$$0 = \sum_{n=0}^{N} (\lambda^m)^n A_n \qquad \text{for } m = 0, 1, 2, \cdots, N.$$

If we let

$$C_N = \begin{pmatrix} 1 & 1 & 1 & \cdots & 1 \\ 1 & \lambda & \lambda^2 & \cdots & \lambda^N \\ 1 & \lambda^2 & (\lambda^2)^2 & \cdots & (\lambda^2)^N \\ \vdots & \vdots & \vdots & \ddots & \vdots \\ 1 & \lambda^N & (\lambda^N)^2 & \cdots & (\lambda^N)^N \end{pmatrix} \tag{3.6}$$

then we have

$$\begin{pmatrix} 0 \\ 0 \\ \vdots \\ 0 \end{pmatrix} = C_N \begin{pmatrix} A_0 \\ A_1 \\ \vdots \\ A_N \end{pmatrix}$$

But $\frac{1}{\sqrt{N+1}} C_N$ is unitary (use the properties of the roots of unity). Thus,

$$A_m = 0 \qquad \text{for all } m = 0, 1, 2, \cdots, N.$$

Hence, since $0 = A_0 = \overline{\mathcal{F}[f](0)}$, we have $\mathcal{F}[f](0) = 0$. Moreover, for $n = 1, 2, \cdots, N$,

$$A_n = \sum_{\ell \in \mathbb{Z}} \overline{\mathcal{F}[f](n + 2^j \ell)} \, b_j(n + 2^j \ell)$$

$$= \sum_{\ell \in \mathbb{Z}} \overline{\mathcal{F}[f](n + 2^j \ell)} \, \frac{(\lambda^n - 1)^2}{(-4\pi^2) 2^{-2j} (n + 2^j \ell)^2}$$

Therefore,

$$\sum_{\ell \in \mathbb{Z}} \frac{\overline{\mathcal{F}[f](n + 2^j \ell)}}{(n + 2^j \ell)^2} = 0 \qquad \text{for } n = 1, 2, \cdots, N.$$

THEOREM 3.7 $f \in C_j$ if and only if $f \in B_{j+1}$, $\mathcal{F}[f](0) = 0$ and

$$\sum_{\ell \in \mathbb{Z}} \frac{\mathcal{F}[f](n + 2^j \ell)}{(n + 2^j \ell)^2} = 0 \qquad for \quad n = 1, 2, \cdots, N(= 2^j - 1).$$

We have just shown the "only if" part; the "if" part follows by reversing the arguments given above.

Theorem 3.7 gives us a characterization of the classes C_j, $j = 0, 1, 2, \cdots$. There is a simple characterization of the classes B_j as well. Since $\{\tilde{\Delta}_{j,m} : m = 0, 1, 2, \cdots, 2^j - 1\}$ is a basis for B_j, if $g \in B_j$, then

$$g = \sum_{m=0}^{N} a_m \tilde{\Delta}_{j,m},$$

and we have

$$\mathcal{F}[g](k) = \sum_{m=0}^{N} a_m \mathcal{F}[\tilde{\Delta}_{j,m}](k) = \sum_{m=0}^{N} a_m \mathcal{F}[\tilde{\Delta}_j](k) e^{-2\pi i 2^{-j} km}$$

$$= \sum_{m=0}^{N} a_m \frac{1}{N+1} b_j(k) e^{-2\pi i 2^{-j} km},$$

where $b_j(k)$ is defined in (3.4). This shows that $k^2 \mathcal{F}[g](k)$ is 2^j-periodic over $\mathbb{Z}$. It turns out that this is a characterization of the space B_j. More precisely, we have

THEOREM 3.8 Suppose $f \in B$, the space of all 1-periodic continuous functions on $\mathbb{R}$. Then $f \in B_j$, $j \geq 0$, if and only if $k^2 \mathcal{F}[f](k)$ is a 2^j-periodic function on $\mathbb{Z}$. In particular, $\mathcal{F}[f](2^j n) = 0$ if $n \neq 0$ and $\mathcal{F}[f](0)$ is arbitrary.

PROOF : If $f \in B_j$ we have seen that $k^2 \mathcal{F}[f](k)$ is 2^j-periodic. Suppose that $k^2 \mathcal{F}[f](k)$ is 2^j-periodic on $\mathbb{Z}$. It is enough to show that f is orthogonal to C_p for all $p \geq j$. Suppose $g \in C_p$, $p \geq j$. By Theorem 3.7 we have $\mathcal{F}[g](0) = 0$ and

$$\sum_{\ell \in \mathbb{Z}} \frac{\mathcal{F}[g](n + 2^p \ell)}{(n + 2^p \ell)^2} = 0 \qquad for \quad n = 1, 2, \cdots, 2^p - 1.$$

Thus, since $\mathcal{F}[g](0)$ and $\mathcal{F}[f](2^p\ell) = 0$ for $\ell \neq 0$, we have

$$(g, f)_{L^2(\mathbb{T})} = \overline{\mathcal{F}[f](0)}\,\mathcal{F}[g](0) + \sum_{k \neq 0} k^2\,\overline{\mathcal{F}[f](k)}\,\frac{\mathcal{F}[g](k)}{k^2}$$

$$= 0 + \sum_{n=1}^{2^p-1} \sum_{\ell \in \mathbb{Z}} (n + 2^p\ell)^2\,\overline{\mathcal{F}[f](n + 2^p\ell)}\,\frac{\mathcal{F}[g](n + 2^p\ell)}{(n + 2^p\ell)^2}$$

$$= \sum_{n=1}^{2^p-1} \sum_{\ell \in \mathbb{Z}} n^2\,\overline{\mathcal{F}[f](n)}\,\frac{\mathcal{F}[g](n + 2^p\ell)}{(n + 2^p\ell)^2} \qquad \text{(since } p \geq j\text{)}$$

$$= \sum_{n=1}^{2^p-1} n^2\,\overline{\mathcal{F}[f](n)} \sum_{\ell \in \mathbb{Z}} \frac{\mathcal{F}[g](n + 2^p\ell)}{(n + 2^p\ell)^2} = 0.$$

∎

We now give a necessary and sufficient condition for the orthonormality of the dyadic translates of a 1-periodic function. This result is similar to Proposition 1.11 of Chapter 2.

THEOREM 3.9 If $f \in L^2(\mathbb{T})$, then $\{f, Uf, \cdots, U^N f\}$, where $U \equiv U_j$, is an orthonormal system if and only if

$$\sum_{\ell \in \mathbb{Z}} \left| \mathcal{F}[f](n + 2^j\ell) \right|^2 = 2^{-j} \qquad \text{for } n = 0, 1, \cdots, N = 2^j - 1. \qquad (3.10)$$

PROOF: Let

$$a_n = \sum_{\ell \in \mathbb{Z}} \left| \mathcal{F}[f](n + 2^j\ell) \right|^2 \qquad \text{for } n = 0, 1, \cdots, N = 2^j - 1,$$

and suppose that $\{f, Uf, \cdots, U^N f\}$ is an orthonormal system. Then, by the Plancherel theorem, the fact that $\|f\|_{L^2(\mathbb{T})} = 1$ is equivalent to

$$\sum_{n=0}^{N} a_n = 1; \qquad (3.11)$$

the orthogonality $(f, U^m f)_{L^2(\mathbb{T})} = 0$ for $m = 1, 2, \cdots, N$, in terms of the Fourier coefficients of f, is equivalent to

$$0 = \sum_{k \in \mathbb{Z}} \mathcal{F}[f](k)\,\overline{\mathcal{F}[f](k)}\,e^{2\pi i 2^{-j}km}$$

$$= \sum_{n=0}^{N} \left\{ \sum_{\ell \in \mathbb{Z}} |\mathcal{F}[f](n + 2^j \ell)|^2 \right\} e^{2\pi i 2^{-j} mn} = \sum_{n=0}^{N} a_n (\overline{\lambda}^m)^n \quad (3.12)$$

for $m = 1, 2, \cdots, N$, where $\lambda = e^{-2\pi i 2^{-j}}$. Thus, if we let A be the column vector

$$\begin{pmatrix} a_0 \\ a_1 \\ \vdots \\ a_N \end{pmatrix},$$

equalities (3.11) and (3.12) can be expressed matricially by the equality

$$C_N^* A = \begin{pmatrix} 1 \\ 0 \\ \vdots \\ 0 \end{pmatrix},$$

where C_N is the matrix given by (3.6) and used in the proof of Theorem 3.7. Since $\frac{1}{\sqrt{N+1}} C_N^*$ is unitary, this means that

$$A = \frac{1}{N+1} C_N \begin{pmatrix} 1 \\ 0 \\ \vdots \\ 0 \end{pmatrix} = 2^{-j} \begin{pmatrix} 1 \\ 1 \\ \vdots \\ 1 \end{pmatrix}.$$

But this is precisely equality (3.10).

Conversely, suppose condition (3.10) is satisfied by an $f \in L^2(\mathbb{T})$. We can "reverse" the equalities we have just derived to obtain (3.11) and (3.12). But these are equivalent to $\|f\|_{L^2(\mathbb{T})}^2 = (f, f)_{L^2(\mathbb{T})} = 1$ and $(f, U^m f)_{L^2(\mathbb{T})} = 0$ for $m = 1, 2, \cdots, N$. Since U is unitary we have

$$\left(U^m f, U^{m+n} f \right)_{L^2(\mathbb{T})} = \left(f, U^n f \right)_{L^2(\mathbb{T})} \quad \text{whenever } 0 \leq m \leq m+n \leq N.$$

This shows that $\{f, Uf, \cdots, U^N f\}$ is an orthonormal system in $L^2(\mathbb{T})$. ∎

We can now describe the promised "wavelet-like" basis for $L^2(\mathbb{T})$. Let ψ be the Franklin wavelet on the real line (see (1.14) in Theorem 1.13). For

$j = 0, 1, 2, \cdots, k = 0, 1, \cdots, 2^j - 1$, let

$$\tilde{\psi}_{j,k}(x) = 2^{\frac{j}{2}} \sum_{\ell \in \mathbb{Z}} \psi(2^j(x + \ell) - k). \tag{3.13}$$

Observe that

$$\tilde{\psi}_{j,k}(x) = \tilde{\psi}_{j,0}(x - 2^{-j}k) = (U_j^k \tilde{\psi}_{j,0})(x) = \sum_{\ell \in \mathbb{Z}} \psi_{j,k}(x + \ell), \tag{3.14}$$

where $\psi_{j,k}(x) = 2^{\frac{j}{2}}\psi(2^j x - k)$. Thus, each $\tilde{\psi}_{j,k}$ is a "periodization" of $\psi_{j,k}$. The exponential decay of ψ (see Proposition 1.16) shows that the series in (3.13) is uniformly convergent on $[0,1]$. Thus, the functions $\tilde{\psi}_{j,k}$ are all continuous.

THEOREM 3.15 *The system*

$$\left\{1, \tilde{\psi}_{j,k} : j = 0, 1, 2, \cdots, k = 0, 1, \cdots, 2^j - 1\right\}$$

is an orthonormal basis for $L^2(\mathbb{T}) \approx L^2([0,1))$.

REMARK: This system is called the **Franklin periodic wavelet** basis.

PROOF: Since

$$\overline{\mathcal{B}} = \mathcal{B}_0 \oplus \left(\bigoplus_{j=0}^{\infty} \mathcal{C}_j\right),$$

where $\mathcal{B}$ is the space of 1-periodic continuous functions on $\mathbb{R}$, and $\mathcal{B}$ is dense in $L^2(\mathbb{T})$, it is enough to show that the system

$$\left\{\tilde{\psi}_{j,k} : k = 0, 1, \cdots, N = 2^j - 1\right\}$$

is an orthonormal basis of $\mathcal{C}_j$, for each $j = 0, 1, 2, \cdots$.

To prove the orthonormality we shall use Theorem 3.9, and, hence, all we need to establish is that

$$\left\{\tilde{\psi}_{j,0}, \tilde{\psi}_{j,1}, \cdots, \tilde{\psi}_{j,N}\right\} = \left\{\tilde{\psi}_{j,0}, U\tilde{\psi}_{j,0}, \cdots, U^N \tilde{\psi}_{j,0}\right\}$$

satisfies (3.10). We claim that

$$\mathcal{F}[\tilde{\psi}_{j,0}](n) = 2^{-\frac{j}{2}} \hat{\psi}(2^{-j} n 2\pi). \tag{3.16}$$

This is shown by a simple periodization argument:

$$\mathcal{F}[\tilde{\psi}_{j,0}](n) = \int_0^1 \left\{ 2^{\frac{j}{2}} \sum_{\ell \in \mathbb{Z}} \psi(2^j(x+\ell)) \right\} e^{-2\pi i n x} \, dx$$

$$= 2^{-\frac{j}{2}} \sum_{\ell \in \mathbb{Z}} \int_{2^j \ell}^{2^j(\ell+1)} \psi(t) e^{-2\pi i n 2^{-j} t} \, dt$$

$$= 2^{-\frac{j}{2}} \int_{-\infty}^{\infty} \psi(t) e^{-2\pi i n 2^{-j} t} \, dt = 2^{-\frac{j}{2}} \hat{\psi}(2^{-j} n 2\pi).$$

Since ψ is an orthonormal wavelet, by Proposition 1.11 of Chapter 2 we have

$$\sum_{k \in \mathbb{Z}} |\hat{\psi}(\xi + 2k\pi)|^2 = 1.$$

Putting $\xi = 2^{-j} n 2\pi$ in this equality and using (3.16) we obtain

$$1 = \sum_{k \in \mathbb{Z}} |\hat{\psi}(2^{-j} n 2\pi + 2k\pi)|^2 = 2^j \sum_{k \in \mathbb{Z}} |\mathcal{F}[\tilde{\psi}_{j,0}](n + 2^j k)|^2,$$

which is equality (3.10) for $\tilde{\psi}_{j,0}$.

The proof of the theorem is not finished yet. We still have to show that the functions $\tilde{\psi}_{j,k}$ are elements of $\mathcal{C}_j$, $j = 0, 1, \cdots$. This can be accomplished by using Theorem 3.7. Because of (3.14) it is enough to show that $\tilde{\psi}_{j,0} \in \mathcal{C}_j$. By (3.16) and (1.14) we have $\mathcal{F}[\tilde{\psi}_{j,0}](0) = 2^{-\frac{j}{2}} \hat{\psi}(0) = 0$, and

$$k^2 \mathcal{F}[\tilde{\psi}_{j,0}](k) = 2^{-\frac{j}{2}} k^2 \hat{\psi}(2^{-j} k 2\pi).$$

A simple computation, together with the fact that $\xi^2 \hat{\psi}(\xi)$ is 4π-periodic (see part (b) of Theorem 1.5), shows that $k^2 \mathcal{F}[\tilde{\psi}_{j,0}](k)$ is 2^{j+1}-periodic. Hence, by Theorem 3.8, $\tilde{\psi}_{j,0} \in \mathcal{B}_{j+1}$.

Finally, it remains to show the equality that appears in Theorem 3.7. Putting $\xi = 2^{-j} m 2\pi$, $m = 1, 2, \cdots, 2^j - 1$, in Theorem 2.19 (with $n = 1$) and using (3.16) again, we obtain

$$0 = \sum_{k \in \mathbb{Z}} \frac{\hat{\psi}(2^{-j} m 2\pi + 2k\pi)}{(2^{-j} m 2\pi + 2k\pi)^2} = \frac{2^{2j} 2^{\frac{j}{2}}}{4\pi^2} \sum_{k \in \mathbb{Z}} \frac{\mathcal{F}[\tilde{\psi}_{j,0}](m + 2^j k)}{(m + 2^j k)^2}.$$

Thus, the last condition in Theorem 3.7 is established. This completes the proof of our theorem. ∎

4.4   Orthonormal bases of periodic splines

The finite dimensional function spaces $\mathcal{B}_j \subset L^2(\mathbb{T})$ of section 4.3 were defined using linear interpolation between the values at the nodes $2^{-j}k$. All the functions in $\mathcal{B}_j$ were contained in the space $\mathcal{B}$ of 1-periodic continuous functions. New scales of spaces are obtained if we use polynomials of higher degree to do this interpolation and assume, globally, a certain degree of smoothness.

More precisely, for $m \geq 1$ and $j \geq 0$, we consider the space $\mathcal{B}_j^{(m)}$ of continuous 1-periodic functions on $\mathbb{R}$ which are in the class C^{m-1} and agree with a polynomial of degree at most m on each interval of the form $[2^{-j}(k-1), 2^{-j}k]$, $k = 1, 2, \cdots, 2^j$. Thus, we have $\mathcal{B}_j^{(1)} = \mathcal{B}_j$ and

$$\mathcal{B}_0^{(m)} \subset \mathcal{B}_1^{(m)} \subset \mathcal{B}_2^{(m)} \subset \cdots \subset \mathcal{B} \subset L^2(\mathbb{T}).$$

Observe that $\mathcal{B}_0^{(m)}$ consists of the constant functions. To see this we let $f \in \mathcal{B}_0^{(m)}$ and write

$$f(x) = a_0 + a_1 x + \cdots + a_m x^m \qquad \text{on } [0,1];$$

since $f^{(m-1)}(x) = (m-1)!\, a_{m-1} + m!\, a_m x$ we must have

$$(m-1)!\, a_{m-1} + m!\, a_m = f^{(m-1)}(1) = f^{(m-1)}(0) = (m-1)!\, a_{m-1}.$$

Thus, $a_m = 0$. The argument can be repeated to prove that $\mathcal{B}_0^{(m)}$ contains only the constant functions.

We can now prove a result that is a generalization of Theorem 3.8. It gives us a characterization of the functions that belong to $\mathcal{B}_j^{(m)}$, and allows us to find the dimension of these spaces.

THEOREM 4.1 Let $m \geq 1$, $j \geq 0$ and $f \in \mathcal{B}$. Then $f \in \mathcal{B}_j^{(m)}$ if and only if $k^{m+1} \mathcal{F}[f](k)$ is a 2^j-periodic function on $\mathbb{Z}$. In particular, $\mathcal{F}[f](2^j n) = 0$ for $n \neq 0$, $\mathcal{F}[f](0)$ is arbitrary and the dimension of $\mathcal{B}_j^{(m)}$ is 2^j.

PROOF : The case $m = 1$ is Theorem 3.8. For a general $m > 1$ we proceed by induction on m. Let us consider the case $m = 2$. If $f \in \mathcal{B}_j^{(2)}$ we can write

$$f(x) = \sum_{k \in \mathbb{Z}} \mathcal{F}[f](k)e^{2\pi ikx}.$$

Thus, $f' \in \mathcal{B}_j^{(1)}$ and

$$f'(x) = \sum_{k \in \mathbb{Z}} k\mathcal{F}[f](k)(2\pi i)e^{2\pi ikx}.$$

By Theorem 3.8, $k^2(k\mathcal{F}[f](k))$ is 2^j-periodic.

Suppose now that $f \in \mathcal{B}$ and $k^3\mathcal{F}[f](k)$ is 2^j-periodic. This implies $\mathcal{F}[f](k) = O(k^{-3})$ as $|k| \to \infty$, and allows us to define

$$h(x) = (2\pi i) \sum_{k \in \mathbb{Z}} k\mathcal{F}[f](k)e^{2\pi ikx};$$

thus, $\mathcal{F}[h](k) = (2\pi i)k\mathcal{F}[f](k)$. Therefore, $k^2\mathcal{F}[h](k) = (2\pi i)k^3\mathcal{F}[f](k)$ is 2^j-periodic. By Theorem 3.8, $h \in \mathcal{B}_j^{(1)}$. But $h = f'$ so that f has one more degree of smoothness and $f \in \mathcal{B}_j^{(2)}$. The induction proof for $m > 2$ is similar.

∎

With the aid of this theorem we can give examples of elements in $\mathcal{B}_j^{(m)}$. Let $\tilde{\chi}_j$ be the 1-periodic function whose restriction to $[0,1)$ is $2^{\frac{j}{2}}\chi_{[0,2^{-j})}$ (see the proof of Lemma 3.3). If $\tilde{\Delta}_j$ is the 1-periodic "triangle" function of section 4.3 we then have $\tilde{\Delta}_j = \tilde{\chi}_j * \tilde{\chi}_j$. For $m \in \mathbb{N}$ we define, by induction,

$$\tilde{\Delta}_j^{(m)} = \tilde{\Delta}_j^{(m-1)} * \tilde{\chi}_j, \qquad m \geq 1,$$

with $\tilde{\Delta}_j^{(0)} = \tilde{\chi}_j$. Since

$$\mathcal{F}[\tilde{\Delta}_j^{(m)}](k) = \begin{cases} 2^{-(m+1)\frac{j}{2}} & \text{if } k = 0, \\ 2^{-(m+1)\frac{j}{2}} \left(\dfrac{e^{-2\pi i2^{-j}k} - 1}{-2\pi i2^{-j}k} \right)^{m+1} & \text{if } k \neq 0. \end{cases} \qquad (4.2)$$

(see the proof of Lemma 3.3), the function $k^{m+1}\mathcal{F}[\tilde{\Delta}_j^{(m)}](k)$ is 2^j-periodic on $\mathbb{Z}$. Thus, $\tilde{\Delta}_j^{(m)} \in \mathcal{B}_j^{(m)}$, $j = 1, 2, \cdots$. It follows that the translations of this function, namely,

$$\tilde{\Delta}_{j,k}^{(m)}(x) = \tilde{\Delta}_j^{(m)}(x - 2^{-j}k), \qquad k = 0, 1, \cdots, 2^j - 1, \qquad (4.3)$$

are also elements of $\mathcal{B}_j^{(m)}$. The collection of elements in (4.3) is a basis for $\mathcal{B}_j^{(m)}$. (We prove this at the end of section 4.6.) There is, however, a simpler way for finding a basis for this space. By Theorem 4.1 the functions

$$g_{j,k}^{(m)}(x) = \begin{cases} 1 & \text{if } k = 0, \\ \displaystyle\sum_{\ell \in \mathbb{Z}} \frac{k^{m+1}}{(k + 2^j \ell)^{m+1}} e^{2\pi i (k + 2^j \ell) x} & \text{for } k = 1, \cdots, 2^j - 1, \end{cases} \quad (4.4)$$

belong to $\mathcal{B}_j^{(m)}$. They are linearly independent since they involve a disjoint collection of characters. Since there are 2^j of them, they form a basis for $\mathcal{B}_j^{(m)}$.

The functions in the spaces $\mathcal{B}_j^{(m)}$ will be called **periodic splines of order** m. We want to find an orthonormal basis for $L^2(\mathbb{T})$ whose elements are periodic splines of order m. The case $m = 1$ was done in section 4.3. For the case $m > 1$ we proceed in a similar manner.

Let $\mathcal{C}_j^{(m)}$ be the orthogonal complement of $\mathcal{B}_j^{(m)}$ in $\mathcal{B}_{j+1}^{(m)}$. Then

$$\mathcal{C}_j^{(m)} \oplus \mathcal{B}_j^{(m)} = \mathcal{B}_{j+1}^{(m)} \qquad \text{and} \qquad \dim(\mathcal{C}_j^{(m)}) = 2^j.$$

Since

$$\mathcal{B}_0^{(m)} \oplus \left(\bigoplus_{j=0}^{\infty} \mathcal{C}_j^{(m)} \right) = \overline{\mathcal{B}},$$

and $\mathcal{B}$ is dense in $L^2(\mathbb{T})$, a basis for $L^2(\mathbb{T})$ can be obtained by finding an orthonormal basis for each $\mathcal{C}_j^{(m)}$, $j = 0, 1, \cdots$. This will be done by an appropriate periodization of the spline wavelet basis of Theorem 2.17.

We begin with the characterization of the space $\mathcal{C}_j^{(m)}$. The result is similar to Theorem 3.7.

THEOREM 4.5 $f \in \mathcal{C}_j^{(m)}$ if and only if $f \in \mathcal{B}_{j+1}^{(m)}$, $\mathcal{F}[f](0) = 0$ and

$$\sum_{\ell \in \mathbb{Z}} \frac{\mathcal{F}[f](n + 2^j \ell)}{(n + 2^j \ell)^{m+1}} = 0 \qquad \text{for } n = 1, 2, \cdots, N(= 2^j - 1).$$

PROOF : A function $f \in \mathcal{C}_j^{(m)}$ if and only if $f \in \mathcal{B}_{j+1}^{(m)}$ and f is orthogonal to all the elements in $\mathcal{B}_j^{(m)}$. The last assertion is equivalent to showing that

$$\sum_{k \in \mathbb{Z}} \mathcal{F}[g_{j,n}^{(m)}](k) \overline{\mathcal{F}[f](k)} = 0 \qquad \text{for } n = 0, 1, \cdots, N,$$

since (4.4) provides us with a basis $\{g_{j,n}^{(m)} : n = 0, 1, \cdots, N\}$ for $\mathcal{B}_j^{(m)}$.

Suppose $f \in \mathcal{C}_j^{(m)}$, then the above condition is satisfied. For $n = 0$ we have $\mathcal{F}[g_{j,0}^{(m)}](k) = \delta_{k,0}$; hence,

$$0 = \sum_{k \in \mathbb{Z}} \mathcal{F}[g_{j,0}^{(m)}](k) \, \overline{\mathcal{F}[f](k)} = \mathcal{F}[f](0).$$

When $1 \leq n \leq N$,

$$\mathcal{F}[g_{j,n}^{(m)}](k) = \begin{cases} \dfrac{n^{m+1}}{(n + 2^j \ell)^{m+1}} & \text{if } k = n + 2^j \ell \text{ for some } \ell \in \mathbb{Z}, \\ 0 & \text{otherwise.} \end{cases}$$

Hence

$$0 = \sum_{k \in \mathbb{Z}} \mathcal{F}[g_{j,n}^{(m)}](k) \, \overline{\mathcal{F}[f](k)} = \sum_{\ell \in \mathbb{Z}} \frac{n^{m+1}}{(n + 2^j \ell)^{m+1}} \, \overline{\mathcal{F}[f](n + 2^j \ell)}.$$

This establishes the required result for $n = 1, 2, \cdots, N$. The converse is obtained by reversing the above argument. $\blacksquare$

We are now ready to describe the promised "wavelet-like" basis for $L^2(\mathbb{T})$ involving splines of order m. Let ψ^m be the spline wavelet of order m on the real line (see formula (2.16) and Theorem 2.17). For $j = 0, 1, 2, \cdots$ and $k = 0, 1, \cdots, 2^j - 1$, let

$$\tilde{\psi}_{j,k}^m(x) = \sum_{\ell \in \mathbb{Z}} \psi_{j,k}^m(x + \ell) \tag{4.6}$$

be the periodization of the elements $\psi_{j,k}^m(x) = 2^{\frac{j}{2}} \psi^m(2^j x - k)$ with j, k ranging as described above. Observe that

$$\tilde{\psi}_{j,k}^m(x) = 2^{\frac{j}{2}} \sum_{\ell \in \mathbb{Z}} \psi^m(2^j(x + \ell) - k)$$

and

$$\tilde{\psi}_{j,k}^m(x) = \tilde{\psi}_{j,0}^m(x - 2^{-j}k) = (U_j^k \tilde{\psi}_{j,0}^m)(x). \tag{4.7}$$

The exponential decay of ψ^m (see Theorem 2.18) shows that the series in (4.6) is uniformly convergent on $[0, 1]$. Thus, each of the functions $\tilde{\psi}_{j,k}^m$ is continuous.

We claim that these functions belong to $C_j^{(m)}$. By (4.7) it is enough to show that $\tilde{\psi}_{j,0}^m \in C_j^{(m)}$. To do this we use Theorem 4.5.

As in (3.16) the relation between the Fourier coefficients of the "periodized" function (as in (3.13) or (4.6)) and the Fourier transform of the original function is given by

$$\mathcal{F}[\tilde{\psi}_{j,0}^m](n) = 2^{-\frac{j}{2}}\widehat{\psi^m}(2^{-j}n2\pi). \qquad (4.8)$$

It follows from the definition of $\widehat{\psi^m}$ (in (2.16)) that $\xi^{m+1}\widehat{\psi^m}(\xi)$ is 4π-periodic. Thus, $k^{m+1}\mathcal{F}[\tilde{\psi}_{j,0}^m](k)$ is 2^{j+1}-periodic on $\mathbb{Z}$:

$$(k + 2^{j+1})^{m+1}\mathcal{F}[\tilde{\psi}_{j,0}^m](k + 2^{j+1})$$

$$= (k + 2^{j+1})^{m+1}2^{-\frac{j}{2}}\widehat{\psi^m}(2^{-j}(k + 2^{j+1})2\pi)$$

$$= (k + 2^{j+1})^{m+1}2^{-\frac{j}{2}}\widehat{\psi^m}(2^{-j}k2\pi + 4\pi)$$

$$= (k + 2^{j+1})^{m+1}2^{-\frac{j}{2}}\widehat{\psi^m}(2^{-j}k2\pi)\frac{(2^{-j}k2\pi)^{m+1}}{(2^{-j}k2\pi + 4\pi)^{m+1}}$$

$$= k^{m+1}\mathcal{F}[\tilde{\psi}_{j,0}^m](k)\,(k + 2^{j+1})^{m+1}\frac{1}{(k + 2^{j+1})^{m+1}} = k^{m+1}\mathcal{F}[\tilde{\psi}_{j,0}^m](k).$$

Theorem 4.1 allows us to conclude that $\tilde{\psi}_{j,0}^m \in \mathcal{B}_{j+1}^{(m)}$.

From (4.8) and the definition of $\widehat{\psi^m}$ (2.16) it follows that $\mathcal{F}[\tilde{\psi}_{j,0}^m](0) = 0$. It remains to show the equality that appears in Theorem 4.5. Putting $\xi = 2^{-j}\ell2\pi$, $\ell = 1, 2, \cdots, 2^j - 1$, in Theorem 2.19 and using (4.8) again, we obtain

$$0 = \sum_{k\in\mathbb{Z}}\frac{\widehat{\psi^m}(2^{-j}\ell2\pi + 2k\pi)}{(2^{-j}\ell2\pi + 2k\pi)^{m+1}} = \frac{2^{\frac{j}{2}}}{(2^{-j}2\pi)^{m+1}}\sum_{k\in\mathbb{Z}}\frac{\mathcal{F}[\tilde{\psi}_{j,0}^m](\ell + 2^jk)}{(\ell + 2^jk)^2}.$$

This finishes the proof of our claim: $\tilde{\psi}_{j,k}^m \in C_j^{(m)}$, $k = 0, 1, \cdots, 2^j - 1$.

THEOREM 4.9 *For each $m \in \mathbb{N}$ the system*

$$\left\{1, \tilde{\psi}_{j,k}^m : j = 0, 1, 2, \cdots, \quad k = 0, 1, \cdots, 2^j - 1\right\}$$

is an orthonormal basis for $L^2(\mathbb{T}) \approx L^2([0, 1))$.

REMARK: Such a basis will be called a basis of **periodic spline wavelets**.

PROOF : Since

$$\overline{\mathcal{B}} = \mathcal{B}_0^{(m)} \oplus \Big(\bigoplus_{j=0}^{\infty} C_j^{(m)} \Big),$$

where $\mathcal{B}$ is the space of all 1-periodic continuous functions on $\mathbb{R}$, and $\mathcal{B}$ is dense in $L^2(\mathbb{T})$, it is enough to show that the system

$$\big\{ \tilde{\psi}_{j,k}^m : k = 0, 1, \cdots, N = 2^j - 1 \big\} \tag{4.10}$$

is an orthonormal basis of $C_j^{(m)}$, for each $j = 0, 1, 2, \cdots$ (recall that $\mathcal{B}_0^{(m)}$ consists of constant functions). The argument preceeding the statement of Theorem 4.9 shows that the elements of the system (4.10) belong to $C_j^{(m)}$. It suffices to show that (4.10) is an orthonormal system. Since each element in the system (4.10) is a translate of the 1-periodic function $\tilde{\psi}_{j,0}^m$, all we need to do is to establish equality (3.10) for this $\tilde{\psi}_{j,0}^m$.

Since ψ^m is an orthonormal wavelet, Proposition 1.11 of Chapter 2 gives us

$$\sum_{k \in \mathbb{Z}} |\widehat{\psi^m}(\xi + 2k\pi)|^2 = 1.$$

Putting $\xi = 2^{-j}n2\pi$, $n = 0, 1, \cdots, 2^j - 1$, in this equality and using (4.8) we obtain

$$1 = \sum_{k \in \mathbb{Z}} |\widehat{\psi^m}(2^{-j}n2\pi + 2k\pi)|^2 = 2^j \sum_{k \in \mathbb{Z}} \big| \mathcal{F}[\tilde{\psi}_{j,0}^m](n + 2^j k) \big|^2.$$

This finishes the proof of our theorem. ∎

Splines are used to approximate functions by interpolation. It is clear that if $f \in \mathcal{B}_j^{(1)} = \mathcal{B}_j$ the values $y_k = f(2^{-j}k)$, $k = 0, 1, \cdots, 2^j - 1$, completely determine f, since this function coincides with the linear interpolation obtained by using y_k, $k = 0, 1, \cdots, 2^j - 1$, as nodes. We shall show that this interpolation property is shared by elements of $\mathcal{B}_j^{(m)}$ **only** when m is odd.

THEOREM 4.11 *Suppose that $y_0, y_1, \cdots, y_N$ are any $2^j = N + 1$ complex numbers and m is an odd positive integer. Then there exists a unique $f \in \mathcal{B}_j^{(m)}$ such that $f(2^{-j}k) = y_k$ for $k = 0, 1, \cdots, N$.*

PROOF : It suffices to show the existence of a unique $\vartheta_j \equiv \vartheta_j^{(m)}$ in $\mathcal{B}_j^{(m)}$ satisfying

$$\vartheta_j(0) = 0, \quad \vartheta_j(2^{-j}) = 1, \quad \text{and} \quad \vartheta_j(2^{-j}k) = 0 \quad \text{for} \quad k = 2, \cdots, N. \quad (4.12)$$

This is clear since, then, $\vartheta_{j,k} = \vartheta_j(x - 2^{-j}k)$, $k = 0, 1, 2, \cdots, N$, is a linearly independent system of 2^j functions in $\mathcal{B}_j^{(m)}$ and, thus, a basis. The desired unique function is

$$f = \sum_{k=0}^{N} y_k \vartheta_{j,k}.$$

We construct ϑ_j first by showing that its Fourier coefficients are uniquely determined and, then, finding them. Let $\gamma(n) = \mathcal{F}[\vartheta_j](n)$, $n \in \mathbb{Z}$, be the nth Fourier coefficient of ϑ_j. Then, since we want $\vartheta_j \in \mathcal{B}_j^{(m)}$, $n^{m+1}\gamma(n)$ has to be 2^j-periodic on $\mathbb{Z}$ (by Theorem 4.1). Hence, we can write

$$\vartheta_j(x) = \gamma(0) + \sum_{n=1}^{N} \sum_{\ell \in \mathbb{Z}} (n + 2^j\ell)^{m+1} \gamma(n + 2^j\ell) \frac{e^{2\pi i(n + 2^j\ell)x}}{(n + 2^j\ell)^{m+1}}$$

$$= \gamma(0) + \sum_{n=1}^{N} n^{m+1}\gamma(n)e^{2\pi inx} \sum_{\ell \in \mathbb{Z}} \frac{e^{2\pi i2^j\ell x}}{(n + 2^j\ell)^{m+1}}.$$

Putting $x = 2^{-j}k$, $k = 0, 1, \cdots, N$, in the above equality we obtain the system of equations

$$0 = \gamma(0) + \sum_{n=1}^{N} n^{m+1}\gamma(n)A_n^{(m)}$$

$$1 = \gamma(0) + \sum_{n=1}^{N} n^{m+1}\gamma(n)\,(\overline{\lambda})^n A_n^{(m)}$$

$$0 = \gamma(0) + \sum_{n=1}^{N} n^{m+1}\gamma(n)\,(\overline{\lambda})^{nk} A_n^{(m)}, \qquad k = 2, 3, \cdots, N,$$

where $\lambda = e^{-2\pi i2^{-j}}$ as before, and

$$A_n^{(m)} = \sum_{\ell \in \mathbb{Z}} \frac{1}{(n + 2^j\ell)^{m+1}}, \qquad n = 1, 2, \cdots, N.$$

Letting $b_n = n^{m+1}\gamma(n)A_n^{(m)}$, $n = 1, 2, \cdots, N$, and $b_0 = \gamma(0)$, the above system of $N + 1$ equations can be written matricially as

$$C_N^* \begin{pmatrix} b_0 \\ b_1 \\ b_2 \\ \vdots \\ b_N \end{pmatrix} = \begin{pmatrix} 0 \\ 1 \\ 0 \\ \vdots \\ 0 \end{pmatrix},$$

where C_N is the same matrix as the one used in the proof of Theorem 3.7 (see (3.6)). As we have seen, $\frac{1}{\sqrt{N+1}}C_N$ is unitary. Therefore, multiplying both sides by $\frac{1}{N+1}C_N = 2^{-j}C_N$ we obtain

$$\begin{pmatrix} b_0 \\ b_1 \\ b_2 \\ \vdots \\ b_N \end{pmatrix} = 2^{-j}C_N \begin{pmatrix} 0 \\ 1 \\ 0 \\ \vdots \\ 0 \end{pmatrix} = 2^{-j} \begin{pmatrix} 1 \\ \lambda \\ \lambda^2 \\ \vdots \\ \lambda^N \end{pmatrix}.$$

Thus,

$$b_n = 2^{-j}\lambda^n \qquad \text{for all } n = 0, 1, \cdots, N.$$

Since m is odd, $m + 1$ is even, and hence $A_n^{(m)} > 0$ for all $n = 1, 2, \cdots, N$. This allows us to write

$$\gamma(n) = \begin{cases} 2^{-j} & \text{if } n = 0, \\ \dfrac{2^{-j}\lambda^n}{n^{m+1}A_n^{(m)}} & \text{for } n = 1, \cdots, N. \end{cases}$$

The 2^j-periodicity of $n^{m+1}\gamma(n)$ determines all the Fourier coefficients of ϑ_j, and hence, ϑ_j is completely determined. This gives us the function ϑ_j satisfying (4.12). $\blacksquare$

The proof of the above theorem allows us to do some explicit calculations. The formula we used to prove Lemma 1.9 gives us, when $m = 1$,

$$A_n^{(1)} = \sum_{\ell \in \mathbb{Z}} \frac{1}{(n + 2^j\ell)^2} = \frac{(2^{-j}\pi)^2}{\sin^2(2^{-j}n\pi)} \qquad \text{for } n = 1, 2, \cdots, N.$$

Hence, $\gamma(0) = 2^{-j}$ and

$$\gamma(n) = 2^{-j}e^{-2\pi i2^{-j}n}\frac{\sin^2(2^{-j}n\pi)}{(2^{-j}n\pi)^2} = -\frac{2^j}{4\pi^2}\left(\frac{e^{-2\pi i2^{-j}n}-1}{n}\right)^2$$

for $n = 1, 2, \cdots, N$. This shows that the function $\vartheta \equiv \vartheta_j^{(1)}$ constructed in the proof of Theorem 4.11 coincides with the function $\tilde{\Delta}_j$ whose graph is given in Figure 4.7 (see Lemma 3.3 to obtain this result).

Lemma 1.9 gives us, for $n = 1, 2, \cdots, N$,

$$A_n^{(3)} = \sum_{\ell \in \mathbb{Z}} \frac{1}{(n + 2^j\ell)^4} = \frac{(2^{-j}\pi)^4}{\sin^4(2^{-j}n\pi)}\left(1 - \tfrac{2}{3}\sin^2(2^{-j}n\pi)\right).$$

Using this formula we can find the Fourier coefficients of the function $\vartheta_j^{(3)}$ in $\mathcal{B}_j^{(3)}$.

The property stated in Theorem 4.11 is not shared by the even order splines in $\mathcal{B}_j^{(2)}$. We can see this by computing $A_n^{(2)}$ for some $1 \le n \le N$. Differentiating once the formula

$$\sum_{n \in \mathbb{Z}} \frac{1}{(x + n)^2} = \frac{\pi^2}{\sin^2(\pi x)}$$

(see the proof of Lemma 1.9) we obtain

$$A_n^{(2)} = (2^{-j}\pi)^3 \frac{\cos(2^{-j}n\pi)}{\sin^3(2^{-j}n\pi)} \qquad n = 1, 2, \cdots, N.$$

For $n = 2^{j-1}$ we obtain $A_{2^{j-1}}^{(2)} = 0$. This forces $b_{2^{j-1}}$ to be 0, where b_n is as in the proof of Theorem 4.11. But we have shown in this theorem that $b_n = 2^{-j}\lambda^n$ with $\lambda = e^{-2\pi i2^{-j}}$ and, thus, they cannot be zero. This shows that the interpolation property is not shared by the elements in $\mathcal{B}_j^{(2)}$. This is a general fact: **the interpolation property fails when m is even.** To see this observe that

$$A_{2^{j-1}}^{(m)} = \sum_{\ell \in \mathbb{Z}} \frac{1}{(2^{j-1} + 2^j\ell)^{m+1}}$$

$$= 2^{-(j-1)(m+1)}\left\{\sum_{\ell=0}^{\infty} \frac{1}{(1 + 2\ell)^{m+1}} + \sum_{\ell=1}^{\infty} \frac{1}{(1 - 2\ell)^{m+1}}\right\}$$

$$= 2^{-(j-1)(m+1)} \left\{ \sum_{\ell=0}^{\infty} \frac{1}{(1+2\ell)^{m+1}} + (-1)^{m+1} \sum_{k=0}^{\infty} \frac{1}{(1+2k)^{m+1}} \right\} = 0$$

when m is even. We notice that for $m = 2$ it is an easy exercise of algebra to show that there is no function $\vartheta_j^{(2)}$ in $\mathcal{B}_j^{(2)}$ that would satisfy condition (4.12) in the proof of Theorem 4.11.

We now consider the case $m = 3$. We shall show that, in this case, the unique function f whose existence is found in Theorem 4.11 is also the one that minimizes a certain functional.

Let $\mathcal{H}^2$ be the Sobolev space of 1-periodic functions with two derivatives in $L^2([0,1))$. Observe that $\mathcal{B}_j^{(3)} \subset \mathcal{H}^2$. On $\mathcal{H}^2$ we consider the functional

$$T(f) = \int_0^1 \left| f''(x) \right|^2 dx. \tag{4.13}$$

THEOREM 4.14 *Suppose that $y_0, y_1, \cdots, y_N$ are any $2^j = N + 1$ complex numbers. Among all $f \in \mathcal{H}^2$ satisfying the specific interpolation property*

$$f(2^{-j}k) = y_k, \qquad k = 0, 1, \cdots, N,$$

the unique element of $\mathcal{B}_j^{(3)}$ satisfying this property minimizes the functional T given by (4.13).

PROOF : Let $\tilde{\Delta}_{j,k}$, $k = 0, 1, \cdots, N$ be the "triangle" functions given at the beginning of section 4.3 (see Figure 4.7), which form a basis of $\mathcal{B}_j^{(1)}$. Suppose that a function $g \in \mathcal{H}^2$ satisfies $g(2^{-j}k) = y_k$ for $k = 0, 1, \cdots, N$. Then

$$\langle g'', \tilde{\Delta}_{j,k} \rangle = \int_0^1 g''(x) \overline{\tilde{\Delta}_{j,k}(x)} \, dx = \int_{2^{-j}k}^{2^{-j}(k+2)} g''(x) \overline{\tilde{\Delta}_{j,k}(x)} \, dx$$

$$= 2^j (y_{k+2} - 2y_{k+1} + y_k)$$

for $k = 0, 1, \cdots, 2^j - 1$ (we put $y_{2^j} = y_0$ and $y_{2^j+1} = y_1$). The last formula can be proved using integration by parts. Since $\{\tilde{\Delta}_{j,k} : k = 0, 1, \cdots, N\}$ is a basis for $\mathcal{B}_j^{(1)}$, this shows that $f'' - g''$ is orthogonal to $\mathcal{B}_j^{(1)}$ for all $f \in \mathcal{H}^2$ satisfying the specific interpolation property. We can restate this as

$$P(f'') = P(g''),$$

where P is the orthogonal projection of L^2 onto $\mathcal{B}_j^{(1)}$. If f is the unique function in $\mathcal{B}_j^{(3)}$ satisfying the specific interpolation property we then have $f'' \in \mathcal{B}_j^{(1)}$ and, hence, $P(f'') = f''$. It follows that

$$\int_0^1 |f''(x)|^2 \, dx = \|f''\|_{L^2(\mathbb{T})}^2 = \|P(f'')\|_{L^2(\mathbb{T})}^2$$

$$= \|P(g'')\|_{L^2(\mathbb{T})}^2 \le \|g''\|_{L^2(\mathbb{T})}^2 = \int_0^1 |g''(x)|^2 \, dx$$

so that the functional $\mathcal{T}$ of (4.13) is minimized by $f \in \mathcal{B}_j^{(3)}$.

∎

4.5   Periodization of wavelets defined on the real line

In sections 4.3 and 4.4 we have constructed periodic Franklin wavelets, as well as more general spline wavelets, by appropriate periodizations of the corresponding wavelets on the real line. We now generalize this procedure to other wavelets.

Throughout this section we shall assume that ψ is an orthonormal wavelet that arises from an MRA with scaling function φ. We shall also assume that both ψ and φ are in $L^1(\mathbb{R})$. This allows us to define, for each $j, k \in \mathbb{Z}$,

$$\widetilde{\varphi_{j,k}}(x) \equiv \sum_{\ell \in \mathbb{Z}} \varphi_{j,k}(x + \ell) = 2^{\frac{j}{2}} \sum_{\ell \in \mathbb{Z}} \varphi(2^j(x + \ell) - k) \qquad (5.1)$$

and

$$\widetilde{\psi_{j,k}}(x) \equiv \sum_{\ell \in \mathbb{Z}} \psi_{j,k}(x + \ell) = 2^{\frac{j}{2}} \sum_{\ell \in \mathbb{Z}} \psi(2^j(x + \ell) - k), \qquad (5.2)$$

which are 1-periodic functions belonging to $L^1([0, 1])$.

It is easy to see that if $j \le 0$, $\widetilde{\varphi_{j,k}} = \widetilde{\varphi_{j,0}}$ and $\widetilde{\psi_{j,k}} = \widetilde{\psi_{j,0}}$ for all $k \in \mathbb{Z}$. We prove this only for the scaling function, since the proof is the same for the wavelet. The main ingredient is the fact that $2^{-j}k \in \mathbb{Z}$ when $j, k \in \mathbb{Z}$ and $j \le 0$:

$$\widetilde{\varphi_{j,k}}(x) = 2^{\frac{j}{2}} \sum_{\ell \in \mathbb{Z}} \varphi(2^j(x + \ell) - k) = 2^{\frac{j}{2}} \sum_{\ell \in \mathbb{Z}} \varphi(2^j(x + \ell - 2^{-j}k))$$

$$= 2^{\frac{j}{2}} \sum_{m \in \mathbb{Z}} \varphi(2^j(x+m)) = \widetilde{\varphi_{j,0}}(x).$$

It turns out that when $j \leq 0$ all these functions are constants. This is proved by computing the Fourier coefficients of $\widetilde{\varphi_{j,0}}$ for $j \leq 0$. Recall that

$$\mathcal{F}[f](k) = \int_0^1 f(x)e^{-2\pi ikx}\, dx.$$

For $j, k \in \mathbb{Z}$, a similar calculation as the one that proves (3.16) gives us

$$\mathcal{F}[\widetilde{\varphi_{j,0}}](k) = 2^{-\frac{j}{2}}\hat{\varphi}(2^{-j}2k\pi), \qquad j,k \in \mathbb{Z}. \tag{5.3}$$

Observe that $\hat{\varphi}$ is continuous since $\varphi \in L^1(\mathbb{R})$, and, hence, it makes sense to write the value of $\hat{\varphi}$ at a particular point $2^{-j}2k\pi$. By Proposition 2.17 of Chapter 2, $\hat{\varphi}(2\ell\pi) = 0$ for all $\ell \neq 0$, $\ell \in \mathbb{Z}$. From (5.3) we deduce that, when $j \leq 0$, all the Fourier coefficients of $\widetilde{\varphi_{j,0}}$ are zero except the one corresponding to $k = 0$. This shows that the functions $\widetilde{\varphi_{j,0}}$ are constants for all $j \leq 0$.

We claim that when $j \leq -1$, all the functions $\widetilde{\psi_{j,0}}$ are zero. A calculation similar to the one that shows (3.16) gives us

$$\mathcal{F}[\widetilde{\psi_{j,0}}](k) = 2^{-\frac{j}{2}}\hat{\psi}(2^{-j}2k\pi), \qquad j,k \in \mathbb{Z}. \tag{5.4}$$

By Proposition 2.17 of Chapter 2, $\hat{\psi}(4\ell\pi) = 0$ for all $\ell \in \mathbb{Z}$. Hence, all the Fourier coefficients of $\widetilde{\psi_{j,0}}$ are zero when $j \leq -1$. This proves the claim.

For $j > 0$, many of the functions in (5.1) coincide; the same is true for the functions in (5.2). We claim that if $j > 0$, $m \in \mathbb{Z}$ and $0 \leq k \leq 2^j - 1$

$$(\varphi_{j,2^j m+k})^\sim = \widetilde{\varphi_{j,k}} \qquad \text{and} \qquad (\psi_{j,2^j m+k})^\sim = \widetilde{\psi_{j,k}}.$$

We write the proof for the wavelet, since the case of the scaling function is proved similarly. Let $k' = 2^j m + k$; then

$$\widetilde{\psi_{j,k'}}(x) = 2^{\frac{j}{2}} \sum_{\ell \in \mathbb{Z}} \psi(2^j(x+\ell-m)-k)$$

$$= 2^{\frac{j}{2}} \sum_{n \in \mathbb{Z}} \psi(2^j(x+n)-k) = \widetilde{\psi_{j,k}}(x).$$

We summarize these results as follows:

LEMMA 5.5 *Let ψ be an orthonormal wavelet that arises from an MRA with a scaling function φ, and suppose that both ψ and φ are elements of $L^1(\mathbb{R})$. If $\widetilde{\varphi_{j,k}}$ and $\widetilde{\psi_{j,k}}$ are defined by (5.1) and (5.2), respectively, we have*

    (a) $\widetilde{\varphi_{j,k}} = \widetilde{\varphi_{j,0}}$ *for all $j \le 0$, $k \in \mathbb{Z}$; moreover $\widetilde{\varphi_{j,0}}$ is constant for $j \le 0$.*

    (b) $\widetilde{\psi_{j,k}} \equiv 0$ *for all $j \le -1$ and $k \in \mathbb{Z}$.*

    (c)   *For $j \ge 0$, $0 \le k \le 2^j - 1$ and $m \in \mathbb{Z}$,*

$$(\varphi_{j,2^j m+k})^{\sim} = \widetilde{\varphi_{j,k}} \qquad and \qquad (\psi_{j,2^j m+k})^{\sim} = \widetilde{\psi_{j,k}}.$$

We have shown that each of the collections of functions given by (5.1) and (5.2) for fixed non-negative j consists of, at most, 2^j distinct functions $\widetilde{\varphi_{j,k}}$ and $\widetilde{\psi_{j,k}}$. Some properties of these functions are inherited from the corresponding properties for the collections $\{\varphi_{j,k}\}$ and $\{\psi_{j,k}\}$, $j, k \in \mathbb{Z}$.

THEOREM 5.6 *For $j \ge 0$,*

    (a)   *the set $\{\widetilde{\varphi_{j,k}} : 0 \le k \le 2^j - 1\}$ is an orthonormal system in $L^2(\mathbb{T})$;*

    (b)   *the set $\{\widetilde{\psi_{j,k}} : 0 \le k \le 2^j - 1\}$ is an orthonormal system in $L^2(\mathbb{T})$;*

    (c)   *$\langle \widetilde{\psi_{j,k}}, \widetilde{\varphi_{j,\ell}} \rangle = 0$ for all $k, \ell \in \mathbb{Z}$ such that $0 \le k, \ell \le 2^j - 1$.*

PROOF : Let us prove (a). Let $k, m \in \mathbb{Z}$ be such that $0 \le k \le 2^j - 1$ and $0 \le m \le 2^j - 1$. Then

$$\langle \widetilde{\varphi_{j,k}}, \widetilde{\varphi_{j,m}} \rangle = \sum_{\ell \in \mathbb{Z}} \int_0^1 \widetilde{\varphi_{j,k}}(x) \, \overline{\varphi_{j,m}(x + \ell)} \, dx$$

$$= \sum_{\ell \in \mathbb{Z}} \int_\ell^{\ell+1} \widetilde{\varphi_{j,k}}(y - \ell) \, \overline{\varphi_{j,m}(y)} \, dy$$

$$= \int_{\mathbb{R}} \widetilde{\varphi_{j,k}}(y) \, \overline{\varphi_{j,m}(y)} \, dy,$$

where the last equality is due to the fact that $\widetilde{\varphi_{j,k}}$ is 1-periodic. Since $j \ge 0$, we can write

$$\widetilde{\varphi_{j,k}}(y) = \sum_{\ell \in \mathbb{Z}} \varphi_{j,k-2^j \ell}(y).$$

This gives

$$\langle \widetilde{\varphi_{j,k}}, \widetilde{\varphi_{j,m}} \rangle = \sum_{\ell \in \mathbb{Z}} \langle \varphi_{j,k-2^j\ell}, \varphi_{j,m} \rangle = \sum_{\ell \in \mathbb{Z}} \delta_{k-2^j\ell, m} .$$

Since $0 \le k, m \le 2^j - 1$, $k - 2^j\ell$ coincides with m only when $\ell = 0$ and, then, $k = m$. Hence $\langle \widetilde{\varphi_{j,k}}, \widetilde{\varphi_{j,m}} \rangle = \delta_{k,m}$, as we wanted to show.

The proofs for (b) and (c) are similar.

∎

REMARK: Parts (a) and (b) of Theorem 5.6 can also be proved by using Theorem 3.9. Observe that for $j \ge 0$ fixed, $\widetilde{\varphi_{j,k}}$ is a translation of $\widetilde{\varphi_{j,0}}$ by $2^{-j}k$, so that in the notation of Theorem 3.9, $\widetilde{\varphi_{j,k}} = U\widetilde{\varphi_{j,0}}$, $0 \le k \le 2^j - 1$.

We want to construct an orthonormal basis for $L^2(\mathbb{T})$ associated with an MRA orthonormal wavelet. For the wavelet ψ and the scaling function φ we shall assume $|\psi(x)| \le R_1(|x|)$ and $|\varphi(x)| \le R_2(|x|)$, where R_1 and R_2 are bounded decreasing functions belonging to $L^1([0, \infty))$. We shall call these kind of functions **radial decreasing L^1-majorants** of ψ and φ, respectively (see also (3.1) of Chapter 5).

For $j = 0, 1, 2, \cdots$, we define $\widetilde{V}_j$ as the subspace of $L^2(\mathbb{T})$ generated by $\{\widetilde{\varphi_{j,k}} : k = 0, 1, \cdots, 2^j - 1\}$; similarly, we define $\widetilde{W}_j$ as the subspace of $L^2(\mathbb{T})$ generated by $\{\widetilde{\psi_{j,k}} : k = 0, 1, \cdots, 2^j - 1\}$. Obviously, $\widetilde{V}_j$ and $\widetilde{W}_j$ have dimension 2^j. Some properties of these subspaces are inherited from the corresponding properties of the subspaces V_j and W_j associated with the MRA we are considering.

THEOREM 5.7 *For $j = 0, 1, 2, \cdots$, we have the following inclusions:*

$$\widetilde{V}_j \subset \widetilde{V}_{j+1} \qquad and \qquad \widetilde{W}_j \subset \widetilde{V}_{j+1}.$$

Moreover,

$$\widetilde{V}_j \oplus \widetilde{W}_j = \widetilde{V}_{j+1}.$$

PROOF : We only prove the inclusion $\widetilde{W}_j \subset \widetilde{V}_{j+1}$. The proof of the other inclusion is similar. Once these results are proved, the equality stated in the theorem follows from part (c) of Theorem 5.6 and the fact that $\dim \widetilde{V}_j = 2^j = \dim \widetilde{W}_j$.

Let

$$\widetilde{\psi_{j,k}}(x) = \sum_{\ell \in \mathbb{Z}} \psi_{j,k}(x + \ell)$$

be one of the functions that generate $\widetilde{W}_j$. Since $\psi_{j,k} \in V_{j+1}$ we can write

$$\psi_{j,k}(x) = \sum_{s \in \mathbb{Z}} \alpha_s^{(j,k)} \varphi_{j+1,s}(x).$$

The conditions imposed on ψ and φ imply that the sequence of coefficients $\{\alpha_s^{(j,k)} : s \in \mathbb{Z}\}$ belongs to $\ell^1(\mathbb{Z})$:

$$\sum_{s \in \mathbb{Z}} |\alpha_s^{(j,k)}| = \sum_{s \in \mathbb{Z}} \left| \int_{\mathbb{R}} \psi_{j,k}(x) \overline{\varphi_{j+1,s}(x)} \, dx \right|$$

$$\leq \|\psi_{j,k}\|_{L^1} \sup_{x \in \mathbb{R}} \sum_{s \in \mathbb{Z}} |\varphi_{j+1,s}(x)| \leq c \|\psi_{j,k}\|_{L^1} \|R_2\|_{L^1([0,\infty))}.$$

Hence,

$$\widetilde{\psi_{j,k}}(x) = \sum_{\ell \in \mathbb{Z}} \sum_{s \in \mathbb{Z}} \alpha_s^{(j,k)} \varphi_{j+1,s}(x + \ell) = \sum_{s \in \mathbb{Z}} \alpha_s^{(j,k)} (\varphi_{j+1,s})^{\sim}(x).$$

Writing $s = n + 2^{j+1}m$, $m \in \mathbb{Z}$, $s = 0, 1, \cdots, 2^{j+1} - 1$ and observing that $(\varphi_{j+1,n+2^j m})^{\sim}(x) = (\varphi_{j+1,n})^{\sim}(x)$ (part (c) of Lemma 5.5), we obtain

$$\widetilde{\psi_{j,k}}(x) = \sum_{n=0}^{2^{j+1}-1} \left\{ \sum_{s \in \mathbb{Z}} \alpha_{n+2^{j+1}m}^{(j,k)} \right\} (\varphi_{j+1,n})^{\sim}(x) \equiv \sum_{n=0}^{2^{j+1}-1} c_n^{(j,k)} (\varphi_{j+1,n})^{\sim}(x).$$

This proves that $\widetilde{\psi_{j,k}} \in \widetilde{V}_{j+1}$ and establishes the desired inclusion. ∎

We have constructed an increasing sequence

$$\widetilde{V}_0 \subset \widetilde{V}_1 \subset \widetilde{V}_2 \subset \cdots$$

of subspaces of $L^2(\mathbb{T})$. To find an orthonormal basis we want to show that

$$\bigcup_{j=0}^{\infty} \widetilde{V}_j$$

is dense in $L^2(\mathbb{T})$. Towards this end we need a result that will be presented in Corollary 3.18 of Chapter 5. This corollary deals with the norm convergence of the projection operators P_j from $L^2(\mathbb{R})$ to V_j (see (3.2) of Chapter 5 for the definition) and requires the scaling function φ to have a radial decreasing L^1-majorant (see the definition in (3.1) of Chapter 5). The reader might find it helpful to look at section 5.3 before continuing with this one.

THEOREM 5.8 *Let ψ be an MRA orthonormal wavelet with scaling function φ such that both ψ and φ have bounded decreasing L^1-majorants. Then, the linear space*

$$\bigcup_{j=0}^{\infty} \tilde{V}_j$$

is dense in the space $L^2(\mathbb{T})$ of all 2π-periodic functions that are square integrable on $\mathbb{T}$.

PROOF : By Corollary 3.18 of Chapter 5,

$$\lim_{j \to \infty} \left\| g - P_j(g) \right\|_{L^\infty} = 0$$

for all bounded uniformly continuous g, where P_j is the orthogonal projection from $L^2(\mathbb{R})$ onto V_j (see (3.2) of Chapter 5). P_j has a natural extension to the bounded functions on $\mathbb{R}$. Thus, if we take $f \in \tilde{V}_j$ we only need to show that $P_j(f)$ belongs to $\tilde{V}_j$ (observe that $\tilde{V}_j \subset L^\infty(\mathbb{R})$ for $j \geq 0$). Since f is 1-periodic,

$$\langle f, \varphi_{j,k+2^j \ell} \rangle = \int_{\mathbb{R}} f(x) \overline{\varphi_{j,k+2^j \ell}(x)} \, dx = \int_{\mathbb{R}} f(y+\ell) \overline{\varphi_{j,k}(y)} \, dy$$

$$= \int_{\mathbb{R}} f(y) \overline{\varphi_{j,k}(y)} \, dy = \langle f, \varphi_{j,k} \rangle.$$

Thus,

$$P_j f(x) = \sum_{k \in \mathbb{Z}} \langle f, \varphi_{j,k} \rangle \varphi_{j,k}(x) = \sum_{n=0}^{2^j - 1} \sum_{\ell \in \mathbb{Z}} \langle f, \varphi_{j,n+2^j \ell} \rangle \varphi_{j,n+2^j \ell}(x)$$

$$= \sum_{n=0}^{2^j - 1} \langle f, \varphi_{j,n} \rangle \sum_{\ell \in \mathbb{Z}} \varphi_{j,n+2^j \ell}(x) = \sum_{n=0}^{2^j - 1} \langle f, \varphi_{j,n} \rangle \widetilde{\varphi_{j,n}}(x).$$

∎

We have all the ingredients needed to construct orthonormal "wavelet" basis for $L^2(\mathbb{T})$.

THEOREM 5.9 *Let ψ be an MRA orthonormal wavelet with scaling function φ such that both ψ and φ have bounded decreasing L^1-majorants. Then the system*

$$\left\{1, \widetilde{\psi_{j,k}} : j = 0, 1, \cdots, \ k = 0, 1, \cdots, 2^j - 1\right\}$$

is an orthonormal basis of $L^2(\mathbb{T})$, where

$$\widetilde{\psi_{j,k}}(x) = \sum_{\ell \in \mathbb{Z}} \psi_{j,k}(x + \ell).$$

This basis is called the **basis of periodic wavelets** *for $L^2(\mathbb{T})$ associated with the wavelet ψ.*

PROOF :   Theorems 5.8 and 5.7 show that

$$L^2(\mathbb{T}) = \widetilde{V}_0 \oplus \left(\bigoplus_{j=0}^{\infty} \widetilde{W}_j\right).$$

The result follows from Theorem 5.6.   ∎

We can apply the procedure described above to the Franklin and spline wavelets of sections 4.1 and 4.2. We obtain the periodized Franklin and spline wavelets of sections 4.3 and 4.4.

Let us apply this procedure to a Lemarié-Meyer wavelet of the form $\hat{\psi}(\xi) = e^{i\frac{\xi}{2}} b(\xi)$ (see Corollary 4.7 of Chapter 1) where $b \in C^\infty$ is an even bell function (as in Theorem 4.5 of Chapter 1). We obtain that the $\widetilde{\psi_{j,k}}$ are trigonometric polynomials. More precisely, for $j = 0$ and $k = 0$, (5.4) gives us

$$\mathcal{F}[\widetilde{\psi_{0,0}}](\ell) = \hat{\psi}(2\ell\pi) = \begin{cases} -\dfrac{1}{\sqrt{2}} & \text{if } \ell = 1 \text{ or } -1, \\[2mm] 0 & \text{otherwise.} \end{cases}$$

Hence,

$$\widetilde{\psi_{0,0}}(x) = -\frac{e^{-2\pi i x} + e^{2\pi i x}}{\sqrt{2}} = -\sqrt{2}\,\cos(2\pi x).$$

In general, if $j \geq 1$, (5.4) gives us

$$\mathcal{F}[\widetilde{\psi_{j,0}}](\ell) = 2^{-\frac{j}{2}} \hat{\psi}(2\pi 2^{-j}\ell) = 2^{-\frac{j}{2}} e^{\pi i 2^{-j}\ell} b(2\pi 2^{-j}\ell).$$

Hence, $\mathcal{F}[\widetilde{\psi_{j,0}}](\ell) = 0$ when $2\pi 2^{-j}\ell \notin \operatorname{supp}(b)$; that is, $|2\pi 2^{-j}\ell| \leq \pi - \varepsilon$ or $|2\pi 2^{-j}\ell| \geq 2\pi + 2\varepsilon$. This shows that $\widetilde{\psi_{j,k}}(x)$ is a finite trigonometric polynomial for these Lemarié-Meyer wavelets.

4.6   Notes and references

1. The Haar wavelet appeared in 1910 ([Haa]). In 1928, Ph. Franklin ([Fra]) introduced a set of continuous orthogonal functions, generalizing the Haar functions, and proved that they form a basis for $L^2([0,1))$. Around 1980, P. Wojtaszczyk ([Wo1]) was able to show that this basis is an unconditional basis for $H^1([0,1))$ and J.O. Strömberg ([Str]) introduced modified Franklin systems with higher degree of smoothness, proving that they are unconditional bases for Hardy spaces over $\mathbb{R}^n$. The subject of unconditionality of wavelet basis will be treated in Chapter 5.

2. Splines have been used for several years in the theory of approximation of functions. The reader is referred to the books of I.J. Schoenberg ([Sch]), C. de Boor ([Boo]), and G. Nürnberger ([Nür]) for the theory and applications of this subject. The connection between wavelets and splines was first made explicit by Y. Meyer ([Me3]); most of the material presented in sections 4.3 and 4.4 has, in fact, appeared in [Me4]. The construction of spline wavelets in $\mathbb{R}$ presented in sections 4.1 and 4.2 is essentially due to G. Battle ([Bat1]) (see also note 3 in this section). There is another way of constructing these wavelets presented by P.G. Lemarié in [Le4]. A comprehensive treatment of splines, spline wavelets and applications can be found in [Chu]. A more detailed study of the graph of the Franklin wavelet is presented in [Ber].

3. The way we constructed the Franklin and spline wavelets in sections 4.1 and 4.2 is related to the notion of an MRA with a Riesz basis, as was introduced after the definition of an MRA in section 2.1. Let

$$V_0 = \overline{\operatorname{span}\{\Delta^n(\cdot - k) : k \in \mathbb{Z}\}},$$

where $\Delta^1 = \chi_{[0,1]} * \chi_{[0,1]}$ and Δ^n is defined inductively in (2.1). For $j \in \mathbb{Z}$, $j \neq 0$, define

$$V_j = \{f \in L^2(\mathbb{R}) : f_{-j,0} \in V_0\},$$

where $f_{-j,0}(x) = 2^{-\frac{j}{2}} f(2^{-j}x)$. Clearly, (1.2) of the definition of an MRA (section 2.1) is satisfied. With the notation of Lemma 1.8 of Chapter 2 and using Proposition 2.7 we obtain

$$\sigma_{\Delta^n}(f) = \left(\sum_{k \in \mathbb{Z}} |\widehat{\Delta^n}(\xi + 2k\pi)|^2\right)^{\frac{1}{2}}$$

$$= \left|2\sin(\tfrac{1}{2}\xi)\right|^{n+1} \left(\sum_{k \in \mathbb{Z}} \frac{1}{(\xi + 2k\pi)^{2(n+1)}}\right)^{\frac{1}{2}} = \sqrt{P_{2n+1}(\tfrac{1}{2}\xi)},$$

since $P_{2n+1}(\xi) > 0$ for all $\xi \in \mathbb{R}$. This last fact and the π-periodicity of P_{2n+1} allows us to conclude $\sqrt{A_n} \leq \sigma_{\Delta^n}(\xi) \leq \sqrt{B_n}$, $0 < A_n \leq B_n < \infty$, for all $\xi \in \mathbb{R}$. Proceeding now as in the argument that follows Lemma 1.8 of Chapter 2 we conclude that $\{\gamma^n(\cdot - k) : k \in \mathbb{Z}\}$ is an orthonormal basis for V_0, where γ^n is defined by

$$\widehat{\gamma^n} = \widehat{\Delta^n}/\sigma_{\Delta^n}.$$

This establishes (1.5) of section 2.1. To verify (1.1) in the definition of an MRA (see Chapter 2), it is enough to show $V_0 \subset V_1$ (sometimes referred to as the **two scale equation**). This will be done by exhibiting a remarkable relation for the basic splines, namely

$$\Delta^n(x) = \sum_{k=0}^{n+1} 2^{-n} \binom{n+1}{k} \Delta^n(2x - k). \tag{6.1}$$

To see this, we use Lemma 2.3 to obtain

$$\frac{\widehat{\Delta^n}(\xi)}{\widehat{\Delta^n}(\tfrac{1}{2}\xi)} = e^{-i\frac{1}{4}(n+1)\xi} \left(\cos(\tfrac{1}{4}\xi)\right)^{n+1} = \left(\frac{1 + e^{-i\frac{\xi}{2}}}{2}\right)^{n+1},$$

so that

$$\widehat{\Delta^n}(\xi) = \frac{1}{2^{n+1}} \sum_{k=0}^{n+1} \binom{n+1}{k} e^{-ik\frac{\xi}{2}} \widehat{\Delta^n}(\tfrac{1}{2}\xi).$$

The proof of (6.1) is finished by taking the inverse Fourier transform. The results proved in section 2.1 show that $\{V_j : j \in \mathbb{Z}\}$ is an MRA and the scaling function is given by γ^n, where

$$\widehat{\gamma^n}(\xi) = \frac{\widehat{\Delta^n}(\xi)}{\sigma_{\Delta^n}(\xi)},$$

which coincides (except for the factor $e^{-i\frac{1}{2}(n+1)\xi}$ if n is odd and $e^{-i\frac{n}{2}\xi}$ if n is even) with $\widehat{\varphi^n}$ given by (2.12).

4. Let us make some observations about the optimal values of A_n and B_n for which $\sqrt{A_n} \leq \sigma_{\Delta^n}(\xi) \leq \sqrt{B_n}$ (see the above paragraph). The optimal B_n is always 1. The best value of A_n is related to the roots of the so-called "Euler-Frobenius polynomials" (see [Chu] for details). From the formula for the polynomial P_3 given in the equality preceding (2.10) and the fact that $[\sigma_{\Delta^n}(\xi)]^2 = P_{2n+1}(\xi/2)$, proved in note 3, we obtain $A_1 = \frac{1}{3}$.

5. Another approach for finding spline wavelets is given in [Le4]. A spline wavelet ψ of order n must belong to $L^2(\mathbb{R}) \cap C^{n-1}(\mathbb{R})$ and is a polynomial of degree at most n on each interval of the form $(\frac{1}{2}k, \frac{1}{2}(k+1))$. An argument similar to the one given in the remark that follows (2.4) leads us to consider

$$\hat{\psi}(\xi) = \xi^{-(n+1)} \mathcal{A}_n(\xi), \tag{6.2}$$

where $\mathcal{A}_n(\xi)$ is a 4π-periodic function. To determine $|\mathcal{A}_n(\xi)|$ we can use equations (1.1) and (1.2) of Chapter 3, that characterize the orthonormality of the system $\{\psi_{j,k} : j, k \in \mathbb{Z}\}$. Writing

$$M_n(\xi) = \sum_{k \in \mathbb{Z}} \frac{1}{|\xi + 4k\pi|^{2(n+1)}}$$

(see Proposition 2.7), equations (1.1) and (1.2) of Chapter 3 here are equivalent to

$$M_n(\xi)\big|\mathcal{A}_n(\xi)\big|^2 + M_n(\xi + 2\pi)\big|\mathcal{A}_n(\xi + 2\pi)\big|^2 = 1 \qquad \text{for } a.e. \ \xi \in \mathbb{R} \tag{6.3}$$

and

$$M_n(\xi)\mathcal{A}_n(\xi) + M_n(\xi + 2\pi)\mathcal{A}_n(\xi + 2\pi) = 0 \qquad \text{for } a.e. \ \xi \in \mathbb{R}, \tag{6.4}$$

respectively. Solving (6.4) for $\mathcal{A}_n(\xi + 2\pi)$ and replacing this value in (6.3) one obtains

$$|\mathcal{A}_n(\xi)| = \frac{\sqrt{M_n(\xi + 2\pi)}}{\sqrt{M_n(\xi)}} \cdot \frac{1}{\sqrt{M_n(\xi) + M_n(\xi + 2\pi)}}.$$

Guided by the general form of a wavelet arising from an MRA (see Proposition 2.13 of Chapter 2), one is led to choose

$$A_n(\xi) = \nu(\xi)e^{i\frac{\xi}{2}} \frac{\sqrt{M_n(\xi + 2\pi)}}{\sqrt{M_n(\xi)}} \cdot \frac{1}{\sqrt{M_n(\xi) + M_n(\xi + 2\pi)}},$$

where $\nu(\xi)$ is a 2π-periodic function such that $|\nu(\xi)| = 1$ for a.e. $\xi \in \mathbb{T}$. Reversing the above argument we easily see that the system

$$\{\psi_{j,k} : j, k \in \mathbb{Z}\},$$

with this choice of ψ, is orthonormal. In particular $\|\psi\|_2 = 1$.

To prove that ψ is an orthonormal wavelet, Theorem 1.1 of Chapter 7 can be used and, hence, all that we need is to show that (1.2) and (1.3) of Chapter 7 are satisfied. In this setting, these two equations are equivalent to

$$\sum_{j \in \mathbb{Z}} 4^{-j(n+1)} |A_n(2^j \xi)|^2 = |\xi|^{2(n+1)} \qquad \text{for a.e. } \xi \in \mathbb{R} \qquad (6.5)$$

and

$$\sum_{j=1}^{\infty} 4^{-j(n+1)} |A_n(2^j \xi)|^2 = -A_n(\xi) \overline{A_n(\xi + 2\pi)} \qquad \text{for a.e. } \xi \in \mathbb{R}, \qquad (6.6)$$

respectively. Equality (6.6) follows by a remarkable equation satisfied by M_n, namely,

$$M_n(\xi) + M_n(\xi + 2\pi) = 4^{n+1} M_n(2\xi). \qquad (6.7)$$

This equation can be proved by starting with the expression for $M_n(2\xi)$ in a series form and summing over the even and odd integers separately. Using (6.7) it can be shown that both the left-hand and the right-hand sides of (6.6) coincide with

$$\frac{1}{M_n(\xi) + M_n(\xi + 2\pi)}.$$

In computing the right-hand side of (6.6) the factor $e^{i\frac{\xi}{2}}$ that appears in the definition of $A_n(\xi)$ is essential. Equation (6.5) now follows from (6.6) and the fact that

$$4^{(n+1)\pi} M_n(2\xi) \sim \xi^{-2(n+1)}$$

in a neighborhood of the origin (to prove this fact use Proposition 2.7 of this chapter).

It can be seen that the wavelet ψ just constructed coincides with the wavelet ψ^n given by (2.16), except for the factor $\nu(\xi)$ when n is odd and $-i\nu(\xi)$ when n is even. All these wavelets arise from an MRA. This is proved directly in [Le4] by exhibiting the corresponding scaling functions. It also follows from Corollary 3.16 of Chapter 7 (observe that ψ has exponential decay at infinity). Another way to prove this result is to show that

$$D_\psi(\xi) = 1 \qquad \text{for a.e. } \xi \in \mathbb{R}$$

(see Theorem 3.2 of Chapter 7). The equality $D_\psi(\xi) = 1$ for a.e. $\xi \in \mathbb{R}$ follows, in this setting, from (6.6) and (6.7).

We have shown that the spline wavelets of order n, ψ^n, have exponential decay at infinity. That is, $|\psi^n(x)| \leq c e^{-\alpha(n)|x|}$. The exponent of decay $\alpha(n)$ depends on n. It is shown in [Le4] that

$$\alpha(n) \sim \frac{\pi^2}{4n} \qquad \text{as } n \to \infty.$$

Another amusing property (also proved in [Le4]) is that for a.e. $\xi \in \mathbb{R}$

$$\lim_{n\to\infty} \beta(n)\widehat{\psi^n}(\xi) = e^{i\frac{\xi}{2}} \chi_{[-2\pi,\pi]\cup[\pi,2\pi]}(\xi),$$

where $\beta(n) = 1$ if n is odd and $\beta(n) = i$ if n is even (see (2.16)). Hence, the Shannon wavelet is recovered as a limit of spline wavelets.

6. There is no C^∞ orthonormal wavelet with exponential decay. To see this suppose that ψ is such a wavelet. By Theorem 3.4 in Chapter 2, all the moments of ψ are zero. On the other hand, the exponential decay of ψ, say, $|\psi(x)| \leq c e^{-\alpha|x|}$ for all $x \in \mathbb{R}$, implies that the function $\hat{\psi}(z)$, defined by

$$\hat{\psi}(z) = \frac{1}{2\pi} \int_{\mathbb{R}} e^{-izx} \psi(x)\, dx,$$

is an analytic function on $|\Im m\, z| < \alpha$. Thus,

$$\frac{d^n}{d\xi^n}\hat{\psi}(0) = 0 \qquad \text{for all } n = 0, 1, 2, \cdots,$$

since all the moments of ψ are zero. The expansion of $\hat{\psi}(z)$ in powers of z around the origin shows that $\hat{\psi} \equiv 0$ in a neighborhood of $z = 0$. Since $\{z \in \mathbb{C} : |\Im m\, z| < \alpha\}$ contains the real line in its interior, ψ must be the zero function on $\mathbb{R}$. This contradiction establishes our claim.

Also, **it is impossible to find a band-limited orthonormal wavelet ψ with exponential decay.** In fact, under these conditions, $\psi \in L^1(\mathbb{R})$ and, thus, $\hat{\psi}$ is continuous. Hence, by Theorem 2.7 of Chapter 3, all the moments of ψ are zero, and the same argument as above shows $\psi \equiv 0$.

On the other hand, if we relax the exponential decay condition, it is possible to construct such wavelets. A function g defined on $\mathbb{R}$ is said to have **subexponential decay** if for every $\varepsilon > 0$, there exist $C_\varepsilon > 0$ and $\alpha_\varepsilon > 0$ such that

$$|g(x)| \leq C_\varepsilon e^{-\alpha_\varepsilon |x|^{1-\varepsilon}}, \qquad x \in \mathbb{R}.$$

The precise result is: there exists a band-limited orthonormal wavelet $\psi \in C^\infty$ such that ψ has subexponential decay. The proof of this result is constructive and the construction is accomplished by using the Gevrey class of functions (see [DH] for details).

7. The set of elements given by (4.3) is a basis for the space $\mathcal{B}_j^{(m)}$, $j = 1, 2, \cdots$, $m = 1, 2, \cdots$. Since the space $\mathcal{B}_j^{(m)}$ has dimension 2^j it is enough to show that the collection

$$\tilde{\Delta}_{j,k}^{(m)}(x) = \tilde{\Delta}_j^{(m)}(x - 2^{-j}k), \qquad k = 0, 1, \cdots, 2^j - 1 \equiv N,$$

is linearly independent. This is proved by induction on m. The case $m = 1$ is clear: see the graphs of the triangle functions $\tilde{\Delta}_{j,k}^{(1)}$ given in Figure 4.7. Let us assume that the result in true for $m - 1$. Our purpose is to show that if

$$a_0 \tilde{\Delta}_j^{(m)}(x) + \cdots + a_N \tilde{\Delta}_j^{(m)}(x - 1 + 2^{-j}) = 0 \qquad (6.8)$$

is valid for all $x \in \mathbb{R}$, we must have $a_0 = a_1 = \cdots = a_N = 0$. Since $\tilde{\Delta}_j^{(m)} = \tilde{\Delta}_j^{(m-1)} * \tilde{\chi}_j$ we deduce

$$\tilde{\Delta}_j^{(m)}(x) = 2^{\frac{j}{2}} \int_{x-2^{-j}}^{x} \tilde{\Delta}_j^{(m-1)}(t)\, dt.$$

Differentiating (6.8), rearranging terms and using the 1-periodicity of $\tilde{\Delta}_j^{(m-1)}$ we obtain

$$2^{\frac{j}{2}}(a_1 - a_0)\tilde{\Delta}_j^{(m-1)}(x - 2^{-j}) + \cdots + 2^{\frac{j}{2}}(a_0 - a_N)\tilde{\Delta}_j^{(m-1)}(x) = 0$$

for all $x \in \mathbb{R}$. By the induction hypothesis, $a_0 = a_1 = \cdots = a_N$. Hence, (6.8) can be written as

$$0 = a_0 \left[\tilde{\Delta}_j^{(m)}{}'(x) + \cdots + \tilde{\Delta}_j^{(m)}{}'(x - 1 + 2^{-j}) \right]$$

$$= a_0 2^{\frac{j}{2}} \int_{x-1}^{x} \tilde{\Delta}_j^{(m-1)}(t) \, dt = a_0 2^{\frac{j}{2}} \int_0^1 \tilde{\Delta}_j^{(m-1)}(t) \, dt,$$

due to the 1-periodicity of $\tilde{\Delta}_j^{(m-1)}$. Since $\tilde{\Delta}_j^{(m-1)}$ is a positive, not identically zero, function, we conclude that $a_0 = 0$, and, consequently,

$$a_1 = \cdots = a_N = a_0 = 0.$$

We are grateful to Pascal Auscher for relating this proof to us.

5

Representation of functions by
wavelets

Since a wavelet system $\{\psi_{j,k} : j, k \in \mathbb{Z}\}$ is an orthonormal basis for the Hilbert space $L^2(\mathbb{R})$, we know that for any $f \in L^2(\mathbb{R})$,

$$f = \sum_{j \in \mathbb{Z}} \sum_{k \in \mathbb{Z}} \langle f, \psi_{j,k} \rangle \psi_{j,k}$$

with convergence in the $L^2(\mathbb{R})$-norm. The main purpose of this chapter is to study if such expansions are well defined and converge in the setting of other function spaces. In particular, we shall consider the spaces L^p and H^1 and shall study convergence in norm, pointwise convergence and unconditional convergence of wavelet expansions on the real line.

We feel that it is useful to begin with a presentation of the results we need from the theory of bases and unconditional bases on Banach spaces. This will be done in the first two sections of this chapter.

5.1 Bases for Banach spaces

Let $\mathbb{B} = (\mathbb{B}, \|\cdot\|)$ be a Banach space over the field $\mathbb{F}$ that is either $\mathbb{R}$ or $\mathbb{C}$. A denumerable set $\mathfrak{B} = \{x_j : j \in \mathbb{N}\} \subset \mathbb{B}$ is said to be a **basis** for $\mathbb{B}$ if and only if for each $x \in \mathbb{B}$ there exists a **unique** sequence $\{\alpha_j\}_{j \in \mathbb{N}} \subset \mathbb{F}$ such that

$$x = \sum_{j \in \mathbb{N}} \alpha_j x_j,$$

where the convergence of the partial sums is in the norm of $\mathbb{B}$; that is,

$$\lim_{n\to\infty}\left\|x - \sum_{j=1}^{n} \alpha_j x_j\right\| = 0.$$

We write $\alpha_j \equiv \alpha_j(x)$ to indicate the dependence of α_j on x. Observe that the x_j's are linearly independent (in particular, non-zero) due to the uniqueness assumption in the representation $x = \sum_{j\in\mathbb{N}} \alpha_j x_j$. We are using the natural numbers, $\mathbb{N}$, as the indexing set for simplicity. We shall see that $\mathbb{Z}$, in the case of Fourier expansions, and $\mathbb{Z} \times \mathbb{Z}$, in the case of wavelet expansions, are also "natural" indexing sets that arise naturally. Associated with a basis $\mathfrak{B} = \{x_j : j \in \mathbb{N}\}$ of $\mathbb{B}$, we have the sequence of linear functionals f_j, $j = 1, 2, \cdots$, defined by

$$f_j(x) = \alpha_j(x), \qquad \text{for all } x \in \mathbb{B}.$$

The first result we will show is that, for each $j \in \mathbb{N}$, the linear functional f_j is continuous; that is, $f_j \in \mathbb{B}^*$, where $\mathbb{B}^*$ is the dual of $\mathbb{B}$. Towards this end we define on $\mathbb{B}$ a new norm and show that it is equivalent to the original one.

LEMMA 1.1 *Let* $\mathbb{B} = (\mathbb{B}, \|\cdot\|)$ *be a Banach space over a field* $\mathbb{F}$ *and* $\mathfrak{B} = \{x_j : j \in \mathbb{N}\}$ *be a basis for* $\mathbb{B}$. *If*

$$x = \sum_{j\in\mathbb{N}} \alpha_j(x)x_j \tag{1.2}$$

is the unique representation of x *in terms of the elements of the basis* $\mathfrak{B}$, *then*

$$\|\|x\|\| \equiv \sup_{n\in\mathbb{N}} \left\|\sum_{j=1}^{n} \alpha_j(x)x_j\right\| \tag{1.3}$$

defines a norm on $\mathbb{B}$ *that is equivalent to* $\|\cdot\|$.

PROOF: Since the representation in (1.2) is norm convergent, $\|\|x\|\| < \infty$; moreover, $\|\|x\|\| \geq \|x\| > 0$ if $x \neq 0$. The other properties of a norm are clearly satisfied by $\|\|\cdot\|\|$. By replacing the elements of the basis $\mathfrak{B}$ by $\|x_j\|^{-1}x_j$ we do not change $\|\|x\|\|$; thus, we may assume that $\|x_j\| = 1$ for all $j \in \mathbb{N}$.

We claim that $(\mathbb{B}, |||\cdot|||)$ is complete. To see this let $\{y^{(k)} : k \in \mathbb{N}\}$ be a Cauchy sequence with respect to $|||\cdot|||$; we then have to exhibit a limit $y \in (\mathbb{B}, |||\cdot|||)$. Given $\varepsilon > 0$, there exists $N(\varepsilon) \in \mathbb{N}$ such that

$$|||y^{(k)} - y^{(m)}||| = \sup_{n \in \mathbb{N}} \left\| \sum_{j=1}^{n} (\alpha_j(y^{(k)}) - \alpha_j(y^{(m)}))x_j \right\| < \varepsilon \qquad (1.4)$$

for all $k, m > N(\varepsilon)$. To keep the notation simple we shall write $\alpha_j^{(k)}$ for $\alpha_j(y^{(k)})$ for $k = 1, 2, 3, \cdots$. We also have, since $\|x_n\| = 1$,

$$\left| \alpha_n^{(k)} - \alpha_n^{(m)} \right| = \left\| (\alpha_n^{(k)} - \alpha_n^{(m)})x_n \right\|$$

$$\leq \left\| \sum_{j=1}^{n} (\alpha_j^{(k)} - \alpha_j^{(m)})x_j \right\| + \left\| \sum_{j=1}^{n-1} (\alpha_j^{(k)} - \alpha_j^{(m)})x_j \right\| < 2\varepsilon$$

for all $k, m > N(\varepsilon)$. This shows that $\{\alpha_n^{(k)}\}_{k \in \mathbb{N}}$ is a Cauchy sequence in $\mathbb{F}$, and, consequently, there exists an $\alpha_n \in \mathbb{F}$ for each $n \in \mathbb{N}$, such that

$$\alpha_n = \lim_{k \to \infty} \alpha_n^{(k)} \qquad \text{for all} \;\; n \in \mathbb{N}.$$

From (1.4) we deduce

$$\left\| \sum_{j=1}^{n} (\alpha_j^{(k)} - \alpha_j^{(m)})x_j \right\| < \varepsilon \qquad \text{when} \;\; k, m > N(\varepsilon);$$

letting $m \to \infty$ we obtain

$$\left\| \sum_{j=1}^{n} (\alpha_j^{(k)} - \alpha_j)x_j \right\| \leq \varepsilon \qquad \text{for} \;\; k > N(\varepsilon) \;\; \text{and} \;\; n \in \mathbb{N}. \qquad (1.5)$$

Since

$$\sum_{j=n+1}^{n+\ell} \alpha_j x_j = \sum_{j=1}^{n+\ell} \alpha_j x_j - \sum_{j=1}^{n} \alpha_j x_j$$

$$= \sum_{j=1}^{n+\ell} (\alpha_j - \alpha_j^{(k)})x_j - \sum_{j=1}^{n} (\alpha_j - \alpha_j^{(k)})x_j + \sum_{j=n+1}^{n+\ell} \alpha_j^{(k)} x_j,$$

an application of (1.5) gives us

$$\left\| \sum_{j=n+1}^{n+\ell} \alpha_j x_j \right\| \le 2\varepsilon + \left\| \sum_{j=n+1}^{n+\ell} \alpha_j^{(k)} x_j \right\| \qquad \text{for } k > N(\varepsilon) \text{ and } n, \ell \in \mathbb{N}.$$

Since $\sum_{j=1}^{\infty} \alpha_j^{(k)} x_j$ converges to $y^{(k)}$, the above inequality and the completeness of $(\mathbb{B}, \|\cdot\|)$ show that the series $\sum_{j=1}^{\infty} \alpha_j x_j$ is convergent as well in the norm $\|\cdot\|$. If we denote this last sum by y and take the supremum over $n \in \mathbb{N}$ on the left-hand side of (1.5) we see that

$$\|| y^{(k)} - y \|| \le \varepsilon \qquad \text{for } k > N(\varepsilon).$$

This shows, as we claimed, that $(\mathbb{B}, \|\cdot\|)$ is a Banach space.

Lemma 1.1 is now easy to prove. Since $(\mathbb{B}, \|\cdot\|)$ and $(\mathbb{B}, \|\cdot\|)$ are two Banach spaces and $\|x\| \le \||x\||$, by the open mapping theorem we also have $\||x\|| \le c\|x\|$ for an appropriate constant c. Thus, $\|\cdot\|$ and $\||\cdot\||$ are equivalent norms on $\mathbb{B}$. ∎

It is now easy to prove that the "coefficient" functionals (or maps) f_j are continuous.

THEOREM 1.6 *The coefficient maps f_i are bounded linear functionals on the Banach space $\mathbb{B}$.*

PROOF : We prove the result for $i = 1$. Writing $x = \sum_{i \in \mathbb{N}} f_i(x) x_i$ we have

$$|f_1(x)| = |f_1(x)| \left\| \frac{x_1}{\|x_1\|} \right\| \le \frac{1}{\|x_1\|} \sup_{n \in \mathbb{N}} \left\| \sum_{i=1}^{n} f_i(x) x_i \right\|$$

$$= \frac{1}{\|x_1\|} \||x\|| \le \frac{c}{\|x_1\|} \|x\|,$$

where the last inequality follows from Lemma 1.1. This shows that f_1 is a bounded linear functional with norm not exceeding $c\|x_1\|^{-1}$. ∎

EXAMPLE A : Any complete orthonormal system in a separable Hilbert space $\mathbb{H}$ is a basis of $\mathbb{H}$. In particular, $\{e^{ikx} : k \in \mathbb{Z}\}$ is a basis for

$L^2(\mathbb{T})$; in this example we can choose $k = 0, 1, -1, 2, -2, \cdots$ to be the (natural) ordering of the basis; however, any other order gives us a basis. If ψ is an orthonormal wavelet, $\{\psi_{j,k} : j, k \in \mathbb{Z}\}$ is a basis for $L^2(\mathbb{R})$; in this example we can choose any order in $\mathbb{Z} \times \mathbb{Z}$. As we shall see, this "independence of order" is not true for general Banach spaces.

EXAMPLE B: Let $\ell^p(\mathbb{N})$, $1 \leq p < \infty$, be the space of complex valued sequences $x = \{x_k\}_{k \in \mathbb{N}}$ such that

$$\|x\|_{\ell^p(\mathbb{N})} \equiv \left(\sum_{k \in \mathbb{N}} |x_k|^p \right)^{\frac{1}{p}} < \infty.$$

Let $\delta_{n,k} = 0$ if $n \neq k$ and $\delta_{n,n} = 1$ for $n, k \in \mathbb{N}$. Consider the standard unit vectors $\mathfrak{e}_n = \{\delta_{n,k}\}_{k \in \mathbb{N}}$ in $\ell^p(\mathbb{N})$. Then $\mathfrak{B} = \{\mathfrak{e}_n : n \in \mathbb{N}\}$ is a basis for $\ell^p(\mathbb{N})$ if $1 \leq p < \infty$. The set $\mathfrak{B}$ is not a basis for $\ell^\infty(\mathbb{N})$, the space of all bounded sequences $x = \{x_k\}_{k \in \mathbb{N}}$ endowed with the norm

$$\|x\|_{\ell^\infty(\mathbb{N})} \equiv \sup_{k \in \mathbb{N}} |x_k|,$$

but it is a basis for c_0, the subspace of all $x = \{x_k\}_{k \in \mathbb{N}} \in \ell^\infty(\mathbb{N})$ such that $\lim_{k \to \infty} |x_k| = 0$.

There are many other examples of bases for the Lebesgue spaces L^p, $1 < p < \infty$, that we have already encountered in the case $p = 2$. The system $\{e^{ikx} : k \in \mathbb{Z}\}$ is a basis for $L^p(\mathbb{T})$, $1 < p < \infty$. If, for $x \in [0, 1]$, we define

$$h(x) = \begin{cases} 1 & \text{if } 0 \leq x \leq \frac{1}{2}, \\ -1 & \text{if } \frac{1}{2} < x \leq 1, \\ 0 & \text{otherwise}, \end{cases}$$

and $h_{j,k}(x) = 2^{\frac{j}{2}} h(2^j x - k)$ for $j = 0, 1, 2, \cdots$, $k = 0, 1, \cdots, 2^j - 1$, then the system $\{h_{j,k} : j = 0, 1, \cdots; k = 0, 1, \cdots, 2^j - 1\}$ is a basis for $L^p([0, 1])$ when $1 \leq p < \infty$ (see the first two paragraphs in section 5.7 for references). In later sections we shall prove that other wavelets can provide bases for several function spaces besides L^2.

5.2   Unconditional bases for Banach spaces

We have seen in the last section that all bases for a Banach space satisfy the property in Theorem 1.6. There are other types of bases in a Banach space $\mathbb{B} = (\mathbb{B}, \|\cdot\|)$; the "nicer" ones for applications are the ones called "unconditional," whose definition is given in the next paragraph. This section is dedicated to obtaining necessary and sufficient conditions for a basis to be unconditional.

Let us consider the class $\mathfrak{N}$ of all finite subsets $\mathcal{N}$ of $\mathbb{N}$. We write

$$\lim_{\mathcal{N} \in \mathfrak{N}} \sum_{i \in \mathcal{N}} y_i = y \qquad (2.1)$$

if and only if for every $\varepsilon > 0$, there exists an $\mathcal{N} = \mathcal{N}(\varepsilon) \in \mathfrak{N}$ such that

$$\left\| y - \sum_{i \in \mathcal{N}'} y_i \right\|_{\mathbb{B}} < \varepsilon \qquad \text{for all } \mathcal{N}' \in \mathfrak{N} \text{ with } \mathcal{N}' \supseteq \mathcal{N}.$$

We say that the series $\sum_{i \in \mathbb{N}} y_i$ **converges unconditionally to** y if and only if (2.1) holds.

A basis $\mathfrak{B} = \{x_j : j \in \mathbb{N}\}$ of a Banach space $\mathbb{B} = (\mathbb{B}, \|\cdot\|)$ is said to be an **unconditional basis** if and only if in the unique representation $x = \sum_{j \in \mathbb{N}} \alpha_j x_j$ the series converges unconditionally.

One should expect that if a series converges unconditionally to y, any permutation of the elements of the series produces a new series converging in norm to the same element y. This is the content of the following lemma.

LEMMA 2.2 *In a Banach space* $\mathbb{B}$, *the series* $\displaystyle\sum_{i \in \mathbb{N}} y_i$ *converges unconditionally to* $y \in \mathbb{B}$ *if and only if* $\displaystyle\sum_{i \in \mathbb{N}} y_{\sigma(i)} = y$ *(convergence in the norm of* $\mathbb{B}$*) for all permutations* σ *of* $\mathbb{N}$.

PROOF: Suppose that $\sum_{i \in \mathbb{N}} y_i$ converges unconditionally to $y \in \mathbb{B}$ and let σ be a permutation of $\mathbb{N}$. Given $\varepsilon > 0$, choose $\mathcal{N} \equiv \mathcal{N}(\varepsilon) \in \mathfrak{N}$ in such a way that

$$\left\| y - \sum_{i \in \mathcal{N}'} y_i \right\|_{\mathbb{B}} < \varepsilon \qquad \text{for all } \mathcal{N}' \supseteq \mathcal{N} \text{ with } \mathcal{N}' \in \mathfrak{N}.$$

Let n_0 be defined by $n_0 = \sup\{j \in \mathbb{N} : \sigma(j) \in \mathcal{N}(\varepsilon)\}$. If $n \geq n_0$ we have

$$\left\| \sum_{i=1}^{n} y_{\sigma(i)} - y \right\|_{\mathbb{B}} = \left\| \sum_{k \in \{\sigma(j):j \leq n\}} y_k - y \right\|_{\mathbb{B}} < \varepsilon$$

since $\{\sigma(j) : j \leq n\} \supseteq \mathcal{N}(\varepsilon)$.

Assume now that $\sum_{j \in \mathbb{N}} y_{\sigma(j)} = y$ for each permutation σ and suppose that

$$\lim_{\mathcal{N} \in \mathfrak{N}} \sum_{j \in \mathcal{N}} y_j \neq y.$$

Then, we can find an $\varepsilon_o > 0$ such that for each $\mathcal{N} \in \mathfrak{N}$ there exists $\mathcal{N}' \supseteq \mathcal{N}$, $\mathcal{N}' \in \mathfrak{N}$, and

$$\left\| \sum_{j \in \mathcal{N}'} y_j - y \right\|_{\mathbb{B}} \geq \varepsilon_o.$$

Let us start with $\mathcal{N}_0 = \{1\}$ and let $\mathcal{N}_0' \supset \mathcal{N}_0$ be such that

$$\left\| \sum_{j \in \mathcal{N}_0'} y_j - y \right\|_{\mathbb{B}} \geq \varepsilon_o.$$

Let $n_1 = \sup\{k \in \mathbb{N} : k \in \mathcal{N}_0'\}$ and put $\mathcal{N}_1 = \{1, 2, \cdots, n_1 + 1\}$. We can find $\mathcal{N}_1' \supset \mathcal{N}_1$ such that

$$\left\| \sum_{j \in \mathcal{N}_1'} y_j - y \right\|_{\mathbb{B}} \geq \varepsilon_o.$$

Continuing in this way we obtain a sequence of elements of $\mathfrak{N}$,

$$\mathcal{N}_0 \subset \mathcal{N}_0' \subset \mathcal{N}_1 \subset \mathcal{N}_1' \subset \cdots \subset \mathcal{N}_k \subset \mathcal{N}_k' \subset \cdots$$

such that

$$\left\| \sum_{j \in \mathcal{N}_k'} y_j - y \right\|_{\mathbb{B}} \geq \varepsilon_o \qquad \text{for all } k = 0, 1, 2, \cdots.$$

Observe that

$$\mathcal{N}_0, \ \mathcal{N}_0' - \mathcal{N}_0, \ \mathcal{N}_1 - \mathcal{N}_0', \ \cdots, \ \mathcal{N}_k' - \mathcal{N}_k, \ \mathcal{N}_{k+1} - \mathcal{N}_k', \ \cdots$$

is a disjoint union which equals $\mathbb{N}$. We can then re-order $\mathbb{N}$ by letting the elements of the nth set of this union precede the elements of the $(n+1)^{\text{th}}$

set, $n = 0, 1, \cdots$. This defines a permutation σ of $\mathbb{N}$ satisfying

$$\sum_{i=1}^{\mathrm{card}(\mathcal{N}_k')} y_{\sigma(i)} = \sum_{j \in \mathcal{N}_k'} y_j.$$

Since $\left\| \sum_{j \in \mathcal{N}_k'} y_j - y \right\|_{\mathbb{B}} \geq \varepsilon_o$, we see that the series $\sum_{j \in \mathbb{N}} y_{\sigma(j)}$ cannot converge to y in the norm of $\mathbb{B}$.

∎

LEMMA 2.3 *Suppose* $\displaystyle\sum_{j \in \mathbb{N}} x_j$ *converges unconditionally to* x; *then*

$$\sum_{j \in \mathbb{N}} |f(x_j)|$$

converges uniformly for all $f \in \mathbb{B}_1^* = \{f \in \mathbb{B}^* : \|f\|_{\mathbb{B}^*} \leq 1\}$. *That is, given* $\varepsilon > 0$, *there exists* $N \equiv N(\varepsilon)$ *such that*

$$\sum_{j=n+1}^{n+k} |f(x_j)| < \varepsilon \qquad whenever \;\; n > N \;\; and \;\; k \in \mathbb{N},$$

for all $f \in \mathbb{B}_1^*$.

PROOF : Suppose $\displaystyle\lim_{\mathcal{N} \in \mathfrak{N}} \sum_{j \in \mathcal{N}} x_j = x$ and ε is an arbitrary positive number. Let $\mathcal{N} = \mathcal{N}(\varepsilon) \in \mathfrak{N}$ be such that

$$\left\| x - \sum_{j \in \mathcal{N}'} x_j \right\|_{\mathbb{B}} < \tfrac{1}{8}\varepsilon \qquad for \; all \;\; \mathcal{N}' \supseteq \mathcal{N} \;\; with \;\; \mathcal{N}' \in \mathfrak{N}.$$

Let $N = N(\varepsilon) = \max\{i : i \in \mathcal{N}(\varepsilon)\}$. For $n \geq N$, $\ell \geq 1$ and $f \in \mathbb{B}_1^*$ we define

$$\mathcal{N}_1(f) = \{i : n + 1 \leq i \leq n + \ell, \; \Re e f(x_i) \geq 0\}, \qquad and$$

$$\mathcal{N}_2(f) = \{i : n + 1 \leq i \leq n + \ell, \; \Re e f(x_i) < 0\}.$$

Then

$$\sum_{i=n+1}^{n+\ell} |\Re e f(x_i)| = \left| \Re e f\Big(\sum_{i \in \mathcal{N}_1(f)} x_i \Big) \right| + \left| \Re e f\Big(\sum_{i \in \mathcal{N}_2(f)} x_i \Big) \right|$$

$$\leq \left\| \sum_{i \in \mathcal{N}_1(f)} x_i \right\|_{\mathbb{B}} + \left\| \sum_{i \in \mathcal{N}_2(f)} x_i \right\|_{\mathbb{B}} \equiv I + II.$$

For I we have

$$I = \left\| x - \sum_{i \in \mathcal{N}(\varepsilon)} x_i - \sum_{i \in \mathcal{N}_1(f)} x_i - x + \sum_{i \in \mathcal{N}(\varepsilon)} x_i \right\|_{\mathbb{B}}$$

$$\leq \left\| x - \sum_{i \in \mathcal{N}(\varepsilon) \cup \mathcal{N}_1(f)} x_i \right\|_{\mathbb{B}} + \left\| x - \sum_{i \in \mathcal{N}(\varepsilon)} x_i \right\|_{\mathbb{B}} \leq \frac{\varepsilon}{8} + \frac{\varepsilon}{8} = \frac{\varepsilon}{4},$$

since both $\mathcal{N}(\varepsilon)$ and $\mathcal{N}(\varepsilon) \cup \mathcal{N}_1(f)$ contain $\mathcal{N}(\varepsilon)$. Similarly, we obtain $II \leq \frac{1}{4}\varepsilon$. We can do the same thing for $\Im m f(x_i)$ and, hence,

$$\sum_{n+1}^{n+\ell} |f(x_i)| \leq \sum_{n+1}^{n+\ell} |\Re e f(x_i)| + \sum_{n+1}^{n+\ell} |\Im m f(x_i)| < \frac{\varepsilon}{2} + \frac{\varepsilon}{2} = \varepsilon.$$

■

This lemma will help us obtain other characterizations of unconditional convergence for series.

THEOREM 2.4 *For a sequence $\{x_i\}_{i \in \mathbb{N}}$ in a Banach space $\mathbb{B}$, the following conditions are equivalent:*

(i)   *The series $\displaystyle\sum_{j \in \mathbb{N}} x_j$ converges unconditionally;*

(ii) *The series $\displaystyle\sum_{j \in \mathbb{N}} \beta_j x_j$ converges for every sequence $\{\beta_j\}_{j \in \mathbb{N}}$ such that $|\beta_j| \leq 1$ for all $j \in \mathbb{N}$;*

(iii) *The series $\displaystyle\sum_{j \in \mathbb{N}} x_{n_j}$ converges for every increasing sequence $\{n_j\}_{j \in \mathbb{N}}$;*

(iv) *The series $\displaystyle\sum_{j \in \mathbb{N}} \varepsilon_j x_j$ converges for every sequence $\{\varepsilon_j\}_{j \in \mathbb{N}}$ such that $\varepsilon_j = \pm 1$.*

PROOF: We start with (i) $\Rightarrow$ (ii). Given $\varepsilon > 0$, Lemma 2.3 allows us to choose an $N \equiv N(\varepsilon)$ such that

$$\sum_{j=n+1}^{n+k} |f(x_j)| < \varepsilon \qquad \text{for all } f \in \mathbb{B}_1^*, \text{ when } n > N \text{ and } k > 1.$$

Then, by duality,

$$\left\| \sum_{j=n+1}^{n+k} \beta_j x_j \right\|_{\mathbb{B}} = \sup_{f \in \mathbb{B}_1^*} \left| f\left(\sum_{j=n+1}^{n+k} \beta_j x_j \right) \right| = \sup_{f \in \mathbb{B}_1^*} \left| \sum_{j=n+1}^{n+k} \beta_j f(x_j) \right|.$$

Since $|\beta_j| \leq 1$, we deduce

$$\left\| \sum_{j=n+1}^{n+k} \beta_j x_j \right\|_{\mathbb{B}} \leq \sup_{f \in \mathbb{B}_1^*} \sum_{j=n+1}^{n+k} |f(x_j)| < \varepsilon.$$

This shows that the partial sums $\sum_{j=1}^{n} \beta_j x_j$ form a Cauchy sequence; the convergence then follows from the completeness of $\mathbb{B}$.

To obtain (iii) from (ii) we select, for an increasing sequence $\{n_j\}_{j \in \mathbb{N}}$,

$$\beta_k = \begin{cases} 1 & \text{if } k = n_j \text{ for some } j, \\ 0 & \text{otherwise,} \end{cases}$$

and apply (ii) to show that the series $\displaystyle\sum_{j \in \mathbb{N}} x_{n_j}$ converges.

Given a sequence $\{\varepsilon_j\}_{j \in \mathbb{N}}$ with $\varepsilon_j = \pm 1$, if we choose

$$\beta_j = \begin{cases} 1 & \text{if } \varepsilon_j = 1, \\ 0 & \text{if } \varepsilon_j = -1, \end{cases}$$

then we have the equality

$$\sum_{j \in \mathbb{N}} \beta_j x_j = \frac{1}{2} \left(\sum_{j \in \mathbb{N}} \varepsilon_j x_j + \sum_{j \in \mathbb{N}} x_j \right)$$

(that is, the partial sums of the series involved satisfy this equality). Thus, if (iii) holds, the series on the left and the second one on the right converge. This shows that (iii) $\Rightarrow$ (iv). This equation, in fact, shows that (iii) and (iv) are equivalent.

Finally we prove that (iii) implies (i). Let us assume that

$$\sum_{i=1}^{\infty} x_{p_i}$$

is convergent for all strictly increasing sequences of natural numbers

$$p_1 < p_2 < \cdots < p_\ell < \cdots.$$

In particular, $\sum_{i=1}^{\infty} x_i$ is a well defined element of our Banach space that converges to an element $x \in \mathbb{B}$. Suppose $\sum_{i \in \mathbb{N}} x_i$ is not unconditionally convergent. In particular, it is not unconditionally convergent to x. Thus, there exists an $\varepsilon > 0$ such that for each $\mathcal{N} \in \mathfrak{N}$ we can find $\mathcal{N}' \in \mathfrak{N}$ with $\mathcal{N} \subset \mathcal{N}'$ and

$$\left\| \sum_{j \in \mathcal{N}'} x_j - x \right\|_{\mathbb{B}} \geq \varepsilon.$$

Let $\mathcal{N}_0 = \{1\}$, then there exists $\mathcal{N}_0' \in \mathfrak{N}$ such that $\mathcal{N}_0 \subset \mathcal{N}_0'$ and

$$\left\| \sum_{j \in \mathcal{N}_0'} x_j - x \right\|_{\mathbb{B}} \geq \varepsilon.$$

Let $n_1 = \max \{j : j \in \mathcal{N}_0'\}$ and $\mathcal{N}_1 = \{1, 2, \cdots, n_1 + 1\}$. Then, there exists $\mathcal{N}_1' \in \mathfrak{N}$ such that $\mathcal{N}_1 \subset \mathcal{N}_1'$ and

$$\left\| \sum_{j \in \mathcal{N}_1'} x_j - x \right\|_{\mathbb{B}} \geq \varepsilon.$$

Continuing this process we obtain an increasing sequence of finite subsets of $\mathbb{N}$

$$\mathcal{N}_0 \subset \mathcal{N}_0' \subset \mathcal{N}_1 \subset \mathcal{N}_1' \subset \cdots \subset \mathcal{N}_{k-1}' \subset \mathcal{N}_k \subset \mathcal{N}_k' \subset \cdots$$

such that $n_k = \max \{j : j \in \mathcal{N}_{k-1}'\}$, $\mathcal{N}_k = \{1, 2, \cdots, n_k + 1\}$, $\mathcal{N}_k \subset \mathcal{N}_k' \in \mathfrak{N}$ and

$$(*) \qquad \left\| \sum_{j \in \mathcal{N}_k'} x_j - x \right\|_{\mathbb{B}} \geq \varepsilon,$$

$k = 1, 2, \cdots$, and $\mathcal{N}_0 = \{1\}$. Let $\mathcal{D}_k = \mathcal{N}_k' - \mathcal{N}_k$, $k = 1, 2, \cdots$. Then $\mathcal{D}_k \subseteq \{n_k + 2, \cdots, n_{k+1} + 1\}$, where $n_k + 2 \leq n_{k+1} + 1$. It follows that the collection $\{\mathcal{D}_k : k = 1, 2, \cdots\}$ is mutually disjoint and

$$\max \{j : j \in \mathcal{D}_k\} < \min \{j : j \in \mathcal{D}_{k+1}\};$$

thus, the union of the $\mathcal{D}_k$'s in their natural order corresponds to a strictly increasing sequence $\{p_i\}_{i \in \mathbb{N}}$ of natural numbers. But, from $(*)$ we have

$$0 < \varepsilon \leq \left\| \left(\sum_{j \in \mathcal{N}_k} x_j - x \right) + \sum_{j \in \mathcal{D}_k} x_j \right\|_{\mathbb{B}}$$

$$\leq \left\| \sum_{j \in \mathcal{N}_k} x_j - x \right\|_{\mathbb{B}} + \left\| \sum_{j \in \mathcal{D}_k} x_j \right\|_{\mathbb{B}} = \left\| \sum_{j=1}^{n_k+1} x_j - x \right\|_{\mathbb{B}} + \left\| \sum_{j \in \mathcal{D}_k} x_j \right\|_{\mathbb{B}}.$$

However, if k is large enough, the first summand does not exceed $\frac{1}{2}\varepsilon$ (since $\sum_{j=1}^{\infty} x_j = x$) while $\sum_{j \in \mathcal{D}_k} x_j$ is the difference of two partial sums of the convergent series $\sum_{i=1}^{\infty} x_{p_i}$ which must tend to 0 as $k \to \infty$ by the Cauchy criterion. This gives us the contradiction

$$\tfrac{1}{2}\varepsilon \leq \left\| \sum_{j \in \mathcal{D}_k} x_j \right\|_{\mathbb{B}} \longrightarrow 0 \qquad \text{as} \ \ k \to \infty$$

and forces us to conclude that $\displaystyle\sum_{j \in \mathbb{N}} x_j$ converges unconditionally.

$\blacksquare$

Recall that a basis $\mathfrak{B} = \{x_j : j \in \mathbb{N}\}$ for a Banach space $\mathbb{B}$ is said to be **an unconditional basis** if in the unique representation (1.2) the series converges unconditionally. The following result is an immediate consequence of Lemmas 2.2 and 2.3 and Theorem 2.4.

THEOREM 2.5 *For a basis* $\mathfrak{B} = \{x_j : j \in \mathbb{N}\}$ *of a Banach space* $\mathbb{B}$, *the following conditions are equivalent:*

(i) $\mathfrak{B}$ *is an unconditional basis for* $\mathbb{B}$;

(ii) *For every permutation* σ *of* $\mathbb{N}$, *the set* $\mathfrak{B}^{\sigma} = \{x_{\sigma(i)} : i \in \mathbb{N}\}$ *is a basis for* $\mathbb{B}$;

(iii) *For every* $x \in \mathbb{B}$ *with the unique representation* $x = \displaystyle\sum_{j \in \mathbb{N}} \alpha_j x_j$ *and for every sequence* $\{\beta_j\}_{j \in \mathbb{N}}$ *such that* $|\beta_j| \leq 1$, *the series* $\displaystyle\sum_{j \in \mathbb{N}} \beta_j \alpha_j x_j$ *converges.*

REMARK: Condition (iii) in Theorem 2.5 follows immediately from condition (ii) of Theorem 2.4. Conditions (iii) and (iv) in Theorem 2.4 obviously lead to two more equivalent conditions.

Let $\mathfrak{B} = \{x_j : j \in \mathbb{N}\}$ be an unconditional basis for a Banach space $\mathbb{B}$ and $\beta = \{\beta_j\}_{j \in \mathbb{N}}$ be a sequence of scalars such that $|\beta_j| \leq 1$ for all $j \in \mathbb{N}$. By definition, the series $\displaystyle\sum_{j \in \mathbb{N}} f_j(x) x_j$, where the f_j's are the coefficient functionals, converges unconditionally. Part (iii) of Theorem 2.5 allows us to define the operator

$$S_\beta(x) = \sum_{j \in \mathbb{N}} \beta_j f_j(x) x_j \qquad \text{for all } x \in \mathbb{B}. \tag{2.6}$$

It is clear that S_β is linear. We claim that S_β is continuous. To prove this we shall use the closed graph theorem. Suppose that $x^{(k)}$ converges to x and $S_\beta(x^{(k)})$ converges to y. We need to show that $S_\beta(x) = y$. Since

$$x = \sum_{j \in \mathbb{N}} f_j(x) x_j, \qquad x^{(k)} = \sum_{j \in \mathbb{N}} f_j(x^{(k)}) x_j, \qquad y = \sum_{j \in \mathbb{N}} f_j(y) x_j,$$

we obtain

$$S_\beta(x^{(k)}) = \sum_{j \in \mathbb{N}} \beta_j f_j(x^{(k)}) x_j.$$

From the convergence of $x^{(k)}$ to x we have, by Theorem 1.6, that for each j fixed, $\beta_j f_j(x^{(k)})$ converges to $\beta_j f_j(x)$ as $k \to \infty$. On the other hand,

$$\beta_j f_j(x^{(k)}) = f_j(\beta_j x^{(k)}) = f_j(S_\beta(x^{(k)})),$$

which converges to $f_j(y)$. Hence $\beta_j f_j(x) = f_j(y)$, which shows that

$$S_\beta(x) = \sum_{j \in \mathbb{N}} \beta_j f_j(x) x_j = \sum_{j \in \mathbb{N}} f_j(y) x_j = y.$$

This proves that S_β is bounded; but the bound may vary with β. We shall show that S_β is uniform; that is, it does not depend on β.

LEMMA 2.7 *For any unconditional basis* $\mathfrak{B} = \{x_j : j \in \mathbb{N}\}$ *of a Banach space* $\mathbb{B}$, *there exists a constant* $C < \infty$ *such that, for all* $x \in \mathbb{B}$,

$$\left\| S_\beta(x) \right\|_{\mathbb{B}} \le C \|x\|_{\mathbb{B}}$$

for all sequences of scalars $\beta = \{\beta_j\}_{j \in \mathbb{N}}$ *with* $|\beta_j| \le 1$.

PROOF : We have just shown that each S_β is a bounded linear operator on $\mathbb{B}$. Let $x = \sum_{j \in \mathbb{N}} f_j(x) x_j \in \mathbb{B}$ and $f \in \mathbb{B}_1^* = \{f \in \mathbb{B}^* : \|f\| \le 1\}$. Applying Lemma 2.3 with $\varepsilon = 1$ we know that there exists an $N \in \mathbb{N}$ such that

$$\sum_{j=N+1}^{\infty} |f_j(x) f(x_j)| \le 1 \qquad \text{for all } f \in \mathbb{B}_1^*.$$

Hence,

$$\sum_{j\in\mathbb{N}}|f_j(x)f(x_j)| \le \sum_{j=1}^{N}|f_j(x)f(x_j)| + 1 \equiv c(x) + 1.$$

For each $f \in \mathbb{B}_1^*$ we define $T_f : \mathbb{B} \longrightarrow \ell^1(\mathbb{N})$ by

$$T_f(x) = \{f(x_j)f_j(x)\}_{j\in\mathbb{N}} \qquad \text{for all } x \in \mathbb{B}.$$

Each T_f is linear. Since $c(x) \le c\|x\|$, the last inequality shows that each T_f is bounded, and that the set $\{\|T_f(x)\|_{\ell^1(\mathbb{N})} : f \in \mathbb{B}_1^*\}$ is bounded by $c(x) + 1$. By the Banach-Steinhaus theorem, there exists a constant $C < \infty$ such that $\|T_f\|_{op} \le C$ for all $f \in \mathbb{B}_1^*$. Hence

$$\|T_f(x)\|_{\ell^1(\mathbb{N})} \le C\|x\|_{\mathbb{B}} \qquad \text{for all } f \in \mathbb{B}_1^*.$$

Therefore,

$$|f(S_\beta(x))| = \left|f\left(\sum_{j\in\mathbb{N}}\beta_j f_j(x)x_j\right)\right| = \left|\sum_{j\in\mathbb{N}}\beta_j f_j(x)f(x_j)\right|$$

$$\le \sum_{j\in\mathbb{N}}|f_j(x)f(x_j)| = \|T_f(x)\|_{\ell^1(\mathbb{N})} \le C\|x\|$$

for all $x \in \mathbb{B}$ and all $f \in \mathbb{B}_1^*$. Taking the supremum over all $f \in \mathbb{B}_1^*$ we obtain

$$\|S_\beta(x)\|_{\mathbb{B}} = \sup_{f\in\mathbb{B}_1^*}|f(S_\beta(x))| \le C\|x\|_{\mathbb{B}},$$

which proves the desired result.

∎

We shall show that the converse of Lemma 2.7 is also true. In fact, we have the following stronger result.

LEMMA 2.8 *Let* $\mathfrak{B} = \{x_j : j \in \mathbb{N}\}$ *be a basis for a Banach space* $\mathbb{B} = (\mathbb{B}, \|\cdot\|)$. *If there exists a constant* $C < \infty$ *such that the operator* S_β *defined by (2.6) satisfies* $\|S_\beta(x)\| \le C\|x\|$ *for all* $x \in \mathbb{B}$ *and finitely non-zero sequences* $\beta = \{\beta_j\}_{j\in\mathbb{N}}$ *with* $\beta_j = 1$ *or* 0, *then the basis* $\mathfrak{B}$ *is unconditional.*

PROOF: Let $x = \sum_{j\in\mathbb{N}} f_j(x)x_j \in \mathbb{B}$ and σ be any permutation of $\mathbb{N}$. By Lemma 2.2 it is enough to show that the series

$$\sum_{j\in\mathbb{N}} f_{\sigma(j)}(x)\, x_{\sigma(j)}$$

converges to x in $\mathbb{B}$. The convergence of $\sum_{j \in \mathbb{N}} f_j(x)x_j$ to x in $\mathbb{B}$ implies that, for given $\varepsilon > 0$, there exists an $N \in \mathbb{N}$ such that

$$\left\| x - \sum_{j=1}^{n} f_j(x)x_j \right\| < \frac{\varepsilon}{2C+1} \qquad \text{for all } n \geq N. \tag{2.9}$$

Let $M = M(\varepsilon, \sigma) \in \mathbb{N}$ be such that $\{1, 2, \cdots, N\} \subseteq \{\sigma(1), \sigma(2), \cdots, \sigma(M)\}$. If $m \geq M$,

$$\left\| x - \sum_{j=1}^{m} f_{\sigma(j)}(x)\, x_{\sigma(j)} \right\| \leq \left\| x - \sum_{j=1}^{N} f_j(x)x_j \right\| + \left\| \sum_{j \leq m, \sigma(j) > N} f_{\sigma(j)}(x)x_{\sigma(j)} \right\|$$

$$= \left\| x - \sum_{j=1}^{N} f_j(x)x_j \right\| + \left\| \sum_{k=N+1}^{M_0} \beta_k f_k(x)x_k \right\|,$$

where $M_0 = \max\{\sigma(1), \sigma(2), \cdots, \sigma(m)\}$ and

$$\beta_k = \begin{cases} 1 & \text{if } k = \sigma(j) \text{ for some } j \leq m, \\ 0 & \text{otherwise.} \end{cases}$$

By (2.9) the first summand is smaller than $(2C+1)^{-1}\varepsilon$. For the second summand we use our hypothesis to obtain

$$\left\| \sum_{k=N+1}^{M_0} \beta_k f_k(x)x_k \right\| \leq C \left\| \sum_{k=N+1}^{M_0} f_k(x)x_k \right\|$$

$$\leq C \left\{ \left\| x - \sum_{k=1}^{M_0} f_k(x)x_k \right\| + \left\| x - \sum_{k=1}^{N} f_k(x)x_k \right\| \right\} \leq C \frac{2\varepsilon}{2C+1}.$$

Hence,

$$\left\| x - \sum_{j=1}^{m} f_{\sigma(j)}(x)\, x_{\sigma(j)} \right\| \leq \varepsilon \qquad \text{for all } m \geq M.$$

∎

The above results give the following characterizations for unconditional bases.

THEOREM 2.10 *For a basis* $\mathfrak{B} = \{x_j : j \in \mathbb{N}\}$ *of a Banach space* $\mathbb{B} = (\mathbb{B}, \|\cdot\|)$, *the following statements are equivalent:*

(i) $\mathfrak{B}$ *is an unconditional basis for* $\mathbb{B}$.

(ii)   *There exists a constant $C > 0$ such that $\|S_\beta(x)\| \leq C\|x\|$ for all sequences $\beta = \{\beta_j\}_{j \in \mathbb{N}}$ with $|\beta_j| \leq 1$.*

(iii)   *There exists a constant $C > 0$ such that $\|S_\varepsilon(x)\| \leq C\|x\|$ for all sequences $\varepsilon = \{\varepsilon_j\}_{j \in \mathbb{N}}$ with $\varepsilon_j = \pm 1$.*

(iv)   *There exists a constant $C > 0$ such that $\|S_\beta(x)\| \leq C\|x\|$ for all finitely non-zero sequences $\beta = \{\beta_j\}_{j \in \mathbb{N}}$ with $\beta_j = 1$ or 0.*

PROOF: The implication (i) $\Rightarrow$ (ii) follows from Lemma 2.7, while the implication (ii) $\Rightarrow$ (iii) is obvious. To prove that (iii) implies (iv) we choose, for given $\beta = \{\beta_j\}_{j \in \mathbb{N}}$ with $\beta_j = 1$ or 0,

$$\varepsilon_j = \begin{cases} 1 & \text{if } \beta_j = 1, \\ -1 & \text{if } \beta_j = 0, \end{cases}$$

so that

$$\sum_{j \in \mathbb{N}} \varepsilon_j f_j(x) x_j = 2 \sum_{j \in \mathbb{N}} \beta_j f_j(x) x_j - \sum_{j \in \mathbb{N}} f_j(x) x_j.$$

The implication (iv) $\Rightarrow$ (i) is Lemma 2.8.   ∎

We conclude this section with some examples of unconditional bases.

EXAMPLE C: The basis $\mathfrak{B} = \{\mathfrak{e}_n : n \in \mathbb{N}\}$ where $\mathfrak{e}_n = \{\delta_{n,k}\}_{k \in \mathbb{N}}$ is an unconditional basis for $\ell^p(\mathbb{N})$, $1 \leq p < \infty$ (see Example B). This follows easily from part (iii) of Theorem 2.10, since if $x = \{x_k\}_{k \in \mathbb{N}} \in \ell^p(\mathbb{N})$

$$\left\|S_\varepsilon(x)\right\|_{\ell^p(\mathbb{N})} = \left\|S_\varepsilon\left(\sum_{k \in \mathbb{N}} x_k \mathfrak{e}_k\right)\right\|_{\ell^p(\mathbb{N})} = \left\|\sum_{k \in \mathbb{N}} \varepsilon_k x_k \mathfrak{e}_k\right\|_{\ell^p(\mathbb{N})}$$

$$= \left(\sum_{k \in \mathbb{N}} |\varepsilon_k x_k|^p\right)^{\frac{1}{p}} = \left(\sum_{k \in \mathbb{N}} |x_k|^p\right)^{\frac{1}{p}} = \|x\|_{\ell^p(\mathbb{N})}$$

EXAMPLE D: Any complete orthonormal system $\{x_k : k \in \mathbb{N}\}$ in a separable Hilbert space $\mathbb{H}$ is an unconditional basis for $\mathbb{H}$. This follows from the fact that if $x \in \mathbb{H}$,

$$\|x\|_{\mathbb{H}}^2 = \sum_{k \in \mathbb{N}} |\langle x, x_k \rangle|^2,$$

and then use Theorem 2.10. More precisely,

$$\left\|S_\varepsilon(x)\right\|_{\mathbb{H}} = \left\|S_\varepsilon\left(\sum_{k\in\mathbb{N}}\langle x, x_k\rangle x_k\right)\right\|_{\mathbb{H}} = \left\|\sum_{k\in\mathbb{N}}\varepsilon_k\langle x, x_k\rangle x_k\right\|_{\mathbb{H}}$$

$$\leq \left(\sum_{k\in\mathbb{N}}\left|\varepsilon_k\langle x, x_k\rangle\right|^2\right)^{\frac{1}{2}} = \left(\sum_{k\in\mathbb{N}}\left|\langle x, x_k\rangle\right|^2\right)^{\frac{1}{2}} = \left\|x\right\|_{\mathbb{H}}.$$

In particular, the trigonometric system is an unconditional basis for $L^2(\mathbb{T})$, and any orthonormal wavelet system is an unconditional basis for $L^2(\mathbb{R})$.

We have mentioned at the end of the preceding section that the trigonometric system $\{e^{ikx} : k \in \mathbb{Z}\}$ is a basis for $L^p(\mathbb{T})$, $1 < p < \infty$. We just observed that this basis is unconditional in $L^2(\mathbb{T})$. It turns out that this basis is **not unconditional** in $L^p(\mathbb{T})$ if $p \neq 2$. One way to see this is to borrow a result from Zygmund's book [Zyg] that says that there exists a function $f \in L^p(\mathbb{T})$, $p \neq 2$, $1 < p < \infty$ such that the function

$$\sum_{k\in\mathbb{Z}}\varepsilon_k c_k e^{ik\xi}, \qquad c_k = \frac{1}{2\pi}\int_{\mathbb{T}} f(x)e^{-ikx}\,dx,$$

is not in $L^p(\mathbb{T})$, $p \neq 2$, for "almost all" choices of $\varepsilon = \{\varepsilon_j\}_{j\in\mathbb{Z}}$ with $\varepsilon_j = \pm 1$. The result then follows from part (iii) of Theorem 2.5.

There are periodic wavelets that provide unconditional bases for $L^p(\mathbb{T})$, $1 < p < \infty$ (see note 15 in section 5.7). When a certain decay at infinity is assumed, the wavelets provide unconditional bases for $L^p(\mathbb{R})$, $1 < p < \infty$, and other spaces (see section 5.6).

5.3   Convergence of wavelet expansions in $L^p(\mathbb{R})$

We know that a function in $L^2(\mathbb{R})$ can be represented by its wavelet expansion with convergence in the $L^2(\mathbb{R})$-norm. In this section we shall prove that, under certain conditions on the wavelet, the wavelet expansion also converges in the $L^p(\mathbb{R})$-norm, $1 \leq p < \infty$, and study the case of $L^\infty(\mathbb{R})$.

All the wavelets we will use in this section are assumed to arise from a multiresolution analysis (MRA) (see Chapter 2). For the MRA we shall assume that the scaling function φ and the wavelet ψ have a "controlled" decrease at infinity.

More precisely, given a function g defined on $\mathbb{R}$, we say that a bounded function $W : [0, \infty) \longrightarrow \mathbb{R}^+$ is a **radial decreasing L^1-majorant** of g if $|g(x)| \leq W(|x|)$ and W satisfies the following conditions:

$$\left. \begin{array}{ll} \text{(a)} & W \in L^1([0, \infty)), \\[2mm] \text{(b)} & W \text{ is decreasing}, \\[2mm] \text{(c)} & W(0) < \infty. \end{array} \right\} \quad (3.1)$$

In fact, the boundedness of W follows from (b) and (c) above.

EXAMPLE E: Two particularly natural choices for W are

$$W(r) = ce^{-\varepsilon r} \qquad \text{for some } \varepsilon > 0,$$

and

$$W(r) = \frac{c}{(1+r)^\alpha} \qquad \text{for some } \alpha > 1.$$

The first example is the majorant that we have encountered when we studied the Franklin and spline wavelets and its scaling functions, since we proved in Chapter 4 that they have exponential decay. The second one is a majorant for any of the Lemarié-Meyer wavelets (see Corollary 4.7 in Chapter 1 and Example D in section 2.2). This second one is also a good majorant for the wavelets described in Theorem 4.1 of Chapter 3 if we assume some regularity for the function $b = |\hat{\psi}|$. Finally, both of the above choices are good majorants for the compactly supported wavelets described in section 2.3.

Suppose that we have a wavelet ψ that arises from an MRA with scaling function φ. Associated with the increasing sequence of subspaces $\{V_j\}_{j \in \mathbb{Z}}$ we have the orthogonal projections of $L^2(\mathbb{R})$ onto V_j given by

$$P_j f = \sum_{k \in \mathbb{Z}} \langle f, \varphi_{j,k} \rangle \varphi_{j,k} \qquad \text{for } f \in L^2(\mathbb{R}). \qquad (3.2)$$

We can also consider the projections Q_j from $L^2(\mathbb{R})$ onto W_j given by

$$Q_j f = \sum_{k \in \mathbb{Z}} \langle f, \psi_{j,k} \rangle \psi_{j,k} \qquad \text{for } f \in L^2(\mathbb{R}). \qquad (3.3)$$

Since $V_j \oplus W_j = V_{j+1}$ it follows that $Q_j = P_{j+1} - P_j$. There is also a natural operator, associated with a wavelet ψ, given by

$$S_{j,k}^{\sigma} f(x) = \sum_{\ell=-\infty}^{j-1} \sum_{m \in \mathbb{Z}} \langle f, \psi_{\ell,m} \rangle \psi_{\ell,m}(x) + \sum_{m=1}^{k} \langle f, \psi_{j,\sigma(m)} \rangle \psi_{j,\sigma(m)}(x), \quad (3.4)$$

where $f \in L^2(\mathbb{R})$, $j, k \in \mathbb{Z}$ and σ is any permutation of $\mathbb{Z}$. This operator is a partial sum of the wavelet expansion of f. Since $V_j = \bigoplus_{\ell=-\infty}^{j-1} W_\ell$ we have

$$S_{j,k}^{\sigma} f(x) = P_j f(x) + \sum_{m=1}^{k} \langle f, \psi_{j,\sigma(m)} \rangle \psi_{j,\sigma(m)}(x). \quad (3.5)$$

The above definitions make sense for any $f \in L^2(\mathbb{R})$. In analogy with the case of the Fourier series we try to see if these operators could be defined in $L^p(\mathbb{R})$ and study their convergence properties as $j \to \infty$.

Writing $\langle f, \varphi_{j,k} \rangle$ as an integral and interchanging (formally) the order of summation and integration, we obtain

$$(P_j f)(x) = \int_{-\infty}^{\infty} 2^j K_\varphi(2^j x, 2^j y) f(y) \, dy, \quad (3.6)$$

where

$$K_\varphi(x, y) = \sum_{k \in \mathbb{Z}} \varphi(x - k) \overline{\varphi(y - k)}. \quad (3.7)$$

Similarly,

$$(Q_j f)(x) = \int_{-\infty}^{\infty} 2^j K_\psi(2^j x, 2^j y) f(y) \, dy, \quad (3.8)$$

where

$$K_\psi(x, y) = \sum_{k \in \mathbb{Z}} \psi(x - k) \overline{\psi(y - k)}. \quad (3.9)$$

For the operator $S_{j,k}^{\sigma}$ we have

$$(S_{j,k}^{\sigma} f)(x) = \int_{-\infty}^{\infty} 2^j K_\varphi(2^j x, 2^j y) f(y) \, dy$$

$$+ \int_{-\infty}^{\infty} 2^j K^{\sigma,k}(2^j x, 2^j y) f(y) \, dy, \quad (3.10)$$

where

$$K^{\sigma,k}(x,y) = \sum_{m=1}^{k} \psi(x - \sigma(m)) \overline{\psi(y - \sigma(m))}. \tag{3.11}$$

In order to justify these interchanges of summation and integration, we need the following lemma.

LEMMA 3.12 *Suppose that W is a function on $[0,\infty)$ that satisfies condition (3.1). Then,*

$$\sum_{k \in \mathbb{Z}} W(|x - k|) W(|y - k|) \le C W\left(\frac{|x - y|}{2}\right), \qquad \text{for all } x, y \in \mathbb{R},$$

where C is a constant which depends only on W.

PROOF : Since $|x - y| \le |x - k| + |y - k|$, either $|x - k| \ge \frac{1}{2}|x - y|$ or $|y - k| \ge \frac{1}{2}|x - y|$. Thus, using (3.1),

$$\sum_{k \in \mathbb{Z}} W(|x - k|) W(|y - k|) \le W\left(\tfrac{|y-x|}{2}\right)\left(\sum_{k \in \mathbb{Z}} W(|x - k|) + \sum_{k \in \mathbb{Z}} W(|y - k|)\right)$$

$$\le C W(\tfrac{1}{2}|x - y|),$$

with C depending only on W. ∎

Let us assume that the scaling function φ and the wavelet ψ have a radial decreasing L^1-majorant W (see (3.1)). We shall use this to extend the operators P_j, Q_j and $S_{j,k}^\sigma$ (see (3.2), (3.3) and (3.4)) to $L^p(\mathbb{R})$, $1 \le p \le \infty$. Observe that if g is bounded by a function W satisfying condition (3.1), then $g_{j,k} \in L^p(\mathbb{R})$ whenever $1 \le p \le \infty$ and $j, k \in \mathbb{Z}$. To see this choose x_0 such that $W(x_0) \le 1$. Hence, since $W(x) \le 1$ for $x \ge x_0$, we have

$$\|g\|_{L^p(\mathbb{R})}^p \le 2\left\{\int_0^{x_0} |W(x)|^p \, dx + \int_{x_0}^\infty |W(x)|^p \, dx\right\}$$

$$\le 2\left\{W(0)^p x_0 + \int_{x_0}^\infty |W(x)| \, dx\right\} < \infty.$$

The result for general $g_{j,k}$ follows by a change of variables. It is clear that the formal interchange of summation and integration needed to obtain (3.6), (3.8) or (3.10) are justified by Lemma 3.12.

PROPOSITION 3.13

(a) *Suppose that φ has a radial decreasing L^1-majorant W; then there exists $C > 0$, independent of j, such that for all $f \in L^2(\mathbb{R}) \cap L^p(\mathbb{R})$, $1 \leq p \leq \infty$, we have*

$$\|P_j f\|_{L^p(\mathbb{R})} \leq C \|W\|_{L^1(0,\infty)} \|f\|_{L^p(\mathbb{R})}.$$

(b) *Suppose that ψ has a radial decreasing L^1-majorant W_1; then there exists $C > 0$, independent of j, such that for all $f \in L^2(\mathbb{R}) \cap L^p(\mathbb{R})$, $1 \leq p \leq \infty$, we have*

$$\|Q_j f\|_{L^p(\mathbb{R})} \leq C \|W_1\|_{L^1(0,\infty)} \|f\|_{L^p(\mathbb{R})}.$$

(c) *Suppose that φ and ψ have radial decreasing L^1-majorants W and W_1, respectively; then there exists $C \equiv C(W, W_1) > 0$, independent of j, k and σ, such that for all $f \in L^2(\mathbb{R}) \cap L^p(\mathbb{R})$, $1 \leq p \leq \infty$,*

$$\|S_{j,k}^\sigma f\|_{L^p(\mathbb{R})} \leq C(W, W_1) \|f\|_{L^p(\mathbb{R})}.$$

PROOF : If $p = \infty$ we have, by using (3.6), (3.7) and Lemma 3.12,

$$|P_j f(x)| \leq C \int_{\mathbb{R}} 2^j W(2^j \tfrac{|x-y|}{2}) |f(y)| \, dy \leq C \|f\|_\infty \|W\|_{L^1(0,\infty)}.$$

This proves (a) for $p = \infty$. For the case $p = 1$ we have

$$\int_{\mathbb{R}} |P_j f(x)| \, dx \leq C \int_{\mathbb{R}} \left(\int_{\mathbb{R}} 2^j W(2^j \tfrac{|x-y|}{2}) |f(y)| \, dy \right) dx$$

$$= C \int_{\mathbb{R}} |f(y)| \left(\int_{\mathbb{R}} 2^j W(2^{j-1} |x - y|) \, dx \right) dy$$

$$\leq C \|W\|_{L^1(0,\infty)} \|f\|_{L^1(\mathbb{R})}.$$

The symbol C denotes different constants in the above equalities and inequalities. The general case of (a) follows by interpolation (see Note and Reference 3 in section 5.7). The proofs for (b) and (c) are similar. ∎

Assuming the conditions of Proposition 3.13, the operators P_j, Q_j and $S_{j,k}^\sigma$ can be extended to $L^p(\mathbb{R})$, $1 \leq p \leq \infty$. We start with some results concerning the scaling function of an MRA. By Theorem 1.7 of Chapter 2 we know $|\hat{\varphi}|(0) = 1$. Thus, multiplying φ by a unimodular constant, we can, and will, assume that $\hat{\varphi}(0) = 1$.

PROPOSITION 3.14 *Suppose that the scaling function φ has a radial decreasing L^1-majorant; then, for almost all $x \in \mathbb{R}$,*

(a) $\quad \sum_{k \in \mathbb{Z}} \varphi(x + k) = 1,$

(b) $\quad \int_{\mathbb{R}} 2^j K_\varphi(2^j x, 2^j y) \, dy = 1 \qquad for\ all\ \ j \in \mathbb{Z}.$

PROOF : Since φ is in $L^1(\mathbb{R})$ the sum in (a) is well defined. Therefore, if

$$A(x) = \sum_{k \in \mathbb{Z}} \varphi(x + k),$$

then $A(x)$ is a 1-periodic function on $\mathbb{R}$ that is integrable on $[0, 1)$. We only need to show that the non-zero indexed Fourier coefficients are zero and the other one is 1. To do this we first observe that

$$\int_0^1 A(x) e^{-2\pi i \ell x} \, dx = \int_{\mathbb{R}} \varphi(x) e^{-2\pi i \ell x} \, dx = \hat{\varphi}(2\ell\pi).$$

Since $\varphi \in L^1(\mathbb{R})$, $\hat{\varphi}$ is continuous (see Chapter 1 of [SW]) and, hence, it makes sense to write $\hat{\varphi}(2\ell\pi)$. By Proposition 2.17 of Chapter 2, it follows that $\hat{\varphi}(2\ell\pi) = 0$ if $\ell \neq 0$. This proves (a). To prove (b) we use (a), an obvious change of variables and the existence of the radial decreasing L^1-majorant to obtain

$$1 = \overline{\hat{\varphi}(0)} = \int_{\mathbb{R}} \overline{\varphi(y)} \, dy = \sum_{k \in \mathbb{Z}} \varphi(2^j x - k) \int_{\mathbb{R}} \overline{\varphi(y)} \, dy$$

$$= \sum_{k \in \mathbb{Z}} \varphi(2^j x - k) \, 2^j \int_{\mathbb{R}} \overline{\varphi(2^j y - k)} \, dy = \int_{\mathbb{R}} 2^j K_\varphi(2^j x, 2^j y) \, dy.$$

∎

In order to study the operators P_j and $S^\sigma_{j,k}$ simultaneously we consider the operator

$$T_j f(x) = \int_{\mathbb{R}} 2^j K(2^j x, 2^j y) \{ f(y) - f(x) \} \, dy, \qquad (3.15)$$

where $|K(x, y)| \leq C\,W\left(\frac{|x - y|}{2}\right)$ and W satisfies (3.1) (see Lemma 3.12).

THEOREM 3.16 *For the operators T_j defined in (3.15) we have*

(a) *If $1 \leq p < \infty$, then $\displaystyle\lim_{j \to \infty} \|T_j f\|_{L^p(\mathbb{R})} = 0$ for all $f \in L^p(\mathbb{R})$.*

(b) $\displaystyle\lim_{j \to \infty} \|T_j f\|_\infty = 0$ *for all bounded uniformly continuous functions f.*

PROOF : To prove (a) we choose an $f \in L^p(\mathbb{R})$. A simple change of variables gives us

$$|T_j f(x)| \leq C \int_{\mathbb{R}} 2^j\,W(2^{j-1}|t|)\,|f(x - t) - f(x)|\,dt. \qquad (3.17)$$

Minkowski's inequality for integrals allows us to conclude

$$\|T_j f\|_{L^p(\mathbb{R})} \leq C \int_{\mathbb{R}} 2^j\,W(2^{j-1}|t|)\left(\int_{\mathbb{R}} |f(x - t) - f(x)|^p\,dx\right)^{\frac{1}{p}} dt$$

$$= C \int_{\mathbb{R}} W(\tfrac{1}{2}|t|)\left(\int_{\mathbb{R}} |f(x - 2^{-j}t) - f(x)|^p\,dx\right)^{\frac{1}{p}} dt.$$

The expression

$$\left(\int_{\mathbb{R}} |f(x - h) - f(x)|^p\,dx\right)^{\frac{1}{p}} \equiv \omega_p(h)$$

is the L^p modulus of continuity of f, which satisfies $\omega_p(h) \to 0$ as $|h| \to 0$ if $1 \leq p < \infty$. By the Lebesgue dominated convergence theorem (the integrand is bounded by the L^1-function $2\|f\|_{L^p(\mathbb{R})} W(\tfrac{1}{2}|t|)$) we obtain the desired result.

To prove (b) let $f \in L^\infty(\mathbb{R})$ be uniformly continuous. Then

$$\omega_\infty(t) \equiv \sup_{x \in \mathbb{R}} |f(x - t) - f(x)| \longrightarrow 0$$

as $t \to 0$ uniformly on x. By (3.17) we obtain

$$\|T_j f\|_\infty \leq C \int_{\mathbb{R}} \omega_\infty(t)\,2^j\,W(2^{j-1}|t|)\,dt,$$

and the proof is finished by applying the Lebesgue dominated convergence theorem. ∎

COROLLARY 3.18 *Suppose that φ is the scaling function of an MRA with a radial decreasing L^1-majorant W.*

(a) *If $1 \leq p < \infty$, then $\lim\limits_{j \to \infty} \|P_j f - f\|_{L^p(\mathbb{R})} = 0$ for all $f \in L^p(\mathbb{R})$.*

(b) *$\lim\limits_{j \to \infty} \|P_j f - f\|_{\infty} = 0$ for all bounded uniformly continuous f.*

PROOF : By part (b) of Proposition 3.14 we can write

$$P_j f(x) - f(x) = \int_{\mathbb{R}} 2^j K_\varphi(2^j x, 2^j y)\{f(y) - f(x)\}\, dy. \tag{3.19}$$

The proof is finished by applying Lemma 3.12 and Theorem 3.16.   ∎

COROLLARY 3.20 *Suppose that the wavelet ψ and the scaling function φ of an MRA have radial decreasing L^1-majorants W_1 and W_2, respectively.*

(a) *If $1 \leq p < \infty$, then $\lim\limits_{j \to \infty} \|S_{j,k}^\sigma f - f\|_{L^p(\mathbb{R})} = 0$ for all $k = 1, 2, \cdots$, and all $f \in L^p(\mathbb{R})$.*

(b) *$\lim\limits_{j \to \infty} \|S_{j,k}^\sigma f - f\|_{\infty} = 0$ for all $k = 1, 2, \cdots$, and all bounded uniformly continuous functions f.*

PROOF : Using (3.5) and $\int_{\mathbb{R}} \psi(y)\, dy = \hat{\psi}(0) = 0$ (see Proposition 2.17 in Chapter 2) we obtain

$$S_{j,k}^\sigma f(x) - f(x)$$

$$= [P_j f(x) - f(x)] + \int_{\mathbb{R}} \left\{ \sum_{m=1}^{k} \psi_{j,\sigma(m)}(x)\, \overline{\psi_{j,\sigma(m)}(y)} \right\} [f(y) - f(x)]\, dy$$

$$= [P_j f(x) - f(x)] + \int_{\mathbb{R}} 2^j K^{\sigma,k}(2^j x, 2^j y) [f(y) - f(x)]\, dy. \tag{3.21}$$

From Corollary 3.18, $\lim\limits_{j \to \infty} \|P_j f - f\|_{L^p(\mathbb{R})} = 0$. By Lemma 3.12,

$$|K^{\sigma,k}(x, y)| \leq C\, W_2(\tfrac{1}{2}|x - y|)$$

(independent of σ and k) so that the corollary follows from Theorem 3.16.   ∎

5.4 Pointwise convergence of wavelet expansions

As we did in section 5.3, all wavelets considered in this section will be supposed to be generated by an MRA. We have shown in the preceding section that both $P_j f$ (see (3.2)) and $S_{j,k}^\sigma f$ (see (3.5)) converge to f in the $L^p(\mathbb{R})$-norm, $1 \le p < \infty$, when $j \to \infty$ and in the $L^\infty(\mathbb{R})$-norm for uniformly continuous functions, when appropriate decay conditions are assumed on the wavelet. This section is devoted to the study of pointwise convergence properties of these operators.

The definitions of the above operators as well as the operators T_j used in the theorem below were given in section 5.3.

THEOREM 4.1 *Let $\{T_j : j \in \mathbb{Z}\}$ be the family of operators defined by (3.15). If $f \in L^p(\mathbb{R})$, $1 \le p \le \infty$, then $\lim_{j \to \infty} T_j f(x) = 0$ for every x in the Lebesgue set of f.*

PROOF: If x is a point in the Lebesgue set of f and $\delta > 0$, there exists an $\eta > 0$ such that

$$\frac{1}{r} \int_{|t| \le r} |f(x-t) - f(x)|\, dt \le \delta \qquad \text{when } 0 < r \le \eta. \qquad (4.2)$$

We have (see (3.17))

$$C^{-1}|T_j f(x)| \le \int_{|t| < \eta} 2^j\, W(2^{j-1}|t|)|f(x-t) - f(x)|\, dt$$

$$+ \int_{|t| \ge \eta} 2^j\, W(2^{j-1}|t|)|f(x-t) - f(x)|\, dt$$

$$= I + II.$$

Since $W(|x|)$ decreases to zero as $|x| \to \infty$ and $W \in L^1([0,\infty))$ we have

$$rW(r) \le \int_{\frac{r}{2} \le |x| \le r} W(|x|)\, dx \longrightarrow 0 \qquad \text{as } r \to \infty.$$

Also, $rW(r) \to 0$ as $r \to 0$ since $W(|x|)$ is continuous at zero. Let

$$g(r) = |f(x-r) - f(x)| \qquad \text{and} \qquad G(r) = \int_0^r g(s)\, ds.$$

From (4.2) we deduce

$$G(r) \le r\delta \qquad \text{when } 0 < r \le \eta. \tag{4.3}$$

Thus, integrating by parts, we obtain

$$
\begin{aligned}
I &= \int_{|t|<\eta} 2^j\, W(2^{j-1}|t|)\, |f(x-t) - f(x)|\, dt \\
&= 2 \int_0^\eta 2^j\, W(2^{j-1}r)\, g(r)\, dr \\
&= 2G(r)2^j\, W(2^{j-1}r)\Big|_0^\eta - 2\int_0^\eta G(r)2^j 2^{j-1}\, W'(2^{j-1}r)\, dr \\
&\le 2r\delta 2^j\, W(2^{j-1}r)\Big|_0^\eta - 2\int_0^{2^j\eta} G(2^{-j}r)2^{j-1}\, W'(\tfrac{1}{2}r)\, dr.
\end{aligned}
$$

Since W is decreasing, $W'(\tfrac{1}{2}r)$ is negative and by (4.3) and the boundedness of $rW(r)$ we obtain

$$I \le C\delta - 2\delta \int_0^{2^j\eta} 2^{-j}\, r\, 2^{j-1}\, W'(\tfrac{1}{2}r)\, dr.$$

In order to estimate the last term we write $A = 2^j\eta$ and observe that

$$
\begin{aligned}
-\int_0^{2^j\eta} 2^{-j}\, r\, 2^{j-1} W'(\tfrac{1}{2}r)\, dr &= -\int_0^A \tfrac{1}{2}u\, W'(\tfrac{1}{2}u)\, du \\
&= -A\,W(\tfrac{1}{2}A) + \int_0^A W(\tfrac{1}{2}v)\, dv = -A\,W(\tfrac{1}{2}A) + 2\int_0^{\frac{1}{2}A} W(u)\, du.
\end{aligned}
$$

As $j \to \infty$ (or, equivalently, $A \to \infty$) the last expression increases to $2\|W\|_{L^1(\mathbb{R})}$ (since $rW(r) \to 0$ as $r \to \infty$). This gives us the estimate

$$I \le C\delta + 4\|W\|_{L^1(\mathbb{R})}\delta = a\delta. \tag{4.4}$$

Observe that the constant a depends only on W and, in particular, does not depend on j.

In order to estimate II we let χ_η be the characteristic function of the set $\{t \in \mathbb{R} : |t| \ge \eta\}$. If p' denotes the conjugate exponent to p (that is

$\frac{1}{p} + \frac{1}{p'} = 1$) we use Hölder's inequality to write

$$II \leq \|f\|_{L^p(\mathbb{R})} \left(\int_{\mathbb{R}} |\chi_\eta(t) 2^j \, W(2^{j-1}|t|)|^{p'} \, dt \right)^{\frac{1}{p'}}$$

$$+ |f(x)| \int_{\mathbb{R}} |\chi_\eta(t) 2^j \, W(2^{j-1}|t|)| \, dt. \qquad (4.5)$$

But

$$\int_{\mathbb{R}} |\chi_\eta(t) 2^j \, W(2^{j-1}|t|)| \, dt = 2 \int_{|s| \geq 2^{j-1}\eta} W(|s|) \, ds$$

tends to zero as $j \to \infty$. Also, the same is true for the first summand of the above inequality since

$$\left(\int_{\mathbb{R}} |\chi_\eta(t) 2^j \, W(2^{j-1}|t|)|^{p'} \, dt \right)^{\frac{1}{p'}}$$

$$= \left(\int_{|t| \geq \eta} |2^j \, W(2^{j-1}|t|)|^{\frac{p'}{p}} |2^j \, W(2^{j-1}|t|)| \, dt \right)^{\frac{1}{p'}}$$

$$\leq \left(\sup_{|t| \geq \eta} |2^j \, W(2^{j-1}|t|)| \right)^{\frac{1}{p}} \left(\int_{|t| \geq \eta} |2^j \, W(2^{j-1}|t|)| \, dt \right)^{\frac{1}{p'}}$$

$$\leq C \left(\|W\|_{L^1(0,1)} \right)^{\frac{1}{p'}} \left(\sup_{|t| \geq \eta} |2^j \, W(2^{j-1}|t|)| \right)^{\frac{1}{p}}.$$

Since $rW(r) \to 0$ as $r \to \infty$ we have that

$$\sup_{|t| \geq \eta} 2^j |W(2^{j-1}|t|)| = 2^j \, W(2^{j-1}\eta)$$

tends to zero as $j \to \infty$. Hence, choosing j large enough we deduce from (4.5) that II can be made smaller than δ. This, together with inequality (4.4), proves the desired result.

∎

COROLLARY 4.6 *Suppose that φ is the scaling function of an MRA and that φ has a radial decreasing L^1-majorant. If $f \in L^p(\mathbb{R})$, $1 \leq p \leq \infty$, then*

$$\lim_{j \to \infty} P_j f(x) = f(x) \qquad \text{for every x in the Lebesgue set of f.}$$

In particular, $\lim_{j \to \infty} P_j f(x) = f(x)$ for almost every $x \in \mathbb{R}$.

PROOF : Apply Theorem 4.1 to $T_j f(x) = P_j f(x) - f(x)$, the expression described in (3.19).

∎

COROLLARY 4.7 *Suppose that the wavelet ψ and the scaling function φ of an MRA each has a radial decreasing L^1-majorant. If $f \in L^p(\mathbb{R})$, $1 \leq p \leq \infty$, then, for $k = 1, 2, \cdots$,*

$$\lim_{j \to \infty} S_{j,k}^\sigma f(x) = f(x) \qquad \text{for all } x \text{ in the Lebesgue set of } f.$$

In particular, the partial sums $S_{j,k}^\sigma f(x)$ converge to $f(x)$ almost everywhere as $j \to \infty$ and $k = 1, 2, \cdots$.

PROOF : Apply Theorem 4.1 to the expression introduced in (3.21).

∎

We remark that Theorem 3.16 and Theorem 4.1 can be found in Chapter 1 of [SW]. We have seen, therefore, how classical results in harmonic analysis are well adapted to treat wavelets.

There are wavelets that do not satisfy the conditions of Corollary 4.6, but the projection operators converge almost everywhere. This is the case for the Shannon wavelet ψ for which $\hat{\psi} = \chi_{[-2\pi,-\pi]\cup[\pi,2\pi]}$. Since the scaling function φ associated with this wavelet satisfies $\hat{\varphi} = \chi_{[-\pi,\pi]}$, we must have

$$\varphi(x) = \frac{\sin(\pi x)}{\pi x},$$

which shows $|\varphi(x)| = O(|x|^{-1})$ at ∞. Thus, φ does not have a radial decreasing L^1-majorant. Nevertheless, the oscillations of φ allow us to obtain pointwise convergence results.

In this particular case we can write explicitly the kernel K_φ defined in (3.7) as

$$K_\varphi(x, y) = \sum_{k \in \mathbb{Z}} \frac{\sin(\pi(x-k))\sin(\pi(y-k))}{\pi(x-k)\pi(y-k)}$$

$$= \frac{\sin(\pi x)\sin(\pi y)}{\pi^2} \sum_{k \in \mathbb{Z}} \frac{1}{(x-k)(y-k)}.$$

To find the value of the last sum we write

$$\frac{1}{(x-k)(y-k)} = \frac{1}{y-x}\left\{\frac{1}{x-k} - \frac{1}{y-k}\right\}$$

and use the identity

$$\sum_{k\in\mathbb{Z}}\frac{1}{(x-k)} = \pi\cot(\pi x).$$

(The easiest way to obtain this last equality is to integrate the equality used at the beginning of the proof of Lemma 1.9 in Chapter 4, after writing $\xi = 2\pi x$.) Simple trigonometric calculation then gives us

$$K_\varphi(x,y) = \frac{\sin(\pi(y-x))}{\pi(y-x)}. \tag{4.8}$$

This shows that, for the Shannon wavelet, the projection operators P_j are given by

$$(P_j f)(x) = \int_{\mathbb{R}} \frac{\sin(2^j\pi(y-x))}{\pi(y-x)} f(y)\,dy. \tag{4.9}$$

Observe that this intergal is the convolution of f with an oscillatory kernel not belonging to $L^1(\mathbb{R})$. In fact, one can show that

$$\lim_{j\to\infty} (P_j f)(x) = f(x) \qquad \text{for a.e. } x \in \mathbb{R} \tag{4.10}$$

when $f \in L^2(\mathbb{R})$. This equality can be regarded as a form of the Fourier inversion formula. To see this observe that

$$\frac{1}{2\pi}\int_{-2^j\pi}^{2^j\pi}\left(\int_{\mathbb{R}} f(y)e^{-i\xi y}\,dy\right)e^{i\xi x}\,d\xi$$

$$= \frac{1}{\pi}\int_{\mathbb{R}} \frac{\sin(2^j\pi(y-x))}{y-x} f(y)\,dy = (P_j f)(x).$$

Letting $j \to \infty$, the left-hand side of this equality gives us $f(x)$ by the Fourier inversion formula for a.e. $x \in \mathbb{R}$. We indicate, in the "Notes and references" section at the end of this chapter, why this result follows from the Carleson-Hunt theorem on the a.e. convergence of Fourier series of functions in $L^p(\mathbb{T})$. This establishes (4.10) for the particular case of Shannon wavelet and $f \in L^p(\mathbb{R})$.

5.5 H^1 and BMO on $\mathbb{R}$

In the section that follows this one we shall obtain results concerning unconditional convergence of wavelet expansions in $L^p(\mathbb{R})$, $1 < p < \infty$, and $H^1(\mathbb{R})$. This section contains a review of the basic facts about $H^1(\mathbb{R})$ as well as some elementary results about its dual $BMO(\mathbb{R})$. (We assume that the reader is familiar with the spaces $L^p(\mathbb{R})$.) We could introduce $H^1(\mathbb{R})$ directly as the "atomic" space characterized by Theorem 5.4 below or the "molecular" space defined by Theorem 5.8. This would make our presentation essentially self-contained. We feel, however, that more readers are likely to have encountered the spaces defined by (5.2). Consequently, we introduce what is necessary to explain how these spaces are related. We shall not, however, present all the proofs since this would lead us far astray from our purpose of presenting the properties of wavelets.

Let $\mathbb{R}_+^2 = \{(x,t) \in \mathbb{R}^2 : t > 0\}$ be the upper half plane. The real part of $\dfrac{i}{\pi z}$, $z = x + it$, is called the **Poisson kernel** for $\mathbb{R}_+^2$, and the imaginary part is called the **conjugate Poisson kernel** for $\mathbb{R}_+^2$. Since

$$\frac{i}{\pi z} = \frac{i}{\pi(x+it)} = \frac{i(x-it)}{\pi(x^2+t^2)} = \frac{1}{\pi}\frac{t}{x^2+t^2} + \frac{i}{\pi}\frac{x}{x^2+t^2},$$

the Poisson kernel and the conjugate Poisson kernel for $\mathbb{R}_+^2$ are

$$P_t(x) = \frac{1}{\pi}\frac{t}{x^2+t^2} \qquad \text{and} \qquad Q_t(x) = \frac{1}{\pi}\frac{x}{x^2+t^2}.$$

Given $f \in L^p(\mathbb{R})$, $1 \le p < \infty$, the **Poisson integral** of f is

$$u(x,t) = (P_t * f)(x) = \frac{1}{\pi}\int_{\mathbb{R}} \frac{t}{(x-y)^2+t^2}\, f(y)\, dy.$$

It is not hard to show that $u(x,t)$ is a harmonic function in $\mathbb{R}_+^2$ that satisfies

$$\lim_{t \to 0^+} u(x,t) = f(x) \qquad \text{for} \quad a.e.\ x \in \mathbb{R}.$$

If we consider

$$v(x,t) = (Q_t * f)(x) = \frac{1}{\pi}\int_{\mathbb{R}} \frac{x-y}{(x-y)^2+t^2}\, f(y)\, dy$$

and write $z = x + it$ we see that

$$F(z) \equiv u(x,t) + iv(x,t) = \frac{i}{\pi} \int_{\mathbb{R}} \frac{f(y)}{z - y}\, dy$$

is an analytic function in $\mathbb{R}^2_+$. The M. Riesz inequality asserts that there exists a constant $A_p > 0$, independent of $f \in L^p(\mathbb{R})$, such that

$$\sup_{t>0} \left(\int_{\mathbb{R}} |F(x + it)|^p\, dx \right)^{\frac{1}{p}} \leq A_p \|f\|_{L^p(\mathbb{R})}, \qquad 1 < p < \infty. \qquad (5.1)$$

When $p = 1$ the inequality (5.1) is no longer true. The space of all real valued functions $f \in L^1(\mathbb{R})$ such that

$$\sup_{t>0} \left\{ \int_{\mathbb{R}} |F(x + it)|\, dx \right\} < \infty \qquad (5.2)$$

is called $ReH^1 \equiv ReH^1(\mathbb{R})$. The expression

$$\|f\|_{ReH^1} = \sup_{t>0} \left\{ \int_{\mathbb{R}} |F(x + it)|\, dx \right\}$$

defines a norm on ReH^1 that makes this space a Banach space over the reals. When ReH^1 is "complexified" we obtain a space that we shall denote by $H^1(\mathbb{R}) \equiv H^1$. That is, $f \in H^1$ if and only if $f = g + ih$ with $g, h \in ReH^1$; in this case

$$\|f\|_{H^1} = \|g\|_{ReH^1} + \|h\|_{ReH^1}$$

defines a Banach space norm on H^1.

The space H^1 we have just defined consists of functions in $L^1(\mathbb{R})$. It is natural to ask if there is a simple description of these functions. There exists, in fact, a characterization of H^1 in terms of elementary particles called "atoms" that is most appropriate for our goal of obtaining wavelet bases for this space. An **atom** is a measurable function a defined on $\mathbb{R}$ that satisfies

$$\begin{cases} \text{(i)} & \text{the support of } a \text{ is contained in a finite interval } I \subset \mathbb{R}, \\[2mm] \text{(ii)} & |a(x)| \leq \dfrac{1}{|I|} \qquad \text{for all } x \in \mathbb{R}, \\[2mm] \text{(iii)} & \displaystyle\int_{\mathbb{R}} a(x)\, dx = 0. \end{cases} \qquad (5.3)$$

The following is a precise statement of the atomic characterization of $H^1 = H^1(\mathbb{R})$:

THEOREM 5.4 *A function f belongs to $H^1(\mathbb{R})$ if and only if f has a decomposition of the form*

$$f = \sum_{j=1}^{\infty} \lambda_j a_j, \tag{5.5}$$

where a_j, $j = 1, 2, 3, \cdots$, are atoms and $\sum_{j=1}^{\infty} |\lambda_j| < \infty$. Moreover,

$$\|f\|_{H^1} \approx \inf \left\{ \sum_{j=1}^{\infty} |\lambda_j| \right\}, \tag{5.6}$$

where the infimum is taken over all decompositions of f of the form (5.5).

There exists a similar characterization of H^1 in terms of certain functions, called molecules, that are somewhat more complicated than atoms. A **molecule** centered at $x_o \in \mathbb{R}$ is a measurable function M defined on $\mathbb{R}$ for which there exists $b > \frac{1}{2}$ such that

$$\begin{cases} \text{(i)} & \int_{\mathbb{R}} M(x)\, dx = 0, \\ \text{(ii)} & \|M\|_2^{1-t} \|M| \cdot -x_o|^b \|_2^t = C < \infty, \end{cases} \tag{5.7}$$

where $t = \frac{1}{2b}$ (so that $0 < t < 1$). The number C is called the **molecular norm** of M. The molecular characterization of H^1 is given by the following result:

THEOREM 5.8 *A function f belongs to $H^1(\mathbb{R})$ if and only if f has a decomposition of the form*

$$f = \sum_{j=1}^{\infty} \lambda_j M_j, \tag{5.9}$$

where each M_j is a molecule with molecular norm 1 and $\sum_{j=1}^{\infty} |\lambda_j| < \infty$. The infimum of all such sums, taken over all representations of f in the form of (5.9), gives us a norm equivalent to $\|f\|_{H^1(\mathbb{R})}$.

It is easy to see that an atom is also a molecule centered at the mid-point of the supporting interval of the atom, with molecular norm not exceeding 1. If x_o is the center of the interval $I \subset \mathbb{R}$ and a is an atom supported in I, we have

$$\|a\|_{L^2(\mathbb{R})} \leq \frac{1}{\sqrt{|I|}},$$

and, for any $b > 0$,

$$\left(\int_I |a(x)|^2 |x - x_o|^{2b} dx \right)^{\frac{1}{2}} \leq \frac{1}{|I|} \left(\int_I (\tfrac{1}{2}|I|)^{2b} dx \right)^{\frac{1}{2}} = \frac{1}{2^b} \frac{|I|^b}{\sqrt{|I|}} \leq \frac{|I|^b}{\sqrt{|I|}}.$$

Hence,

$$\|a\|_2^{1-t} \|a| \cdot - x_o|^b\|_2^t \leq \left(\frac{1}{\sqrt{|I|}} \right)^{1-t} \frac{|I|^{bt}}{(\sqrt{|I|})^t} = 1,$$

since $bt = \frac{1}{2}$. This shows that the atom a is a molecule.

One of the useful properties of molecules is that many important operators in analysis map atoms into molecules, having bounded molecular norms. This is the case for the "conjugate operator" (also known as the **Hilbert transform**) which is the operator $\mathcal{H}$ defined by

$$(\mathcal{H}f)(x) = \lim_{t \to 0+} v(x,t) = \lim_{t \to 0+} \frac{1}{\pi} \int_{\mathbb{R}} \frac{x-y}{(x-y)^2 + t^2} f(y)\, dy$$

$$= \lim_{t \to 0+} \frac{1}{\pi} \int_{|x-y|>t} \frac{f(y)}{x-y}\, dy.$$

This operator is the prototype of the Calderón-Zygmund operators. In the next section we shall show that these operators do, indeed, map atoms into molecules with bounded molecular norms. As we shall see, the basic role played by an appropriate wavelet ψ is that the kernels giving us the various partial sums of the wavelet expansion

$$\sum_{j \in \mathbb{Z}} \sum_{k \in \mathbb{Z}} \langle f, \psi_{j,k} \rangle \psi_{j,k}$$

are Calderón-Zygmund operators that map atoms into molecules with uniformly bounded norms. This will permit us to use Theorem 2.10 in order to show that $\{\psi_{j,k} : j, k \in \mathbb{Z}\}$ is an unconditional basis for the spaces we are studying.

Of course the importance of the Hilbert transform to the study of the space $H^1(\mathbb{R})$ is paramount. In fact, the definition of $ReH^1(\mathbb{R})$ given by inequality (5.2) is equivalent to the assertion that $f \in ReH^1(\mathbb{R})$ if and only if f and $\mathcal{H}f$ belong to $L^1(\mathbb{R})$ and f is real-valued. Since $\mathcal{H}$ is an operator that commutes with translations it has the property that there exists a bounded measurable function $m(\xi)$ such that

$$(\mathcal{H}f)^\wedge(\xi) = m(\xi)\hat{f}(\xi) \qquad a.e. \qquad \text{for all } f \in L^1 \cap L^2.$$

The "multiplier" m is easily seen to be

$$m(\xi) = -i\,\mathrm{sgn}\,\xi = -i \begin{cases} 1 & \text{if } \xi > 0, \\ 0 & \text{if } \xi = 0, \\ -1 & \text{if } \xi < 0. \end{cases}$$

That wavelets are most natural for the study of the space $H^1(\mathbb{R})$ follows from the fact that ψ **is an orthonormal wavelet if and only if** $\mathcal{H}\psi$ **is an orthonormal wavelet.** This follows immediately from the fact that equations (2.1) and (2.2) in Chapter 3 completely characterize an orthonormal wavelet ψ. (This general result is established in full generality in Chapter 7.)

We shall also consider the space dual to $H^1(\mathbb{R})$. In order to do this we introduce the **sharp maximal function** of a locally integrable function f on $\mathbb{R}$:

$$f^\sharp(x) \equiv \sup \frac{1}{|I|} \int_I |f(y) - f_I|\, dy,$$

where the supremum is taken over all intervals $I = (a, b)$ containing x and

$$f_I = \frac{1}{|I|} \int_I f(y)\, dy.$$

On the set of all locally integrable functions f for which $f^\sharp \in L^\infty(\mathbb{R})$ we define

$$\|f\|_* = \|f^\sharp\|_\infty.$$

We obtain a seminorm $\|\cdot\|_*$ such that $\|f\|_* = 0$ if and only if f equals a constant almost everywhere. The quotient of this space modulo these constant functions is called the space of functions of **bounded mean oscillation** and is denoted by $BMO \equiv BMO(\mathbb{R})$; moreover, $\|\cdot\|_*$ is a norm on this quotient space such that $(BMO, \|\cdot\|_*)$ is a Banach space. We shall also say that a function f belongs to BMO when it determines one of the equivalence classes in this quotient space. The following remarks are designed to make it plausible to the reader that the space dual to $H^1(\mathbb{R})$ is equivalent to $BMO(\mathbb{R})$.

Suppose that f is a locally integrable function and there exists a constant $C > 0$ such that for each interval I

$$\frac{1}{|I|} \int_I |f(x) - a_I| \, dx \leq C \tag{5.10}$$

for a constant a_I, then $f \in BMO$. In fact,

$$\frac{1}{|I|} \int_I |f(x) - f_I| \, dx \leq \frac{1}{|I|} \int_I |f(x) - a_I| \, dx + \frac{1}{|I|} \int_I |a_I - f_I| \, dx$$

$$\leq C + \left| \int_I (a_I - f(y)) \frac{dy}{|I|} \right| \leq 2C.$$

This shows that $\|f\|_* \leq 2C$. Now suppose $g \in BMO$. Since

$$\big| |g(x)| - |g_I| \big| \leq |g(x) - g_I|,$$

$f = |g|$ satisfies (5.10) with $a_I = |g_I|$ and $C = \|g\|_*$. Thus, $|g| \in BMO$ and its norm does not exceed $2\|g\|_*$. We see, therefore, that the space of BMO functions is a lattice: if f and g are two such functions, then

$$f \wedge g = \min\{f, g\} = \frac{f + g - |f - g|}{2} \qquad \text{and}$$

$$f \vee g = \max\{f, g\} = \frac{f + g + |f - g|}{2}$$

belong to BMO and

$$\|f \wedge g\|_* \leq \tfrac{3}{2}(\|f\|_* + \|g\|_*), \qquad \|f \vee g\|_* \leq \tfrac{3}{2}(\|f\|_* + \|g\|_*). \tag{5.11}$$

In particular, if $b \in BMO$, then

$$b_N(x) = \begin{cases} N & \text{if } b(x) > N \\ b(x) & \text{if } -N \leq b(x) \leq N \\ -N & \text{if } b(x) < -N \end{cases}$$

also belongs to BMO. Moreover, letting N denote the constant function having value N and using the fact that constants have BMO norm 0, we have, from (5.11),

$$\left\|b_N\right\|_* = \left\|(b \wedge N) \vee (-N)\right\|_* \leq \tfrac{3}{2}\left\|b \wedge N\right\|_* \leq \tfrac{9}{4}\left\|b\right\|_*. \tag{5.12}$$

Suppose a is an atom supported in an interval I and $b \in BMO$. Then, using (5.3) (ii) and (iii), we have

$$\left|\int_{\mathbb{R}} a(x)b(x)\,dx\right| = \left|\int_I a(x)(b(x) - b_I)\,dx\right| \leq \frac{1}{|I|}\int_I |b(x) - b_I|\,dx \leq \|b\|_*.$$

Thus, if b is a bounded function in BMO and $f = \sum_j \lambda_j a_j \in H^1(\mathbb{R})$, then $fb \in L^1(\mathbb{R})$ and

$$\left|\int_{\mathbb{R}} f(x)b(x)\,dx\right| = \left|\sum_j \lambda_j \int_{I_j} a_j(x)b(x)\,dx\right| \leq \left(\sum_j |\lambda_j|\right)\|b\|_*,$$

where the a_j's are atoms supported in the intervals I_j. Taking the infimum over all sums $\sum_j |\lambda_j|$ such that $f = \sum_j \lambda_j a_j$, with a_j an atom, we have shown that the linear functional $\mathcal{L}_b$ defined by

$$\mathcal{L}_b(f) = \int_{\mathbb{R}} f(x)b(x)\,dx \tag{5.13}$$

satisfies

$$|\mathcal{L}_b(f)| \leq \|b\|_*\|f\|_{H^1} \tag{5.14}$$

when $b \in BMO \cap L^\infty$. That is, $\mathcal{L}_b \in (H^1)^*$, the dual of H^1. We cannot define $\mathcal{L}_b$ directly by the integral in (5.13) for a general $b \in BMO$ since fb is not necessarily integrable for $f \in H^1$. But $\mathcal{L}_{b_N}$ can be so defined, for each $N = 1, 2, 3, \cdots$. Combining (5.12) with (5.14) we have

$$|\mathcal{L}_{b_N}(f)| \leq \tfrac{9}{4}\|b\|_*\|f\|_{H^1}. \tag{5.15}$$

Let $\mathbb{V}$ be the subspace of H^1 consisting of all f that are finite linear combinations of atoms. If $f \in \mathbb{V}$, then f is both bounded and compactly supported; moreover, $\mathbb{V}$ is easily seen to be dense in H^1. The integral in equality (5.13) is well defined for $f \in \mathbb{V}$ and a general $b \in BMO$. What is not a priori clear is that the linear functional defined by (5.13) on $\mathbb{V}$ is bounded. But this is an easy consequence of (5.15) and the fact that

$$\int_{\mathbb{R}} b_N(x)f(x)\,dx \longrightarrow \int_{\mathbb{R}} b(x)f(x)\,dx \qquad \text{when } N \to \infty$$

for $f \in \mathbb{V}$.

This shows that $\mathcal{L}_b(f) = \int_{\mathbb{R}} b(x)f(x)\,dx$ is, not only well defined on $\mathbb{V}$, but the linear functional $\mathcal{L}_b$ is bounded on $\mathbb{V}$:

$$|\mathcal{L}_b(f)| \leq \tfrac{9}{4}\|b\|_*\|f\|_{H^1}. \tag{5.16}$$

We then define $\mathcal{L}_b$ on all of H^1 to be the unique linear extension to H^1 of this functional originally defined on the dense space $\mathbb{V}$. We shall not prove that all linear functionals $\mathcal{L} \in (H^1)^*$ have the form $\mathcal{L} = \mathcal{L}_b$ for some $b \in BMO$; however, as stated above, these observations make this statement quite plausible.

There is one more technique that shall be used in the next section: the "interpolation of operators." In very general terms this technique can be described as follows: suppose T is an operator on a linear topological space $\mathbb{W}$ containing two Banach spaces $\mathbb{X}_0$ and $\mathbb{X}_1$ such that T is bounded from $\mathbb{X}_j$ to $\mathbb{X}_j$, $j = 0, 1$, then there exists a scale of spaces $\mathbb{X}_\theta$, $\theta \in [0, 1]$, such that $\mathbb{X}_\theta \subseteq \mathbb{W}$ and $T : \mathbb{X}_\theta \longrightarrow \mathbb{X}_\theta$ boundedly. In fact, considerably more general results are true, but we shall only apply this technique to the case $\mathbb{X}_0 = H^1(\mathbb{R})$, $\mathbb{X}_1 = L^2(\mathbb{R})$ where the "intermediate" spaces are the spaces $L^{p_\theta}(\mathbb{R})$, $p_\theta = 2/(2-\theta)$. We shall not prove this result but will cite complete references in the seventh section of this chapter.

5.6 Wavelets as unconditional bases for $H^1(\mathbb{R})$ and $L^p(\mathbb{R})$ with $1 < p < \infty$

We shall prove that if an orthonormal wavelet ψ satisfies certain regularity conditions and has an appropriate decay at infinity, then the family $\{\psi_{j,k} : j, k \in \mathbb{Z}\}$ is an unconditional basis for $H^1(\mathbb{R})$ and $L^p(\mathbb{R})$, $1 < p < \infty$.

Guided by Theorem 2.10 we are led to examine the boundedness of the operator

$$T_\beta f = \sum_{j \in \mathbb{Z}} \sum_{k \in \mathbb{Z}} \beta_{j,k} \langle f, \psi_{j,k} \rangle \psi_{j,k} \tag{6.1}$$

(in $H^1(\mathbb{R})$ and in $L^p(\mathbb{R})$, $1 < p < \infty$), where $\beta = \{\beta_{j,k}\}$ is a sequence such that $\beta_{j,k} = 1$ for a **finite** number of indices and $\beta_{j,k} = 0$ for the remaining indices. The operator T_β can be written as an integral operator of the form

$$(T_\beta f)(x) = \int_{\mathbb{R}} K_\beta(x,y)f(y)\,dy, \qquad (6.2)$$

where

$$K_\beta(x,y) = \sum_{j\in\mathbb{Z}}\sum_{k\in\mathbb{Z}} \beta_{j,k}\psi_{j,k}(x)\,\overline{\psi_{j,k}(y)}. \qquad (6.3)$$

If the wavelet ψ is bounded by a radial decreasing L^1-majorant W, we can use Lemma 3.12 to obtain the following estimate:

$$|K_\beta(x,y)| \le \sum_{j\in\mathbb{Z}} 2^j \sum_{k\in\mathbb{Z}} |\psi(2^j x - k)\psi(2^j y - k)|$$

$$\le \sum_{j\in\mathbb{Z}} 2^j \sum_{k\in\mathbb{Z}} W(|2^j x - k|)W(|2^j y - k|)$$

$$\le C \sum_{j\in\mathbb{Z}} 2^j W(2^{j-1}|x - y|), \qquad (6.4)$$

where C depends only on W. We shall use this estimate to show that the operator T_β is a Calderón-Zygmund operator.

In the one-dimensional case we are considering, a **Calderón-Zygmund operator** T is a bounded linear operator on $L^2(\mathbb{R})$ such that, for $x \notin$ supp (f),

$$(Tf)(x) = \int_{\mathbb{R}} K(x,y)f(y)\,dy,$$

where the kernel K is a jointly measurable function satisfying

$$|K(x,y)| \le \frac{C_1}{|x - y|}, \qquad (6.5)$$

$$|K(x_o,y) - K(x,y)| \le \frac{C_2|x - x_o|}{|x - y|^2} \quad \text{if } |x - x_o| \le \tfrac{1}{2}|x - y|, \quad (6.6)$$

$$|K(x,y_o) - K(x,y)| \le \frac{C_3|y - y_o|}{|x - y|^2} \quad \text{if } |y - y_o| \le \tfrac{1}{2}|x - y|. \quad (6.7)$$

The Hilbert transform $\mathcal{H}$, which we introduced in the last section, is easily seen to be a Calderón-Zygmund operator. The kernel associated with $\mathcal{H}$ is

$$K(x,y) = \frac{1}{\pi}\frac{1}{x - y}$$

and it is immediate that (6.5), (6.6) and (6.7) are satisfied by $K(x, y)$. The fact that $\mathcal{H}$ is bounded on $L^2(\mathbb{R})$ is an elementary fact; perhaps the simplest way of seeing this is to use the fact that

$$(\mathcal{H}f)^\wedge(\xi) = -i\,(\text{sgn}\,\xi)\hat{f}(\xi).$$

THEOREM 6.8 *Suppose that T is a Calderón-Zygmund operator such that*

$$\int_{\mathbb{R}} Tf(x)\,dx = 0 \qquad and \qquad \int_{\mathbb{R}} T^*f(x)\,dx = 0$$

whenever $f \in L^2(\mathbb{R}) \cap L^\infty(\mathbb{R})$ and $\int_{\mathbb{R}} f(x)\,dx = 0$, where T^* is the dual of T. Then T extends to a bounded operator on $H^1(\mathbb{R})$, $BMO(\mathbb{R})$ and on $L^p(\mathbb{R})$, $1 < p < \infty$, with operator norm depending only on $\|T\|_{L^2(\mathbb{R})}$ and the constants that appear in the inequalities (6.5), (6.6) and (6.7).

PROOF: It is enough to show that the operator T maps atoms into molecules with molecular norms that are bounded independently of the atoms. (This follows from Theorem 5.4 and Theorem 5.8 announced in the last section.) Let a be an atom with supp $(a) \subset I$. The boundedness of T on $L^2(\mathbb{R})$ implies

$$\|Ta\|_{L^2(\mathbb{R})} \le C\|a\|_{L^2(\mathbb{R})} \le C|I|^{-\frac{1}{2}}. \tag{6.9}$$

Let $x \notin 3I$ and $y \in I$, so that, if y_o is the center of I, $|y - y_o| \le \frac{1}{2}|x - y|$. Using (5.3) (iii) and (6.7) we obtain

$$|Ta(x)| = \left| \int_I [K(x, y) - K(x, y_o)]a(y)\,dy \right| \le C_3 \int_I \frac{|y - y_o|}{|x - y|^2}\,|a(y)|\,dy.$$

The inequality $|y - y_o| \le \frac{1}{2}|x - y|$ implies

$$\tfrac{3}{2}|x - y| \ge |x - y| + |y - y_o| \ge |x - y_o|.$$

Hence, the Schwarz inequality gives us

$$|Ta(x)| \le C \frac{C_3}{|x - y_o|^2} \int_I |y - y_o|\,|a(y)|\,dy$$

$$\le C\frac{C_3}{|x - y_o|^2}\|a\|_{L^2(\mathbb{R})}\left(\int_I |y - y_o|^2\,dy\right)^{\frac{1}{2}} \le \frac{C}{|x - y_o|^2}|I| \tag{6.10}$$

when $x \notin 3I$ (here, as well as in other cases where it is not a source of confusion, C may denote different constants). Choose $\frac{1}{2} < b < \frac{3}{2}$ and consider

$$\left\| |\cdot - y_0|^b Ta \right\|_{L^2(\mathbb{R})}^2 = \int_{\mathbb{R}} |y - y_0|^{2b} |Ta(y)|^2 \, dy$$

$$= \left(\int_{3I} + \int_{(3I)^c} \right) |y - y_0|^{2b} |Ta(y)|^2 \, dy.$$

Inequalities (6.9) and (6.10) then imply

$$\left\| |\cdot - y_0|^b Ta \right\|_{L^2(\mathbb{R})}^2 \leq C|I|^{2b-1} + C \left(\int_{(3I)^c} |y - y_0|^{2b-4} \, dy \right) |I|^2$$

$$\leq C|I|^{2b-1}. \qquad (6.11)$$

This inequality, together with (6.9) and $t = \dfrac{1}{2b}$, gives us

$$\left\| Ta \right\|_{L^2(\mathbb{R})}^{1-t} \left\| |\cdot - y_0|^b Ta \right\|_{L^2(\mathbb{R})}^{t} \leq C|I|^{-\frac{1-t}{2}} |I|^{\frac{1-t}{2}} = C,$$

which is part (ii) of (5.7). The moment condition (5.7) (i) for Ta follows from $\int_{\mathbb{R}} a(x) \, dx = 0$ and our assumption. This finishes the proof that the operator T is bounded on $H^1(\mathbb{R})$.

The case $1 < p \leq 2$ follows from interpolation between $H^1(\mathbb{R})$ and $L^2(\mathbb{R})$ (see the last paragraph of section 5.5 and the Notes and References at the end of this chapter). The cases $2 < p < \infty$ and BMO can be reduced to the one just established if we show that the adjoint operator T^* is bounded on $H^1(\mathbb{R})$. But this is proved as we just did for T, if we use (6.6) instead of (6.7). ∎

We now return to the main purpose of this section, the construction of unconditional wavelet bases. We begin with the following result:

THEOREM 6.12 *Suppose that ψ is an orthonormal wavelet such that ψ and ψ' have a common radial decreasing L^1-majorant W satisfying*

$$\int_0^\infty s W(s) \, ds < \infty.$$

Then, the operators T_β defined by (6.2) and (6.3) are bounded on $H^1(\mathbb{R})$, $BMO(\mathbb{R})$ and $L^p(\mathbb{R})$, $1 < p < \infty$, with norm bounded by a constant independent of the finitely non-zero sequence β consisting of zeroes and ones.

PROOF : Clearly T_β is bounded on $L^2(\mathbb{R})$ since the system $\{\psi_{j,k} : j, k \in \mathbb{Z}\}$ is an orthonormal basis for $L^2(\mathbb{R})$:

$$\|T_\beta f\|_{L^2(\mathbb{R})}^2 = \sum_{j,k \in \mathbb{Z}} |\beta_{j,k} \langle f, \psi_{j,k}\rangle|^2 \leq \sum_{j,k \in \mathbb{Z}} |\langle f, \psi_{j,k}\rangle|^2 = \|f\|_{L^2(\mathbb{R})}^2.$$

For any $f \in L^2(\mathbb{R})$,

$$\int_{\mathbb{R}} T_\beta f(x)\, dx = 0 = \int_{\mathbb{R}} T_\beta^* f(x)\, dx,$$

since $T_\beta f$ and $T_\beta^* f$ are finite linear combinations of the $\psi_{j,k}$'s and, moreover, $0 = \hat{\psi}(0) = \int_{\mathbb{R}} \psi(x)\, dx$ (see Proposition 2.1 in section 7.2). It remains to show (6.5), (6.6) and (6.7) so that Theorem 6.8 applies. To prove (6.5) we use (6.4) and obvious estimates, to obtain

$$|K_\beta(x,y)| \leq C \sum_{j \in \mathbb{Z}} 2^j W(2^{j-1}|x-y|)$$

$$= C \sum_{j=-\infty}^{0} 2^j W(2^{j-1}|x-y|) + C \sum_{j=1}^{\infty} 2^j W(2^{j-1}|x-y|)$$

$$\leq 2C \int_0^1 W(\tfrac{1}{2}t|x-y|)\, dt + 2C \int_1^\infty W(\tfrac{1}{2}t|x-y|)\, dt$$

$$= 2C \int_0^\infty \frac{2}{|x-y|} W(s)\, ds = \frac{4C}{|x-y|} \|W\|_{L^1(0,\infty)}.$$

To prove (6.6) we assume $x_o < x$. We shall show

$$\left|\frac{\partial K_\beta}{\partial x}(x,y)\right| \leq \frac{C}{|x-y|^2} \qquad \text{for } y \in \mathbb{R}. \tag{6.13}$$

It is easily seen that inequality (6.13) implies (6.6). To see this apply the Mean Value Theorem to obtain a point $x' \in (x_o, x)$ such that

$$|K_\beta(x_o,y) - K_\beta(x,y)| \leq |x_o - x| \cdot \left|\frac{\partial K_\beta(x',y)}{\partial x}\right| \leq \frac{C|x_o - x|}{|x' - y|^2}.$$

Observe that (6.6) implies that $y \notin (x_o, x)$. If $y \geq x$, it is clear that

$$|x' - y| \geq |x - y| \geq \tfrac{1}{2}|x - y|.$$

If $y \leq x_o$ we use (6.6) to obtain

$$|x' - y| \geq |x_o - y| \geq |x - y| - |x - x_o| \geq \tfrac{1}{2}|x - y|.$$

Hence,

$$|K_\beta(x_o, y) - K_\beta(x, y)| \leq \frac{4C|x_o - x|}{|x - y|^2} \quad \text{if } |x - x_o| \leq \frac{1}{2}|x - y|.$$

Thus, we need to show (6.13). We use Lemma 3.12 and the fact that W is decreasing to obtain

$$\left| \frac{\partial K_\beta}{\partial x}(x, y) \right| = \left| \sum_{j,k \in \mathbb{Z}} \beta_{j,k} 2^j 2^j \psi'(2^j x - k) \overline{\psi(2^j y - k)} \right|$$

$$\leq \sum_{j \in \mathbb{Z}} \sum_{k \in \mathbb{Z}} 2^{2j} W(|2^j x - k|) W(|2^j y - k|)$$

$$\leq C \sum_{j \in \mathbb{Z}} 2^{2j} W(2^{j-1}|x - y|) \leq 2C \int_0^\infty t W(\tfrac{1}{2}t|x - y|)\, dt$$

$$= 2C \int_0^\infty \frac{2s}{|x - y|} W(s) \frac{2ds}{|x - y|} = \frac{8C}{|x - y|^2} \int_0^\infty s W(s)\, ds.$$

This proves (6.13) and, consequently, (6.6). Inequality (6.7) follows from a similar argument. This finishes the proof of the theorem. ∎

Now we can prove the unconditionality of some wavelet bases for $L^p(\mathbb{R})$, $1 < p < \infty$.

THEOREM 6.14 *Suppose that ψ is an orthonormal wavelet such that ψ and ψ' have a common radial decreasing L^1-majorant W satisfying*

$$\int_0^\infty s W(s)\, ds < \infty.$$

Then, the system $\{\psi_{j,k} : j, k \in \mathbb{Z}\}$ is an unconditional basis for $L^p(\mathbb{R})$, $1 < p < \infty$.

PROOF: We start by showing that the system considered is a basis for $L^p(\mathbb{R})$, $1 < p < \infty$. To do this let $S_{m,n}f$ be the "rectangular" partial sum of the wavelet expansion of f; that is,

$$S_{m,n}f = \sum_{|j| \leq m} \sum_{|k| \leq n} \langle f, \psi_{j,k} \rangle \psi_{j,k} \tag{6.15}$$

for $f \in L^p(\mathbb{R})$, $1 < p < \infty$. This operator is well defined by Theorem 6.12. We shall show, first, that given $f \in L^p(\mathbb{R})$, $1 < p < \infty$, and $\varepsilon > 0$ we can find m and n large enough so that

$$\left\| f - S_{m,n}f \right\|_{L^p(\mathbb{R})} < \varepsilon.$$

Let $C = \sup \|T_\beta\| < \infty$, where the T_β's are the operators of Theorem 6.12 and the supremum is taken over all admissible sequences $\beta = \{\beta_{j,k}\}$ considered in Theorem 6.12. Since $L^2(\mathbb{R}) \cap L^p(\mathbb{R})$ is dense in $L^p(\mathbb{R})$ we can find $g \in L^2(\mathbb{R}) \cap L^p(\mathbb{R})$ such that

$$\left\| f - g \right\|_{L^p(\mathbb{R})} < \frac{\varepsilon}{C+3}.$$

We can write

$$\left\| f - S_{m,n}f \right\|_p \leq \left\| f - g \right\|_p + \left\| g - S_{m,n}g \right\|_p + \left\| S_{m,n}(g - f) \right\|_p \tag{6.16}$$

The first summand on the right hand side of (6.16) is smaller than $\frac{\varepsilon}{C+3}$; by Theorem 6.12, the last summand is smaller than $\frac{\varepsilon C}{C+3}$. It remains to estimate $\left\| g - S_{m,n}g \right\|_p$ for $g \in L^2(\mathbb{R}) \cap L^p(\mathbb{R})$. By duality, and the density of $L^2(\mathbb{R}) \cap L^{p'}(\mathbb{R})$ in $L^{p'}(\mathbb{R})$ (where $\frac{1}{p} + \frac{1}{p'} = 1$), we can find $h \in L^2(\mathbb{R}) \cap L^{p'}(\mathbb{R})$ such that

$$\left\| g - S_{m,n}g \right\|_p \leq \left| \int_{\mathbb{R}} [g(x) - S_{m,n}g(x)] \, \overline{h(x)} \, dx \right| + \frac{\varepsilon}{C+3}. \tag{6.17}$$

Using the Schwarz inequality we deduce

$$\left| \int_{\mathbb{R}} [g(x) - S_{m,n}g(x)] \, \overline{h(x)} \, dx \right| = \left| \int_{\mathbb{R}} g(x) \left[\overline{h(x)} - \overline{S_{m,n}h(x)} \right] dx \right|$$

$$\leq \left\| g \right\|_2 \left\| h - S_{m,n}h \right\|_2.$$

Since $\{\psi_{j,k} : j, k \in \mathbb{Z}\}$ is an orthonormal basis for $L^2(\mathbb{R})$, we can find m and n large enough so that

$$\left\| h - S_{m,n}h \right\|_2 < \frac{\varepsilon}{\|g\|_2 (C+3)}.$$

Hence,

$$\left\|f - S_{m,n}f\right\|_p \le \frac{\varepsilon}{C+3} + \left\|g\right\|_2 \frac{\varepsilon}{\left\|g\right\|_2(C+3)} + \frac{\varepsilon}{C+3} + \frac{\varepsilon C}{C+3} = \varepsilon.$$

The uniqueness of the representation

$$f = \sum_{j \in \mathbb{Z}} \sum_{k \in \mathbb{Z}} c_{j,k} \psi_{j,k} \qquad (6.18)$$

with convergence in $L^p(\mathbb{R})$, $1 < p < \infty$, follows from the orthonormality of the system $\{\psi_{j,k} : j, k \in \mathbb{Z}\}$: if (6.18) is true, multiply both sides by $\overline{\psi_{m,n}}$ and integrate to obtain $c_{m,n} = \langle f, \psi_{m,n} \rangle$. The unconditionality of the wavelet basis follows from Theorem 6.12 and Theorem 2.10. ∎

THEOREM 6.19 *Suppose that ψ is an orthonormal wavelet for $L^2(\mathbb{R})$ such that ψ and ψ' have a common radial decreasing L^1-majorant W satisfying*

$$\int_0^\infty sW(s) \, ds < \infty.$$

Then, the system $\{\psi_{j,k} : j, k \in \mathbb{Z}\}$ is an unconditional basis for $H^1(\mathbb{R})$.

PROOF: The proof is very similar to that of Theorem 6.14. We need to show that this system is a basis for $H^1(\mathbb{R})$. Inequality (6.16) is true with the L^p-norm replaced by the H^1-norm and choosing g to be a finite linear combination of atoms. Since $(H^1(\mathbb{R}))^* = BMO(\mathbb{R})$ we can find a bounded function $h \in BMO(\mathbb{R})$ such that (6.17) is true with $\|\cdot\|_p$ replaced by $\|\cdot\|_{H^1(\mathbb{R})}$. By choosing M large enough we have

$$\left\|g - S_{m,n}g\right\|_{H^1} \le \left| \int_{\mathbb{R}} [g(x) - S_{m,n}g(x)] \chi_{[-M,M]}(x) \, \overline{h(x)} \, dx \right| + \frac{\varepsilon}{C+3}. \quad (6.20)$$

Observe that $\chi_{[-M,M]} h \in L^2(\mathbb{R})$ and, thus, the proof can be finished as was the proof of Theorem 6.14. ∎

Theorem 6.14 and Theorem 6.19 apply to many wavelets. They apply to all sufficiently smooth (on the Fourier transform side) Lemarié-Meyer wavelets, to all differentiable compactly supported wavelets and to all spline wavelets of order $n \geq 2$. These theorems do not apply to the Shannon wavelet $\psi_{\mathfrak{s}}$ (since $\psi_{\mathfrak{s}}(x) \sim |x|^{-1}$ as $|x| \to \infty$). They cannot be applied to the Haar wavelet and the Franklin wavelet ψ^1; this is due to the fact that we have assumed that the wavelet is differentiable in the proof of Theorem 6.12. In fact, we obtained condition (6.6) for a Calderón-Zygmund operator as a consequence of (6.13) and the Mean Value Theorem.

If we want to extend these results for the Franklin wavelet ψ^1, we would like to show (6.6) directly, without the aid of (6.13). First notice that, at the points x where ψ^1 is differentiable,

$$\left|\frac{d}{dx}\psi^1(x)\right| \leq Ce^{-\varepsilon|x|} \equiv CW(|x|). \tag{6.21}$$

This can be seen by using the exponential decay of ψ^1 and the linearity of ψ^1 on intervals of the form $[\frac{1}{2}\ell, \frac{1}{2}(\ell+1)]$, $\ell \in \mathbb{Z}$. Alternatively, we can apply the argument that was used in the proof of Proposition 1.16 of Chapter 4.

For $j \in \mathbb{Z}$, let

$$K_j(x,y) = \sum_{k\in\mathbb{Z}} \beta_{j,k}\psi_{j,k}(x)\,\overline{\psi_{j,k}(y)} = 2^j \sum_{k\in\mathbb{Z}} \beta_{j,k}\psi(2^jx - k)\,\overline{\psi(2^jy - k)}.$$

Thus,

$$K_\beta(x,y) = \sum_{j\in\mathbb{Z}} K_j(x,y).$$

Suppose that $x_o, x \in \left[\dfrac{\ell}{2^{j+1}}, \dfrac{\ell+1}{2^{j+1}}\right]$; then

$$2^jx_o - k, 2^jx - k \in \left[\frac{\ell}{2} - k, \frac{\ell+1}{2} - k\right].$$

With this choice of x_o and x we can apply (6.21) to obtain

$$|K_j(x_o,y) - K_j(x,y)| \leq 2^j \sum_{k\in\mathbb{Z}} |\psi(2^jx_o - k) - \psi(2^jx - k)|\cdot|\psi(2^jy - k)|$$

$$\leq 2^{2j} \sum_{k\in\mathbb{Z}} \left|\frac{d}{dx}\psi(2^jx(j) - k)\right|\cdot|x - x_o|\cdot|\psi(2^jy - k)|$$

$$\leq C2^{2j}|x - x_o| \sum_{k\in\mathbb{Z}} W(|2^jx(j) - k|)W(|2^jy - k|),$$

where $x(j) \in (x_o, x)$. By Lemma 3.12 we then obtain

$$|K_j(x_o,y) - K_j(x,y)| \leq C2^{2j}|x - x_o|W(2^{j-1}|x(j) - y|).$$

For $x(j) \in (x_o, x)$ and $|x - y| \geq 2|x - x_o|$ we have $|x(j) - y| \geq \frac{1}{2}|x - y|$. Since W is decreasing, we obtain

$$|K_j(x_o, y) - K_j(x, y)| \leq C 2^{2j}|x - x_o| W(2^{j-2}|x - y|).$$

It follows from this inequality, as in the proof of (6.13), that

$$|K_\beta(x_o, y) - K_\beta(x, y)| \leq C|x - x_o| \sum_{j \in \mathbb{Z}} 2^{2j} W(2^{j-2}|x - y|)$$

$$\leq \frac{16C}{|x - y|^2}|x - x_o| \, \|W\|_{L^1(0,\infty)}.$$

This shows (6.6) when $x, x_o \in \left[\dfrac{\ell}{2^{j+1}}, \dfrac{\ell+1}{2^{j+1}}\right]$ for some $\ell \in \mathbb{Z}$.

If $x_o \in \left[\dfrac{\ell}{2^{j+1}}, \dfrac{\ell+1}{2^{j+1}}\right]$ and $x \in \left[\dfrac{\ell+1}{2^{j+1}}, \dfrac{\ell+2}{2^{j+1}}\right]$ we write

$$|\psi(2^j x_o - k) - \psi(2^j x - k)|$$

$$\leq \left|\psi(2^j x_o - k) - \psi\left(2^j \frac{\ell+1}{2^{j+1}} - k\right)\right| + \left|\psi\left(2^j \frac{\ell+1}{2^{j+1}} - k\right) - \psi(2^j x - k)\right|$$

and proceed as before to obtain

$$|K_\beta(x_o, y) - K_\beta(x, y)| \leq \frac{C}{|x - y|^2}\left\{\left|\frac{\ell+1}{2^{j+1}} - x_o\right| + \left|x - \frac{\ell+1}{2^{j+1}}\right|\right\}$$

$$= \frac{C|x - x_o|}{|x - y|^2} \tag{6.22}$$

if $|x - x_o| \leq \frac{1}{2}|x - y|$.

Finally, if x and x_o are arbitrary we write

$$x_o \equiv x_\ell < x_{\ell+1} < \cdots < x_{m-1} < x_m \equiv x,$$

where $x_n \in \left[\dfrac{n}{2^{j+1}}, \dfrac{n+1}{2^{j+1}}\right]$, $n = \ell, \cdots, m-1$. By applying (6.22) $m - \ell$ times we obtain

$$|K_\beta(x_o, y) - K_\beta(x, y)| \leq \frac{C}{|x - y|^2} \sum_{n=\ell}^{m-1} |x_{n+1} - x_n| = \frac{C|x - x_o|}{|x - y|^2}.$$

Arguments similar to the ones that established Theorem 6.14 and Theorem 6.19 give us the following result:

THEOREM 6.23 *If ψ^1 is the Franklin wavelet for the real line, then the system $\{\psi^1_{j,k} : j, k \in \mathbb{Z}\}$ is an unconditional basis for $H^1(\mathbb{R})$ and $L^p(\mathbb{R})$, $1 < p < \infty$.*

5.7 Notes and references

1. A basis can be defined on any general topological vector space $\mathbb{B}$; in this case the convergence of $x = \sum_{i \in \mathbb{N}} \alpha_i(x)x_i$ is in the topology of $\mathbb{B}$. In this general setting it is not always true that the coefficient maps f_i are continuous (see [Sin] page 153, or [Mar] page 127). The book of I. Singer ([Sin]) is a standard reference for matters related to bases on topological vector spaces. If for a basis $\mathfrak{B}$ these coefficient functionals are all continuous, $\mathfrak{B}$ is called a **Schauder basis** for the topological vector space $\mathbb{B}$. Theorem 1.6 shows that all bases in a Banach space are Schauder bases.

2. That the system $\{e^{ikx} : k \in \mathbb{Z}\}$ is a basis for $L^p(\mathbb{T})$, $1 < p < \infty$, can be found in [Kat] page 50. The non-unconditionality of this basis in $L^p(\mathbb{T})$, if $p \neq 2$, was first published by Paley and Zygmund (see [PZ]). The basis $\{h_{j,k} : j = 0, 1, 2, \cdots, k = 0, 1, \cdots, 2^j - 1\}$ for $L^p([0,1])$, $1 \leq p < \infty$, presended at the end of section 5.1, is often called the (periodic) Haar basis. The proof that it is a basis can be found in [Wo2].

3. We have used several results from the theory of interpolation of operators. For the one used in the proof of Proposition 3.13 see Theorem 1.3 in Chapter V of the book [SW] by Stein and Weiss. For the interpolation between $H^1(\mathbb{R})$ and $L^2(\mathbb{R})$, mentioned at the end of section 5.5 and used in the proof of Theorem 6.8 see [FS1]. (Another good reference is Theorem 6.1 of Chapter III in [GR].)

4. Important articles in the development of the modern theory of $H^1(\mathbb{R})$ are [BGS], [Coi] and [FS1]. Details of the results stated in section 5.5 for $H^1(\mathbb{R})$ and $BMO(\mathbb{R})$ can be found in [GR] and [CW2]. The duality theorem between $H^1(\mathbb{R})$ and $BMO(\mathbb{R})$ was originally proved by C. Fefferman in [Fef]. Extensions of some of these results to more general settings can be found in [TW].

5. The Carleson-Hunt theorem asserts that the Fourier series of a function

$f \in L^p(\mathbb{T})$ converges pointwise to $f(x)$ for a.e. $x \in \mathbb{T}$, $p > 1$. This result is true for $L^p(\mathbb{R})$, $1 < p < \infty$. More precisely,

$$\frac{1}{2\pi} \int_{-N}^{N} \left\{ \int_{\mathbb{R}} f(y) e^{-i\xi y} \, dy \right\} e^{ix\xi} \, d\xi \longrightarrow f(x) \quad a.e. \quad \text{as } N \to \infty$$

for such f. The proof of this (as well as the proper definition of the integral involved) can be obtained from a "transference" argument (see [BPW] and [KT]).

6. Results about norm and pointwise convergence of wavelet expansions were first published in [Me1] under more restrictive hypotheses. Our treatment of these two topics is essentially similar to the one published in [KKR1] and [KKR2]. (The n-dimensional case is also included in these articles.) These references also contain necessary and sufficient conditions for given rates of convergence of wavelet expansions in the L^∞-norm. More on pointwise convergence of wavelet expansions can be found in [Wal1]. It is worthwhile to observe that the pointwise convergence of wavelet expansions that we considered in this chapter is relatively easy to prove when compared with the convergence of the partial sums of Fourier series (the celebrated Carleson-Hunt theorem). In fact, the convergence we are studying for wavelet expansions is really an analog of summability results for Fourier series in which the Lebesgue points of f play an important role.

7. Condition (6.5) was not used in the proof of Theorem 6.8. In fact, (6.5) is used in the Calderón-Zygmund operator theory to justify the definition of T. On the other hand, under very weak assumptions on T (namely that T satisfies the so-called **weak boundedness** property), conditions (6.6) and (6.7) imply (6.5). The interested reader can find a proof in [HS].

8. The subject of unconditionality of wavelet bases for Lebesgue spaces $L^p(\mathbb{R})$, $1 < p < \infty$, is treated in [Me1] for wavelets with polynomial decay. A result with weaker hypotheses, which are similar to the ones in Theorem 6.14, is given in [Gri2].

9. The story of unconditional bases for the Hardy space $H^1(\mathbb{R})$ is an interesting one. Around 1980, B. Maurey gave a non-constructive proof of the existence of an unconditional basis for $H^1(\mathbb{R})$ (see [Mau]). Soon thereafter, constructive proofs were given by L. Carleson ([Car2]) and P. Wojtaszczyk ([Wo1]). The example of an unconditional basis for H^1 given in [Wo1] is the Franklin system. The fact that a large class of wavelets are unconditional bases for $H^1(\mathbb{R})$ was discovered by Y. Meyer (see [Me1]). Our approach in section 5.6, using the atomic decomposition of H^1 described in [Coi] and the Calderón-Zygmund theory for boundedness of certain operators, is essentially the approach of [Me1].

10. To keep our exposition simple we have not treated the case of the H^p spaces for $0 < p < 1$. It was J.O. Strömberg ([Str]) who first described unconditional bases for these spaces: they are spline systems of higher order. It is worthwhile noticing that Strömberg's spline systems are spline wavelets as described in Chapter 4. The relation of his construction to wavelets was not observed until well after the development of the general theory of wavelets began. The approach described in section 5.6 for H^1 and L^p also works, with minor modifications, for H^p, $0 < p < 1$. One needs atoms with higher zero moments to be able to have "atomic" and "molecular" theories for these spaces, and, consequently, one needs wavelets with increasing regularities as $p \to 0$ (see Theorem 3.4 of Chapter 2) (see [Me1] for details). Because of this, Y. Meyer often refers to J.O. Strömberg as the "creator" of the modern concept of a wavelet.

11. The existence of unconditional bases on weighted Lebesgue and Hardy spaces was treated in [GK1], [GK2] and [GK3]. They found conditions on a weight w for spline wavelets to be unconditional bases for $L^p(w)$, $1 < p < \infty$, and $H^p(w)$, $0 < p \leq 1$. For the case $L^p(w)$, $1 < p < \infty$, the condition is $w \in A_p$, while for the space $H^p(w)$, $0 < p \leq 1$, the regularity of the spline wavelet and the critical index of the weight play a role.

The subject of unconditional bases of wavelets for Sobolev spaces is treated in [Gri3] (see also the end of section 6.6). Chapter 6 also provides results on unconditionality of wavelet bases for many fundamental spaces in Analysis (see the end of sections 6.4, 6.5 and 6.6, and also the notes in section 6.8 concerning the Triebel-Lizorkin spaces).

12. The Haar system is an unconditional basis for $L^p(\mathbb{R})$, $1 < p < \infty$ (see, for example, [Wo2]). We sketch a proof of this result below. As in (6.1) we are led to examine the boundedness of the operator

$$T_\beta f = \sum_{j \in \mathbb{Z}} \sum_{k \in \mathbb{Z}} \beta_{j,k} \langle f, h_{j,k} \rangle h_{j,k} \tag{7.1}$$

on $L^p(\mathbb{R})$, $1 < p < \infty$, where h denotes the Haar wavelet

$$h(x) = \chi_{[0,\frac{1}{2})}(x) - \chi_{[\frac{1}{2},1)}(x),$$

and $\beta = \{\beta_{j,k}\}$ is a sequence such that $\beta_{j,k} = 1$ for a finite number of indices and $\beta_{j,k} = 0$ for the remaining indices. Thus, as in (6.2) and (6.3), we need to study the operator

$$(T_\beta f)(x) = \int_{\mathbb{R}} K_\beta(x, y) f(y) \, dy,$$

where

$$K_\beta(x, y) = \sum_{j \in \mathbb{Z}} \sum_{k \in \mathbb{Z}} \beta_{j,k} h_{j,k}(x) \overline{h_{j,k}(y)}.$$

Due to the lack of regularity of the Haar wavelet our approach in section 5.6 fails, since we cannot prove (6.13). Instead, it can be proved directly that our operator T_β is of **weak type** $(1, 1)$; that is,

$$\left| \{ x \in \mathbb{R} : |(T_\beta f)(x)| > \alpha \} \right| \leq \frac{C}{\alpha} \|f\|_{L^1} \qquad (7.2)$$

for all $f \in L^1(\mathbb{R})$ and $\alpha > 0$. Using the boundedness of T_β on $L^2(\mathbb{R})$ and the Marcinkiewicz interpolation theorem (see Chapter I of [Ste1]), the unconditionality of the Haar basis for $L^p(\mathbb{R})$, $1 < p < \infty$, follows as in the proof of Theorem 6.14.

Inequality (7.2) is proved by using a technique called the **Calderón-Zygmund decomposition** of a function (see [CW2], [Ste1] or [GR]). Given $f \in L^1(\mathbb{R})$ and $\alpha > 0$, f can be written as a sum of a "good" part g and a "bad" part $b = \sum_i b_i$; moreover, the b_i's are supported in disjoint dyadic cubes Q_i such that if $\Omega = \bigcup_i Q_i$ we have

$$
\begin{array}{lll}
\text{i)} & f(x) = g(x) + b(x) & \text{for a.e. } x \in \mathbb{R}, \\[2mm]
\text{ii)} & |g(x)| = |f(x)| \leq \alpha & \text{for a.e. } x \notin \Omega, \\[2mm]
\text{iii)} & |g(x)| \leq 2\alpha & \text{for all } x \in \Omega, \\[2mm]
\text{iv)} & |\Omega| \leq \dfrac{1}{\alpha} \|f\|_{L^1}, & \\[2mm]
\text{v)} & \displaystyle\int_{Q_i} b_i(x) \, dx = 0 & \text{for all } i.
\end{array}
$$

From i) we have

$$\left| \{ x \in \mathbb{R} : |(T_\beta f)(x)| > \alpha \} \right|$$

$$\leq \left| \{ x \in \mathbb{R} : |(T_\beta g)(x)| > \tfrac{1}{2}\alpha \} \right| + \left| \{ x \in \mathbb{R} : |(T_\beta b)(x)| > \tfrac{1}{2}\alpha \} \right|.$$

Since g is the "good" part, the first summand on the right-hand side of the above inequality is easy to estimate. In fact, using the boundedness of T_β on $L^2(\mathbb{R})$ together with ii), iii) and iv), we obtain

$$\left|\{x \in \mathbb{R} : |(T_\beta g)(x)| > \tfrac{1}{2}\alpha\}\right| \leq \frac{4}{\alpha^2} \int_\mathbb{R} |(T_\beta g)(x)|^2 \, dx \leq \frac{4}{\alpha^2} \int_\mathbb{R} |g(x)|^2 \, dx$$

$$\leq \frac{4}{\alpha^2} \left\{ \int_{\mathbb{R}\setminus\Omega} |g(x)|^2 \, dx + \int_\Omega |g(x)|^2 \, dx \right\}$$

$$\leq \frac{4}{\alpha^2} \left\{ \int_{\mathbb{R}\setminus\Omega} \alpha |f(x)| \, dx + 4\alpha^2 |\Omega| \right\} \leq \frac{C}{\alpha} \|f\|_{L^1}.$$

The "bad" part is not so bad, after all. The key observation is that if $x \notin Q_i$ (one of the dyadic cubes in the Calderón-Zygmund decomposition) then

$$(T_\beta b_i)(x) = \sum_{j\in\mathbb{Z}} \sum_{k\in\mathbb{Z}} \beta_{j,k} h_{j,k}(x) \int_{Q_i} b_i(y) \, h_{j,k}(y) \, dy = 0. \qquad (7.3)$$

To prove (7.3) recall that given two dyadic intervals Q_i and $Q_{j,k}$, if they are not disjoint, one of them must be contained in the other. Thus, if $Q_{j,k} \subset Q_i$, then $x \notin Q_{j,k}$ and it follows that $h_{j,k}(x) = 0$. On the other hand, if $Q_i \subset Q_{j,k}$ and $Q_i \neq Q_{j,k}$, $h_{j,k}$ is constant in Q_i and, using v), we deduce

$$\int_{Q_i} b_i(y) \, h_{j,k}(y) \, dy = 0.$$

This establishes (7.3). Hence,

$$\left|\{x \in \mathbb{R} : |(T_\beta b)(x)| > \tfrac{1}{2}\alpha\}\right| \leq |\Omega| \leq \frac{1}{\alpha} \|f\|_{L^1},$$

and the proof of (7.2) is finished.

13. Theorem 5.9 of Chapter 4 exhibits a basis for $L^2(\mathbb{T})$ obtained by periodizing an orthonormal wavelet ψ on $\mathbb{R}$ that arises from an MRA, assuming a certain decay at infinity. Since

$$\widetilde{V}_m = \widetilde{V}_0 \oplus \left(\bigoplus_{j=0}^{m-1} \widetilde{W}_j \right),$$

the projection operator $\widetilde{P}_m$ from $L^2(\mathbb{T})$ onto $\widetilde{V}_m$ is given by

$$(\widetilde{P}_m f)(x) = \int_0^1 f(y)\,dy + \sum_{j=0}^{m-1} \sum_{k=0}^{2^j-1} \langle f, \tilde{\psi}_{j,k} \rangle \, \tilde{\psi}_{j,k}(x) \qquad (7.4)$$

for $f \in L^2(\mathbb{T})$, where $\langle f, \tilde{\psi}_{j,k} \rangle$ denotes the inner product in $L^2(\mathbb{T})$. Since the system $\{\tilde{\varphi}_{m,k} : 0 \leq k \leq 2^m - 1\}$ is an orthonormal basis of $\widetilde{V}_m$ (see Theorem 5.6 of Chapter 4) we can also write the projection operator as follows:

$$(\widetilde{P}_m f)(x) = \sum_{k=0}^{2^m-1} \langle f, \tilde{\varphi}_{j,k} \rangle \, \tilde{\varphi}_{j,k}(x).$$

It can be shown, in analog with Proposition 3.13, that $\widetilde{P}_m$ is a bounded operator on $L^p(\mathbb{T})$, $1 \leq p \leq \infty$. When $1 \leq p < \infty$ we have

$$\lim_{m \to \infty} \left\| \widetilde{P}_m f - f \right\|_{L^p(\mathbb{T})} = 0 \qquad \text{for } f \in L^p(\mathbb{T}),$$

and when $p = \infty$ we have

$$\lim_{m \to \infty} \left\| \widetilde{P}_m f - f \right\|_{L^\infty(\mathbb{T})} = 0$$

for all 1-periodic continuous functions f on $\mathbb{R}$. These results can be proved by using Theorem 5.8 of Chapter 4. They correspond, in the periodic case, to Corollary 3.18.

Another operator associated with the periodic wavelets described in section 4.5 is the **partial sum operator** (very similar to the one associated with series), given by

$$(\widetilde{S}_q f)(x) = (\widetilde{P}_j f)(x) + \sum_{k=0}^{n} \langle f, \tilde{\psi}_{j,k} \rangle \, \tilde{\psi}_{j,k}(x), \qquad (7.5)$$

where $q = 2^j + n$, $j = 0, 1, 2, \cdots$ and $n = 0, 1, 2, \cdots, 2^j - 1$. When $1 \leq p < \infty$,

$$\lim_{q \to \infty} \left\| \widetilde{S}_q f - f \right\|_{L^p(\mathbb{T})} = 0 \qquad \text{for all } f \in L^p(\mathbb{T}),$$

and when $p = \infty$,   .

$$\lim_{q \to \infty} \left\| \widetilde{S}_m f - f \right\|_{L^\infty(\mathbb{T})} = 0$$

for all 1-periodic continuous functions f on $\mathbb{R}$. These results correspond, in the periodic case, to Corollary 3.20. They show that, in this periodic case, wavelet expansions behave better than Fourier series in $L^1(\mathbb{T})$ since it is not true that if $f \in L^1(\mathbb{T})$ the general partial sums of the Fourier series converge to f in $L^1(\mathbb{T})$. (This observation applies to the case $p = \infty$ as well.) All the results stated here require the assumptions made for the wavelet in Theorem 5.9 of Chapter 4.

14. Similar to section 5.4, it is natural to study the pointwise convergence properties of the operators $\widetilde{P}_m$ and $\widetilde{S}_q \equiv \widetilde{S}_{2^j+n}$ (see (7.4) and (7.5)) defined for periodic wavelets. We have, if $1 \le p \le \infty$ and $f \in L^p(\mathbb{T})$,

$$\lim_{m\to\infty} (\widetilde{P}_m f)(x) = f(x) \qquad \text{for a.e. } x \in \mathbb{T},$$

and, more generally,

$$\lim_{j\to\infty} (\widetilde{S}_{2^j+n}f)(x) = f(x) \qquad \text{for a.e. } x \in \mathbb{T}, \ n \in \mathbb{N}.$$

The idea of the proof is to show $\widetilde{P}_m f = P_m \tilde{f}$ for $m \ge 0$, where

$$\tilde{f}(x) = \sum_{\ell \in \mathbb{Z}} f(x + \ell)$$

is the 1-periodic version of f and P_m is defined in (3.2). This allows us to obtain these convergence properties for $(\widetilde{P}_m f)(x)$ by using the convergence results proved in section 5.4. The assumptions we require on the wavelets are as in Corollary 4.6 and Corollary 4.7.

15. The unconditionality of periodic wavelet bases for $L^p(\mathbb{T})$, $1 < p < \infty$, and for $H^1(\mathbb{T})$ can also be obtained. The results are similar to the main results in section 5.6; that is, a periodic wavelet associated with a wavelet ψ on $\mathbb{R}$ having enough decay at infinity, provides an unconditional basis in each of these spaces. One approach for showing this result is to study the boundedness of the operator

$$\widetilde{T}_\beta f = \beta_0 \int_0^1 f(y)\,dy + \sum_{j=0}^{\infty} \sum_{k=0}^{2^j-1} \beta_{j,k} \langle f, \tilde{\psi}_{j,k} \rangle \, \tilde{\psi}_{j,k}$$

on the above spaces, where the sequence $\beta = \{\beta_{j,k}\}$ is made up of zeros and ones and is finitely non-zero. Another approach is to prove, for the periodized case, results similar to the ones contained in sections 6.4 and 6.5 for wavelets on the real line (see also [Me1], section 3.11, 6.2 and 6.3).

6

Characterizations of function spaces using wavelets

For an orthonormal wavelet ψ and a function $f \in L^2(\mathbb{R})$ we have

$$\|f\|_{L^2(\mathbb{R})} = \left(\sum_{j \in \mathbb{Z}} \sum_{k \in \mathbb{Z}} |\langle f, \psi_{j,k} \rangle|^2 \right)^{\frac{1}{2}}.$$

Thus, $L^2(\mathbb{R})$ can be characterized as the space of functions f such that the sequence $\{\langle f, \psi_{j,k} \rangle : j, k \in \mathbb{Z}\}$ belongs to $\ell^2(\mathbb{Z})$. The purpose of this chapter is to obtain characterizations for other function spaces: the Lebesgue spaces $L^p(\mathbb{R})$, $1 < p < \infty$, the Hardy space $H^1(\mathbb{R})$, the Sobolev space $L^{p,s}(\mathbb{R})$, $1 < p < \infty$, $s \in \mathbb{Z}$, as well as some Lipschitz and Besov spaces. These characterizations will show, in particular, that appropriate wavelets generate unconditional bases for these spaces. Though some of these results were obtained in the last chapter, the techniques we shall use here are quite different. These characterizations are accomplished by using well known techniques and notions in Fourier analysis that were developed during this century, before the modern development of the theory of wavelets. These are the Littlewood-Paley theory, vector-valued Calderón-Zygmund operators, and inequalities for general maximal functions similar to the Hardy-Littlewood maximal function. A complete description of all these subjects will lead us far astray from the main purpose of this chapter; hence, we present a summary of the relevant results, and appropriate references are given in the last section of this chapter.

6.1 Wavelets and sampling theorems

The orthonormality of the wavelet basis $\{\psi_{j,k} : j, k \in \mathbb{Z}\}$ allows us to write

$$f = \sum_{j \in \mathbb{Z}} \sum_{k \in \mathbb{Z}} \langle f, \psi_{j,k} \rangle \psi_{j,k} \tag{1.1}$$

with convergence in $L^2(\mathbb{R})$. We have seen in Chapter 5 that expansions related to (1.1) are valid for functions in $L^p(\mathbb{R})$ when the wavelet basis arises from a multiresolution analysis and the corresponding scaling function satisfies a "mild" condition (see sections 5.3 and 5.4 for details).

Expansions like (1.1) are related to **sampling theorems**. Let us rewrite (1.1) in a form that will help us explain what is meant by this terminology. Assume that (1.1) holds; the precise convergence of this formula is not a matter of concern at this moment. For a function g defined on $\mathbb{R}$, we let $\tilde{g}(x) = \overline{g(-x)}$. Then, we have (formally)

$$\langle f, \psi_{j,k} \rangle = \int_{\mathbb{R}} f(x) \overline{\psi_{j,k}(x)} \, dx = \int_{\mathbb{R}} f(x) 2^{\frac{j}{2}} \overline{\psi(2^j x - k)} \, dx$$

$$= \int_{\mathbb{R}} f(x) \overline{\psi_{j,0}(x - 2^{-j} k)} \, dx = \int_{\mathbb{R}} f(x) \widetilde{\psi_{j,0}}(2^{-j} k - x) \, dx$$

$$= (f * \widetilde{\psi_{j,0}})(2^{-j} k). \tag{1.2}$$

Thus, from (1.1) we obtain the representation

$$f = \sum_{j \in \mathbb{Z}} \sum_{k \in \mathbb{Z}} (f * \widetilde{\psi_{j,0}})(2^{-j} k) \, \psi_{j,k}. \tag{1.3}$$

In (1.3) the values of f are recovered by using the family $\{\psi_{j,k} : j, k \in \mathbb{Z}\}$ and the values of $f * \widetilde{\psi_{j,0}}$ at the points $2^{-j} k$, $j, k \in \mathbb{Z}$. The latter are often referred to as **samples** of this function (clearly related to f) and the system $\{\psi_{j,k} : j, k \in \mathbb{Z}\}$ is used to reconstruct f from these samples.

The classical sampling theorem of C. Shannon uses samples of an appropriate function f:

THEOREM 1.4 (Shannon Sampling Theorem)

Suppose that $f \in L^1(\mathbb{R})$ and supp$(\hat{f}) \subseteq [-B\pi, B\pi]$. Then,

$$f(x) = \sum_{k \in \mathbb{Z}} f\left(\frac{k}{B}\right) \frac{\sin \pi(Bx - k)}{\pi(Bx - k)}, \tag{1.5}$$

where the symmetric partial sums of this series converge to f in the L^2-norm.

REMARK 1: We are really assuming f to be continuous. This is justified by our assumptions. In fact, the Paley-Wiener theorem allows us to consider f to be the restriction to $\mathbb{R}$ of an entire function of exponential type. Hence, the values $f(k/B)$, $k \in \mathbb{Z}$, are well defined.

REMARK 2: Observe that the Shannon sampling theorem states that a band-limited function in $L^1(\mathbb{R})$ such that supp$(\hat{f}) \subseteq [-B\pi, B\pi]$ is uniquely determined by sampling f at the points $\{B^{-1}k : k \in \mathbb{Z}\}$ and the reconstruction is performed by the family of functions

$$\left\{ \frac{\sin \pi(Bx - k)}{\pi(Bx - k)} : k \in \mathbb{Z} \right\}.$$

This family is related to the function

$$\varphi(x) = \begin{cases} \dfrac{\sin \pi x}{\pi x} & \text{if } x \neq 0, \\ 1 & \text{if } x = 0 \end{cases}$$

(usually denoted by sinc$\,x$), which is the scaling function of the Shannon wavelet (Example C of Chapter 2).

REMARK 3: For simplicity let $B = 1$. Observe that $f = f * \varphi$ when supp$(\hat{f}) \subseteq [-\pi, \pi]$ and $\varphi = $ sinc. To see this we use the fact that $\hat{\varphi} = \chi_{[-\pi,\pi]}$ and, thus, $\hat{f} = \hat{f}\hat{\varphi} = (f * \varphi)^{\wedge}$. This shows that the sampling of f in Theorem 1.4 is a special case of sampling a convolution in a way that is similar to that done in (1.3).

PROOF: We expand $\hat{f}$ in its Fourier series on $[-B\pi, B\pi]$ to obtain

$$\hat{f}(\xi) = \sum_{k \in \mathbb{Z}} c_{-k} e^{-ikB^{-1}\xi}, \tag{1.6}$$

where

$$c_{-k} = \frac{1}{2\pi B} \int_{-B\pi}^{B\pi} \hat{f}(\xi) e^{ikB^{-1}\xi}\, d\xi.$$

By the Fourier inversion formula we obtain

$$c_{-k} = \frac{1}{B}(\hat{f})^{\vee}(B^{-1}k) = B^{-1}f(B^{-1}k).$$

Using these values in (1.6) and, again, using the inversion formula, we obtain

$$f(x) = \frac{1}{2\pi}\int_{-B\pi}^{B\pi} \hat{f}(\xi) e^{ix\xi}\, d\xi = \frac{1}{2\pi}\int_{-B\pi}^{B\pi}\sum_{k\in\mathbb{Z}} \frac{f(B^{-1}k)}{B} e^{-ikB^{-1}\xi} e^{ix\xi}\, d\xi$$

$$= \sum_{k\in\mathbb{Z}} \frac{f(B^{-1}k)}{2\pi B} \int_{-B\pi}^{B\pi} e^{i\xi(x - kB^{-1})}\, d\xi$$

$$= \sum_{k\in\mathbb{Z}} \frac{f(B^{-1}k)}{2\pi B} \frac{2\sin\pi(Bx - k)}{x - kB^{-1}} = \sum_{k\in\mathbb{Z}} f(B^{-1}k)\frac{\sin\pi(Bx - k)}{\pi(Bx - k)}.$$

The convergence is in $L^2(\mathbb{R})$; in fact, the coefficients are in $\ell^2(\mathbb{Z})$ and $\{\mathrm{sinc}\,(\cdot - k)\}_{k\in\mathbb{Z}}$ is an orthonormal sequence.   ∎

If $\mathrm{supp}\,(\hat{f}) \subseteq [-B\pi, B\pi]$ and $B' > B$, formula (1.5) is valid for f with B replaced by B', since $\mathrm{supp}\,(\hat{f}) \subseteq [-B'\pi, B'\pi]$. This means that f is also determined by its values on the "denser" grid $\{k/B' : k \in \mathbb{Z}\}$. But, in general, we cannot replace the grid $\{k/B : k \in \mathbb{Z}\}$ by a "larger" one; that is, one obtained when B is replaced by B' with $B' < B$. To see this consider the following example. Let

$$f(x) = g(x)\frac{\sin\pi B'x}{\pi B'x}$$

where $B' > 0$, $g \in \mathcal{S}$, $g \not\equiv 0$, $\mathrm{supp}\,(g) = [-\varepsilon\pi, \varepsilon\pi]$ with $\varepsilon = B - B'$ and $g(0) = 0$. From

$$\left(\frac{1}{B'}\chi_{[-B'\pi, B'\pi]}\right)^{\wedge}(x) = \frac{\sin(B'\pi x)}{B'\pi x},$$

we obtain

$$\hat{f}(\xi) = \left(\hat{g} * \frac{1}{B'}\chi_{[-B'\pi, B'\pi]}\right)(\xi) = \frac{1}{B'}\int_{\xi - B'\pi}^{\xi + B'\pi} \hat{g}(y)\, dy.$$

Thus, $\operatorname{supp}(\hat{f}) \subseteq [-(B' + \varepsilon)\pi, (B' + \varepsilon)\pi] \subseteq [-B\pi, B\pi]$. But,

$$f\left(\frac{k}{B'}\right) = g\left(\frac{k}{B'}\right)\frac{\sin(k\pi)}{k\pi} = 0 \qquad \text{for all } k \in \mathbb{Z}, \ k \neq 0.$$

Since $g(0) = 0$ we have $f(k/B') = 0$ for all $k \in \mathbb{Z}$, so that formula (1.5) cannot be valid with B replaced by B', since f is not the zero function.

The idea of the proof of the Shannon sampling theorem (Theorem 1.4) is to expand $\hat{f}$ in a Fourier series and then use the Fourier inversion formula. The same idea can be used (at least formally) to obtain expressions similar to (1.1) for a large family of functions related to wavelets.

Let Φ be a function in $L^2(\mathbb{R})$ such that $\operatorname{supp}(\hat{\Phi}) \subseteq [-\pi, \pi]$ and

$$\sum_{j \in \mathbb{Z}} \left|\hat{\Phi}(2^j \xi)\right|^2 = 1 \qquad \text{for a.e. } \xi \in \mathbb{R} \setminus \{0\}. \tag{1.7}$$

The existence of such functions is obtained immediately from the examples of band-limited wavelets presented in Chapter 3. In fact, if ψ is a band-limited orthonormal wavelet with $\operatorname{supp}(\hat{\psi}) \subseteq [-2^J\pi, 2^J\pi]$, the function $\Phi \equiv \Phi^J$ given by $\Phi^J(x) = 2^{-J}\psi(2^{-J}x)$ (thus, $\widehat{\Phi^J}(\xi) = \hat{\psi}(2^J\xi)$) satisfies (1.7) by Theorem 2.4 of Chapter 3.

From (1.7) we deduce

$$1 = \sum_{j \in \mathbb{Z}} \left|\hat{\Phi}(2^{-j}\xi)\right|^2 = \sum_{j \in \mathbb{Z}} \hat{\Phi}(2^{-j}\xi)\,\overline{\hat{\Phi}(2^{-j}\xi)} \tag{1.8}$$

almost everywhere on $\mathbb{R} \setminus \{0\}$. Easy calculations show that

$$\hat{\Phi}(2^{-j}\xi) = 2^{\frac{j}{2}}(\Phi_{j,0})^{\wedge}(\xi) \qquad \text{and} \qquad \overline{\hat{\Phi}(2^{-j}\xi)} = 2^{\frac{j}{2}}(\tilde{\Phi}_{j,0})^{\wedge}(\xi), \tag{1.9}$$

where, as before, $\tilde{g}(x) = \overline{g(-x)}$. Thus, if f is a **tempered distribution**, we obtain (formally) from (1.8) and (1.9) the equalities

$$\hat{f}(\xi) = 1 \cdot \hat{f}(\xi) = \sum_{j \in \mathbb{Z}} 2^j (\Phi_{j,0})^{\wedge}(\xi)(\tilde{\Phi}_{j,0})^{\wedge}(\xi)\hat{f}(\xi)$$

$$= \sum_{j \in \mathbb{Z}} 2^j (\Phi_{j,0})^{\wedge}(\xi)(\tilde{\Phi}_{j,0} * f)^{\wedge}(\xi). \tag{1.10}$$

The function $(\tilde{\Phi}_{j,0} * f)^{\wedge} = (\tilde{\Phi}_{j,0})^{\wedge}\hat{f}$ has support contained in $[-2^j\pi, 2^j\pi]$ because $\operatorname{supp}((\tilde{\Phi})^{\wedge}) \subseteq [-\pi, \pi]$ and (1.9) holds. We can expand $(\tilde{\Phi}_{j,0} * f)^{\wedge}$

in a Fourier series on $[-2^j\pi, 2^j\pi]$ to obtain

$$(\widetilde{\Phi}_{j,0} * f)^\wedge(\xi) = \sum_{k \in \mathbb{Z}} \frac{1}{2^j 2\pi} \left(\int_{-2^j\pi}^{2^j\pi} (\widetilde{\Phi}_{j,0} * f)^\wedge(\eta) e^{i2^{-j}k\eta} \, d\eta \right) e^{-i2^{-j}k\xi}$$

$$= \sum_{k \in \mathbb{Z}} 2^{-j} (\widetilde{\Phi}_{j,0} * f)(2^{-j}k) e^{-i2^{-j}k\xi}$$

on $[-2^j\pi, 2^j\pi]$. Observe that the right-hand side of the above equality is a $2^j 2\pi$-periodic function, while the left-hand side is zero outside the period interval $[-2^j\pi, 2^j\pi]$.

Substituting in (1.10) we deduce

$$f(x) = \left(\sum_{j \in \mathbb{Z}} 2^j (\Phi_{j,0})^\wedge(\xi) \sum_{k \in \mathbb{Z}} 2^{-j} (\widetilde{\Phi}_{j,0} * f)(2^{-j}k) \, e^{-i2^{-j}k\xi} \right)^\vee (x)$$

$$= \sum_{j \in \mathbb{Z}} \sum_{k \in \mathbb{Z}} (\widetilde{\Phi}_{j,0} * f)(2^{-j}k) \left((\Phi_{j,0})^\wedge(\xi) \, e^{-i2^{-j}k\xi} \right)^\vee (x)$$

$$= \sum_{j \in \mathbb{Z}} \sum_{k \in \mathbb{Z}} (\widetilde{\Phi}_{j,0} * f)(2^{-j}k) \Phi_{j,k}(x),$$

since $(\Phi_{j,k})^\wedge(\xi) = (\Phi_{j,0})^\wedge(\xi) \, e^{-i2^{-j}k\xi}$. Finally, equality (1.2) applied to Φ instead of ψ allows us to conclude

$$f(x) = \sum_{j \in \mathbb{Z}} \sum_{k \in \mathbb{Z}} \langle f, \Phi_{j,k} \rangle \Phi_{j,k}(x), \tag{1.11}$$

which is valid, at least formally, for all functions $\Phi \in L^2(\mathbb{R})$ such that $\mathrm{supp}\,(\widehat{\Phi}) \subseteq [-\pi, \pi]$ and satisfying (1.7).

We shall encounter expressions similar to (1.11) often in the next sections.

6.2 Littlewood-Paley theory

The trigonometric system $\{e^{ikx} : k \in \mathbb{Z}\}$ is an orthonormal basis of $L^2(\mathbb{T}, \frac{dx}{2\pi})$. Therefore, a function $f \in L^2(\mathbb{T})$ if and only if its Fourier coefficients

$$c_k = \int_{\mathbb{T}} f(x)e^{-ikx} \frac{dx}{2\pi}, \qquad k \in \mathbb{Z},$$

satisfy $\sum_{k \in \mathbb{Z}} |c_k|^2 < \infty$. In this case,

$$\|f\|_{L^2(\mathbb{T}, \frac{dx}{2\pi})} = \left(\sum_{k \in \mathbb{Z}} |c_k|^2 \right)^{\frac{1}{2}}. \tag{2.1}$$

It is well known that this result is a general fact about orthonormal bases in Hilbert spaces. Formula (2.1) can be interpreted in several different ways. Here are two which are appropriate for our subsequent development. On one hand, we can view (2.1) as stating that the operator that sends f to the sequence $\{c_k\}_{k \in \mathbb{Z}}$, of its Fourier coefficients, is an isometry from $L^2(\mathbb{T}, \frac{dx}{2\pi})$ into $\ell^2(\mathbb{Z})$. On the other hand, (2.1) provides us with a characterization of the elements of $L^2(\mathbb{T})$ which depends only on the magnitude of the coefficients c_k.

Similar characterizations cannot be obtained for functions $f \in L^p(\mathbb{T})$, $p \neq 2$. Although $\{e^{ikx} : k \in \mathbb{Z}\}$ is a basis for $L^p(\mathbb{T})$, $1 < p < \infty$, we have pointed out at the end of section 5.2 that it is not unconditional. If a characterization like (2.1), involving only the magnitude of the coefficients c_k were true for functions in $L^p(\mathbb{T})$, Theorem 2.10 of Chapter 5 would allow us to conclude that the basis is unconditional.

The important feature of the Littlewood-Paley theory is that a characterization of the spaces $L^p(\mathbb{T})$, $1 < p < \infty$, can be obtained if we group the terms of a Fourier series in dyadic blocks. More precisely, suppose that $\sum_{k \in \mathbb{Z}} c_k e^{ikx}$ is the Fourier series of a function $f \in L^1(\mathbb{T})$. For $N \geq 0$, consider the dyadic partial sums of f, namely

$$(\Delta_{N+1}f)(x) = \sum_{2^N \leq |k| < 2^{N+1}} c_k e^{ikx}, \tag{2.2}$$

and $\Delta_0 = c_0$. Littlewood and Paley showed that there exist two positive constants A and B (depending on p) such that

$$A\|f\|_{L^p(\mathbb{T})} \leq \left\| \left(\sum_{N=0}^{\infty} |\Delta_N f|^2 \right)^{\frac{1}{2}} \right\|_{L^p(\mathbb{T})} \leq B\|f\|_{L^p(\mathbb{T})}, \qquad 1 < p < \infty. \tag{2.3}$$

The function that appears within the norm in the middle term is called the (classical) **Littlewood-Paley** function,

$$d(f)(x) = \Big(\sum_{N=0}^{\infty} |\Delta_N f(x)|^2\Big)^{\frac{1}{2}}. \tag{2.4}$$

Thus, when $f \in L^p(\mathbb{T})$, $1 < p < \infty$, $d(f) \in L^p(\mathbb{T})$. Conversely, $d(f) \in L^p(\mathbb{T})$ implies $f \in L^p(\mathbb{T})$.

There are many situations in which these ideas have been used. A "continuous" version of the results we just described involves the Abel means of Fourier series. We introduce the **Littlewood-Paley G-function** by defining

$$G(f)(x) = \Big(\int_0^1 (1-r)\Big|\big(f * \frac{dP_r}{dr}\big)(x)\Big|^2 dr\Big)^{\frac{1}{2}},$$

where

$$P_r(x) = \frac{1-r^2}{1 - 2r\cos x + r^2}, \qquad 0 \le r < 1,$$

is the Poisson kernel for the unit disk.

There is also a version of the Littlewood-Paley G-function associated with $\mathbb{R}$. This involves the Poisson kernel

$$P_t(x) = \frac{1}{\pi}\frac{t}{x^2 + t^2}, \qquad t > 0, \ x \in \mathbb{R},$$

for the upper half plane $\mathbb{R}_+^2 = \{(x,t) \in \mathbb{R}^2 : t > 0\}$ (see also section 5.5). In this situation, the Littlewood-Paley G-function is given by

$$G(f)(x) = \Big(\int_0^{\infty} \Big|t\big(\frac{\partial P_t}{\partial t} * f\big)(x)\Big|^2 \frac{dt}{t}\Big)^{\frac{1}{2}}, \tag{2.5}$$

and it can be used to characterize those functions f which belong to $L^p(\mathbb{R})$, $1 < p < \infty$. More precisely, $f \in L^p(\mathbb{R})$, $1 < p < \infty$, if and only if $G(f) \in L^p(\mathbb{R})$ and there exist positive constants A_p and B_p such that

$$A_p\|f\|_{L^p(\mathbb{R})} \le \|G(f)\|_{L^p(\mathbb{R})} \le B_p\|f\|_{L^p(\mathbb{R})}. \tag{2.6}$$

Observing that $t\frac{\partial}{\partial t}P_t(x) = \varphi_t(x)$, where

$$\varphi(x) = \frac{1}{\pi}\frac{x^2 - 1}{(x^2 + 1)^2} \qquad \text{and} \qquad \varphi_t(x) = \frac{1}{t}\varphi\big(\frac{x}{t}\big),$$

we can write (2.5) as

$$G(f)(x) = \left(\int_0^\infty |\varphi_t * f(x)|^2 \frac{dt}{t} \right)^{\frac{1}{2}}. \tag{2.7}$$

This suggests the use of a more general version of this G-function that depends on an appropriate smooth function φ having mean zero, as is the case for the function $\varphi(x) = \frac{1}{\pi} \frac{x^2-1}{(x^2+1)^2}$. We shall develop this approach below in order to obtain a characterization of the spaces $L^p(\mathbb{R})$, but we shall use, instead, a discrete version of (2.7). To motivate the introduction of this discrete version, suppose that estimates for φ_t are all of the same order on the dyadic intervals $2^{-k} \le t \le 2^{-k+1}$, $k \in \mathbb{Z}$. Then

$$[G(f)(x)]^2 = \sum_{k\in\mathbb{Z}} \int_{2^{-k}}^{2^{-k+1}} |\varphi_t * f(x)|^2 \frac{dt}{t}$$

$$\sim \sum_{k\in\mathbb{Z}} |\varphi_{2^{-k}} * f(x)|^2 \int_{2^{-k}}^{2^{-k+1}} \frac{dt}{t} = C \sum_{k\in\mathbb{Z}} |\varphi_{2^{-k}} * f(x)|^2.$$

In the sequel we shall use these discrete versions of the Littlewood-Paley function; that is, we write

$$\mathfrak{g}(f)(x) = \left(\sum_{k\in\mathbb{Z}} |\varphi_{2^{-k}} * f(x)|^2 \right)^{\frac{1}{2}} \tag{2.8}$$

and shall show that these expressions can be used to characterize the spaces $L^p(\mathbb{R})$, $1 < p < \infty$, when appropriate functions φ are used.

Here is a precise result:

THEOREM 2.9 *Let φ be an integrable function on $\mathbb{R}$ such that*

$$\hat{\varphi}(0) = \int_{\mathbb{R}} \varphi(x) \, dx = 0,$$

and assume that, for some $\alpha > 0$, satisfies

$$|\varphi(x)| \le C \frac{1}{(1+|x|)^{1+\alpha}}, \qquad x \in \mathbb{R}, \tag{2.10}$$

and

$$\int_{\mathbb{R}} |\varphi(x+h) - \varphi(x)| \, dx \le C|h|^\alpha, \qquad h \in \mathbb{R}. \tag{2.11}$$

Then, the operators

$$\mathfrak{g}(f)(x) = \left(\sum_{j \in \mathbb{Z}} |\varphi_{2^{-j}} * f(x)|^2 \right)^{\frac{1}{2}}$$

and

$$G(f)(x) = \left(\int_0^\infty |\varphi_t * f(x)|^2 \, \frac{dt}{t} \right)^{\frac{1}{2}}$$

are bounded in $L^p(\mathbb{R})$, $1 < p < \infty$. They are also bounded operators from $H^1(\mathbb{R})$ to $L^1(\mathbb{R})$.

We say that a function φ defined on $\mathbb{R}$ belongs to the regularity class $\mathfrak{R}^0$ if there exist constants C_0, C_1, $\gamma > 0$ and $\varepsilon > 0$ such that

$$
\begin{cases}
\text{i)} & \displaystyle\int_{\mathbb{R}} \varphi(x) \, dx = 0; \\[2ex]
\text{ii)} & |\varphi(x)| \le \dfrac{C_0}{(1 + |x|)^{2+\gamma}} \qquad \text{for all } x \in \mathbb{R}; \\[2ex]
\text{iii)} & |\varphi'(x)| \le \dfrac{C_1}{(1 + |x|)^{1+\varepsilon}} \qquad \text{for all } x \in \mathbb{R}.
\end{cases}
\qquad (2.12)
$$

REMARK: Observe that if $\varphi \in \mathfrak{R}^0$ with constants $C_0, C_1, \gamma > 0$ and $\varepsilon > 0$ then condition iii) in (2.12) holds for any $\varepsilon' < \varepsilon$. Hence, in definition (2.12) we can always take $0 < \varepsilon < \gamma$. This fact will be used in some of the proofs that follow.

Clearly, every $\varphi \in \mathfrak{R}^0$ satisfies the conditions of Theorem 2.9, and, hence, its associated Littlewood-Paley G- and $\mathfrak{g}$-functions are bounded operators on $L^p(\mathbb{R})$, $1 < p < \infty$, and are also bounded as maps from $H^1(\mathbb{R})$ to $L^1(\mathbb{R})$. The class $\mathfrak{R}^0$, and others that require more "regularity," that we shall introduce later, are the ones we shall use to characterize function spaces using wavelet coefficients.

We have stated Theorem 2.9 for functions defined on $\mathbb{R}$, since we only deal with wavelets defined in one dimension. There is a similar result for functions defined on $\mathbb{R}^n$, as stated in the reference given in note 3 of section 6.8.

Theorem 2.9 is obtained in [GR] as a consequence of results concerning "vector-valued" Calderón-Zygmund operators. We have encountered

Calderón-Zygmund operators in section 5.6 as useful tools for the proof that some wavelet bases are unconditional for $L^p(\mathbb{R})$, $1 < p < \infty$, and $H^1(\mathbb{R})$.

The words "vector-valued" that appear in the previous paragraph need some explanation. A "vector-valued" function is a function defined on $\mathbb{R}$ with values in a Banach space. We can also consider a Calderón-Zygmund kernel to be a function k defined on $\mathbb{R} \times \mathbb{R}$ with values in $\mathcal{L}(\mathbb{A}, \mathbb{B})$, the space of bounded linear operators from the Banach space $\mathbb{A}$ into the Banach space $\mathbb{B}$, satisfying inequalities similar to those in (6.5), (6.6) and (6.7) of Chapter 5, obtained by replacing the absolute values on the left-hand side of each inequality by $\|\cdot\|_{\mathcal{L}(\mathbb{A},\mathbb{B})}$.

Vector-valued functions of this type arise naturally in Theorem 2.9. In fact, we can consider $\mathfrak{g}(f)$ as the $\ell^2(\mathbb{Z})$-norm of the vector-valued function whose value at x is the sequence

$$(T_\varphi f)(x) \equiv \{\varphi_{2^{-j}} * f(x) : j \in \mathbb{Z}\};$$

that is,

$$\mathfrak{g}(f)(x) = \left\|\{\varphi_{2^{-j}} * f(x)\}_{j \in \mathbb{Z}}\right\|_{\ell^2(\mathbb{Z})} = \left\|(T_\varphi f)(x)\right\|_{\ell^2(\mathbb{Z})}.$$

In this case, the Banach space considered is $\ell^2(\mathbb{Z})$. For the function $G(f)$ we use the Banach space $L^2\big((0, \infty), \frac{dt}{t}\big)$.

We shall concentrate on the discrete version, $\mathfrak{g}(f)$, of the Littlewood-Paley function. We want to obtain equivalences of norms such as the one in (2.6). What is asserted in Theorem 2.9 is that, under appropriate conditions on φ,

$$\left\|\mathfrak{g}(f)\right\|_{L^p(\mathbb{R})} \leq C\|f\|_{L^p(\mathbb{R})}, \qquad 1 < p < \infty. \tag{2.13}$$

We are, thus, led to find another condition on φ that guarantees an inequality that is the reverse of inequality (2.13). The following result contains such a condition:

COROLLARY 2.14 *Let $\varphi \in \mathfrak{R}^0$ and suppose that*

$$\sum_{j \in \mathbb{Z}} |\hat{\varphi}(2^j \xi)|^2 = M \qquad \text{for a.e. } \xi \in \mathbb{R}. \tag{2.15}$$

Then, there exist constants A_p and B_p, $0 < A_p \leq B_p < \infty$, such that

$$A_p\|f\|_{L^p(\mathbb{R})} \leq \left\|\mathfrak{g}(f)\right\|_{L^p(\mathbb{R})} \leq B_p\|f\|_{L^p(\mathbb{R})}$$

for every $f \in L^p(\mathbb{R})$, $1 < p < \infty$, *where* $\mathfrak{g}(f)$ *is defined in* Theorem 2.9.

PROOF: We only need to show the first inequality in the conclusion of the corollary, since the second one is a consequence of Theorem 2.9 (see (2.13)). Equality (2.15) allows us to show

$$\left\|\mathfrak{g}(f)\right\|_{L^2(\mathbb{R})} = \sqrt{M}\left\|f\right\|_{L^2(\mathbb{R})}, \qquad f \in L^2(\mathbb{R}). \tag{2.16}$$

In fact, using Plancherel's theorem and (2.15), we obtain

$$\left\|\mathfrak{g}(f)\right\|_{L^2(\mathbb{R})}^2 = \int_{\mathbb{R}} \sum_{j \in \mathbb{Z}} \left|\varphi_{2^{-j}} * f(x)\right|^2 dx = \sum_{j \in \mathbb{Z}} \frac{1}{2\pi} \int_{\mathbb{R}} \left|\hat{\varphi}(2^j \xi)\hat{f}(\xi)\right|^2 d\xi$$

$$= \frac{M}{2\pi} \int_{\mathbb{R}} \left|\hat{f}(\xi)\right|^2 d\xi = M\left\|f\right\|_{L^2(\mathbb{R})}^2.$$

We now use the just established equality (2.16) to "reverse" (2.13). Applying the polarization identity to (2.16), and letting $\langle \cdot, \cdot \rangle$ denote the inner product in $\ell^2(\mathbb{Z})$, we obtain

$$M\left|\int_{\mathbb{R}} f(x)h(x)\,dx\right| = \left|\int_{\mathbb{R}} \langle T_\varphi f(x), T_\varphi h(x)\rangle\,dx\right|$$

$$\leq \int_{\mathbb{R}} \mathfrak{g}(f)(x)\,\mathfrak{g}(h)(x)\,dx \leq \left\|\mathfrak{g}(f)\right\|_{L^p(\mathbb{R})}\left\|\mathfrak{g}(h)\right\|_{L^{p'}(\mathbb{R})}$$

for $f \in L^p(\mathbb{R}) \cap L^2(\mathbb{R})$, $h \in L^{p'}(\mathbb{R}) \cap L^2(\mathbb{R})$, $1 < p < \infty$ and $\dfrac{1}{p} + \dfrac{1}{p'} = 1$ (a density argument allows us to drop the assumption that $f, h \in L^2(\mathbb{R})$). By taking the supremum over all h such that $\|h\|_{L^{p'}(\mathbb{R})} \leq 1$ and using (2.13) with p replaced by p' we deduce

$$\left\|f\right\|_{L^p(\mathbb{R})} = \sup_{\|h\|_{p'} \leq 1} \left|\int_{\mathbb{R}} f(x)h(x)\,dx\right|$$

$$\leq \frac{1}{M}\left\|\mathfrak{g}(f)\right\|_{L^p(\mathbb{R})}\left\|\mathfrak{g}(h)\right\|_{L^{p'}(\mathbb{R})} \leq \frac{C_p}{M}\left\|\mathfrak{g}(f)\right\|_{L^p(\mathbb{R})},$$

which proves the desired result.   ∎

REMARK 1: Though we do not make use of this fact, weaker conditions on φ still give us the conclusions in Corollary 2.14; these are the conditions that φ satisfies in Theorem 2.9, together with (2.15).

The conditions presented in Corollary 2.14 are natural when considering orthonormal wavelets ψ. All the examples of wavelets we have given satisfy $\hat{\psi}(0) = 0$. Condition (2.15) is satisfied, with $M = 1$, for all orthonormal wavelets. So far, we have proved this result for band-limited wavelets (see Theorem 2.4 of Chapter 3). In Chapter 7 we shall prove that this is true for every orthonormal wavelet (see Theorem 1.1 of Chapter 7). Conditions ii) and iii) in the definition of the regularity class $\mathfrak{R}^0$ (see (2.12)) are verified by many wavelets: the Lemarié-Meyer wavelets of Chapter 1, the differentiable compactly supported wavelets of Daubechies developed in section 2.3, and the spline wavelets of Chapter 4. They are not verified, however, by the oldest known wavelets: the Haar wavelet and the Shannon wavelet.

REMARK 2: It is worthwhile to point out that when ψ is an orthonormal wavelet and $\psi \in \mathfrak{R}^0$, conditions i) of (2.12) follows from ii). This is a consequence of Proposition 3.6 in Chapter 2.

6.3   Necessary tools

The approach that we shall follow in order to obtain wavelet characterizations of several function spaces requires the use of certain "maximal functions" and estimates for the convolution of two functions. These tools are developed in this section. The results presented below are true on $\mathbb{R}^n$ (with obvious modifications), but we state them for the case $n = 1$ since we are only considering wavelets defined on $\mathbb{R}$.

For a function g defined on $\mathbb{R}$ and for a real number $\lambda > 0$ we consider the **maximal function**

$$g_\lambda^*(x) = \sup_{y \in \mathbb{R}} \frac{|g(x - y)|}{(1 + |y|)^\lambda}, \qquad x \in \mathbb{R}. \tag{3.1}$$

LEMMA 3.2 *Let $g \in L^1(\mathbb{R})$ be such that $\hat{g}$ is compactly supported. Then, for any real $\lambda > 0$, there exists a c_λ such that*

$$(g')_\lambda^*(x) \le c_\lambda g_\lambda^*(x), \qquad x \in \mathbb{R}.$$

PROOF: Suppose that $\text{supp}(\hat{g}) \subseteq \{\xi \in \mathbb{R} : |\xi| \le L\}$ for some $L > 0$. Let γ be a function in the Schwartz class such that $\hat{\gamma}(\xi) = 1$ if $|\xi| \le L$. Then,

$\hat{\gamma}(\xi)\hat{g}(\xi) = \hat{g}(\xi)$ for all $\xi \in \mathbb{R}$. Hence, $\gamma * g = g$ and $g' = \gamma' * g$. This proves that g is differentiable. Moreover,

$$|g'(x - y)| = \left|\int_{\mathbb{R}} \gamma'(x - y - z)g(z)\, dz\right| = \left|\int_{\mathbb{R}} \gamma'(w - y)g(x - w)\, dw\right|$$

$$\leq \int_{\mathbb{R}} |\gamma'(w - y)|\, (1 + |w - y|)^{\lambda}(1 + |y|)^{\lambda}\, \frac{|g(x - w)|}{(1 + |w|)^{\lambda}}\, dw,$$

where the last inequality is due to the following simple calculations:

$$1 + |w| \leq 1 + |w - y| + |y| \leq (1 + |w - y|)(1 + |y|).$$

Therefore,

$$|g'(x - y)| \leq g_{\lambda}^{*}(x)\, (1 + |y|)^{\lambda} \int_{\mathbb{R}} |\gamma'(w - y)|\, (1 + |w - y|)^{\lambda}\, dw.$$

Since γ' also belongs to the Schwartz class, the last integral equals a finite constant c_{λ}, independent of y, and we obtain

$$|g'(x - y)| \leq c_{\lambda}\, g_{\lambda}^{*}(x)\, (1 + |y|)^{\lambda}.$$

The desired result now follows immediately.   ∎

Another tool that we shall need is the **Hardy-Littlewood maximal function**, $\mathcal{M}f(x)$, given by

$$\mathcal{M}f(x) = \sup_{r > 0} \frac{1}{2r} \int_{|y - x| \leq r} |f(y)|\, dy \tag{3.3}$$

for a locally integrable function f on $\mathbb{R}$. It is well known that $\mathcal{M}$ is bounded on $L^{r}(\mathbb{R})$, $1 < p \leq \infty$. An important property of $\mathcal{M}$ that we shall need is the following vector-valued inequality:

THEOREM 3.4 *Suppose* $1 < p, q < \infty$; *then, there exists a constant* $C_{p,q}$ *such that*

$$\left\|\left\{\sum_{i=1}^{\infty}(\mathcal{M}f_{i})^{q}\right\}^{\frac{1}{q}}\right\|_{L^{p}(\mathbb{R})} \leq C_{p,q}\left\|\left\{\sum_{i=1}^{\infty}|f_{i}|^{q}\right\}^{\frac{1}{q}}\right\|_{L^{p}(\mathbb{R})}$$

for any sequence $\{f_{i} : i = 1, 2, \cdots\}$ *of locally integrable functions.*

The following result establishes an inequality between the two maximal functions we have just defined.

LEMMA 3.5 *Let g be a band-limited function on $\mathbb{R}$ such that $g_\lambda^*(x) < \infty$ for all $x \in \mathbb{R}$. Then, there exists a constant C_λ such that*

$$g_\lambda^*(x) \leq C_\lambda \left\{ \mathcal{M}(|g|^{\frac{1}{\lambda}})(x) \right\}^\lambda, \qquad x \in \mathbb{R}. \tag{3.6}$$

PROOF : Since g is band-limited, g is differentiable on $\mathbb{R}$ (by the Paley-Wiener theorem), so that it makes sense to consider the pointwise values of g. Let $x, y \in \mathbb{R}$ and $0 < \delta < 1$. Choose $z \in \mathbb{R}$ such that $|(x-y)-z| < \delta$. We apply the mean value theorem to g and the endpoints $x-y$ and z to obtain

$$|g(x-y)| \leq |g(z)| + \delta \sup_{\{w:|(x-y)-w|<\delta\}} |g'(w)|.$$

Taking the $(\frac{1}{\lambda})^{\text{th}}$ power and integrating with respect to the variable z over $[x-y-\delta, x-y+\delta]$ we obtain

$$|g(x-y)|^{\frac{1}{\lambda}} \leq \frac{\tilde{c}_\lambda}{2\delta} \int_{x-y-\delta}^{x-y+\delta} |g(z)|^{\frac{1}{\lambda}} \, dz + \tilde{c}_\lambda \delta^{\frac{1}{\lambda}} \sup_{\{w:|(x-y)-w|<\delta\}} |g'(w)|^{\frac{1}{\lambda}}, \tag{3.7}$$

here we have used the inequalities

$$(a+b)^s \leq a^s + b^s \qquad \text{if } 0 < s \leq 1, \qquad \text{and}$$
$$(a+b)^s \leq 2^{s-1}(a^s + b^s) \qquad \text{if } 1 \leq s < \infty.$$

Since $[x-y-\delta, x-y+\delta]$ is contained in $[x-|y|-\delta, x+|y|+\delta]$, we have

$$\int_{x-y-\delta}^{x-y+\delta} |g(z)|^{\frac{1}{\lambda}} \, dz \leq \int_{x-|y|-\delta}^{x+|y|+\delta} |g(z)|^{\frac{1}{\lambda}} \, dz \leq 2(\delta + |y|)\mathcal{M}(|g|^{\frac{1}{\lambda}})(x),$$

and

$$\sup_{\{w:|(x-y)-w|<\delta\}} |g'(w)|^{\frac{1}{\lambda}} \leq \sup_{\{w:|x-w|<|y|+\delta\}} |g'(w)|^{\frac{1}{\lambda}}$$

$$= \sup_{\{t:|t|<|y|+\delta\}} |g'(x+t)|^{\frac{1}{\lambda}} \leq (1+|y|+\delta)\left[(g')_\lambda^*(x)\right]^{\frac{1}{\lambda}}.$$

Substituting these last two inequalities in (3.7) we obtain

$$|g(x-y)|^{\frac{1}{\lambda}} \leq \tilde{c}_\lambda \frac{\delta + |y|}{\delta} \mathcal{M}(|g|^{\frac{1}{\lambda}})(x) + \tilde{c}_\lambda \delta^{\frac{1}{\lambda}}(1+|y|+\delta)\left[(g')^*_\lambda(x)\right]^{\frac{1}{\lambda}}$$

$$\leq \tilde{c}_\lambda \left\{ \frac{1}{\delta}\mathcal{M}(|g|^{\frac{1}{\lambda}})(x) + \delta^{\frac{1}{\lambda}}\left[(g')^*_\lambda(x)\right]^{\frac{1}{\lambda}} \right\}(1+|y|+\delta).$$

Since $\delta < 1$ we have $(1+|y|+\delta) \leq 2(1+|y|)$, so that, taking the λth power, we deduce

$$\frac{|g(x-y)|}{(1+|y|)^\lambda} \leq \overline{c}_\lambda \left\{ \frac{1}{\delta^\lambda}\left[\mathcal{M}(|g|^{\frac{1}{\lambda}})(x)\right]^\lambda + \delta\left[(g')^*_\lambda(x)\right] \right\}$$

(where we have used, again, the same inequalities employed to obtain (3.7)). Hence,

$$g^*_\lambda(x) \leq \overline{c}_\lambda \left\{ \frac{1}{\delta^\lambda}\left[\mathcal{M}(|g|^{\frac{1}{\lambda}})(x)\right]^\lambda + \delta\left[(g')^*_\lambda(x)\right] \right\}.$$

Using Lemma 3.2 and taking δ small enough so that $\overline{c}_\lambda c_\lambda \delta < \frac{1}{2}$ (where c_λ is the constant that appears in Lemma 3.2) we obtain

$$g^*_\lambda(x) \leq \frac{\overline{c}_\lambda}{\delta^\lambda}\left[\mathcal{M}(|g|^{\frac{1}{\lambda}})(x)\right]^\lambda + \tfrac{1}{2}g^*_\lambda(x).$$

Since $g^*_\lambda(x) < \infty$ we obtain the desired result by subtracting the last term from the first one in the above inequality.

∎

We shall always apply Lemma 3.5 with band-limited $g \in L^p(\mathbb{R})$, $1 \leq p \leq \infty$, and, in this case, the condition $g^*_\lambda(x) < \infty$ for all $x \in \mathbb{R}$ is always fulfilled. To show this we need the following inequality (see 4 of section 6.8 for references and comments):

THEOREM 3.8 **(Plancherel-Polya Inequality)**

Suppose $0 < p \leq \infty$ and $j \in \mathbb{Z}$. Let $g \in \mathcal{S}'$ and

$$\operatorname{supp}(\hat{g}) \subseteq \{\xi \in \mathbb{R} : |\xi| \leq 2^{j+1}\}.$$

Then, if $g \in L^p(\mathbb{R})$, there exists a constant C_p such that

$$\left\{ \sum_{k\in\mathbb{Z}} \sup_{z\in I_{j,k}} |g(z)|^p \right\}^{\frac{1}{p}} \leq C_p 2^{\frac{j}{p}}\|g\|_{L^p(\mathbb{R})},$$

where $I_{j,k} = [2^{-j}k, 2^{-j}(k+1)]$.

Corollary 3.9 *If g is a band-limited function defined on $\mathbb{R}$ such that $g \in L^p(\mathbb{R})$, $0 < p \leq \infty$, then we have $g_\lambda^*(x) < \infty$ for all $x \in \mathbb{R}$.*

Proof : Applying Theorem 3.8 with j large enough so that

$$\text{supp}\,(\hat{g}) \subseteq \{\xi \in \mathbb{R} : |\xi| \leq 2^{j+1}\},$$

we obtain

$$\sup_{z \in \mathbb{R}} |g(z)| \leq \Big\{\sum_{k \in \mathbb{Z}} \sup_{z \in I_{j,k}} |g(z)|^p \Big\}^{\frac{1}{p}} \leq C_p 2^{\frac{j}{p}} \|g\|_{L^p(\mathbb{R})}.$$

The result now follows by using the definition of g_λ^* given in (3.1). ∎

We shall apply Lemma 3.5 for functions of the form $\varphi_{2^{-j}} * f$ with φ band-limited – recall that functions of the above form appear in the definition of $\mathfrak{g}(f)$ (see Theorem 2.9).

Lemma 3.10 *Let φ be band-limited, $f \in \mathcal{S}'$ and $0 < p \leq \infty$ such that $\varphi_{2^{-j}} * f \in L^p(\mathbb{R})$ for all $j \in \mathbb{Z}$. Then, for any real $\lambda > 0$, there exists a constant C_λ such that*

$$(\varphi_{j,\lambda}^{**}f)(x) \leq C_\lambda \big\{\mathcal{M}\big(|\varphi_{2^{-j}} * f|^{\frac{1}{\lambda}}\big)(x)\big\}^\lambda, \qquad x \in \mathbb{R},$$

where

$$(\varphi_{j,\lambda}^{**}f)(x) \equiv \sup_{y \in \mathbb{R}} \frac{|(\varphi_{2^{-j}} * f)(x - y)|}{(1 + 2^j|y|)^\lambda},$$

and $\varphi_t(x) = \frac{1}{t}\varphi(\frac{x}{t})$.

Proof : Let $g(x) = (\varphi_{2^{-j}} * f)(2^{-j}x)$, so that g belongs to $L^p(\mathbb{R})$. By Corollary 3.9, $g_\lambda^*(x) < \infty$ for all $x \in \mathbb{R}$. Since $(\varphi_{2^{-j}})^\wedge(\xi) = \hat{\varphi}(2^{-j}\xi)$ and φ is band-limited, so is g. By Lemma 3.5,

$$g_\lambda^*(t) \leq C_\lambda \big\{\mathcal{M}\big(|g|^{\frac{1}{\lambda}}\big)(t)\big\}^\lambda, \qquad t \in \mathbb{R}. \tag{3.11}$$

But

$$g_\lambda^*(t) = \sup_{y \in \mathbb{R}} \frac{|g(t - y)|}{(1 + |y|)^\lambda} = \sup_{y \in \mathbb{R}} \frac{|(\varphi_{2^{-j}} * f)(2^{-j}t - 2^{-j}y)|}{(1 + |y|)^\lambda}$$

$$= \sup_{z \in \mathbb{R}} \frac{|(\varphi_{2^{-j}} * f)(2^{-j}t - z)|}{(1 + 2^j|z|)^\lambda} = (\varphi_{j,\lambda}^{**}f)(2^{-j}t),$$

and

$$M\left(|g|^{\frac{1}{\lambda}}\right)(t) = \sup_{r>0} \frac{1}{2r} \int_{t-r}^{t+r} \left|(\varphi_{2^{-j}} * f)(2^{-j}y)\right|^{\frac{1}{\lambda}} dy$$

$$= \sup_{r>0} \frac{2^j}{2r} \int_{2^{-j}t-2^{-j}r}^{2^{-j}t+2^{-j}r} \left|(\varphi_{2^{-j}} * f)(z)\right|^{\frac{1}{\lambda}} dz$$

$$= M\left(|\varphi_{2^{-j}} * f|^{\frac{1}{\lambda}}\right)(2^{-j}t).$$

Thus, the result follows from (3.11) and these last two equalities by taking $x = 2^{-j}t$.

■

We also need a result that allows us to estimate the convolution of two functions if appropriate inequalities for each one of them are assumed. The simplest case is contained in the following lemma:

LEMMA 3.12 *Let $\varepsilon > 0$. Suppose that g and h satisfy*

a) $|g(x)| \le \dfrac{C_1}{(1+|x|)^{1+\varepsilon}}$ *for all $x \in \mathbb{R}$,*   *and*

b) $|h(x)| \le \dfrac{C_2}{(1+|x|)^{1+\varepsilon}}$ *for all $x \in \mathbb{R}$,*

with C_1 and C_2 independent of $x \in \mathbb{R}$. Then, there exists a constant C such that for all $j, k, \ell, m \in \mathbb{Z}$ and $j \le \ell$ we have

$$\left|(g_{j,k} * h_{\ell,m})(x)\right| \le \dfrac{C2^{\frac{1}{2}(j-\ell)}}{\left(1 + 2^j|x - 2^{-j}k - 2^{-\ell}m|\right)^{1+\varepsilon}} \quad \text{for all } x \in \mathbb{R}.$$

REMARK: One can interpret the above lemma as follows: the functions g and h are "localized" at the origin; the functions $g_{j,k}$ and $h_{\ell,m}$ are then "localized" at $2^{-j}k$ and $2^{-\ell}m$, having a "localization width" that is roughly 2^{-j} and $2^{-\ell}$ and a height $2^{\frac{j}{2}}$ and $2^{\frac{\ell}{2}}$, respectively. Lemma 3.12 tells us that the convolution $g_{j,k} * h_{\ell,m}$ is localized at $2^{-j}k + 2^{-\ell}m$ (the sum of the localization points of each factor), its width is 2^{-j} (the bigger width), and its height is $2^{\frac{1}{2}(j-\ell)}$.

PROOF : By dilating and translating it is enough to give a proof for the case $j = k = m = 0$; that is, for $\ell \geq 0$,

$$\left|(g_{0,0} * h_{\ell,0})(x)\right| \leq \frac{2^{-\frac{\ell}{2}} C}{\left(1 + |x|\right)^{1+\varepsilon}} \qquad \text{for all } x \in \mathbb{R}. \tag{3.13}$$

This holds from the equality $(g_{j,k} * h_{\ell,m})(x) = (g_{0,0} * h_{\ell-j,0})(2^j x - k - 2^{j-\ell} m)$. Let

$$E_1 = \{y \in \mathbb{R} : |y - x| \leq 3\},$$

$$E_2 = \{y \in \mathbb{R} : |y - x| > 3 \text{ and } |y| \leq \tfrac{1}{2}|x|\},$$

$$E_3 = \{y \in \mathbb{R} : |y - x| > 3 \text{ and } |y| > \tfrac{1}{2}|x|\}.$$

If $y \in E_1$ we have $1 + |x| \leq 1 + |x - y| + |y| \leq 4 + |y| \leq 4(1 + |y|)$. If $y \in E_3$ we have $1 + |x| \leq 1 + 2|y| \leq 2(1 + |y|)$. Hence, for $y \in E_1 \cup E_3$, we have $1 + |x| \leq 4(1 + |y|)$, and obtain

$$\int_{E_1 \cup E_3} |g(y)| \cdot |h_{\ell,0}(x - y)| \, dy \leq \frac{4C_1}{\left(1 + |x|\right)^{1+\varepsilon}} \int_{\mathbb{R}} \frac{2^{\frac{\ell}{2}} C_2}{\left(1 + 2^\ell |x - y|\right)^{1+\varepsilon}} \, dy$$

$$\leq C \, 2^{-\frac{\ell}{2}} \frac{1}{\left(1 + |x|\right)^{1+\varepsilon}}$$

for some $C > 0$, since $\varepsilon > 0$. If $y \in E_2$, $\tfrac{1}{2}|x| \leq |y - x| \leq \tfrac{3}{2}|x|$ and $4|x - y| = |x - y| + 3|x - y| > 3 + \tfrac{3}{2}|x| > 1 + |x|$. Thus,

$$1 + 2^\ell |x - y| \geq 2^\ell |x - y| \geq 2^{\ell-2}(1 + |x|) \qquad \text{for } y \in E_2.$$

Therefore,

$$\int_{E_2} |g(y)| \cdot |h_{\ell,0}(x - y)| \, dy \leq \frac{2^{(2-\ell)(1+\varepsilon)} 2^{\frac{\ell}{2}} C_2}{\left(1 + |x|\right)^{1+\varepsilon}} \int_{\mathbb{R}} \frac{C_1}{\left(1 + |y|\right)^{1+\varepsilon}} \, dy$$

$$\leq C \, 2^{-\frac{\ell}{2}} \frac{1}{\left(1 + |x|\right)^{1+\varepsilon}}$$

for some $C > 0$, since $\ell > 0$ and $\varepsilon > 0$. Combining the above two estimates we obtain the desired result.

∎

In Lemma 3.12 the "height" of the convolution is $2^{\frac{1}{2}(j-\ell)}$. If we want "least height" we need to assume that h has some moments equal to zero; the price that we pay is that in this case we need to control not only g, but also some of its derivatives.

LEMMA 3.14 Let $r \geq \varepsilon > 0$ and $N \in \mathbb{N}$. Suppose that g and h satisfy

(a) $\left|\dfrac{d^n g}{dx^n}(x)\right| \leq \dfrac{C_{n,1}}{(1+|x|)^{1+\varepsilon}}$ for all $x \in \mathbb{R}$ and $0 \leq n \leq N+1$;

(b) $\displaystyle\int_{\mathbb{R}} x^n h(x)\, dx = 0$ for all $0 \leq n \leq N$;

(c) $|h(x)| \leq \dfrac{C_2}{(1+|x|)^{2+N+r}}$ for all $x \in \mathbb{R}$;

with $C_{n,1}$, $0 \leq n \leq N+1$, and C_2 independent of $x \in \mathbb{R}$. Then, there exists a constant C such that for all $j, k, \ell, m \in \mathbb{Z}$ and $j \leq \ell$ we have

$$\left|(g_{j,k} * h_{\ell,m})(x)\right| \leq \frac{C 2^{(j-\ell)(\frac{1}{2}+N+1)}}{\left(1 + 2^j|x - 2^{-j}k - 2^{-\ell}m|\right)^{1+\varepsilon}} \quad \text{for all } x \in \mathbb{R}.$$

PROOF : As in the proof of Lemma 3.12, it is enough to consider the case $j = k = m = 0$; that is, for $\ell \geq 0$,

$$\left|(g_{0,0} * h_{\ell,0})(x)\right| \leq \frac{2^{-\ell(\frac{1}{2}+N+1)}C}{\left(1+|x|\right)^{1+\varepsilon}} \quad \text{for all } x \in \mathbb{R}. \tag{3.15}$$

We let E_1, E_2 and E_3 be the sets introduced in the proof of Lemma 3.12. Then, by (b) we can write

$$\left|(g_{0,0} * h_{\ell,0})(x)\right| = \left|\int_{\mathbb{R}} \left\{g(y) - \sum_{n=0}^{N} \frac{1}{n!} \frac{d^n g}{dx^n}(x)(y-x)^n\right\} h_{\ell,0}(x-y)\, dy\right|$$

$$\leq \left\{\int_{E_1} + \int_{E_2} + \int_{E_3}\right\} \left\{\left|g(y) - \sum_{n=0}^{N} \frac{1}{n!} \frac{d^n g}{dx^n}(x)(y-x)^n\right| \cdot \left|h_{\ell,0}(x-y)\right|\right\} dy$$

$$\equiv I + II + III.$$

For $y \in E_1$, we use (a) with $n = N+1$ to obtain, for z between y and x,

$$\left|g(y) - \sum_{n=0}^{N} \frac{1}{n!} \frac{d^n g}{dx^n}(x)(y-x)^n\right| \leq C \left|\frac{d^{N+1} g}{dx^{N+1}}(z)\right| \cdot |y-x|^{N+1}$$

$$\le \frac{C}{\left(1+|z|\right)^{1+\varepsilon}}|y-x|^{N+1}.$$

Since z lies between x and y and $|y-x| \le 3$ when $y \in E_1$, we have $1+|x| \le 1+|x-z|+|z| \le 4+|z| \le 4(1+|z|)$. Thus, we obtain, using (c), that

$$I \le \frac{C\,2^{\frac{\ell}{2}}}{\left(1+|x|\right)^{1+\varepsilon}} \int_{E_1} |x-y|^{N+1} \frac{1}{\left(1+2^{\ell}|x-y|\right)^{2+N+r}}\,dy$$

$$\le \frac{2^{\frac{\ell}{2}}C}{\left(1+|x|\right)^{1+\varepsilon}} \int_0^{3\cdot 2^{\ell}} 2^{-\ell(N+1)}t^{N+1} \frac{1}{\left(1+t\right)^{2+N+r}}\,2^{-\ell}dt$$

$$\le \frac{2^{-\ell(\frac{1}{2}+N+1)}C}{\left(1+|x|\right)^{1+\varepsilon}}$$

since $r > 0$. For $y \in E_2$ we have $1+2^{\ell}|x-y| \ge 2^{\ell-2}(1+|x|)$ (as in the proof of Lemma 3.12). Hence, using (a) and (c) we obtain

$$II \le C \int_{E_2} \left\{ \frac{1}{\left(1+|y|\right)^{1+\varepsilon}} + \sum_{n=0}^{N} \frac{|x-y|^n}{\left(1+|x|\right)^{1+\varepsilon}} \right\} \frac{2^{\frac{\ell}{2}}}{\left(1+2^{\ell}|x-y|\right)^{2+N+r}}\,dy$$

$$\le C \int_{E_2} \left\{ \frac{1}{\left(1+|y|\right)^{1+\varepsilon}} + \sum_{n=0}^{N} \frac{|x-y|^n}{\left(1+|x|\right)^{1+\varepsilon}} \right\} \frac{2^{\frac{\ell}{2}}2^{-\ell(2+N+r)}}{\left(1+|x|\right)^{2+N+r}}\,dy.$$

For $y \in E_2$, $|y-x| > 3$, so that $|x-y|^n \le |x-y|^N$ for $n = 0, 1, \cdots, N$. Also, in this case, $|x-y| \le |x|+|y| \le \frac{3}{2}|x|$. Hence,

$$II \le \frac{2^{\frac{\ell}{2}}2^{-\ell(2+N+r)}C}{\left(1+|x|\right)^{2+N+r}} \left\{ \int_{E_2} \frac{1}{\left(1+|y|\right)^{1+\varepsilon}}\,dy + \frac{|x|^N}{\left(1+|x|\right)^{1+\varepsilon}} \int_{|y|\le \frac{1}{2}|x|}\,dy \right\}$$

$$\le \frac{2^{-\ell(\frac{3}{2}+N+r)}C}{\left(1+|x|\right)^{2+N+r}} \left\{ 1 + \frac{|x|^{N+1}}{\left(1+|x|\right)^{1+\varepsilon}} \right\}$$

$$\le \frac{2^{-\ell(\frac{1}{2}+N+1)}C}{\left(1+|x|\right)^{1+r}} \le \frac{2^{-\ell(\frac{1}{2}+N+1)}C}{\left(1+|x|\right)^{1+\varepsilon}}$$

since $r \ge \varepsilon > 0$. For $y \in E_3$, $1+|x| \le 2(1+|y|)$ (as in the proof of

Lemma 3.12). Using (a) and (c) again we obtain

$$
III \leq C \int_{E_3} \left\{ \frac{1}{(1+|y|)^{1+\varepsilon}} + \sum_{n=0}^{N} \frac{|x-y|^n}{(1+|x|)^{1+\varepsilon}} \right\} \frac{2^{\frac{\ell}{2}}}{(1+2^\ell|x-y|)^{2+N+r}} \, dy
$$

$$
\leq C \frac{2^{\frac{\ell}{2}}}{(1+|x|)^{1+\varepsilon}} \int_3^\infty \frac{t^N}{(1+2^\ell|t|)^{2+N+r}} \, dt
$$

$$
= C \frac{2^{\frac{\ell}{2}-\ell N}}{(1+|x|)^{1+\varepsilon}} \int_{3\cdot 2^\ell}^\infty \frac{2^{-\ell} w^N}{(1+w)^{2+N+r}} \, dw
$$

$$
\leq C \, 2^{-\ell(\frac{1}{2}+N)} \frac{2^{-\ell(1+r)}}{(1+|x|)^{1+\varepsilon}} \leq \frac{2^{-\ell(\frac{1}{2}+N+1)} C}{(1+|x|)^{1+\varepsilon}}.
$$

These three estimates give us the result.

∎

REMARK: In Lemma 3.14 we have assumed that the $(N+1)^{\text{th}}$ derivative of g exists and satisfies (a). But all that we have used in the proof of the lemma related to this condition involves the application of the Mean-Value theorem to $\frac{d^N}{dx^N} g$ (see the estimate for I). Hence, in Lemma 3.14 one could replace the condition on $\frac{d^{N+1}}{dx^{N+1}} g$ by a Lipschitz condition on $\frac{d^N}{dx^N} g$ and still obtain a similar conclusion. More precisely, if we replace the assumption on $\frac{d^{N+1}}{dx^{N+1}} g$ of Lemma 3.14 by

$$
\left| \frac{d^N g}{dx^N}(x) - \frac{d^N g}{dx^N}(y) \right| \leq C_N |x-y|^\delta \sup_{|z| \leq |x-y|} \frac{1}{(1+|z-x|)^{1+\varepsilon}}
$$

for all $x, y \in \mathbb{R}$, $0 < \delta \leq 1$, then the conclusion will be

$$
\left| (g_{j,k} * h_{\ell,m})(x) \right| \leq \frac{2^{(j-\ell)(\frac{1}{2}+N+\delta)} C}{(1+2^j|x-2^{-j}k-2^{-\ell}m|)^{1+\varepsilon}}
$$

for all $x \in \mathbb{R}$ and $j \leq \ell$.

We need one more tool. This is a pair of inequalities which are of similar nature to the one stated in Lemma 3.5. In the next lemma we write, as before, $I_{j,k} = [2^{-j}k, 2^{-j}(k+1)]$ for $j, k \in \mathbb{Z}$.

LEMMA 3.16 *Given $\varepsilon > 0$ and $1 \leq r < 1 + \varepsilon$, there exists a constant C such that for all sequences $\{s_{j,k} : j, k \in \mathbb{Z}\}$ of complex numbers and all $x \in I_{j,k}$,*

(a) $\quad \displaystyle\sum_{m \in \mathbb{Z}} \frac{|s_{\ell,m}|}{\left(1 + 2^\ell |2^{-j}k - 2^{-\ell}m|\right)^{1+\varepsilon}} \leq C \left[\mathcal{M}\left(\sum_{m \in \mathbb{Z}} |s_{\ell,m}|^{\frac{1}{r}} \chi_{I_{\ell,m}} \right)(x) \right]^r$

$\quad$ *if $\ell \leq j$, and*

(b) $\quad \displaystyle\sum_{m \in \mathbb{Z}} \frac{|s_{\ell,m}|}{\left(1 + 2^j |2^{-\ell}m - 2^{-j}k|\right)^{1+\varepsilon}} \leq C 2^{(\ell-j)r} \left[\mathcal{M}\left(\sum_{m \in \mathbb{Z}} |s_{\ell,m}|^{\frac{1}{r}} \chi_{I_{\ell,m}} \right)(x) \right]^r$

$\quad$ *if $\ell \geq j$;*

where $\mathcal{M}$ is the Hardy-Littlewood maximal function defined by (3.3).

PROOF : We begin by proving (a). Write

$$E_0 = \{m \in \mathbb{Z} : |2^{\ell-j}k - m| \leq 1\} \qquad \text{and}$$

$$E_n = \{m \in \mathbb{Z} : 2^{n-1} < |2^{\ell-j}k - m| \leq 2^n\} \qquad \text{for } n = 1, 2, 3, \cdots.$$

Then, we have

$$\sum_{m \in \mathbb{Z}} \frac{|s_{\ell,m}|}{\left(1 + 2^\ell |2^{-j}k - 2^{-\ell}m|\right)^{1+\varepsilon}} \leq \sum_{n=0}^{\infty} \sum_{m \in E_n} \frac{|s_{\ell,m}|}{\left(1 + |2^{\ell-j}k - m|\right)^{1+\varepsilon}}$$

$$\leq C_\varepsilon \sum_{n=0}^{\infty} 2^{-n(1+\varepsilon)} \sum_{m \in E_n} |s_{\ell,m}|.$$

Since $0 < \frac{1}{r} \leq 1$, the space $\ell^{\frac{1}{r}}$ is continuously contained in ℓ^1. Thus,

$$\sum_{m \in E_n} |s_{\ell,m}| \leq \left(\sum_{m \in E_n} |s_{\ell,m}|^{\frac{1}{r}} \right)^r.$$

Since $I_{\ell,m} \subseteq [x - 2^{n-\ell+1}, x + 2^{n-\ell+1}]$ when $x \in I_{j,k}$, $\ell \leq j$, and $m \in E_n$, we have

$$\sum_{m \in E_n} |s_{\ell,m}|^{\frac{1}{r}} = 2^\ell \int_{x-2^{n-\ell+1}}^{x+2^{n-\ell+1}} \left(\sum_{m \in E_n} |s_{\ell,m}|^{\frac{1}{r}} \chi_{I_{\ell,m}}(y) \right) dy.$$

Therefore, when $x \in I_{j,k}$, we can write

$$
\sum_{m \in \mathbb{Z}} \frac{|s_{\ell,m}|}{\left(1 + 2^{\ell}|2^{-j}k - 2^{-\ell}m|\right)^{1+\varepsilon}}
$$

$$
\leq C_{\varepsilon} \sum_{n=0}^{\infty} 2^{-n(1+\varepsilon)} \left[2^{\ell} \int_{x - 2^{n-\ell+1}}^{x + 2^{n-\ell+1}} \left(\sum_{m \in E_n} |s_{\ell,m}|^{\frac{1}{r}} \chi_{I_{\ell,m}}(y) \right) dy \right]^{r}
$$

$$
\leq C_{\varepsilon,r} \sum_{n=0}^{\infty} 2^{-n(1+\varepsilon)} 2^{nr} \left[\mathcal{M} \left(\sum_{m \in E_n} |s_{\ell,m}|^{\frac{1}{r}} \chi_{I_{\ell,m}} \right)(x) \right]^{r}
$$

$$
\leq C_{\varepsilon,r} \left[\mathcal{M} \left(\sum_{m \in \mathbb{Z}} |s_{\ell,m}|^{\frac{1}{r}} \chi_{I_{\ell,m}} \right)(x) \right]^{r},
$$

since $\displaystyle\sum_{n=0}^{\infty} 2^{-n(1+\varepsilon-r)} < \infty$, due to the fact that $1 + \varepsilon - r > 0$. This proves part (a) of the lemma.

We now prove part (b). Let

$$
F_0 = \{ m \in \mathbb{Z} : |2^{j-\ell}m - k| \leq 1 \} \qquad \text{and}
$$

$$
F_n = \{ m \in \mathbb{Z} : 2^{n-1} < |2^{j-\ell}m - k| \leq 2^n \} \qquad \text{for} \ \ n = 1, 2, 3, \cdots.
$$

As in the proof of part (a), we have

$$
\sum_{m \in \mathbb{Z}} \frac{|s_{\ell,m}|}{\left(1 + 2^{j}|2^{-\ell}m - 2^{-j}k|\right)^{1+\varepsilon}} \leq C_{\varepsilon} \sum_{n=0}^{\infty} 2^{-n(1+\varepsilon)} \left(\sum_{m \in F_n} |s_{\ell,m}|^{\frac{1}{r}} \right)^{r},
$$

where we have already used that $r \geq 1$. When $x \in I_{j,k}$, $\ell \geq j$, and $m \in F_n$, we have $I_{\ell,m} \subseteq [x - 2^{n-j+1}, x + 2^{n-j+1}]$. Thus

$$
\sum_{m \in F_n} |s_{\ell,m}|^{\frac{1}{r}} = 2^{\ell} \int_{x - 2^{n-j+1}}^{x + 2^{n-j+1}} \left(\sum_{m \in F_n} |s_{\ell,m}|^{\frac{1}{r}} \chi_{I_{\ell,m}}(y) \right) dy.
$$

Therefore, when $x \in I_{j,k}$, we can write

$$
\sum_{m \in \mathbb{Z}} \frac{|s_{\ell,m}|}{\left(1 + 2^{j}|2^{-\ell}m - 2^{-j}k|\right)^{1+\varepsilon}}
$$

$$
\leq C_{\varepsilon} \sum_{n=0}^{\infty} 2^{-n(1+\varepsilon)} \left[2^{\ell} \int_{x - 2^{n-j+1}}^{x + 2^{n-j+1}} \left(\sum_{m \in F_n} |s_{\ell,m}|^{\frac{1}{r}} \chi_{I_{\ell,m}}(y) \right) dy \right]^{r}
$$

$$\leq C_{\varepsilon,r} \sum_{n=0}^{\infty} 2^{-n(1+\varepsilon)} 2^{(\ell-j+n)r} \left[\mathcal{M} \Big(\sum_{m \in F_n} |s_{\ell,m}|^{\frac{1}{r}} \chi_{I_{\ell,m}} \Big)(x) \right]^r$$

$$\leq C_{\varepsilon,r} 2^{(\ell-j)r} \left[\mathcal{M} \Big(\sum_{m \in \mathbb{Z}} |s_{\ell,m}|^{\frac{1}{r}} \chi_{I_{\ell,m}} \Big)(x) \right]^r,$$

since $1 + \varepsilon - r > 0$.

∎

6.4 The Lebesgue spaces $L^p(\mathbb{R})$ with $1 < p < \infty$

In section 6.2 we have given a characterization of $L^p(\mathbb{R})$, $1 < p < \infty$, in terms of a (discrete) Littlewood-Paley $\mathfrak{g}$ function. In this section we shall give a characterization of these spaces using wavelet coefficients. We start by giving another characterization of $L^p(\mathbb{R})$, $1 < p < \infty$, this time using the maximal functions $\varphi_{j,\lambda}^{**} f$ defined in Lemma 3.10. We shall use the class $\mathfrak{R}^0$ whose definition is given in (2.12).

THEOREM 4.1 *Let $\varphi \in \mathfrak{R}^0$ be a band-limited function. Given a real number $\lambda \geq 1$ and $1 < p < \infty$, there exists a constant $A_{p,\lambda} < \infty$ such that*

$$\left\| \left\{ \sum_{j \in \mathbb{Z}} |\varphi_{j,\lambda}^{**} f|^2 \right\}^{\frac{1}{2}} \right\|_{L^p(\mathbb{R})} \leq A_{p,\lambda} \|f\|_{L^p(\mathbb{R})} \qquad \text{for all } f \in L^p(\mathbb{R}).$$

PROOF : Let $f \in L^p(\mathbb{R})$, $1 < p < \infty$. Since $\varphi_{2^{-j}} \in L^1(\mathbb{R})$, $\varphi_{2^{-j}} * f \in L^p(\mathbb{R})$ for all $j \in \mathbb{Z}$; thus, we can use Lemma 3.10 to obtain

$$\left\| \left\{ \sum_{j \in \mathbb{Z}} |\varphi_{j,\lambda}^{**} f|^2 \right\}^{\frac{1}{2}} \right\|_{L^p(\mathbb{R})} \leq C_\lambda \left\| \left\{ \sum_{j \in \mathbb{Z}} [\mathcal{M}(|\varphi_{2^{-j}} * f|^{\frac{1}{\lambda}})]^{2\lambda} \right\}^{\frac{1}{2}} \right\|_{L^p(\mathbb{R})}$$

$$= C_\lambda \left\| \left\{ \sum_{j \in \mathbb{Z}} [\mathcal{M}(|\varphi_{2^{-j}} * f|^{\frac{1}{\lambda}})]^{2\lambda} \right\}^{\frac{1}{2\lambda}} \right\|_{L^{p\lambda}(\mathbb{R})}^{\lambda}.$$

We apply Theorem 3.4 to $p\lambda > 1$ and $q = 2\lambda \geq 2 > 1$ and obtain

$$\left\| \left\{ \sum_{j \in \mathbb{Z}} |\varphi_{j,\lambda}^{**} f|^2 \right\}^{\frac{1}{2}} \right\|_{L^p(\mathbb{R})} \leq C_{p,\lambda} \left\| \left\{ \sum_{j \in \mathbb{Z}} |\varphi_{2^{-j}} * f|^2 \right\}^{\frac{1}{2\lambda}} \right\|_{L^{p\lambda}(\mathbb{R})}^{\lambda}$$

$$= C_{p,\lambda} \left\| \left\{ \sum_{j \in \mathbb{Z}} |\varphi_{2^{-j}} * f|^2 \right\}^{\frac{1}{2}} \right\|_{L^p(\mathbb{R})} \le A_{p,\lambda} \|f\|_{L^p(\mathbb{R})},$$

where the last inequality follows from Corollary 2.14. (The fact that we are dealing with a sequence indexed by $\mathbb{Z}$, instead of the natural numbers, as in the case in Theorem 3.4, is clearly not a problem.) $\blacksquare$

REMARK: If in addition to assuming that $\varphi \in \mathfrak{R}^0$ and φ is band-limited we also assume that φ satisfies

$$\sum_{j \in \mathbb{Z}} |\hat{\varphi}(2^j \xi)|^2 = M \qquad \text{for a.e. } \xi \in \mathbb{R},$$

we have the promised characterization of $L^p(\mathbb{R})$, $1 < p < \infty$, in terms of the functions $\varphi_{j,\lambda}^{**} f$. That is,

$$A\|f\|_{L^p(\mathbb{R})} \le \left\| \left\{ \sum_{j \in \mathbb{Z}} |\varphi_{j,\lambda}^{**} f|^2 \right\}^{\frac{1}{2}} \right\|_{L^p(\mathbb{R})} \le B\|f\|_{L^p(\mathbb{R})},$$

with A and B depending only on p and λ, and $\lambda \ge 1$. By Theorem 4.1 it remains to show the left-hand side inequality. This follows easily from Corollary 2.14 and the trivial inequality $|(\varphi_{2^{-j}} * f)(x)| \le (\varphi_{j,\lambda}^{**} f)(x)$.

THEOREM 4.2 *Let $\psi \in \mathfrak{R}^0$ be a band-limited function. For $1 < p < \infty$ and $f \in L^p(\mathbb{R})$, we have*

$$\left\| \left\{ \sum_{j \in \mathbb{Z}} \sum_{k \in \mathbb{Z}} |\langle f, \psi_{j,k} \rangle|^2 2^j \chi_{[2^{-j}k, 2^{-j}(k+1)]} \right\}^{\frac{1}{2}} \right\|_{L^p(\mathbb{R})} \le C\|f\|_{L^p(\mathbb{R})}, \qquad (4.3)$$

with C independent of f.

PROOF : We start by noticing that, for $f \in L^p(\mathbb{R})$, the numbers $\langle f, \psi_{j,k} \rangle$ make sense since $\psi \in L^{p'}(\mathbb{R})$. In fact,

$$|\langle f, \psi_{j,k} \rangle| \le 2^{j(\frac{1}{p} - \frac{1}{2})} \|\psi\|_{L^{p'}(\mathbb{R})} \|f\|_{L^p(\mathbb{R})}.$$

We have

$$|\langle f, \psi_{j,k} \rangle| = 2^{\frac{j}{2}} \left| \int_{\mathbb{R}} f(x) \overline{\psi(2^j x - k)} \, dx \right| = 2^{-\frac{j}{2}} \left| \int_{\mathbb{R}} f(x) \overline{\tilde{\psi}_{2^{-j}}(x - 2^{-j}k)} \, dx \right|$$

$$= 2^{-\frac{j}{2}} |(\tilde{\psi}_{2^{-j}} * f)(2^{-j}k)| \le 2^{-\frac{j}{2}} \sup_{y \in I_{j,k}} |(\tilde{\psi}_{2^{-j}} * f)(y)|,$$

where $I_{j,k} = [2^{-j}k, 2^{-j}(k+1)]$ and $\tilde{\psi}(y) = \overline{\psi(-y)}$. Fixing $j \in \mathbb{Z}$ we have

$$\sum_{k \in \mathbb{Z}} |\langle f, \psi_{j,k} \rangle|^2 2^j \chi_{[2^{-j}k, 2^{-j}(k+1)]}(x)$$

$$\leq \sum_{k \in \mathbb{Z}} \left\{ \sup_{y \in I_{j,k}} |(\tilde{\psi}_{2^{-j}} * f)(y)| \right\}^2 \chi_{I_{j,k}}(x) \leq \left\{ \sup_{|z| \leq 2^{-j}} |(\tilde{\psi}_{2^{-j}} * f)(x-z)| \right\}^2$$

$$= \left\{ \sup_{|z| \leq 2^{-j}} \frac{|(\tilde{\psi}_{2^{-j}} * f)(x-z)|}{(1 + 2^j|z|)^\lambda} \right\}^2 (1 + 2^j|z|)^{2\lambda} \leq 2^{2\lambda} [(\psi_{j,\lambda}^{**} f)(x)]^2$$

for any $\lambda > 0$. Inequality (4.3) now follows by applying Theorem 4.1 with any $\lambda \geq 1$. ∎

To obtain the reverse inequality to (4.3) we shall assume ψ to be an orthonormal wavelet and we derive it with an argument similar to the one described in Corollary 2.14. This will give us the promised characterization of $L^p(\mathbb{R})$, $1 < p < \infty$, in terms of wavelet coefficients.

We shall use the following notation. Given two functions f and ψ for which $\langle f, \psi \rangle$ makes sense, we define

$$(\mathcal{W}_\psi f)(x) = \left\{ \sum_{j \in \mathbb{Z}} \sum_{k \in \mathbb{Z}} |\langle f, \psi_{j,k} \rangle|^2 2^j \chi_{[2^{-j}k, 2^{-j}(k+1)]}(x) \right\}^{\frac{1}{2}}. \tag{4.4}$$

Observe that if T_ψ is the operator mapping f into the $\ell^2(\mathbb{Z} \times \mathbb{Z})$ valued function

$$(T_\psi f)(x) = \left\{ \langle f, \psi_{j,k} \rangle 2^{\frac{j}{2}} \chi_{[2^{-j}k, 2^{-j}(k+1)]}(x) : j \in \mathbb{Z}, k \in \mathbb{Z} \right\},$$

we have

$$(\mathcal{W}_\psi f)(x) = \sqrt{(T_\psi f)(x) \cdot (T_\psi f)(x)},$$

where $\cdot$ denotes the dot product in $\ell^2(\mathbb{Z} \times \mathbb{Z})$.

THEOREM 4.5 *Let ψ be a band-limited wavelet such that $\psi \in \mathfrak{R}^0$. Given $p \in (1, \infty)$, there exist two constants $0 < A_p \leq B_p < \infty$ such that*

$$A_p \|f\|_{L^p(\mathbb{R})} \leq \|\mathcal{W}_\psi f\|_{L^p(\mathbb{R})} \leq B_p \|f\|_{L^p(\mathbb{R})} \tag{4.6}$$

for all $f \in L^p(\mathbb{R})$.

PROOF: We first observe that the coefficients $\langle f, \psi_{j,k}\rangle$ are well defined since $\psi \in L^{p'}(\mathbb{R})$, where $\frac{1}{p'} + \frac{1}{p} = 1$. The inequality on the right in (4.6) is precisely inequality (4.3). Thus, we have already obtained a constant $B_p < \infty$ such that

$$\|\mathcal{W}_\psi f\|_{L^p(\mathbb{R})} \le B_p \|f\|_{L^p(\mathbb{R})}. \tag{4.7}$$

For $p = 2$ we have equality (with $B_p = 1$) because ψ is an orthonormal wavelet:

$$\int_{\mathbb{R}} (T_\psi f)(x) \cdot (T_\psi f)(x)\, dx = \|\mathcal{W}_\psi f\|_{L^2(\mathbb{R})}^2$$

$$= \int_{\mathbb{R}} \sum_{j \in \mathbb{Z}} \sum_{k \in \mathbb{Z}} |\langle f, \psi_{j,k}\rangle|^2 2^j \chi_{I_{j,k}}(x)\, dx$$

$$= \sum_{j \in \mathbb{Z}} \sum_{k \in \mathbb{Z}} |\langle f, \psi_{j,k}\rangle|^2 = \|f\|_{L^2(\mathbb{R})}^2,$$

where, as in the previous section, $I_{j,k} = [2^{-j}k, 2^{-j}(k+1)]$. From this equality, the polarization identity and a density argument we obtain

$$\int_{\mathbb{R}} f(x)g(x)\, dx = \int_{\mathbb{R}} (T_\psi f)(x) \cdot (T_\psi g)(x)\, dx$$

for $f \in L^p(\mathbb{R})$, $g \in L^{p'}(\mathbb{R})$, where p' is the conjugate exponent to p. Using a duality argument, together with Hölder's inequality and (4.7) for $L^{p'}(\mathbb{R})$ we deduce

$$\|f\|_{L^p(\mathbb{R})} = \sup_{\|g\|_{p'} \le 1} \left| \int_{\mathbb{R}} f(x)g(x)\, dx \right|$$

$$\le \sup_{\|g\|_{p'} \le 1} \|\mathcal{W}_\psi f\|_{L^p(\mathbb{R})} \|\mathcal{W}_\psi g\|_{L^{p'}(\mathbb{R})} \le B_{p'} \|\mathcal{W}_\psi f\|_{L^p(\mathbb{R})}.$$

∎

Theorem 4.5 can clearly be applied to the Lemarié-Meyer wavelets given in Corollary 4.7 of Chapter 1 and Example D of Chapter 2, if we choose b sufficiently smooth, since they are band-limited. The drawback is that it does not apply to the compactly supported wavelets of Daubechies (section 2.3) or to the spline wavelets of Chapter 4, because they are not band-limited. We shall now show how to overcome this problem. The idea is to show that for any two wavelets ψ and φ, with "mild" decreasing conditions, the expressions $\left\|\mathcal{W}_\psi f\right\|_{L^p(\mathbb{R})}$ and $\left\|\mathcal{W}_\varphi f\right\|_{L^p(\mathbb{R})}$ are equivalent for $f \in L^p(\mathbb{R})$, $1 < p < \infty$.

We start with a result that follows from Lemma 3.14.

PROPOSITION 4.8 *Let $\psi, \varphi \in \mathfrak{R}^0$ and $j, k, \ell, m \in \mathbb{Z}$.*

(a) *There exist constants $C < \infty$ and $\varepsilon > 0$ such that for $j \geq \ell$ we have*

$$\left|\langle \psi_{j,k}, \varphi_{\ell,m}\rangle\right| \leq \frac{2^{\frac{3}{2}(\ell-j)}C}{\left(1 + 2^\ell|2^{-j}k - 2^{-\ell}m|\right)^{1+\varepsilon}}.$$

(b) *There exist constants $C < \infty$ and $\varepsilon > 0$ such that for $j \leq \ell$ we have*

$$\left|\langle \psi_{j,k}, \varphi_{\ell,m}\rangle\right| \leq \frac{2^{\frac{3}{2}(j-\ell)}C}{\left(1 + 2^j|2^{-\ell}m - 2^{-j}k|\right)^{1+\varepsilon}}.$$

PROOF: Suppose that ψ is associated with the constants $C_0^1, C_1^1, \gamma^1 > 0$ and $\varepsilon^1 > 0$ and that φ is associated with the constants $C_0^2, C_1^2, \gamma^2 > 0$ and $\varepsilon^2 > 0$ in definition (2.12). We choose

$$C = \max\left\{C_0^1, C_1^1, C_0^2, C_1^2\right\}, \quad \gamma = \min\left\{\gamma^1, \gamma^2\right\} \quad \text{and} \quad \varepsilon = \min\left\{\varepsilon^1, \varepsilon^2, \gamma\right\}.$$

Then, ψ and φ satisfy the condition (2.12) with C, γ and ε ($\gamma \geq \varepsilon$).

For a function g defined on $\mathbb{R}$ we write $\tilde{g}(x) = \overline{g(-x)}$. Then, we have

$$\overline{\langle \psi_{j,k}, \varphi_{\ell,m}\rangle} = \langle \varphi_{\ell,m}, \psi_{j,k}\rangle = (\varphi_{\ell,m} * \tilde{\psi}_{j,-k})(0).$$

Part (a) follows from Lemma 3.14 with $N = 0$. Part (b) follows from (a) or, again, by an application of Lemma 3.14. ∎

THEOREM 4.9 *Suppose that $\psi, \varphi \in \mathfrak{R}^0$ and φ is an orthonormal wavelet. If $1 < p < \infty$, there exists a constant $0 < C_p < \infty$ such that*

$$\left\|\mathcal{W}_\psi f\right\|_{L^p(\mathbb{R})} \leq C_p \left\|\mathcal{W}_\varphi f\right\|_{L^p(\mathbb{R})} \tag{4.10}$$

for all $f \in L^p(\mathbb{R})$.

PROOF : Since φ is an orthonormal wavelet we can write

$$\psi_{j,k}(x) = \sum_{\ell \in \mathbb{Z}} \sum_{m \in \mathbb{Z}} \langle \psi_{j,k}, \varphi_{\ell,m} \rangle \varphi_{\ell,m}(x),$$

with convergence in $L^2(\mathbb{R})$ and, hence, also in the sense of distributions. Then,

$$(\mathcal{W}_\psi f)(x) = \left\{ \sum_{j \in \mathbb{Z}} \sum_{k \in \mathbb{Z}} |\langle f, \psi_{j,k} \rangle|^2 2^j \chi_{I_{j,k}}(x) \right\}^{\frac{1}{2}}$$

$$= \left\{ \sum_{j \in \mathbb{Z}} \sum_{k \in \mathbb{Z}} \left| \sum_{\ell \in \mathbb{Z}} \sum_{m \in \mathbb{Z}} \langle f, \varphi_{\ell,m} \rangle \overline{\langle \psi_{j,k}, \varphi_{\ell,m} \rangle} \right|^2 2^j \chi_{I_{j,k}}(x) \right\}^{\frac{1}{2}},$$

where $I_{j,k} = [2^{-j}k, 2^{-j}(k+1)]$. Writing

$$A_1(j,k) = \sum_{\ell \leq j} \sum_{m \in \mathbb{Z}} \langle f, \varphi_{\ell,m} \rangle \overline{\langle \psi_{j,k}, \varphi_{\ell,m} \rangle}$$

and

$$A_2(j,k) = \sum_{\ell > j} \sum_{m \in \mathbb{Z}} \langle f, \varphi_{\ell,m} \rangle \overline{\langle \psi_{j,k}, \varphi_{\ell,m} \rangle},$$

we have

$$(\mathcal{W}_\psi f)(x) \leq \left\{ \sum_{j \in \mathbb{Z}} \sum_{k \in \mathbb{Z}} |A_1(j,k)|^2 2^j \chi_{I_{j,k}}(x) \right\}^{\frac{1}{2}}$$

$$+ \left\{ \sum_{j \in \mathbb{Z}} \sum_{k \in \mathbb{Z}} |A_2(j,k)|^2 2^j \chi_{I_{j,k}}(x) \right\}^{\frac{1}{2}}. \qquad (4.11)$$

To estimate A_1 we use part (a) of Proposition 4.8 to obtain

$$|A_1(j,k)| \leq \sum_{\ell \leq j} \sum_{m \in \mathbb{Z}} |\langle f, \varphi_{\ell,m} \rangle| \frac{2^{\frac{3}{2}(\ell-j)} C}{(1 + |2^{\ell-j}k - m|)^{1+\varepsilon}}$$

$$= C \sum_{\ell \leq j} 2^{\frac{3}{2}(\ell-j)} \left\{ \sum_{m \in \mathbb{Z}} \frac{|\langle f, \varphi_{\ell,m} \rangle|}{(1 + |2^{\ell-j}k - m|)^{1+\varepsilon}} \right\}.$$

By part (a) of Lemma 3.16 (with $r = 1$) we deduce

$$|A_1(j,k)| \leq C \sum_{\ell \leq j} 2^{\frac{3}{2}(\ell-j)} \left[\mathcal{M} \left(\sum_{m \in \mathbb{Z}} |\langle f, \varphi_{\ell,m} \rangle| \chi_{I_{\ell,m}} \right)(x) \right]$$

for all $x \in I_{j,k}$. Therefore,

$$\left\|\left\{\sum_{j\in\mathbb{Z}}\sum_{k\in\mathbb{Z}}|A_1(j,k)|^2 2^j \chi_{I_{j,k}}\right\}^{\frac{1}{2}}\right\|_{L^p(\mathbb{R})}$$

$$\leq C\left\|\left\{\sum_{j\in\mathbb{Z}} 2^j \left[\sum_{\ell\leq j} 2^{\frac{3}{2}(\ell-j)} \mathcal{M}\left(\sum_{m\in\mathbb{Z}}|\langle f,\varphi_{\ell,m}\rangle|\chi_{I_{\ell,m}}\right)\right]^2\right\}^{\frac{1}{2}}\right\|_{L^p(\mathbb{R})}$$

$$= C\left\|\left\{\sum_{j\in\mathbb{Z}}\left[\sum_{\ell\leq j} 2^{\ell-j} \mathcal{M}\left(\sum_{m\in\mathbb{Z}}|\langle f,\varphi_{\ell,m}\rangle|2^{\frac{\ell}{2}}\chi_{I_{\ell,m}}\right)\right]^2\right\}^{\frac{1}{2}}\right\|_{L^p(\mathbb{R})}$$

$$\leq C\left\|\left\{\sum_{j=0}^{\infty} 2^{-j}\right\}\left\{\sum_{\ell\in\mathbb{Z}}\left[\mathcal{M}\left(\sum_{m\in\mathbb{Z}}|\langle f,\varphi_{\ell,m}\rangle|2^{\frac{\ell}{2}}\chi_{I_{\ell,m}}\right)\right]^2\right\}^{\frac{1}{2}}\right\|_{L^p(\mathbb{R})}$$

$$\leq C\left\|\left\{\sum_{\ell\in\mathbb{Z}}\left[\mathcal{M}\left(\sum_{m\in\mathbb{Z}}|\langle f,\varphi_{\ell,m}\rangle|2^{\frac{\ell}{2}}\chi_{I_{\ell,m}}\right)\right]^2\right\}^{\frac{1}{2}}\right\|_{L^p(\mathbb{R})},$$

where we have used Young's inequality for convolutions

$$\|\{a_j\} * \{b_\ell\}\|_{\ell^2} \equiv \left\|\left\{\sum_\ell a_{j-\ell} b_\ell\right\}_j\right\|_{\ell^2} \leq \|\{a_j\}\|_{\ell^1}\|\{b_\ell\}\|_{\ell^2} \qquad (4.12)$$

with

$$a_j = \begin{cases} 2^{-j} & \text{if } j \geq 0, \\ 0 & \text{if } j < 0, \end{cases}$$

and

$$b_\ell = \mathcal{M}\left(\sum_{m\in\mathbb{Z}}|\langle f,\varphi_{\ell,m}\rangle|2^{\frac{\ell}{2}}\chi_{I_{\ell,m}}\right)(x).$$

We now use the vector-valued inequality for the Hardy-Littlewood maximal function (Theorem 3.4) with $q = 2$, to obtain

$$\left\|\left\{\sum_{j\in\mathbb{Z}}\sum_{k\in\mathbb{Z}}|A_1(j,k)|^2 2^j \chi_{I_{j,k}}\right\}^{\frac{1}{2}}\right\|_{L^p(\mathbb{R})}$$

$$\leq C_p\left\|\left\{\sum_{\ell\in\mathbb{Z}}\left[\sum_{m\in\mathbb{Z}}|\langle f,\varphi_{\ell,m}\rangle|2^{\frac{\ell}{2}}\chi_{I_{\ell,m}}\right]^2\right\}^{\frac{1}{2}}\right\|_{L^p(\mathbb{R})}$$

$$= C_p\left\|\left\{\sum_{\ell\in\mathbb{Z}}\sum_{m\in\mathbb{Z}}|\langle f,\varphi_{\ell,m}\rangle|^2 2^\ell \chi_{I_{\ell,m}}\right\}^{\frac{1}{2}}\right\|_{L^p(\mathbb{R})}$$

$$= C_p \left\| \mathcal{W}_\varphi f \right\|_{L^p(\mathbb{R})}. \tag{4.13}$$

To estimate A_2 we use part (b) of Proposition 4.8 together with part (b) of Lemma 3.16 with $r = 1$, to obtain

$$|A_2(j,k)| \leq \sum_{\ell > j} \sum_{m \in \mathbb{Z}} |\langle f, \varphi_{\ell,m} \rangle| \frac{2^{\frac{3}{2}(j-\ell)} C}{\left(1 + |2^{j-\ell} m - k|\right)^{1+\varepsilon}}$$

$$\leq C \sum_{\ell > j} 2^{\frac{1}{2}(j-\ell)} \left[\mathcal{M} \Big(\sum_{m \in \mathbb{Z}} |\langle f, \varphi_{\ell,m} \rangle| \chi_{I_{\ell,m}} \Big)(x) \right]$$

for all $x \in I_{j,k}$. Therefore, since $\{I_{j,k} : k \in \mathbb{Z}\}$ is a disjoint collection of dyadic intervals, we have

$$\left\| \Big\{ \sum_{j \in \mathbb{Z}} \sum_{k \in \mathbb{Z}} |A_2(j,k)|^2 2^j \chi_{I_{j,k}} \Big\}^{\frac{1}{2}} \right\|_{L^p(\mathbb{R})}$$

$$\leq C \left\| \Big\{ \sum_{j \in \mathbb{Z}} 2^j \Big[\sum_{\ell > j} 2^{\frac{1}{2}(j-\ell)} \mathcal{M} \Big(\sum_{m \in \mathbb{Z}} |\langle f, \varphi_{\ell,m} \rangle| \chi_{I_{\ell,m}} \Big) \Big]^2 \Big\}^{\frac{1}{2}} \right\|_{L^p(\mathbb{R})}$$

$$= C \left\| \Big\{ \sum_{j \in \mathbb{Z}} \Big[\sum_{\ell > j} 2^{j-\ell} \mathcal{M} \Big(\sum_{m \in \mathbb{Z}} |\langle f, \varphi_{\ell,m} \rangle| 2^{\frac{\ell}{2}} \chi_{I_{\ell,m}} \Big) \Big]^2 \Big\}^{\frac{1}{2}} \right\|_{L^p(\mathbb{R})}.$$

Since $\sum_{\ell > j} 2^{j-\ell} = 1$, the estimate for A_2 is now obtained as was the estimate for A_1, so that

$$\left\| \Big\{ \sum_{j \in \mathbb{Z}} \sum_{k \in \mathbb{Z}} |A_2(j,k)|^2 2^j \chi_{I_{j,k}} \Big\}^{\frac{1}{2}} \right\|_{L^p(\mathbb{R})} \leq C_p \left\| \mathcal{W}_\varphi f \right\|_{L^p(\mathbb{R})}. \tag{4.14}$$

Inequalities (4.13) and (4.14), together with (4.11), prove (4.10), which is the desired result. ∎

Theorem 4.9 is an important achievement. It tells us that, as soon as we prove the equivalence

$$\|f\|_{L^p(\mathbb{R})} \approx \|\mathcal{W}_\psi f\|_{L^p(\mathbb{R})},$$

$1 < p < \infty$, $f \in L^p(\mathbb{R})$, for a **single wavelet** that belongs to the regularity class $\mathfrak{R}^0$ (see (2.12)), the equivalence is true for all wavelets belonging to the class $\mathfrak{R}^0$. Thus, combining Theorem 4.5 and Theorem 4.9 we deduce

THEOREM 4.15 *Let ψ be an orthonormal wavelet such that $\psi \in \mathfrak{R}^0$. If $1 < p < \infty$, there exist constants A_p and B_p, $0 < A_p \leq B_p < \infty$, such that*

$$A_p \|f\|_{L^p(\mathbb{R})} \leq \|\mathcal{W}_\psi f\|_{L^p(\mathbb{R})} \leq B_p \|f\|_{L^p(\mathbb{R})}$$

for all $f \in L^p(\mathbb{R})$.

We have already mentioned (see the beginning of the proof of Theorem 4.9) that the convergence

$$f \sim \sum_{j \in \mathbb{Z}} \sum_{k \in \mathbb{Z}} \langle f, \psi_{j,k} \rangle \psi_{j,k}$$

is in $\mathcal{S}'$. It is now easy to prove that if $f \in L^p(\mathbb{R})$, $1 < p < \infty$, and $\psi \in \mathfrak{R}^0$ is an orthonormal wavelet, the above convergence is in $L^p(\mathbb{R})$. To see this we observe that, by the left side of the inequality in Theorem 4.15,

$$\left\| f - \sum_{|j| \leq N} \sum_{k \in \mathbb{Z}} \langle f, \psi_{j,k} \rangle \psi_{j,k} \right\|_{L^p(\mathbb{R})} = \left\| \sum_{|j| > N} \sum_{k \in \mathbb{Z}} \langle f, \psi_{j,k} \rangle \psi_{j,k} \right\|_{L^p(\mathbb{R})}$$

$$\leq C \left\| \left\{ \sum_{|j| > N} \sum_{k \in \mathbb{Z}} |\langle f, \psi_{j,k} \rangle|^2 2^j \chi_{I_{j,k}} \right\}^{\frac{1}{2}} \right\|_{L^p(\mathbb{R})} \longrightarrow 0, \qquad \text{as } N \to \infty,$$

by the Lebesgue Dominated Convergence Theorem. This shows that, if ψ is an orthonormal wavelet and $\psi \in \mathfrak{R}^0$, $\{\psi_{j,k} : j, k \in \mathbb{Z}\}$ is a basis for $L^p(\mathbb{R})$, $1 < p < \infty$. Theorem 4.15 together with Theorem 2.10 of Chapter 5 allow us to conclude that it is also an unconditional basis for $L^p(\mathbb{R})$, $1 < p < \infty$. One should compare this result with Theorem 6.14 of Chapter 5.

6.5 The Hardy space $H^1(\mathbb{R})$

In this section we shall give a characterization of the space $H^1(\mathbb{R})$ in terms of wavelet coefficients. This characterization is very similar to the one given in section 6.4 for the Lebesgue spaces. The definition of $H^1(\mathbb{R})$ was given in section 5.5; we shall use the atomic characterization given by Theorem 5.4 of Chapter 5.

Although we shall use methods similar to the ones developed in sections 6.2, 6.3 and 6.4, there are two main difficulties to overcome. First, Theorem 3.4 on vector-valued inequalities for the Hardy-Littlewood maximal function is not true for $p = 1$. Second, the space $H^1(\mathbb{R})$ is not reflexive and a duality argument like the one used to establish the left side of inequality (4.6) is not available to us. Instead, we shall make use of the atomic characterization of $H^1(\mathbb{R})$. Some of the arguments given in this section are similar to the ones described in the previous section, to which the reader should refer for details.

The first result concerns the Littlewood-Paley function, and it is a corollary to Theorem 2.9.

PROPOSITION 5.1 *Suppose* $\varphi \in \mathfrak{R}^0$, *the class defined by* (2.12). *Then, there exists a constant* A, $0 < A < \infty$, *such that*

$$\|\mathfrak{g}(f)\|_{L^1(\mathbb{R})} \leq A\|f\|_{H^1(\mathbb{R})} \qquad \text{for all } f \in H^1(\mathbb{R}),$$

where $\mathfrak{g}(f)$ *is defined in* Theorem 2.9.

We now prove a result for $H^1(\mathbb{R})$ which is similar to Theorem 4.1.

PROPOSITION 5.2 *Let* $\varphi \in \mathfrak{R}^0$ *be a band-limited function. Given a real number* $\lambda > 1$, *there exists a constant* A_λ, $0 < A_\lambda < \infty$, *such that*

$$\left\|\left\{\sum_{j\in\mathbb{Z}}|\varphi_{j,\lambda}^{**}f|^2\right\}^{\frac{1}{2}}\right\|_{L^1(\mathbb{R})} \leq A_\lambda\|f\|_{H^1(\mathbb{R})} \qquad \text{for all } f \in H^1(\mathbb{R}),$$

where $\varphi_{j,\lambda}^{**}$ *is the map defined in* Lemma 3.10.

PROOF: Suppose $f \in H^1(\mathbb{R})$. By Proposition 5.1, $\varphi_{2^{-j}} * f \in L^1(\mathbb{R})$ for all $j \in \mathbb{Z}$; thus, we can use Lemma 3.10 to obtain

$$\left\|\left\{\sum_{j\in\mathbb{Z}}|\varphi_{j,\lambda}^{**}f|^2\right\}^{\frac{1}{2}}\right\|_{L^1(\mathbb{R})} \leq C_\lambda\left\|\left\{\sum_{j\in\mathbb{Z}}[\mathcal{M}(|\varphi_{2^{-j}} * f|^{\frac{1}{\lambda}})]^{2\lambda}\right\}^{\frac{1}{2}}\right\|_{L^1(\mathbb{R})}$$

$$= C_\lambda\left\|\left\{\sum_{j\in\mathbb{Z}}[\mathcal{M}(|\varphi_{2^{-j}} * f|^{\frac{1}{\lambda}})]^{2\lambda}\right\}^{\frac{1}{2\lambda}}\right\|_{L^\lambda(\mathbb{R})}^\lambda.$$

Since $\lambda > 1$ we can apply Theorem 3.4 with $p = \lambda > 1$ and $q = 2\lambda > 2 > 1$ to obtain

$$\left\| \left\{ \sum_{j \in \mathbb{Z}} |\varphi_{j,\lambda}^{**} f|^2 \right\}^{\frac{1}{2}} \right\|_{L^1(\mathbb{R})} \leq C_\lambda \left\| \left\{ \sum_{j \in \mathbb{Z}} |\varphi_{2^{-j}} * f|^2 \right\}^{\frac{1}{2\lambda}} \right\|_{L^\lambda(\mathbb{R})}^\lambda$$

$$= C_\lambda \left\| \left\{ \sum_{j \in \mathbb{Z}} |\varphi_{2^{-j}} * f|^2 \right\}^{\frac{1}{2}} \right\|_{L^1(\mathbb{R})} \leq A_\lambda \|f\|_{H^1(\mathbb{R})},$$

where the last inequality follows from Proposition 5.1. ∎

The next result is an analogue to Theorem 4.2 for $H^1(\mathbb{R})$ instead of $L^p(\mathbb{R})$, $1 < p < \infty$.

PROPOSITION 5.3 *Let $\psi \in \mathfrak{R}^0$ be a band-limited function. There exists a constant C, $0 < C < \infty$, such that*

$$\left\| \left\{ \sum_{j \in \mathbb{Z}} \sum_{k \in \mathbb{Z}} |\langle f, \psi_{j,k} \rangle|^2 2^j \chi_{[2^{-j}k, 2^{-j}(k+1)]} \right\}^{\frac{1}{2}} \right\|_{L^1(\mathbb{R})} \leq C \|f\|_{H^1(\mathbb{R})}$$

for all $f \in H^1(\mathbb{R})$.

PROOF : As in the proof of Theorem 4.2 we have

$$\sum_{k \in \mathbb{Z}} |\langle f, \psi_{j,k} \rangle|^2 2^j \chi_{[2^{-j}k, 2^{-j}(k+1)]}(x) \leq 2^{2\lambda} \left[(\psi_{j,\lambda}^{**} f)(x) \right]^2$$

for any $j \in \mathbb{Z}$ and any $\lambda > 0$. We now apply Proposition 5.2 with any $\lambda > 1$ to obtain the desired result. ∎

Proposition 5.3 is proved only for band-limited functions. But the result is true for more general functions, not necessarily band-limited. What we need to show is an analogue to Theorem 4.9 for $p = 1$ (recall the definition of $\mathcal{W}_\psi f$ given in (4.4)):

THEOREM 5.4 *Suppose that $\psi, \varphi \in \mathfrak{R}^0$ and φ is an orthonormal wavelet. There exists a constant C, $0 < C < \infty$, such that*

$$\|\mathcal{W}_\psi f\|_{L^1(\mathbb{R})} \leq C \|\mathcal{W}_\varphi f\|_{L^1(\mathbb{R})} \qquad \text{for all } f \in H^1(\mathbb{R}).$$

PROOF : As in the proof of Theorem 4.9 (see (4.11)) we have

$$(\mathcal{W}_\psi f)(x) \le \left\{ \sum_{j\in\mathbb{Z}}\sum_{k\in\mathbb{Z}} |A_1(j,k)|^2 2^j \chi_{I_{j,k}}(x) \right\}^{\frac{1}{2}}$$

$$+ \left\{ \sum_{j\in\mathbb{Z}}\sum_{k\in\mathbb{Z}} |A_2(j,k)|^2 2^j \chi_{I_{j,k}}(x) \right\}^{\frac{1}{2}} \qquad (5.5)$$

for $x \in \mathbb{R}$, where

$$A_1(j,k) = \sum_{\ell \le j}\sum_{m\in\mathbb{Z}} \langle f, \varphi_{\ell,m}\rangle \overline{\langle \psi_{j,k}, \varphi_{\ell,m}\rangle}$$

and

$$A_2(j,k) = \sum_{\ell > j}\sum_{m\in\mathbb{Z}} \langle f, \varphi_{\ell,m}\rangle \overline{\langle \psi_{j,k}, \varphi_{\ell,m}\rangle}.$$

Observe that in writing (5.5) we are using the fact that φ is an orthonormal wavelet. Proceeding as in the proof of Theorem 2.9, but using part (a) of Lemma 3.16 with $1 < r < 1 + \varepsilon$, we obtain

$$\left\| \left\{ \sum_{j\in\mathbb{Z}}\sum_{k\in\mathbb{Z}} |A_1(j,k)|^2 2^j \chi_{I_{j,k}} \right\}^{\frac{1}{2}} \right\|_{L^1(\mathbb{R})}$$

$$\le C \left\| \left\{ \sum_{j\in\mathbb{Z}} \left[\sum_{\ell \le j} 2^{\ell-j} \mathcal{M}\left(\sum_{m\in\mathbb{Z}} |\langle f,\varphi_{\ell,m}\rangle|^{\frac{1}{r}} 2^{\frac{\ell}{2r}} \chi_{I_{\ell,m}} \right)^r \right]^2 \right\}^{\frac{1}{2}} \right\|_{L^1(\mathbb{R})}$$

$$\le C \left\| \left\{ \sum_{\ell\in\mathbb{Z}} \left[\mathcal{M}\left(\sum_{m\in\mathbb{Z}} |\langle f,\varphi_{\ell,m}\rangle|^{\frac{1}{r}} 2^{\frac{\ell}{2r}} \chi_{I_{\ell,m}} \right) \right]^{2r} \right\}^{\frac{1}{2}} \right\|_{L^1(\mathbb{R})}$$

$$= C \left\| \left\{ \sum_{\ell\in\mathbb{Z}} \left[\mathcal{M}\left(\sum_{m\in\mathbb{Z}} |\langle f,\varphi_{\ell,m}\rangle|^{\frac{1}{r}} 2^{\frac{\ell}{2r}} \chi_{I_{\ell,m}} \right) \right]^{2r} \right\}^{\frac{1}{2r}} \right\|_{L^r(\mathbb{R})}^r.$$

Applying Theorem 3.4 with $p = r > 1$ and $q = 2r > 2 > 1$ we obtain

$$\left\| \left\{ \sum_{j\in\mathbb{Z}}\sum_{k\in\mathbb{Z}} |A_1(j,k)|^2 2^j \chi_{I_{j,k}} \right\}^{\frac{1}{2}} \right\|_{L^1(\mathbb{R})}$$

$$\le C_r \left\| \left\{ \sum_{\ell\in\mathbb{Z}} \left[\sum_{m\in\mathbb{Z}} |\langle f,\varphi_{\ell,m}\rangle|^{\frac{1}{r}} 2^{\frac{\ell}{2r}} \chi_{I_{\ell,m}} \right]^{2r} \right\}^{\frac{1}{2r}} \right\|_{L^r(\mathbb{R})}^r$$

$$= C_r \left\| \left\{ \sum_{\ell\in\mathbb{Z}}\sum_{m\in\mathbb{Z}} |\langle f,\varphi_{\ell,m}\rangle|^2 2^\ell \chi_{I_{\ell,m}} \right\}^{\frac{1}{2}} \right\|_{L^1(\mathbb{R})} = C_r \left\| \mathcal{W}_\varphi f \right\|_{L^1(\mathbb{R})}. \qquad (5.6)$$

To handle the second summand on the right-hand side of (5.5) we proceed as above, but using part (b), instead of part (a), of Lemma 3.16 with $1 < r < \min\{2, 1+\varepsilon\}$. Thus, we obtain

$$\left\|\left\{\sum_{j\in\mathbb{Z}}\sum_{k\in\mathbb{Z}}|A_2(j,k)|^2 2^j \chi_{I_{j,k}}\right\}^{\frac{1}{2}}\right\|_{L^1(\mathbb{R})}$$

$$\leq C \left\|\left\{\sum_{j\in\mathbb{Z}} 2^j \left[\sum_{\ell>j} 2^{\frac{3}{2}(j-\ell)} 2^{(\ell-j)r} \mathcal{M}\left(\sum_{m\in\mathbb{Z}}|\langle f,\varphi_{\ell,m}\rangle|^{\frac{1}{r}}\chi_{I_{\ell,m}}\right)^r\right]^2\right\}^{\frac{1}{2}}\right\|_{L^1(\mathbb{R})}$$

$$= C \left\|\left\{\sum_{j\in\mathbb{Z}} \left[\sum_{\ell>j} 2^{(j-\ell)(2-r)} \mathcal{M}\left(\sum_{m\in\mathbb{Z}}|\langle f,\varphi_{\ell,m}\rangle|^{\frac{1}{r}} 2^{\frac{\ell}{2r}}\chi_{I_{\ell,m}}\right)^r\right]^2\right\}^{\frac{1}{2}}\right\|_{L^1(\mathbb{R})}$$

$$\leq C \left\|\left\{\sum_{s=0}^{\infty} 2^{-s(2-r)}\right\}\left\{\sum_{\ell\in\mathbb{Z}}\left[\mathcal{M}\left(\sum_{m\in\mathbb{Z}}|\langle f,\varphi_{\ell,m}\rangle|^{\frac{1}{r}} 2^{\frac{\ell}{2r}}\chi_{I_{\ell,m}}\right)\right]^{2r}\right\}^{\frac{1}{2}}\right\|_{L^1(\mathbb{R})},$$

where we have used Young's convolution inequality for sequences. Since $2 - r > 0$, the series $\sum_{s=0}^{\infty} 2^{-s(2-r)}$ converges, and, hence,

$$\left\|\left\{\sum_{j\in\mathbb{Z}}\sum_{k\in\mathbb{Z}}|A_2(j,k)|^2 2^j \chi_{I_{j,k}}\right\}^{\frac{1}{2}}\right\|_{L^1(\mathbb{R})}$$

$$\leq C_r \left\|\left\{\sum_{\ell\in\mathbb{Z}}\left[\mathcal{M}\left(\sum_{m\in\mathbb{Z}}|\langle f,\varphi_{\ell,m}\rangle|^{\frac{1}{r}} 2^{\frac{\ell}{2r}}\chi_{I_{\ell,m}}\right)\right]^{2r}\right\}^{\frac{1}{2}}\right\|_{L^1(\mathbb{R})}$$

$$= C_r \left\|\left\{\sum_{\ell\in\mathbb{Z}}\left[\mathcal{M}\left(\sum_{m\in\mathbb{Z}}|\langle f,\varphi_{\ell,m}\rangle|^{\frac{1}{r}} 2^{\frac{\ell}{2r}}\chi_{I_{\ell,m}}\right)\right]^{2r}\right\}^{\frac{1}{2r}}\right\|_{L^r(\mathbb{R})}^r.$$

We apply Theorem 3.4 with $p = r > 1$ and $q = 2r > 2 > 1$ to obtain (see the details that led to (5.6))

$$\left\|\left\{\sum_{j\in\mathbb{Z}}\sum_{k\in\mathbb{Z}}|A_2(j,k)|^2 2^j \chi_{I_{j,k}}\right\}^{\frac{1}{2}}\right\|_{L^1(\mathbb{R})} \leq C_r \left\|\mathcal{W}_\varphi f\right\|_{L^1(\mathbb{R})}. \tag{5.7}$$

Combining (5.5), (5.6) and (5.7) we obtain the desired result. ∎

Combining Proposition 5.3 and Theorem 5.4 we obtain the boundedness of $\mathcal{W}_\psi$ from $H^1(\mathbb{R})$ to $L^1(\mathbb{R})$ for functions ψ that are not necessarily band-limited.

THEOREM 5.8 *For any $\psi \in \mathfrak{R}^0$, there exists a constant C, $0 < C < \infty$, such that*

$$\left\|\mathcal{W}_\psi f\right\|_{L^1(\mathbb{R})} \leq C \|f\|_{H^1(\mathbb{R})} \qquad \text{for all } f \in H^1(\mathbb{R}).$$

We are interested in a characterization of $H^1(\mathbb{R})$ similar to the one we gave for $L^p(\mathbb{R})$, $1 < p < \infty$, in Theorem 4.15. Thus, we need to show the reverse inequality of the one that appears in Theorem 5.8. As mentioned before, we cannot use duality as we did for the case $L^p(\mathbb{R})$, $1 < p < \infty$.

The idea is to exploit the atomic characterization of $H^1(\mathbb{R})$. We shall show that, for compactly supported wavelets ψ, the representation

$$f \sim \sum_{j \in \mathbb{Z}} \sum_{k \in \mathbb{Z}} \langle f, \psi_{j,k} \rangle \psi_{j,k}$$

can be rearranged to produce an atomic decomposition of $f \in H^1(\mathbb{R})$. For the interested reader we notice that if the wavelet ψ does not have a compact support, the above representation can be rearranged to produce a molecular decomposition of f, making the computations more complicated. Once the reverse inequality of the one that appears in Theorem 5.8 is proved, we shall use Theorem 5.4 to obtain the desired characterization of $H^1(\mathbb{R})$ with wavelets that belong to the class $\mathfrak{R}^0$.

In section 5.5 we have described the atomic decomposition of $H^1(\mathbb{R})$ in terms of atoms that are in $L^\infty(\mathbb{R})$ (see (5.3) of Chapter 5). For our purpose in this section, it is more advantageous to consider the atomic decomposition of $H^1(\mathbb{R})$ in terms of atoms that are in $L^2(\mathbb{R})$. More precisely, a **2-atom** a is a measurable function defined on $\mathbb{R}$ that satisfies

$$\begin{cases} \text{(i)} \quad \text{the support of } a \text{ is contained in a finite interval } I \subset \mathbb{R}, \\[2mm] \text{(ii)} \quad \|a\|_{L^2(\mathbb{R})} \leq \dfrac{1}{|I|^{\frac{1}{2}}} \left(\text{equivalently, } \left(\int_I |a(x)|^2 \dfrac{dx}{|I|} \right)^{\frac{1}{2}} \leq \dfrac{1}{|I|} \right), \\[2mm] \text{(iii)} \quad \displaystyle\int_\mathbb{R} a(x)\, dx = 0. \end{cases} \qquad (5.9)$$

We shall use the following result that tells us that the atomic characterization of $H^1(\mathbb{R})$ can be achieved with 2-atoms replacing the atoms introduced in the last chapter:

THEOREM 5.10 *A function f belongs to $H^1(\mathbb{R})$ if and only if f has a decomposition of the form*

$$f = \sum_{j=1}^{\infty} \lambda_j a_j, \tag{5.11}$$

where each a_j, $j = 1, 2, \cdots$, is a 2-atom and $\sum_{j=1}^{\infty} |\lambda_j| < \infty$. Moreover,

$$\|f\|_{H^1(\mathbb{R})} \approx \inf\Big\{\sum_{j=1}^{\infty} |\lambda_j|\Big\},$$

where the infimum is taken over all decompositions of f of the form (5.11).

The proof of this result can be found in [CW2] and [GR].

THEOREM 5.12 *Let ψ be a compactly supported wavelet (see section 2.3). If f is a function such that $\|\mathcal{W}_\psi f\|_{L^1(\mathbb{R})} < \infty$, then $f \in H^1(\mathbb{R})$. Moreover, there exists a constant B, $0 < B < \infty$, independent of f, such that*

$$\|f\|_{H^1(\mathbb{R})} \le B\|\mathcal{W}_\psi f\|_{L^1(\mathbb{R})}.$$

PROOF : To simplify our notation we shall write $\psi_I = \psi_{I_{j,k}}$ for $\psi_{j,k}$ and let $\mathcal{D}$ denote the family of all dyadic intervals on $\mathbb{R}$. Thus,

$$(\mathcal{W}_\psi f)(x) = \Big(\sum_{I \in \mathcal{D}} |\langle f, \psi_I \rangle|^2 |I|^{-1} \chi_I(x)\Big)^{\frac{1}{2}}.$$

For $k \in \mathbb{Z}$ let $\Omega_k = \{x \in \mathbb{R} : (\mathcal{W}_\psi f)(x) > 2^k\}$. We have $\Omega_k \supseteq \Omega_{k+1}$ and a summation by parts argument shows

$$\sum_{k \in \mathbb{Z}} 2^k |\Omega_k| \le 2\|\mathcal{W}_\psi f\|_{L^1(\mathbb{R})}. \tag{5.13}$$

Let

$$\mathcal{A}_k = \{I \in \mathcal{D} : |I \cap \Omega_k| \ge \tfrac{1}{2}|I|\}, \qquad k \in \mathbb{Z}.$$

Denote by $\tilde{I}_k^i$ the maximal dyadic intervals contained in $\mathcal{A}_k$. Let $\mathcal{B}_k = \mathcal{A}_k \setminus \mathcal{A}_{k+1}$ and $\{\tilde{I}_k^i : i \in \Lambda_k\}$ the collection of all such maximal dyadic intervals contained in $\mathcal{B}_k$, where the index set Λ_k depends on k. Let $\widetilde{\mathcal{D}}$ be collection of all $I \in \mathcal{D}$ such that $I \in \mathcal{A}_k$ for some $k \in \mathbb{Z}$. Then, we have the following almost disjoint partition of $\widetilde{\mathcal{D}}$:

$$\widetilde{\mathcal{D}} = \bigcup_k \bigcup_{i \in \Lambda_k} \{I : I \subset \tilde{I}_k^i, I \in \mathcal{B}_k\}. \tag{5.14}$$

If $\langle f, \psi_I \rangle \neq 0$, there exists $k_0 \in \mathbb{Z}$ such that $|\langle f, \psi_I \rangle| \cdot |I|^{-\frac{1}{2}} > 2^{k_0}$; hence, $\mathcal{W}_\psi f(x) > 2^{k_0}$ for all $x \in I$. This implies that if $\langle f, \psi_I \rangle \neq 0$, $I \subset \Omega_{k_0}$ and, consequently, $I \in \widetilde{\mathcal{D}}$. Thus, we have

$$f(x) = \sum_{j \in \mathbb{Z}} \sum_{k \in \mathbb{Z}} \langle f, \psi_{j,k} \rangle \psi_{j,k}(x) = \sum_{I \in \widetilde{\mathcal{D}}} \langle f, \psi_I \rangle \psi_I(x),$$

since $\langle f, \psi_I \rangle = 0$ if $I \notin \widetilde{\mathcal{D}}$. Using (5.14) the above representation can be written as

$$f(x) = \sum_{k \in \mathbb{Z}} \sum_{i \in \Lambda_k} \left\{ \sum_{\substack{I \subset \tilde{I}_k^i \\ I \in \mathcal{B}_k}} \langle f, \psi_I \rangle \psi_I(x) \right\}. \tag{5.15}$$

The fact that the wavelet ψ is compactly supported now plays an important role. By translations, if necessary, we can always assume that there exists $m \in \mathbb{N}$ such that $\text{supp}(\psi) \subset [0, m]$. Thus, the support of $\psi_{I_{j,k}} \equiv \psi_{j,k}$ is contained in $[2^{-j}k, 2^{-j}(k + m)] \equiv I_{j,k}[m]$, and $|I_{j,k}[m]| = 2^{-j}m = m|I_{j,k}|$. Let

$$\lambda(k, i) = |\tilde{I}_k^i|^{\frac{1}{2}} m^{\frac{1}{2}} \left(\sum_{\substack{I \subset \tilde{I}_k^i \\ I \in \mathcal{B}_k}} |\langle f, \psi_I \rangle|^2 \right)^{\frac{1}{2}}.$$

Define

$$a_{(k,i)}(x) = \begin{cases} \dfrac{1}{\lambda(k, i)} \displaystyle\sum_{\substack{I \subset \tilde{I}_k^i \\ I \in \mathcal{B}_k}} \langle f, \psi_I \rangle \psi_I(x) & \text{if } \lambda(k, i) \neq 0, \\[6mm] 0 & \text{if } \lambda(k, i) = 0. \end{cases}$$

Then, (5.15) can be rewritten in the form

$$f(x) = \sum_{k \in \mathbb{Z}} \sum_{i \in \Lambda_k} \lambda(k, i) \, a_{(k,i)}(x), \tag{5.16}$$

which is the promised decomposition of f in terms of 2-atoms. Indeed, the support of each $a_{(k,i)}$ is contained in $\tilde{I}_k^i[m]$. Moreover, since ψ is an orthonormal wavelet, the above definitions give us

$$\left\|a_{(k,i)}\right\|_{L^2(\mathbb{R})}^2 = \frac{1}{|\lambda(k,i)|^2} \sum_{\substack{I \subset \tilde{I}_k^i \\ I \in \mathcal{B}_k}} |\langle f, \psi_I \rangle|^2 = \frac{1}{m|\tilde{I}_k^i|} = \frac{1}{|\tilde{I}_k^i[m]|} . \qquad (5.17)$$

This last equality implies that the series which defines $a_{(k,i)}$ also converges in $L^1(\mathbb{R})$. Since $\hat{\psi}$ is continuous, by Proposition 2.1 in Chapter 7 we have $\int_{\mathbb{R}} \psi_I(x)\,dx = 0$. Thus, we deduce that $\int_{\mathbb{R}} a_{(k,i)}(x)\,dx = 0$ by integrating term by term. This shows that (5.9) holds for each $a_{(k,i)}$ and, hence, these functions are 2-atoms as promised.

Finally, we need to estimate $\sum_{k \in \mathbb{Z}} \sum_{i \in \Lambda_k} |\lambda(k,i)|$. If $I \in \mathcal{B}_k$, then $I \notin \mathcal{A}_{k+1}$. Thus, $|I \cap \Omega_{k+1}| < \frac{1}{2}|I|$ and

$$|I \setminus \Omega_{k+1}| = |I| - |I \cap \Omega_{k+1}| > |I| - \tfrac{1}{2}|I| = \tfrac{1}{2}|I|.$$

Hence,

$$\sum_{\substack{I \subset \tilde{I}_k^i \\ I \in \mathcal{B}_k}} |\langle f, \psi_I \rangle|^2 \leq 2 \sum_{\substack{I \subset \tilde{I}_k^i \\ I \in \mathcal{B}_k}} |\langle f, \psi_I \rangle|^2 |I|^{-1} |I \setminus \Omega_{k+1}|$$

$$\leq 2 \int_{\tilde{I}_k^i \setminus \Omega_{k+1}} \sum_{I \in \mathcal{D}} |\langle f, \psi_I \rangle|^2 |I|^{-1} \chi_I(x)\,dx$$

$$= 2 \int_{\tilde{I}_k^i \setminus \Omega_{k+1}} \left[(\mathcal{W}_\psi f)(x) \right]^2 dx \leq 2 \cdot 2^{2(k+1)} |\tilde{I}_k^i|,$$

since outside of Ω_{k+1} we have $(\mathcal{W}_\psi f)(x) \leq 2^{k+1}$. Thus,

$$\sum_{k \in \mathbb{Z}} \sum_{i \in \Lambda_k} |\lambda(k,i)| \leq 2m^{\frac{1}{2}} \sqrt{2} \sum_{k \in \mathbb{Z}} \sum_{i \in \Lambda_k} 2^k |\tilde{I}_k^i|^{\frac{1}{2}} |\tilde{I}_k^i|^{\frac{1}{2}} = 2\sqrt{2m} \sum_{k \in \mathbb{Z}} \sum_{i \in \Lambda_k} 2^k |\tilde{I}_k^i|.$$

Since $\tilde{I}_k^i \in \mathcal{B}_k$ we have $|\tilde{I}_k^i| \leq 2|\tilde{I}_k^i \cap \Omega_k|$, and, since the $\tilde{I}_k^i$ are disjoint, we obtain

$$\sum_{k \in \mathbb{Z}} \sum_{i \in \Lambda_k} |\lambda(k,i)| \leq 4\sqrt{2m} \sum_{k \in \mathbb{Z}} 2^k \sum_{i \in \Lambda_k} |\tilde{I}_k^i \cap \Omega_k|$$

$$\leq 4\sqrt{2m} \sum_{k \in \mathbb{Z}} 2^k |\Omega_k| \leq 8\sqrt{2m} \left\| \mathcal{W}_\psi f \right\|_{L^1(\mathbb{R})},$$

where the last inequality is due to (5.13). By Theorem 5.10 we have that $f \in H^1(\mathbb{R})$ and

$$\|f\|_{H^1(\mathbb{R})} \leq \sum_{k \in \mathbb{Z}} \sum_{i \in \Lambda_k} |\lambda(k,i)| \leq 8\sqrt{2m}\, \|\mathcal{W}_\psi f\|_{L^1(\mathbb{R})}.$$

∎

Combining Theorem 5.4, Theorem 5.8 and Theorem 5.12 we obtain the characterization of $H^1(\mathbb{R})$ in terms of wavelet coefficients.

THEOREM 5.18 *Let ψ be an orthonormal wavelet in $\mathfrak{R}^0$. There exist constants A and B, $0 < A \leq B < \infty$, such that*

$$A\|f\|_{H^1(\mathbb{R})} \leq \|\mathcal{W}_\psi f\|_{L^1(\mathbb{R})} \leq B\|f\|_{H^1(\mathbb{R})}$$

for all $f \in H^1(\mathbb{R})$.

As in the case of $L^p(\mathbb{R})$, $1 < p < \infty$, it can now be shown that the convergence of

$$\sum_{j \in \mathbb{Z}} \sum_{k \in \mathbb{Z}} \langle f, \psi_{j,k} \rangle \psi_{j,k}$$

to f is in the $H^1(\mathbb{R})$-norm. Hence, if $\psi \in \mathfrak{R}^0$, the system $\{\psi_{j,k} : j, k \in \mathbb{Z}\}$ is a basis for $H^1(\mathbb{R})$. Theorem 5.18 together with Theorem 2.10 of Chapter 5 allow us to conclude that it is also an unconditional basis for $H^1(\mathbb{R})$. The important feature here is that Theorem 5.18 involves a condition on the coefficients of an expansion in terms of the system $\{\psi_{j,k} : j, k \in \mathbb{Z}\}$ that uses only the absolute values of these coefficients. We also would like to point out that the result we just obtained is comparable to Theorem 6.19 of Chapter 5.

6.6   The Sobolev spaces $L^{p,s}(\mathbb{R})$, $1 < p < \infty$, $s = 1, 2, 3, \cdots$

The Sobolev spaces $L^{p,s}$ were first studied in the context of finding regularity properties of the solutions of differential equations. For $1 < p < \infty$ and $s = 1, 2, 3, \cdots$, we define the space $L^{p,s}(\mathbb{R}) \equiv L^{p,s}$ to be the space of all functions $f \in L^p(\mathbb{R})$ such that, for all $n = 1, 2, \cdots, s$, the nth derivative of f also belongs to $L^p(\mathbb{R})$. The nth derivative of a function $f \in L^p(\mathbb{R})$ is considered here in the sense of distributions; that is, it is a function $D^n f$ such that

$$\int_{\mathbb{R}} (D^n f)(x) \varphi(x)\, dx = (-1)^n \int_{\mathbb{R}} f(x) D^n \varphi(x)\, dx$$

for every test function $\varphi \in \mathcal{S}$. The quantity

$$\|f\|_{L^{p,s}} = \|f\|_{L^p} + \sum_{n=1}^{s} \|D^n f\|_{L^p} \tag{6.1}$$

is a norm on the space $L^{p,s}$, with respect to which it is a Banach space.

There are equivalent definitions of the spaces $L^{p,s}(\mathbb{R})$. One of them, which we shall describe below, involves the multiplier $(1 + |\xi|^2)^{\frac{s}{2}}$. We shall use the following "multiplier theorem" that can be found in [Ste1] (Chapter IV).

THEOREM 6.2 *Suppose that $m(\xi)$ is in $C^1(\mathbb{R} \backslash \{0\})$, and there is a constant B such that*

$$\left| D^j m(\xi) \right| \leq B \frac{1}{|\xi|^j} \qquad for \ \ j = 0, 1. \tag{6.3}$$

Then, the operator T_m given by

$$(T_m)^{\wedge}(\xi) = m(\xi) \hat{f}(\xi),$$

originally defined for $f \in L^p(\mathbb{R}) \cap L^2(\mathbb{R})$, is bounded on $L^p(\mathbb{R})$, $1 < p < \infty$; that is, there exists a positive A such that

$$\left\| T_m f \right\|_{L^p} \leq A \|f\|_{L^p} \qquad for \ all \ \ f \in L^p(\mathbb{R}).$$

For a function $f \in L^p(\mathbb{R})$ we consider the quantity

$$\|f\|_{W^{p,s}} = \left\| \left[(1 + |\cdot|^2)^{\frac{s}{2}} \hat{f}(\cdot) \right]^{\vee} \right\|_{L^p}, \qquad 1 < p < \infty, \ s = 1, 2, \cdots,$$

where $\wedge$ and $\vee$ denote, as usual, the Fourier transform and the inverse Fourier transform; though here we consider them as acting on distributions (see [SW] for details).

THEOREM 6.4 *For $1 < p < \infty$ and $s \in \mathbb{N}$, there exist two constants A and B, $0 < A \leq B < \infty$, such that*

$$A\|f\|_{L^{p,s}} \leq \|f\|_{W^{p,s}} \leq B\|f\|_{L^{p,s}} \qquad \text{for all } f \in L^{p,s}(\mathbb{R}).$$

Thus, $\|\cdot\|_{L^{p,s}}$ and $\|\cdot\|_{W^{p,s}}$ define equivalent norms on the space $L^{p,s}(\mathbb{R})$.

PROOF: For an integer n, $0 \leq n \leq s$, the function

$$m(\xi) \equiv m_n(\xi) = \frac{\xi^n}{(1 + |\xi|^2)^{\frac{s}{2}}}$$

is in the class $C^1(\mathbb{R} \setminus \{0\})$ and satisfies (6.3). Hence, by Theorem 6.2, we have, for $f \in \mathcal{S}$,

$$\|D^n f\|_{L^p} = C\left\| [(\cdot)^n \hat{f}(\cdot)]^\vee \right\|_{L^p}$$

$$= C\left\| \left\{ \frac{(\cdot)^n}{(1 + |\cdot|^2)^{\frac{s}{2}}} (1 + |\cdot|^2)^{\frac{s}{2}} \hat{f}(\cdot) \right\}^\vee \right\|_{L^p}$$

$$\leq CA\left\| [(1 + |\cdot|^2)^{\frac{s}{2}} \hat{f}(\cdot)]^\vee \right\|_{L^p} = CA\|f\|_{W^{p,s}}.$$

The first inequality in Theorem 6.4 now follows immediately for $f \in \mathcal{S}$. A density argument extends this result to all $f \in L^p(\mathbb{R})$.

To establish the other inequality we start with $f \in L^{p,s}(\mathbb{R})$. Consider $\rho \in C^\infty(\mathbb{R})$ such that $\rho(t) = \rho(-t)$, $\rho(t) \geq 0$ for all $t \in \mathbb{R}$, and

$$\rho(t) = \begin{cases} 0 & \text{for } |t| \leq \frac{1}{2}, \\ 1 & \text{for } |t| \geq 1. \end{cases}$$

Then, the function

$$m(\xi) = \frac{(1 + |\xi|^2)^{\frac{s}{2}}}{(1 + [\rho(\xi)\xi]^s)}$$

satisfies (6.3). By Theorem 6.2 we obtain

$$\|f\|_{W^{p,s}} = \left\| [(1 + |\cdot|^2)^{\frac{s}{2}} \hat{f}(\cdot)]^\vee \right\|_{L^p}$$

$$= \left\| \left\{ \frac{(1 + |\cdot|^2)^{\frac{s}{2}}}{(1 + [(\cdot)\rho(\cdot)]^s)} (1 + [(\cdot)\rho(\cdot)]^s) \hat{f}(\cdot) \right\}^\vee \right\|_{L^p}$$

$$\leq A \left\| \left\{ \left(1 + [(\cdot)\rho(\cdot)]^s\right) \hat{f}(\cdot) \right\}^{\vee} \right\|_{L^p}$$

$$\leq CA \|f\|_{L^p} + CA \left\| \left\{ [(\cdot)\rho(\cdot)]^s \hat{f}(\cdot) \right\}^{\vee} \right\|_{L^p}$$

Since the function $[\rho(\xi)]^s$ also satisfies (6.3) and $\{(\cdot)^s \hat{f}(\cdot)\}^{\vee}(x) = c\,(D^s f)(x)$ we obtain

$$\|f\|_{W^{p,s}} \leq C_1 \|f\|_{L^p} + C_2 \|D^s f\|_{L^p} \leq C \|f\|_{L^{p,s}}. \tag{6.5}$$

∎

In the proof of the last theorem we have found another equivalent norm for the spaces $L^{p,s}(\mathbb{R})$, namely,

$$\|f\|^*_{L^{p,s}} \equiv \|f\|_{L^p} + \|D^s f\|_{L^p}. \tag{6.6}$$

It is clear that $\|f\|^*_{L^{p,s}} \leq \|f\|_{L^{p,s}}$, and the reverse inequality follows from Theorem 6.4 and the first inequality in (6.5). Thus, we have

$$\|f\|^*_{L^{p,s}} \leq \|f\|_{L^{p,s}} \leq C \|f\|^*_{L^{p,s}} \tag{6.7}$$

for all $f \in L^{p,s}(\mathbb{R})$, $1 < p < \infty$, $s = 1, 2, \cdots$.

There is also a Littlewood-Paley type characterizations of the Sobolev spaces, as was the case for the Lebesgue spaces $L^p(\mathbb{R})$, $1 < p < \infty$. To obtain this we need to invoke a Hilbert space valued version of the multiplier theorem we used, Theorem 6.2. This is sometimes called the Hörmander-Mihlin multiplier theorem. The proof can be found in [Tr2] (section 2.2.4) or [BL] (Theorem 6.1.6). We state this result as is done in [BL].

Given two Hilbert spaces $\mathbb{H}_0$ and $\mathbb{H}_1$, we denote by $\mathcal{L}(\mathbb{H}_0, \mathbb{H}_1)$ the set of all bounded linear operators T from $\mathbb{H}_0$ to $\mathbb{H}_1$, and write $\|T\|_{\mathcal{L}(\mathbb{H}_0, \mathbb{H}_1)}$ for the operator norm of T.

THEOREM 6.8 (Hörmander-Mihlin Multiplier Theorem)

Let $\mathbb{H}_0$ and $\mathbb{H}_1$ be two Hilbert spaces. Assume that m is a function defined on $\mathbb{R}$ with values in $\mathcal{L}(\mathbb{H}_0, \mathbb{H}_1)$ such that

$$\left\| (D^j m)(\xi) \right\|_{\mathcal{L}(\mathbb{H}_0, \mathbb{H}_1)} \leq B \frac{1}{|\xi|^j}, \qquad j = 0, 1, \tag{6.9}$$

for some positive constant $B < \infty$. Then, the operator T_m given by

$$(T_m f)^\wedge(\xi) = m(\xi)\, \hat{f}(\xi) \qquad \text{for all } f \in S(\mathbb{H}_0),$$

can be extended to a bounded linear operator from $L^p(\mathbb{R}; \mathbb{H}_0)$ to $L^p(\mathbb{R}; \mathbb{H}_1)$, $1 < p < \infty$. That is, there exists a constant $0 < C < \infty$ such that

$$\left\| T_m f \right\|_{L^p(\mathbb{R}; \mathbb{H}_1)} \leq C \left\| f \right\|_{L^p(\mathbb{R}; \mathbb{H}_0)} \qquad \text{for all } f \in L^p(\mathbb{R}; \mathbb{H}_0).$$

REMARK: This theorem will be applied with $\mathbb{H}_0$ and $\mathbb{H}_1$ equal to $\mathbb{C}$ or $\ell^2(\mathbb{Z})$; in the last case the function m can be viewed as a matrix and the operator norm that appears on the left-hand side of inequality (6.9) can be taken to be the sum of the absolute values of the entries of this matrix. This is the way the theorem is stated in [Tr2].

The version of the Littlewood-Paley function we need is the following: for $s \in \mathbb{N}$ define

$$\mathfrak{g}^s(f)(x) = \left\{ \sum_{j \in \mathbb{Z}} \left(2^{js} |\varphi_{2-j} * f(x)| \right)^2 \right\}^{\frac{1}{2}}, \tag{6.10}$$

where φ is a band-limited function in S with a Fourier transform supported in $\{\xi \in \mathbb{R} : 2^{-N} \leq |\xi| \leq 2^N\}$ for some $N \in \mathbb{N}$ and

$$\sum_{j \in \mathbb{Z}} |\hat{\varphi}(2^j \xi)|^2 = 1 \qquad \text{for a.e. } \xi \in \mathbb{R}. \tag{6.11}$$

REMARK: One can assume weaker hypotheses on φ in the definition of $\mathfrak{g}^s(f)$, but the ones we are considering here will make our proofs simpler, and they fit perfectly into the frame of band-limited wavelets. Although the definition of $\mathfrak{g}^s(f)$ depends on φ, it will become obvious that different φ's give equivalent characterizations of the Sobolev spaces.

THEOREM 6.12 *Let $\varphi \in S$ be such that*

$$\operatorname{supp}(\hat{\varphi}) \subset \{\xi \in \mathbb{R} : 2^{-N} \leq |\xi| \leq 2^N\} \qquad \text{for some } N \in \mathbb{N},$$

and (6.11) is satisfied. Then, for $1 < p < \infty$ and $s = 1, 2, \cdots$, $f \in L^{p,s}(\mathbb{R})$ if and only if $f \in L^p(\mathbb{R})$ and $\mathfrak{g}^s(f) \in L^p(\mathbb{R})$; moreover,

$$\left\| f \right\|_{L^p} + \left\| \mathfrak{g}^s(f) \right\|_{L^p}$$

defines a norm for $L^{p,s}(\mathbb{R})$ that is equivalent to $\|\cdot\|_{L^{p,s}}$.

PROOF: Assume $f \in L^{p,s}(\mathbb{R})$, then,

$$\left(2^{js}\,\varphi_{2-j} * f\right)^\wedge(\xi) = C\,2^{js}(\varphi_{2-j})^\wedge(\xi)\hat{f}(\xi) = C\,2^{js}\xi^{-s}(\varphi_{2-j})^\wedge(\xi)\xi^s\hat{f}(\xi).$$

The function $m : \mathbb{R} \longrightarrow \ell^2(\mathbb{Z}) \equiv \mathcal{L}(\mathbb{C}, \ell^2(\mathbb{Z}))$ given by

$$m(\xi) = \{2^{js}\xi^{-s}(\varphi_{2-j})^\wedge(\xi) : j \in \mathbb{Z}\}$$

satisfies (6.9) with $\mathbb{H}_0 = \mathbb{C}$ and $\mathbb{H}_1 = \ell^2(\mathbb{Z})$. To see this we have to estimate $\|m(\xi)\|_{\ell^2(\mathbb{Z})}$ and $\|Dm(\xi)\|_{\ell^2(\mathbb{Z})}$. Since $(\varphi_{2-j})^\wedge(\xi) = \hat{\varphi}(2^{-j}\xi)$, the assumption we are making about the support of $\hat{\varphi}$ implies that

$$\sum_{j\in\mathbb{Z}}\left|2^{js}\xi^{-s}(\varphi_{2-j})^\wedge(\xi)\right|^2 \quad \text{and} \quad \sum_{j\in\mathbb{Z}}\left|2^{js}D(\xi^{-s}(\varphi_{2-j})^\wedge(\xi))\right|^2$$

have only a finite number (at most 2^{2N}) of non-zero terms. Thus, (6.9) will be proved if we show

$$\begin{cases} \text{i)} & 2^{js}\left|\xi^{-s}(\varphi_{2-j})^\wedge(\xi)\right| \le B_0; \\[2mm] \text{ii)} & 2^{js}\left|D(\xi^{-s}(\varphi_{2-j})^\wedge(\xi))\right| \le B_1|\xi|^{-1}. \end{cases} \qquad (6.13)$$

Since on the support of $(\varphi_{2-j})^\wedge$ we have $2^{-N} \le |2^{-j}\xi| \le 2^N$, we deduce

$$2^{js}\left|\xi^{-s}(\varphi_{2-j})^\wedge(\xi)\right| = 2^{js}\left|\xi^{-s}\hat{\varphi}(2^{-j}\xi)\right| \le C\,2^{js}2^{(N-j)s} = C\,2^{Ns}$$

and

$$\begin{aligned} 2^{js}\left|D(\xi^{-s}(\varphi_{2-j})^\wedge(\xi))\right| &= 2^{js}\left|D(\xi^{-s}\hat{\varphi}(2^{-j}\xi))\right| \\[2mm] &\le 2^{js}\left[C_1|\xi^{-s-1}\hat{\varphi}(2^{-j}\xi)| + C_2|\xi^{-s}2^{-j}(D\hat{\varphi})(2^{-j}\xi)|\right] \\[2mm] &\le C\,2^{js}\left[2^{(N-j)s}\frac{1}{|\xi|} + 2^{N(s+1)}2^{-sj}\frac{1}{|\xi|}\right] \\[2mm] &\le \frac{C}{|\xi|}\left[2^{Ns} + 2^{N(s+1)}\right] = \frac{C}{|\xi|}, \end{aligned}$$

which proves (6.13).

By the Hörmander-Mihlin multiplier theorem (Theorem 6.8) we obtain

$$\left\|\mathfrak{g}^s(f)\right\|_{L^p(\mathbb{R})} = C\left\|\left\{\sum_{j\in\mathbb{Z}}\left|(2^{js}[(\cdot)^{-s}(\varphi_{2^{-j}})^\wedge](\cdot)^s\hat{f})^\vee\right|^2\right\}^{\frac{1}{2}}\right\|_{L^p(\mathbb{R})}$$

$$\leq C\left\|\left((\cdot)^s\hat{f}\right)^\vee\right\|_{L^p(\mathbb{R})} = C\|D^s f\|_{L^p(\mathbb{R})}.$$

Thus, $\mathfrak{g}^s(f) \in L^p(\mathbb{R})$ and

$$\|f\|_{L^p} + \|\mathfrak{g}^s(f)\|_{L^p} \leq \|f\|_{L^p} + C\|D^s f\|_{L^p} \leq C\|f\|_{L^{p,s}},$$

where the last inequality is due to (6.7). Observe that to show this inequality we have not used (6.11). This equality will be needed to prove the converse result.

Assume now that $f \in L^p(\mathbb{R})$ and $\mathfrak{g}^s(f) \in L^p(\mathbb{R})$. By (6.7) it suffices to prove that $D^s f \in L^p(\mathbb{R})$. Equality (6.11) allows us to write

$$\hat{f}(\xi) = \sum_{j\in\mathbb{Z}}(\varphi_{2^{-j}})^\wedge(\xi)\,(\tilde{\varphi}_{2^{-j}})^\wedge(\xi)\,\hat{f}(\xi),$$

where $\tilde{\varphi}(x) = \overline{\varphi(-x)}$. Then,

$$D^s f(x) = C\left((\cdot)^s\hat{f}(\cdot)\right)^\vee(x) = C\left(\sum_{j\in\mathbb{Z}}(\cdot)^s(\tilde{\varphi}_{2^{-j}})^\wedge(\cdot)\,(\varphi_{2^{-j}})^\wedge(\cdot)\,\hat{f}(\cdot)\right)^\vee(x)$$

$$= C\sum_{j\in\mathbb{Z}}(2^{-js}(\cdot)^s(\tilde{\varphi}_{2^{-j}})^\wedge(\cdot)\,2^{sj}(\varphi_{2^{-j}})^\wedge(\cdot)\,\hat{f}(\cdot))^\vee(x).$$

The function $m : \mathbb{R} \longrightarrow \mathcal{L}(\ell^2(\mathbb{Z}), \mathbb{C})$ given by

$$m(\xi)(\mathfrak{a}) = \sum_{j\in\mathbb{Z}}2^{-js}\xi^s(\tilde{\varphi}_{2^{-j}})^\wedge(\xi)a_j \qquad \text{for} \quad \mathfrak{a} = \{a_j\}_{j\in\mathbb{Z}} \in \ell^2(\mathbb{Z})$$

satisfies (6.9) with $\mathbb{H}_0 = \ell^2(\mathbb{Z})$ and $\mathbb{H}_1 = \mathbb{C}$. As before, we only need to show

$$\begin{cases} \text{i)} & 2^{-js}\left|\xi^s(\tilde{\varphi}_{2^{-j}})^\wedge(\xi)\right| \leq B_0; \\ \text{ii)} & 2^{-js}\left|D(\xi^s(\tilde{\varphi}_{2^{-j}})^\wedge(\xi))\right| \leq B_1|\xi|^{-1}, \end{cases} \tag{6.14}$$

taking into account our assumption on the support of $\hat{\varphi}$. The proof of (6.14) is similar to the one of (6.13). More precisely,

$$2^{-js}\left|\xi^s(\tilde{\varphi}_{2^{-j}})^\wedge(\xi)\right| = 2^{-js}\left|\xi^s\widehat{\tilde{\varphi}}(2^{-j}\xi)\right| \leq C\,2^{-js}2^{(N+j)s} = C\,2^{Ns}$$

and

$$2^{-js}\left|D(\xi^s(\tilde{\varphi}_{2-j})^\wedge(\xi))\right| = 2^{-js}\left|D(\xi^s\widehat{\tilde{\varphi}}(2^{-j}\xi))\right|$$

$$\leq 2^{-js}\left[C_1|\xi^{s-1}\widehat{\tilde{\varphi}}(2^{-j}\xi)| + C_2|\xi^s 2^{-j}(D\widehat{\tilde{\varphi}})(2^{-j}\xi)|\right]$$

$$\leq 2^{-js}\left[2^{(N+j)s}\frac{C_1}{|\xi|} + 2^{(N+j)(s+1)}2^{-j}\frac{C_2}{|\xi|}\right]$$

$$\leq \frac{C}{|\xi|}\left[2^{Ns} + 2^{N(s+1)}\right] = \frac{C}{|\xi|}.$$

We now again use the Hörmander-Mihlin multiplier theorem (Theorem 6.8), with $\mathbb{H}_0 = \ell^2(\mathbb{Z})$ and $\mathbb{H}_1 = \mathbb{C}$ in this case, to obtain

$$\left\|D^s f\right\|_{L^p(\mathbb{R})} = C\left\|\sum_{j \in \mathbb{Z}}\left(\{2^{-js}(\cdot)^s(\tilde{\varphi}_{2-j})^\wedge\}2^{sj}(\varphi_{2-j})^\wedge \hat{f}\right)^\vee\right\|_{L^p(\mathbb{R})}$$

$$\leq C\left\|\left\{\sum_{j \in \mathbb{Z}}(2^{sj}|(\varphi_{2-j})^\wedge \hat{f}|)^\vee\right\}^{\frac{1}{2}}\right\|_{L^p(\mathbb{R})}$$

$$= C\left\|\left\{\sum_{j \in \mathbb{Z}}(2^{js}|\varphi_{2-j} * f|)^2\right\}^{\frac{1}{2}}\right\|_{L^p(\mathbb{R})} = C\left\|\mathfrak{g}^s(f)\right\|_{L^p(\mathbb{R})}.$$

This shows that $D^s f \in L^p(\mathbb{R})$; moreover, by (6.7),

$$\left\|f\right\|_{L^{p,s}} \leq C(\left\|f\right\|_{L^p} + \left\|D^s f\right\|_{L^p}) \leq C(\left\|f\right\|_{L^p} + \left\|\mathfrak{g}^s(f)\right\|_{L^p}).$$

∎

In order to characterize the Sobolev spaces using wavelet coefficients we start with a result similar to Theorem 4.1. For simplicity we shall denote by $\mathfrak{B}$ the space of all $\psi \in \mathcal{S}$ such that there exists an $N \in \mathbb{N}$ for which $\text{supp}(\hat{\psi}) \subset \{\xi \in \mathbb{R} : 2^{-N} \leq |\xi| \leq 2^N\}$ and

$$\sum_{j \in \mathbb{Z}}\left|\hat{\psi}(2^j\xi)\right|^2 = 1 \qquad \text{for a.e. } \xi \in \mathbb{R}.$$

THEOREM 6.15 *Suppose $\psi \in \mathfrak{B}$. Given a real number $\lambda \geq 1$, a natural number $s \geq 1$ and $1 < p < \infty$, there exist two constants $A = A_{p,\lambda,s}$ and $B = B_{p,\lambda,s}$, $0 < A \leq B < \infty$, such that*

$$A\left\|f\right\|_{L^{p,s}} \leq \left\|f\right\|_{L^p} + \left\|\left\{\sum_{j \in \mathbb{Z}}[2^{js}(\psi_{j,\lambda}^{**}f)]^2\right\}^{\frac{1}{2}}\right\|_{L^p} \leq B\left\|f\right\|_{L^{p,s}}$$

for all $f \in L^{p,s}(\mathbb{R})$, where $\psi_{j,\lambda}^{**}$ is defined in Lemma 3.10.

PROOF: If $f \in L^{p,s}(\mathbb{R})$, then $\psi_{2^{-j}} * f \in L^p(\mathbb{R})$ for all $j \in \mathbb{Z}$, and we can use Lemma 3.10 and the vector-valued maximal inequality (given in Theorem 3.4) with $p = p\lambda > 1$ (since $\lambda \geq 1$) and $q = 2\lambda$ to obtain

$$\left\|\left\{\sum_{j\in\mathbb{Z}}|2^{js}(\psi_{j,\lambda}^{**}f)|^2\right\}^{\frac{1}{2}}\right\|_{L^p} \leq C_\lambda \left\|\left\{\sum_{j\in\mathbb{Z}}2^{2js}[\mathcal{M}(|\psi_{2^{-j}} * f|^{\frac{1}{\lambda}})]^{2\lambda}\right\}^{\frac{1}{2}}\right\|_{L^p}$$

$$= C_\lambda \left\|\left\{\sum_{j\in\mathbb{Z}}2^{2js}[\mathcal{M}(|\psi_{2^{-j}} * f|^{\frac{1}{\lambda}})]^{2\lambda}\right\}^{\frac{1}{2\lambda}}\right\|_{L^{p\lambda}}$$

$$\leq C_{p,\lambda} \left\|\left\{\sum_{j\in\mathbb{Z}}2^{2js}|\psi_{2^{-j}} * f|^2\right\}^{\frac{1}{2\lambda}}\right\|_{L^{p\lambda}}$$

$$= C_{p,\lambda} \left\|\left\{\sum_{j\in\mathbb{Z}}|2^{js}\psi_{2^{-j}} * f|^2\right\}^{\frac{1}{2}}\right\|_{L^p}$$

$$= C_{p,\lambda}\left\|\mathfrak{g}^s(f)\right\|_{L^p}.$$

The right-hand side of the inequalities we want to prove follows immediately from here. The left-hand side inequality follows from the fact

$$|\psi_{2^{-j}} * f(x)| \leq (\psi_{j,\lambda}^{**}f)(x)$$

and Theorem 6.12.

∎

THEOREM 6.16 *Let $\psi \in \mathfrak{B}$. For $1 < p < \infty$ and $s = 1, 2, \cdots$, there exists a constant $C_{p,s}$, $0 < C_{p,s} < \infty$, such that*

$$\left\|\left\{\sum_{j\in\mathbb{Z}}\sum_{k\in\mathbb{Z}}|\langle f, \psi_{j,k}\rangle|^2(1+2^{2js})2^j\chi_{[2^{-j}k,2^{-j}(k+1)]}\right\}^{\frac{1}{2}}\right\|_{L^p} \leq C_{p,s}\|f\|_{L^{p,s}}$$

for all $f \in L^{p,s}(\mathbb{R})$.

PROOF: As in the proof of Theorem 4.2 we have

$$|\langle f, \psi_{j,k}\rangle| \leq 2^{-\frac{j}{2}}\sup_{y\in I_{j,k}}|(\tilde{\psi}_{2^{-j}} * f)(y)|,$$

where $I_{j,k} = [2^{-j}k, 2^{-j}(k+1)]$ and $\tilde{\psi}(y) = \overline{\psi(-y)}$. Again, as before, we obtain, for each fixed $j \in \mathbb{Z}$,

$$\sum_{k \in \mathbb{Z}} |\langle f, \psi_{j,k} \rangle|^2 2^j \chi_{I_{j,k}}(x) \leq 2^{2\lambda} \left[(\psi_{j,\lambda}^{**} f)(x) \right]^2$$

for any $\lambda > 0$. We now apply Theorem 4.2 and Theorem 6.15 with $\lambda \geq 1$ to obtain

$$\left\| \left\{ \sum_{j \in \mathbb{Z}} \sum_{k \in \mathbb{Z}} |\langle f, \psi_{j,k} \rangle|^2 (1 + 2^{2js}) 2^j \chi_{I_{j,k}} \right\}^{\frac{1}{2}} \right\|_{L^p}$$

$$\leq \left\| \left\{ \sum_{j \in \mathbb{Z}} \sum_{k \in \mathbb{Z}} |\langle f, \psi_{j,k} \rangle|^2 2^j \chi_{I_{j,k}} \right\}^{\frac{1}{2}} \right\|_{L^p}$$

$$+ \left\| \left\{ \sum_{j \in \mathbb{Z}} \sum_{k \in \mathbb{Z}} |\langle f, \psi_{j,k} \rangle|^2 2^{2js} 2^j \chi_{I_{j,k}} \right\}^{\frac{1}{2}} \right\|_{L^p}$$

$$\leq C \|f\|_{L^p} + C_\lambda \left\| \left\{ \sum_{j \in \mathbb{Z}} 2^{2js} |(\psi_{j,\lambda}^{**} f)|^2 \right\}^{\frac{1}{2}} \right\|_{L^p} \leq C \|f\|_{L^{p,s}}.$$

$\blacksquare$

The reverse inequality to the one in Theorem 6.16 can be proved using duality. We shall use the following notation

$$(\mathcal{W}_\psi^s f)(x) = \left\{ \sum_{j \in \mathbb{Z}} \sum_{k \in \mathbb{Z}} |\langle f, \psi_{j,k} \rangle|^2 (1 + 2^{2js}) 2^j \chi_{[2^{-j}k, 2^{-j}(k+1)]}(x) \right\}^{\frac{1}{2}} \quad (6.17)$$

related to the previous theorem and the next one.

THEOREM 6.18 *Let $\psi \in \mathcal{S}$ be a band-limited orthonormal wavelet. For $1 < p < \infty$ and $s = 1, 2, \cdots$, there exist two constants $A_{p,s}$ and $B_{p,s}$, $0 < A_{p,s} \leq B_{p,s} < \infty$, such that*

$$A_{p,s} \|f\|_{L^{p,s}(\mathbb{R})} \leq \|\mathcal{W}_\psi^s f\|_{L^p(\mathbb{R})} \leq B_{p,s} \|f\|_{L^{p,s}} \quad (6.19)$$

for all $f \in L^{p,s}(\mathbb{R})$.

PROOF: By Theorem 6.16 (notice that $\hat{\psi}$ has support away from the origin due to Theorem 2.7 in Chapter 3) we only need to establish the left-hand side inequality of (6.19). For $f, g \in \mathcal{S}$ (which is dense in $L^{p,s}(\mathbb{R})$) we have (since ψ is an orthonormal wavelet)

$$\int_{\mathbb{R}} (D^s f)(x)\, g(x)\, dx = C \int_{\mathbb{R}} f(x)\, (D^s g)(x)\, dx$$

$$= C \int_{\mathbb{R}} \Big\{ \sum_{j \in \mathbb{Z}} \sum_{k \in \mathbb{Z}} \langle f, \psi_{j,k} \rangle \psi_{j,k}(x) \Big\} \Big\{ \sum_{\ell \in \mathbb{Z}} \sum_{m \in \mathbb{Z}} \langle D^s g, \psi_{\ell,m} \rangle \psi_{\ell,m}(x) \Big\}\, dx$$

$$= C \sum_{j \in \mathbb{Z}} \sum_{k \in \mathbb{Z}} \langle f, \psi_{j,k} \rangle \, \langle D^s g, \psi_{j,k} \rangle$$

$$= C \int_{\mathbb{R}} \sum_{j \in \mathbb{Z}} \sum_{k \in \mathbb{Z}} \langle f, \psi_{j,k} \rangle \, 2^{js} 2^{\frac{j}{2}} \langle D^s g, \psi_{j,k} \rangle \, 2^{-js} 2^{\frac{j}{2}} \chi_{I_{j,k}}(x)\, dx.$$

Using the Cauchy-Schwartz inequality for $\ell^2(\mathbb{Z} \times \mathbb{Z})$ we obtain

$$\left| \int_{\mathbb{R}} (D^s f)(x)\, g(x)\, dx \right|$$

$$\leq C \int_{\mathbb{R}} \Big(\sum_{j \in \mathbb{Z}} \sum_{k \in \mathbb{Z}} |\langle f, \psi_{j,k} \rangle|^2 2^{2js} 2^j \chi_{I_{j,k}}(x) \Big)^{\frac{1}{2}}$$

$$\cdot \Big(\sum_{j \in \mathbb{Z}} \sum_{k \in \mathbb{Z}} |\langle D^s g, \psi_{j,k} \rangle|^2 2^{-2js} 2^j \chi_{I_{j,k}}(x) \Big)^{\frac{1}{2}}\, dx$$

$$\leq C \int_{\mathbb{R}} (\mathcal{W}_\psi^s f)(x) \Big(\sum_{j \in \mathbb{Z}} \sum_{k \in \mathbb{Z}} |\langle D^s g, \psi_{j,k} \rangle \, 2^{-js}|^2 2^j \chi_{I_{j,k}}(x) \Big)^{\frac{1}{2}}\, dx.$$

Notice that $\langle D^s g, \psi_{j,k} \rangle \, 2^{-js} = c 2^{-js} \langle g, D^s \psi_{j,k} \rangle = c \langle g, (D^s \psi)_{j,k} \rangle$. Thus,

$$\left| \int_{\mathbb{R}} (D^s f)(x)\, g(x)\, dx \right| \leq C \int_{\mathbb{R}} (\mathcal{W}_\psi^s f)(x)\, (\mathcal{W}_{D^s \psi} g)(x)\, dx.$$

Since $\psi \in \mathcal{S}$ is a band-limited orthonormal wavelet, we can apply Hölder's inequality, Theorem 4.9 with $\psi = D^s \psi$ and $\varphi = \psi$, and Theorem 4.5 to obtain

$$\left| \int_{\mathbb{R}} (D^s f)(x)\, g(x)\, dx \right| \leq C \big\| \mathcal{W}_\psi^s f \big\|_{L^p} \big\| \mathcal{W}_{D^s \psi} g \big\|_{L^{p'}}$$

$$\leq C \big\| \mathcal{W}_\psi^s f \big\|_{L^p} \big\| \mathcal{W}_\psi g \big\|_{L^{p'}} \leq C \big\| \mathcal{W}_\psi^s f \big\|_{L^p} \big\| g \big\|_{L^{p'}}.$$

Taking the supremum over all $g \in \mathcal{S}$ such that $\|g\|_{L^{p'}} \leq 1$ we deduce

$$\big\| D^s f \big\|_{L^p} \leq C \big\| \mathcal{W}_\psi^s f \big\|_{L^p}.$$

Obviously, $(\mathcal{W}_\psi f)(x) \le (\mathcal{W}_\psi^s f)(x)$ since $1 \le (1 + 2^{2js})$ for all $j \in \mathbb{Z}$. Hence, by Theorem 4.5,

$$\|f\|_{L^p} \le C \|\mathcal{W}_\psi f\|_{L^p} \le C \|\mathcal{W}_\psi^s f\|_{L^p} \, .$$

Therefore, the left-hand side inequality of (6.19) is proved using (6.7).   ■

Theorem 6.18 gives a characterization of the $L^{p,s}(\mathbb{R})$ spaces, $1 < p < \infty$, $s = 1, 2, \cdots$, using band-limited orthonormal wavelets that belong to the Schwartz class. This theorem can be extended to more general wavelets. The ideas are similar to the ones used to prove Theorem 4.9. We need to use Lemma 3.12 and Lemma 3.14 with $N = s$.

In order to simplify the writing of the statements that follow we shall use the following notations. For $N \in \mathbb{N} \cup \{-1\}$, let $\mathfrak{D}^N$ be the set of all functions f defined on $\mathbb{R}$ for which there exist constants $\varepsilon > 0$ and $c_n < \infty$, $n = 0, 1, \cdots, N + 1$, such that

$$|D^n f(x)| \le \frac{c_n}{(1 + |x|)^{1+\varepsilon}} \qquad \text{for all } x \in \mathbb{R} \text{ and } 0 \le n \le N + 1.$$

We write $\mathfrak{M}^N$ for the set of all functions f defined on $\mathbb{R}$ for which there exist constants $\gamma > 0$ and $C < \infty$ such that

$$\int_\mathbb{R} x^n f(x) \, dx = 0 \qquad \text{for } n = 0, 1, \cdots, N,$$

and

$$|f(x)| \le C \frac{1}{(1 + |x|)^{2+N+\gamma}} \qquad \text{for all } x \in \mathbb{R}.$$

The reader should notice that these definitions are adapted to the conditions of Lemma 3.14 and Lemma 3.12.

PROPOSITION 6.20 *Let* $s = 1, 2, \cdots$ *and* $j, k, \ell, m \in \mathbb{Z}$.

(a)   *If* $\varphi \in \mathfrak{D}^s$ *and* $\psi \in \mathfrak{M}^s$, *there exist constants* $C < \infty$ *and* $\varepsilon > 0$ *such that*

$$|\langle \psi_{j,k}, \varphi_{\ell,m} \rangle| \le \frac{C \, 2^{(\ell-j)(\frac{1}{2}+s+1)}}{(1 + 2^\ell |2^{-j}k - 2^{-\ell}m|)^{1+\varepsilon}} \qquad \text{for } j \ge \ell.$$

(b)   If $\varphi, \psi \in \mathfrak{D}^{-1}$, there exist constants $C < \infty$ and $\varepsilon > 0$ such that

$$\left|\langle \psi_{j,k}, \varphi_{\ell,m} \rangle\right| \le \frac{C\, 2^{\frac{1}{2}(j-\ell)}}{\left(1 + 2^j |2^{-j}k - 2^{-\ell}m|\right)^{1+\varepsilon}} \qquad \text{for } j \le \ell.$$

PROOF: Suppose that φ is associated with the constants $\varepsilon' > 0$ and c'_n, $n = 0, 1, \cdots, N+1$, and ψ is associated with the constants $\gamma > 0$ and $C' < \infty$. We choose

$$C = \max\{c'_0, \cdots, c'_{N+1}, C'\} \qquad \text{and} \qquad \varepsilon = \min\{\varepsilon', \gamma\}.$$

Then $\varphi \in \mathfrak{D}^s$ with constants C for all $n = 0, 1, \cdots, N+1$ and $\varepsilon > 0$ and $\psi \in \mathfrak{M}^s$ with constant C and $\gamma \ge \varepsilon$.

For a function g defined on $\mathbb{R}$ we write $\tilde{g}(x) = \overline{g(-x)}$. Then, we have

$$\overline{\langle \psi_{j,k}, \varphi_{\ell,m} \rangle} = \langle \varphi_{\ell,m}, \psi_{j,k} \rangle = (\varphi_{\ell,m} * \tilde{\psi}_{j,-k})(0).$$

To prove (a) apply Lemma 3.14 with the roles of j and ℓ reversed and $N = s$.

To prove (b) apply Lemma 3.12 together with

$$\langle \psi_{j,k}, \varphi_{\ell,m} \rangle = (\psi_{j,k} * \tilde{\varphi}_{\ell,-m})(0).$$

∎

THEOREM 6.21 Let $s = 1, 2, 3, \cdots$ and $\psi, \varphi \in \mathfrak{D}^s \cap \mathfrak{M}^s$. Assume that φ is also an orthonormal wavelet. Then, for $1 < p < \infty$, there exists a constant $C_{p,s}$, $0 < C_{p,s} < \infty$, such that

$$\left\| \mathcal{W}_\psi^s f \right\|_{L^p(\mathbb{R})} \le C_{p,s} \left\| \mathcal{W}_\varphi^s f \right\|_{L^p(\mathbb{R})} \qquad \text{for all } f \in L^{p,s}(\mathbb{R}),$$

where $\mathcal{W}_\psi^s f$ and $\mathcal{W}_\varphi^s f$ are defined by (6.17).

PROOF: It is enough to prove the result for $\widetilde{\mathcal{W}}_\varphi^s f$ and $\widetilde{\mathcal{W}}_\psi^s f$ instead of $\mathcal{W}_\varphi^s f$ and $\mathcal{W}_\psi^s f$, where

$$(\widetilde{\mathcal{W}}_\psi^s f)(x) = \left\{ \sum_{j \in \mathbb{Z}} \sum_{k \in \mathbb{Z}} |\langle f, \psi_{j,k} \rangle|^2 2^{2js} 2^j \chi_{[2^{-j}k, 2^{-j}(k+1)]}(x) \right\}^{\frac{1}{2}};$$

that is,

$$\left\| \widetilde{\mathcal{W}}_{\psi}^s f \right\|_{L^p(\mathbb{R})} \leq C_{p,s} \left\| \widetilde{\mathcal{W}}_{\varphi}^s f \right\|_{L^p(\mathbb{R})} \qquad \text{for all } f \in L^{p,s}(\mathbb{R}). \tag{6.22}$$

In fact, assuming that (6.22) is true, we have, by Theorem 4.9,

$$\left\| \mathcal{W}_{\psi}^s f \right\|_{L^p} \leq \left\| \mathcal{W}_{\psi} f \right\|_{L^p} + \left\| \widetilde{\mathcal{W}}_{\psi}^s f \right\|_{L^p} \leq C_1 \left\| \mathcal{W}_{\varphi} f \right\|_{L^p} + C_2 \left\| \widetilde{\mathcal{W}}_{\varphi}^s f \right\|_{L^p}$$

$$\leq C \{ \left\| \mathcal{W}_{\varphi} f \right\|_{L^p} + \left\| \mathcal{W}_{\varphi}^s f \right\|_{L^p} \} = 2C \left\| \mathcal{W}_{\varphi}^s f \right\|_{L^p}.$$

To prove (6.22) we follow the same ideas that we used in the proof of Theorem 4.9. Since φ is an orthonormal wavelet we can write

$$\psi_{j,k}(x) = \sum_{\ell \in \mathbb{Z}} \sum_{m \in \mathbb{Z}} \langle \psi_{j,k}, \varphi_{\ell,m} \rangle \varphi_{\ell,m}(x), \qquad j, k \in \mathbb{Z}.$$

Thus,

$$(\widetilde{\mathcal{W}}_{\psi}^s f)(x) = \left\{ \sum_{j \in \mathbb{Z}} \sum_{k \in \mathbb{Z}} \left| \sum_{\ell \in \mathbb{Z}} \sum_{m \in \mathbb{Z}} \langle f, \varphi_{\ell,m} \rangle \overline{\langle \psi_{j,k}, \varphi_{\ell,m} \rangle} \right|^2 2^{2js} 2^j \chi_{I_{j,k}}(x) \right\}^{\frac{1}{2}},$$

where $I_{j,k} = [2^{-j}k, 2^{-j}(k+1)]$. Writing

$$A_1(j,k) = \sum_{\ell \leq j} \sum_{m \in \mathbb{Z}} \langle f, \varphi_{\ell,m} \rangle \overline{\langle \psi_{j,k}, \varphi_{\ell,m} \rangle}$$

and

$$A_2(j,k) = \sum_{\ell > j} \sum_{m \in \mathbb{Z}} \langle f, \varphi_{\ell,m} \rangle \overline{\langle \psi_{j,k}, \varphi_{\ell,m} \rangle},$$

we have

$$(\widetilde{\mathcal{W}}_{\psi}^s f)(x) \leq \left\{ \sum_{j \in \mathbb{Z}} \sum_{k \in \mathbb{Z}} \left| A_1(j,k) \right|^2 2^{2js} 2^j \chi_{I_{j,k}}(x) \right\}^{\frac{1}{2}}$$

$$+ \left\{ \sum_{j \in \mathbb{Z}} \sum_{k \in \mathbb{Z}} \left| A_2(j,k) \right|^2 2^{2js} 2^j \chi_{I_{j,k}}(x) \right\}^{\frac{1}{2}}. \tag{6.23}$$

To estimate A_1 we use part (a) of Proposition 6.20 to obtain

$$\left| A_1(j,k) \right| \leq C \sum_{\ell \leq j} \sum_{m \in \mathbb{Z}} \left| \langle f, \varphi_{\ell,m} \rangle \right| \frac{2^{(\ell-j)(\frac{1}{2}+s+1)}}{(1 + 2^\ell |2^{-j}k - 2^{-\ell}m|)^{1+\varepsilon}}$$

$$= C \sum_{\ell \le j} 2^{(\ell-j)(\frac{1}{2}+s+1)} \Big\{ \sum_{m \in \mathbb{Z}} \frac{|\langle f, \varphi_{\ell,m} \rangle|}{(1 + 2^{\ell}|2^{-j}k - 2^{-\ell}m|)^{1+\varepsilon}} \Big\},$$

for some $C < \infty$ and $\varepsilon > 0$. We now apply part (a) of Lemma 3.16 with $r = 1$ to deduce

$$|A_1(j,k)| \le C \sum_{\ell \le j} 2^{(\ell-j)(\frac{1}{2}+s+1)} \Big[\mathcal{M}\Big(\sum_{m \in \mathbb{Z}} |\langle f, \varphi_{\ell,m} \rangle| \chi_{I_{\ell,m}} \Big)(x) \Big]$$

for all $x \in I_{j,k}$. Therefore, since $\{I_{j,k} : k \in \mathbb{Z}\}$ is a collection of disjoint dyadic intervals, we have

$$\Big\| \Big\{ \sum_{j \in \mathbb{Z}} \sum_{k \in \mathbb{Z}} |A_1(j,k)|^2 2^{2js} 2^j \chi_{I_{j,k}} \Big\}^{\frac{1}{2}} \Big\|_{L^p}$$

$$\le C \Big\| \Big\{ \sum_{j \in \mathbb{Z}} 2^{2js} 2^j \Big[\sum_{\ell \le j} 2^{(\ell-j)(\frac{1}{2}+s+1)} \mathcal{M}\Big(\sum_{m \in \mathbb{Z}} |\langle f, \varphi_{\ell,m} \rangle| \chi_{I_{\ell,m}} \Big) \Big]^2 \Big\}^{\frac{1}{2}} \Big\|_{L^p}$$

$$= C \Big\| \Big\{ \sum_{j \in \mathbb{Z}} \Big[\sum_{\ell \le j} 2^{\ell-j} \mathcal{M}\Big(\sum_{m \in \mathbb{Z}} |\langle f, \varphi_{\ell,m} \rangle| 2^{\ell s} 2^{\frac{\ell}{2}} \chi_{I_{\ell,m}} \Big) \Big]^2 \Big\}^{\frac{1}{2}} \Big\|_{L^p}$$

$$\le C \Big\| \Big\{ \sum_{\ell \in \mathbb{Z}} \Big[\mathcal{M}\Big(\sum_{m \in \mathbb{Z}} |\langle f, \varphi_{\ell,m} \rangle| 2^{\ell s} 2^{\frac{\ell}{2}} \chi_{I_{\ell,m}} \Big) \Big]^2 \Big\}^{\frac{1}{2}} \Big\|_{L^p},$$

where the last inequality is due to Young's inequality for convolutions, as in the proof of Theorem 4.9. Using the vector-valued inequality for the Hardy-Littlewood maximal function (Theorem 3.4) with $q = 2$, we obtain

$$\Big\| \Big\{ \sum_{j \in \mathbb{Z}} \sum_{k \in \mathbb{Z}} |A_1(j,k)|^2 2^{2js} 2^j \chi_{I_{j,k}} \Big\}^{\frac{1}{2}} \Big\|_{L^p}$$

$$\le C_p \Big\| \Big\{ \sum_{\ell \in \mathbb{Z}} \sum_{m \in \mathbb{Z}} |\langle f, \varphi_{\ell,m} \rangle|^2 2^{2\ell s} 2^{\ell} \chi_{I_{\ell,m}} \Big\}^{\frac{1}{2}} \Big\|_{L^p}$$

$$= C_p \big\| \widetilde{\mathcal{W}}_{\varphi}^s f \big\|_{L^p}. \tag{6.24}$$

To estimate A_2 we use part (b) of Proposition 6.20 (observe that $\mathfrak{D}^s$ and $\mathfrak{M}^s$ are contained in $\mathfrak{D}^{-1}$), together with part (b) of Lemma 3.16 with

$r = 1$, to obtain

$$|A_2(j,k)| \leq C \sum_{\ell > j} \sum_{m \in \mathbb{Z}} |\langle f, \varphi_{\ell,m} \rangle| \frac{2^{\frac{1}{2}(j-\ell)}}{\left(1 + 2^j |2^{-j} k - 2^{-\ell} m|\right)^{1+\varepsilon}}$$

$$\leq C \sum_{\ell > j} 2^{\frac{1}{2}(j-\ell)} 2^{\ell - j} \left[\mathcal{M}\left(\sum_{m \in \mathbb{Z}} |\langle f, \varphi_{\ell,m} \rangle| \chi_{I_{\ell,m}} \right)(x) \right]$$

for some $C < \infty$, $\varepsilon > 0$, and for all $x \in I_{j,k}$. Therefore, since $\{I_{j,k} : k \in \mathbb{Z}\}$ is a collection of disjoint dyadic intervals, we have

$$\left\| \left\{ \sum_{j \in \mathbb{Z}} \sum_{k \in \mathbb{Z}} |A_2(j,k)|^2 2^{2js} 2^j \chi_{I_{j,k}} \right\}^{\frac{1}{2}} \right\|_{L^p}$$

$$\leq C \left\| \left\{ \sum_{j \in \mathbb{Z}} 2^{2js} 2^j \left[\sum_{\ell > j} 2^{-\frac{1}{2}(j-\ell)} \mathcal{M}\left(\sum_{m \in \mathbb{Z}} |\langle f, \varphi_{\ell,m} \rangle| \chi_{I_{\ell,m}} \right) \right]^2 \right\}^{\frac{1}{2}} \right\|_{L^p}$$

$$= C \left\| \left\{ \sum_{j \in \mathbb{Z}} \left[\sum_{\ell > j} 2^{(j-\ell)s} \mathcal{M}\left(\sum_{m \in \mathbb{Z}} |\langle f, \varphi_{\ell,m} \rangle| 2^{\ell s} 2^{\frac{\ell}{2}} \chi_{I_{\ell,m}} \right) \right]^2 \right\}^{\frac{1}{2}} \right\|_{L^p}.$$

Since $s \geq 1$ the series $\sum_{\ell > j} 2^{(j-\ell)s}$ converges, so that using Young's inequality for convolutions and the vector-valued inequality for the Hardy-Littlewood maximal function (Theorem 3.4) with $q = 2$, we obtain

$$\left\| \left\{ \sum_{j \in \mathbb{Z}} \sum_{k \in \mathbb{Z}} |A_2(j,k)|^2 2^{2js} 2^j \chi_{I_{j,k}} \right\}^{\frac{1}{2}} \right\|_{L^p(\mathbb{R})}$$

$$\leq C \left\| \left\{ \sum_{\ell \in \mathbb{Z}} \left[\mathcal{M}\left(\sum_{m \in \mathbb{Z}} |\langle f, \varphi_{\ell,m} \rangle| 2^{\ell s} 2^{\frac{\ell}{2}} \chi_{I_{\ell,m}} \right) \right]^2 \right\}^{\frac{1}{2}} \right\|_{L^p}$$

$$\leq C \left\| \left\{ \sum_{\ell \in \mathbb{Z}} \sum_{m \in \mathbb{Z}} |\langle f, \varphi_{\ell,m} \rangle|^2 2^{2\ell s} 2^\ell \chi_{I_{\ell,m}} \right\}^{\frac{1}{2}} \right\|_{L^p} = C \left\| \widetilde{W}_\varphi^s f \right\|_{L^p}. \tag{6.25}$$

Finally, inequality (6.22) follows from (6.23), (6.24) and (6.25).   ∎

We now have the necessary ingredients to prove Theorem 6.18 for more general orthonormal wavelets, not necessarily band-limited and belonging to the Schwartz class. For a non-negative integer s, let $\mathfrak{R}^s = \mathfrak{D}^s \cap \mathfrak{M}^s$; that is, $f \in \mathfrak{R}^s$ if there exist constants $\varepsilon > 0$, $\gamma > 0$, $C < \infty$ and $C_n < \infty$,

$n = 1, 2, \cdots, s + 1$, such that

$$
\begin{cases}
\text{i)} \quad \displaystyle\int_{\mathbb{R}} x^n f(x)\, dx = 0 & \text{for } n = 0, 1, \cdots, s; \\[2mm]
\text{ii)} \quad |f(x)| \leq \dfrac{C}{\left(1 + |x|\right)^{2+s+\gamma}} & \text{for } x \in \mathbb{R}; \\[2mm]
\text{iii)} \quad |D^n f(x)| \leq \dfrac{C_n}{\left(1 + |x|\right)^{1+\varepsilon}} & \text{for } x \in \mathbb{R}, \ n = 1, 2, \cdots, s + 1.
\end{cases}
\tag{6.26}
$$

Observe that the class $\mathfrak{R}^s$ contains all the s-regular wavelets described in [Me1], and all functions in the Schwartz class for which i) holds.

THEOREM 6.27 *Let $s = 1, 2, \cdots$, and suppose that ψ is an orthonormal wavelet such that $\psi \in \mathfrak{R}^s$. Then, for $1 < p < \infty$, there exist two constants $A_{p,s}$ and $B_{p,s}$, $0 < A_{p,s} \leq B_{p,s} < \infty$, such that*

$$
A_{p,s} \big\| f \big\|_{L^{p,s}} \leq \big\| \mathcal{W}^s_\psi f \big\|_{L^p} \leq B_{p,s} \big\| f \big\|_{L^{p,s}} \qquad \text{for all } f \in L^{p,s}(\mathbb{R}),
$$

where $\mathcal{W}^s_\psi f$ is defined in (6.17).

PROOF : Apply Theorem 6.18 and Theorem 6.21. Observe that all band-limited wavelets which belong to the Schwartz class $\mathcal{S}$ are contained in $\mathfrak{R}^s$.

∎

Theorem 6.27 allows us to prove that if $s = 1, 2, \cdots$, and $1 < p < \infty$, any orthonormal wavelet $\psi \in \mathfrak{R}^s$ provides an unconditional basis for the Sobolev space $L^{p,s}(\mathbb{R})$, since $\mathcal{W}^s_\psi f$ is a condition that involves the modulus of the coefficients of an expansion in terms of the system $\{\psi_{j,k} : j, k \in \mathbb{Z}\}$. We leave the details to the readers.

REMARK: When ψ is an orthonormal wavelet belonging to $\mathfrak{R}^s$ for an $s \in \mathbb{N}$, conditions ii) and iii) of (6.26) imply i) (see Theorem 3.4 of Chapter 2).

6.7   The Lipschitz spaces $\Lambda_\alpha(\mathbb{R})$, $0 < \alpha < 1$, and the Zygmund class $\Lambda_*(\mathbb{R})$

For $0 < \alpha < 1$ we define the **Lipschitz space** $\Lambda_\alpha \equiv \Lambda_\alpha(\mathbb{R})$ as the set of all $f \in L^\infty(\mathbb{R})$ such that

$$\sup_{x \in \mathbb{R}} |f(x + h) - f(x)| \leq C|h|^\alpha. \tag{7.1}$$

The norm in Λ_α is given by

$$\|f\|_{\Lambda_\alpha} = \|f\|_{L^\infty} + \sup_{|h|>0} \frac{\|f(\cdot + h) - f(\cdot)\|_{L^\infty}}{|h|^\alpha}, \tag{7.2}$$

and, with this norm, Λ_α is a Banach space. We observe that the functions in Λ_α may be taken to be continuous (see [Ste1], section 4.1 of Chapter V).

We start with a characterization of the spaces Λ_α, $0 < \alpha < 1$. Recall the definition of $\mathfrak{R}^0$ given in (2.12) and (6.26): $f \in \mathfrak{R}^0$ if and only if there exist constants $\varepsilon > 0$, $\gamma > 0$, $C_0 < \infty$ and $C_1 < \infty$ such that

$$\begin{cases} \text{i)} & \displaystyle\int_{\mathbb{R}} f(x)\,dx = 0; \\[2ex] \text{ii)} & |f(x)| \leq \dfrac{C_0}{\left(1 + |x|\right)^{2+\gamma}} \quad \text{for all } x \in \mathbb{R}; \\[2ex] \text{iii)} & |Df(x)| \leq \dfrac{C_1}{\left(1 + |x|\right)^{1+\varepsilon}} \quad \text{for all } x \in \mathbb{R}. \end{cases} \tag{7.3}$$

As we shall see in the proof of the next result, condition ii) is stronger than what is needed to characterize the space Λ_α, but it is necessary for the space Λ_* which will be defined later (see the remark that follows the proof of the next result).

Let $\varphi \in \mathfrak{R}^0$ be a function which satisfies

$$\sum_{j \in \mathbb{Z}} |\hat{\varphi}(2^j \xi)|^2 = 1 \quad \text{for } a.e.\ \xi \in \mathbb{R}. \tag{7.4}$$

It is easy to produce such functions (even with $\hat{\varphi} \in \mathcal{S}$ and φ band-limited). In our context we can show the existence by taking φ to be one of the Lemarié-Meyer wavelets.

Define $\mathbb{B}^\alpha \equiv \mathbb{B}^\alpha(\mathbb{R}) \equiv \mathbb{B}_\varphi^\alpha$ to be the space of all $f \in L^\infty(\mathbb{R})$ such that

$$\sup_{j \in \mathbb{Z}} 2^{j\alpha} \|\varphi_{2^{-j}} * f\|_{L^\infty(\mathbb{R})} < \infty. \tag{7.5}$$

In $\mathbb{B}^\alpha$ a norm is given by

$$\|f\|_{\mathbb{B}^\alpha} = \|f\|_{L^\infty} + \sup_{j \in \mathbb{Z}} 2^{j\alpha} \|\varphi_{2^{-j}} * f\|_{L^\infty}. \tag{7.6}$$

THEOREM 7.7 *For $0 < \alpha < 1$, $\Lambda_\alpha = \mathbb{B}^\alpha$ and their norms are equivalent.*

PROOF: Suppose that $f \in \Lambda_\alpha$. Since $\varphi \in \mathfrak{R}^0$ we have $\int_\mathbb{R} \varphi(y)\,dy = 0$ and, hence,

$$\left|\varphi_{2^{-j}} * f(x)\right| = \left|\int_\mathbb{R} \varphi_{2^{-j}}(y)\{f(x-y) - f(x)\}\,dy\right|.$$

By (7.2) we then have

$$\left|\varphi_{2^{-j}} * f(x)\right| \leq \|f\|_{\Lambda_\alpha} \int_\mathbb{R} |\varphi_{2^{-j}}(y)| \cdot |y|^\alpha\,dy$$

$$= \|f\|_{\Lambda_\alpha} 2^{-j\alpha} \int_\mathbb{R} |\varphi(u)| \cdot |u|^\alpha\,du \leq C\,2^{-j\alpha}\|f\|_{\Lambda_\alpha}.$$

Observe that

$$\int_\mathbb{R} |\varphi(u)| \cdot |u|^\alpha\,du \leq \int_\mathbb{R} \frac{C\,|u|^\alpha}{(1+|u|)^{2+\gamma}}\,du < \infty. \tag{7.8}$$

It follows from this inequality and (7.5) that $f \in \mathbb{B}^\alpha$ and $\|f\|_{\mathbb{B}^\alpha} \leq C\|f\|_{\Lambda_\alpha}$.

Now we assume $f \in \mathbb{B}^\alpha$. By (7.4) we can write

$$f(x+h) - f(x) = \sum_{j \in \mathbb{Z}} \{\tilde\varphi_{2^{-j}} * \varphi_{2^{-j}} * f(x+h) - \tilde\varphi_{2^{-j}} * \varphi_{2^{-j}} * f(x)\},$$

where $\tilde\varphi(x) = \overline{\varphi(-x)}$. After taking absolute values inside the sum and separating the summation into two parts, we have

$$\left|f(x+h) - f(x)\right|$$

$$\leq \left\{ \sum_{j \geq -\log_2 |h|} + \sum_{j < -\log_2 |h|} \right\} \left|\tilde\varphi_{2^{-j}} * \varphi_{2^{-j}} * f(x+h) - \tilde\varphi_{2^{-j}} * \varphi_{2^{-j}} * f(x)\right|$$

$$= I + II.$$

Since

$$\left\|\tilde{\varphi}_{2^{-j}} * \varphi_{2^{-j}} * f\right\|_{L^\infty} \leq \left\|\tilde{\varphi}_{2^{-j}}\right\|_{L^1}\left\|\varphi_{2^{-j}} * f\right\|_{L^\infty} = \left\|\varphi\right\|_{L^1}\left\|\varphi_{2^{-j}} * f\right\|_{L^\infty},$$

we have

$$I \leq 2\|\varphi\|_{L^1}\|f\|_{\mathbb{B}^\alpha} \sum_{j \geq -\log_2|h|} 2^{-j\alpha} = C\|f\|_{\mathbb{B}^\alpha}|h|^\alpha.$$

To estimate II we need to use the condition $\alpha < 1$. Since $f \in \mathbb{B}^\alpha$, we have

$$\left|\tilde{\varphi}_{2^{-j}} * \varphi_{2^{-j}} * f(x+h) - \tilde{\varphi}_{2^{-j}} * \varphi_{2^{-j}} * f(x)\right|$$

$$\leq \int_{\mathbb{R}} \left|\tilde{\varphi}_{2^{-j}}(x+h-z) - \tilde{\varphi}_{2^{-j}}(x-z)\right| \cdot \left|\varphi_{2^{-j}} * f(z)\right| dz$$

$$\leq 2^{-j\alpha}\|f\|_{\mathbb{B}^\alpha} \int_{\mathbb{R}} \left|\varphi(2^j z - 2^j x - 2^j h) - \varphi(2^j z - 2^j x)\right| 2^j dz$$

$$= 2^{-j\alpha}\|f\|_{\mathbb{B}^\alpha} \int_{\mathbb{R}} \left|\varphi(w - 2^j h) - \varphi(w)\right| dw$$

$$\leq 2^{-j\alpha}\|f\|_{\mathbb{B}^\alpha} 2^j|h| \int_{\mathbb{R}} \left(\sup_{|t| \leq 2^j|h|} \left|\varphi'(w+t)\right|\right) dw,$$

where the last inequality is due to the Mean-Value theorem. Since we want to estimate II we have $j < -\log_2|h|$, or equivalently, $2^j|h| < 1$. In this case,

$$\sup_{|t| \leq 2^j|h|} \left|\varphi'(w+t)\right| \leq \sup_{|t| \leq 1} \left|\varphi'(w+t)\right| \leq \frac{C}{\left(1+|w|\right)^{1+\varepsilon}}.$$

Thus, since $\alpha < 1$, we have

$$II \leq C \sum_{j < -\log_2|h|} 2^{j(1-\alpha)}|h|\, \|f\|_{\mathbb{B}^\alpha}$$

$$= C\|f\|_{\mathbb{B}^\alpha}|h| \cdot |h|^{\alpha-1} = C\|f\|_{\mathbb{B}^\alpha}|h|^\alpha. \qquad (7.9)$$

The estimates we have found for I and II show that $f \in \Lambda_\alpha$ and

$$\|f\|_{\Lambda_\alpha} \leq C\|f\|_{\mathbb{B}^\alpha}.$$

∎

REMARK 1: As the reader can see by looking at (7.8), it is sufficient to assume that

$$|\varphi(x)| \leq \frac{C_0}{(1 + |x|)^{1+\alpha+\gamma}} \qquad \text{for some } \gamma > 0,$$

instead of condition ii) of (7.3), to have the characterization given in Theorem 7.7.

REMARK 2: Originally, the definition of $\mathbb{B}^\alpha \equiv \mathbb{B}^\alpha_\varphi$, $0 < \alpha < 1$, depends on the choice of the function φ. Theorem 7.7 shows that this definition is, in fact, independent of φ. Hence, any two functions φ^1 and φ^2 satisfying (7.3) and (7.4) give rise to the same space $\mathbb{B}^\alpha$, $0 < \alpha < 1$, with equivalent norms. We shall use this freedom in the following.

It makes sense to define Λ_1 by replacing α by 1 in (7.1) and (7.2). Also, it makes sense to define $\mathbb{B}^1$ by replacing α by 1 in (7.5) and (7.6). The proof of Theorem 7.7 shows that if φ satisfies (7.3) and (7.4), then $\Lambda_1 \subset \mathbb{B}^1$. But the proof does not work to prove the reverse inclusion since the series that appears in (7.9) is not convergent when $\alpha = 1$. In fact, we shall observe below that Λ_1 and $\mathbb{B}^1$ do not coincide. As we shall prove below, $\mathbb{B}^1$ coincides with the **Zygmund class** of functions, $\Lambda_* \equiv \Lambda_*(\mathbb{R})$, defined as the set of all $f \in L^\infty(\mathbb{R})$ such that

$$\|f\|_{\Lambda_*} = \|f\|_{L^\infty} + \sup_{|h|>0} \frac{\|f(\cdot + h) + f(\cdot - h) - 2f(\cdot)\|_{L^\infty}}{|h|} < \infty. \qquad (7.10)$$

THEOREM 7.11 *If $\varphi \in \mathfrak{R}^1$ and satisfies (7.4), then $\Lambda_* = \mathbb{B}^1_\varphi$, and their norms are equivalent.*

PROOF: Suppose that $f \in \Lambda_*$, and assume, for now, that φ is even. Then, $\int_\mathbb{R} \varphi(y)\,dy = 0$ and the evenness of φ allow us to write

$$(\varphi_{2^{-j}} * f)(x) = \tfrac{1}{2} \int_\mathbb{R} \varphi_{2^{-j}}(y)\{f(x + y) + f(x - y) - 2f(x)\}\,dy.$$

Using (7.10) we obtain

$$|\varphi_{2^{-j}} * f(x)| \leq \tfrac{1}{2}\|f\|_{\Lambda_*} \int_\mathbb{R} |\varphi_{2^{-j}}(y)| \cdot |y|\,dy$$

$$= \tfrac{1}{2}\|f\|_{\Lambda_*} 2^{-j} \int_\mathbb{R} |\varphi(u)| \cdot |u|\,du \leq C\,2^{-j}\|f\|_{\Lambda_*},$$

where the last inequality is due to part ii) of (7.3). Thus, when φ is even and belongs to $\mathfrak{R}^1$ we have $\Lambda_* \subseteq \mathbb{B}^1 \equiv \mathbb{B}_\varphi^1$ and

$$\|f\|_{\mathbb{B}^1} \leq C\|f\|_{\Lambda_*}. \tag{7.12}$$

We now must show that (7.12) is true when $\mathbb{B}^1$ is defined by a function φ not necessarily even. Choose ψ satisfying, in addition to being even and belonging to $\mathfrak{R}^1$, condition (7.4); then

$$(\varphi_{2^{-j}} * f)(x) = \sum_{\ell \in \mathbb{Z}} (\varphi_{2^{-j}} * \tilde{\psi}_{2^{-\ell}} * \psi_{2^{-\ell}} * f)(x),$$

where $\tilde{\varphi}(x) = \overline{\varphi(-x)} = \overline{\psi(x)}$. Thus, by (7.12),

$$2^j \left\|\varphi_{2^{-j}} * f\right\|_{L^\infty} \leq 2^j \sum_{\ell \in \mathbb{Z}} \left\|\varphi_{2^{-j}} * \tilde{\psi}_{2^{-\ell}}\right\|_{L^1} \left\|\psi_{2^{-\ell}} * f\right\|_{L^\infty}$$

$$\leq 2^j \left\{ \sum_{\ell \leq j} + \sum_{\ell > j} \right\} \left\{ \left\|\varphi_{2^{-j}} * \tilde{\psi}_{2^{-\ell}}\right\|_{L^1} 2^{-\ell} C \|f\|_{\Lambda_*} \right\}$$

$$= 2^j \left\{ \sum_{\ell \leq j} + \sum_{\ell > j} \right\} \left\{ 2^{\frac{j}{2}} 2^{\frac{\ell}{2}} \left\|\varphi_{j,0} * \tilde{\psi}_{\ell,0}\right\|_{L^1} 2^{-\ell} C \|f\|_{\Lambda_*} \right\}$$

$$\equiv I + II.$$

To estimate I we use Lemma 3.14 with $g = \tilde{\psi}$ and $h = \varphi$ ($N = 1$) to obtain

$$I \leq C\|f\|_{\Lambda_*} \sum_{\ell \leq j} 2^{j-\ell} 2^{\frac{1}{2}(\ell+j)} 2^{(\ell-j)(\frac{1}{2}+2)} \int_{\mathbb{R}} \frac{1}{\left(1 + 2^\ell|x|\right)^{1+\varepsilon}} \, dx$$

$$= C\|f\|_{\Lambda_*} \sum_{\ell \leq j} 2^{\ell-j} = C\|f\|_{\Lambda_*}.$$

To estimate II we use Lemma 3.12 with $g = \varphi$ and $h = \tilde{\psi}$ to obtain

$$II \leq C\|f\|_{\Lambda_*} \sum_{\ell > j} 2^{j-\ell} 2^{\frac{1}{2}(\ell+j)} 2^{\frac{1}{2}(j-\ell)} \int_{\mathbb{R}} \frac{1}{\left(1 + 2^j|x|\right)^{1+\varepsilon}} \, dx$$

$$= C\|f\|_{\Lambda_*} \sum_{\ell > j} 2^{j-\ell} = C\|f\|_{\Lambda_*}.$$

This shows that $\Lambda_* \subseteq \mathbb{B}^1 \equiv \mathbb{B}_\varphi^1$ and $\|f\|_{\mathbb{B}^1} \leq C\|f\|_{\Lambda_*}$ when $\varphi \in \mathfrak{R}^1$.

Suppose now that $f \in \mathbb{B}^1$. By (7.4) we can write

$$f(x+h) + f(x-h) - 2f(x)$$

$$= \sum_{j \in \mathbb{Z}} \Big\{ \tilde{\varphi}_{2^{-j}} * \varphi_{2^{-j}} * f(x+h)$$

$$+ \tilde{\varphi}_{2^{-j}} * \varphi_{2^{-j}} * f(x-h) - 2\tilde{\varphi}_{2^{-j}} * \varphi_{2^{-j}} * f(x) \Big\}$$

$$= \Big\{ \sum_{j < -\log_2 |h|} + \sum_{j \geq -\log_2 |h|} \Big\} \Big\{ \tilde{\varphi}_{2^{-j}} * \varphi_{2^{-j}} * f(x+h)$$

$$+ \tilde{\varphi}_{2^{-j}} * \varphi_{2^{-j}} * f(x-h) - 2\tilde{\varphi}_{2^{-j}} * \varphi_{2^{-j}} * f(x) \Big\}$$

$$= I + II.$$

To estimate $|II|$ we have

$$|II| \leq 4 \|\varphi\|_{L^1} \sum_{j \geq -\log_2 |h|} \|\varphi_{2^{-j}} * f\|_{L^\infty} \leq 4\|\varphi\|_{L^1} \sum_{j \geq -\log_2 |h|} 2^{-j} \|f\|_{\mathbb{B}^1}$$

$$= C\|f\|_{\mathbb{B}^1} |h|.$$

We now estimate $|I|$. Since $f \in \mathbb{B}^1$ we have

$$|I| \leq \sum_{j < -\log_2 |h|} \int_{\mathbb{R}} \Big\{ \big| \tilde{\varphi}_{2^{-j}}(x+h-z) + \tilde{\varphi}_{2^{-j}}(x-h-z)$$

$$- 2\tilde{\varphi}_{2^{-j}}(x-z) \big| \Big\} 2^{-j} \|f\|_{\mathbb{B}^1} \, dz$$

$$= \|f\|_{\mathbb{B}^1} \sum_{j < -\log_2 |h|} 2^{-j} \int_{\mathbb{R}} \big| \tilde{\varphi}(w+2^j h) + \tilde{\varphi}(w-2^j h) - 2\tilde{\varphi}(w) \big| \, dw.$$

Using the Mean-Value theorem twice and taking into account the fact that $j < -\log_2 |h|$, or equivalently, $2^j |h| < 1$, we have

$$|I| \leq C\|f\|_{\mathbb{B}^1} \sum_{j < -\log_2 |h|} 2^j |h|^2 \int_{\mathbb{R}} \Big\{ \sup_{|t| \leq 1} \big| D^2 \varphi(w+t) \big| \Big\} \, dw.$$

Since $\left|D^2\varphi(x)\right| \leq \dfrac{C}{(1+|x|)^{1+\varepsilon}}$ we deduce

$$|I| \leq C\|f\|_{\mathbb{B}^1}|h|^2 \sum_{j<-\log_2|h|} 2^j = C\|f\|_{\mathbb{B}^1}|h|.$$

These two estimates for $|I|$ and $|II|$ show that $\mathbb{B}^1 \subseteq \Lambda_*$ and

$$\|f\|_{\Lambda_*} \leq C\|f\|_{\mathbb{B}^1}.$$

∎

REMARK 3: Observe that for these spaces the estimates when j is large are more difficult than the corresponding ones for j small (see the difference between the estimates for I and II in the above proof). This is due to the fact that in the definition of $\mathbb{B}^\alpha$ only the large values of j are important. In fact, if $j \leq 0$ the condition $f \in L^\infty$ already implies $2^{j\alpha}\|\varphi_{2^{-j}} * f\|_{L^\infty} \leq A$:

$$2^{j\alpha}\|\varphi_{2^{-j}} * f\|_{L^\infty} \leq 2^{j\alpha}\|\varphi_{2^{-j}}\|_{L^1}\|f\|_{L^\infty} = 2^{j\alpha}\|f\|_{L^\infty} \leq \|f\|_{L^\infty}.$$

We shall encounter this situation again in the next result, in which a wavelet characterization of the spaces $\mathbb{B}^\alpha$, $\alpha > 0$, is given.

REMARK 4: In section V.4 of [Ste1] there is an example of a function $f \in L^\infty(\mathbb{R})$ such that $f \in \Lambda_*$, but the inequality

$$\|f(\cdot + t) - f(\cdot)\|_{L^\infty} \leq C|t|$$

fails for all C. This shows that Λ_1 does not coincide with $\mathbb{B}^1$ (recall that the inclusion $\Lambda_1 \subseteq \mathbb{B}^1$ is always true; see the comments before Theorem 7.11).

Let $\alpha > 0$ and write $\alpha = N + s$ for an $N \in \mathbb{Z}^+$ and $0 \leq s < 1$. Given $\varphi \in \mathfrak{R}^N$ and satisfying (7.4) we define $\mathbb{B}^\alpha \equiv \mathbb{B}^\alpha(\mathbb{R}) \equiv \mathbb{B}^\alpha_\varphi$ as the space of all functions $f \in L^\infty(\mathbb{R})$ such that

$$\sup_{j\in\mathbb{Z}} 2^{j\alpha}\|\varphi_{2^{-j}} * f\|_{L^\infty} < \infty, \tag{7.13}$$

and the norm in $\mathbb{B}^\alpha$ is given by

$$\|f\|_{\mathbb{B}^\alpha} = \|f\|_{L^\infty} + \sup_{j\in\mathbb{Z}} 2^{j\alpha}\|\varphi_{2^{-j}} * f\|_{L^\infty}. \tag{7.14}$$

The next proposition shows that the definition of $\mathbb{B}^\alpha$ is independent of the choice of $\varphi \in \mathfrak{R}^N$ satisfying (7.4).

PROPOSITION 7.15 *Let $\alpha = N + s > 0$ for some $N \in \mathbb{Z}^+$ and $0 \le s < 1$. If $\varphi, \psi \in \mathfrak{R}^N$ and ψ satisfies (7.4), then, there exists a constant $0 < C < \infty$ such that*

$$\sup_{j\in\mathbb{Z}} 2^{j\alpha}\|\varphi_{2^{-j}} * f\|_{L^\infty} \le C \sup_{j\in\mathbb{Z}} 2^{j\alpha}\|\psi_{2^{-j}} * f\|_{L^\infty}$$

for all $f \in \mathbb{B}^\alpha$. In particular, the definition of $\mathbb{B}^\alpha$ is independent of the choice of φ.

PROOF : By (7.4) we can write

$$(\varphi_{2^{-j}} * f)(x) = \sum_{\ell\in\mathbb{Z}}(\varphi_{2^{-j}} * \tilde{\psi}_{2^{-\ell}} * \psi_{2^{-\ell}} * f)(x),$$

where $\tilde{\psi}(x) = \overline{\psi(-x)}$. Thus,

$$2^{j\alpha}\|\varphi_{2^{-j}} * f\|_{L^\infty}$$

$$\le 2^{j\alpha}\sum_{\ell\in\mathbb{Z}}\|\varphi_{2^{-j}} * \tilde{\psi}_{2^{-\ell}}\|_{L^1}\|\psi_{2^{-\ell}} * f\|_{L^\infty}$$

$$\le 2^{j\alpha}\Big\{\sum_{\ell\le j}+\sum_{\ell>j}\Big\}\Big\{2^{-\ell\alpha}\Big(\sup_{\ell\in\mathbb{Z}}2^{\ell\alpha}\|\psi_{2^{-\ell}} * f\|_{L^\infty}\Big)\|\varphi_{2^{-j}} * \tilde{\psi}_{2^{-\ell}}\|_{L^1}\Big\}$$

$$\le \Big(\sup_{\ell\in\mathbb{Z}}2^{\ell\alpha}\|\psi_{2^{-\ell}} * f\|_{L^\infty}\Big)\Big\{\sum_{\ell\le j}+\sum_{\ell>j}\Big\}\Big\{2^{(j-\ell)\alpha}2^{\frac{1}{2}(j+\ell)}\|\varphi_{j,0} * \tilde{\psi}_{\ell,0}\|_{L^1}\Big\}$$

$$\le \Big(\sup_{\ell\in\mathbb{Z}}2^{\ell\alpha}\|\psi_{2^{-\ell}} * f\|_{L^\infty}\Big)\Big\{I + II\Big\}.$$

The estimate for II is easy. By Lemma 3.12 with $g = \varphi$ and $h = \tilde{\psi}$ we deduce

$$II \le C\sum_{\ell>j}2^{(j-\ell)\alpha}2^{\frac{1}{2}(j+\ell)}2^{\frac{1}{2}(j-\ell)}\int_{\mathbb{R}}\frac{1}{\left(1+2^j|x|\right)^{1+\varepsilon}}\,dx$$

$$\le C\sum_{\ell>j}2^{(j-\ell)\alpha} = C.$$

To estimate I we use Lemma 3.14 with $g = \tilde{\psi}$ and $h = \varphi$ to obtain

$$I \leq C \sum_{\ell \leq j} 2^{(j-\ell)\alpha} 2^{\frac{1}{2}(j+\ell)} 2^{(\ell-j)(\frac{1}{2}+N+1)} \int_{\mathbb{R}} \frac{1}{\left(1 + 2^\ell |x|\right)^{1+\varepsilon}} \, dx$$

$$\leq C \sum_{\ell \leq j} 2^{(\ell-j)(N+1-\alpha)} = C,$$

since $N + 1 - \alpha = 1 - s > 0$. This proves the desired result. ∎

We now give the characterization of $\mathbb{B}^\alpha$, $\alpha > 0$, in terms of wavelet coefficients.

THEOREM 7.16 *Let $\alpha = N + s > 0$ for some $N \in \mathbb{Z}^+$ and $0 \leq s < 1$, and $\psi \in \mathfrak{R}^N$ be an orthonormal wavelet. Then, $f \in \mathbb{B}^\alpha$ if and only if $f \in L^\infty(\mathbb{R})$ and*

$$\left|\langle f, \psi_{j,k}\rangle\right| \leq C \, 2^{-j(\alpha+\frac{1}{2})} \qquad \text{for all } k \in \mathbb{Z}, \; j = 1, 2, \cdots. \tag{7.17}$$

PROOF : Assume that $f \in \mathbb{B}^\alpha$. Since

$$\langle f, \psi_{j,k}\rangle = 2^{-\frac{j}{2}} f * \tilde{\psi}_{2^{-j}}(2^{-j}k),$$

we have

$$\left|\langle f, \psi_{j,k}\rangle\right| \leq \left(\sup_{j \in \mathbb{Z}} 2^{j\alpha} \left\| f * \tilde{\psi}_{2^{-j}} \right\|_{L^\infty}\right) 2^{-j\alpha} 2^{-\frac{j}{2}} = C \, 2^{-j(\alpha+\frac{1}{2})}$$

for all $k \in \mathbb{Z}$, $j \in \mathbb{Z}$. Observe that this inequality is valid for all $j \in \mathbb{Z}$ but, similar to Remark 3 that follows the proof of Theorem 7.11, if $j \leq 0$ this follows from $f \in L^\infty(\mathbb{R})$.

Suppose now that $f \in L^\infty(\mathbb{R})$ and satisfies (7.17). Take φ to be any function belonging to $\mathfrak{R}^N$. Since ψ is an orthonormal wavelet,

$$(\varphi_{2^{-j}} * f)(x) = \sum_{\ell \in \mathbb{Z}} \sum_{m \in \mathbb{Z}} \langle f, \psi_{\ell,m}\rangle \, (\varphi_{2^{-j}} * \psi_{\ell,m})(x), \qquad j \in \mathbb{Z}.$$

Thus, if $j = 1, 2, \cdots$,

$$2^{j\alpha} \left\| \varphi_{2^{-j}} * f \right\|_{L^\infty} \leq C \sum_{\ell \in \mathbb{Z}} \sum_{m \in \mathbb{Z}} 2^{-\frac{\ell}{2}} 2^{(j-\ell)\alpha} \left\| \varphi_{2^{-j}} * \psi_{\ell,m} \right\|_{L^\infty}$$

$$= C \sum_{\ell \in \mathbb{Z}} \sum_{m \in \mathbb{Z}} 2^{\frac{1}{2}(j-\ell)} 2^{(j-\ell)\alpha} \left\| \varphi_{j,0} * \psi_{\ell,m} \right\|_{L^\infty}$$

$$= C \left\{ \sum_{\ell \leq j} + \sum_{\ell > j} \right\} \left\{ \sum_{m \in \mathbb{Z}} 2^{(j-\ell)(\alpha+\frac{1}{2})} \left\| \varphi_{j,0} * \psi_{\ell,m} \right\|_{L^\infty} \right\}$$

$$= C \left\{ I + II \right\}.$$

To estimate I we use Lemma 3.14 with $g = \psi$ and $h = \varphi$ to obtain

$$I \leq C \sum_{\ell \leq j} 2^{(j-\ell)(\alpha+\frac{1}{2})} 2^{(\ell-j)(\frac{1}{2}+N+1)} = C \sum_{\ell \leq j} 2^{(\ell-j)(N+1-\alpha)} = C,$$

since $N + 1 - \alpha = 1 - s > 0$.

To estimate II we use Lemma 3.12 with $g = \varphi$ and $h = \psi$ to obtain

$$II \leq C \sum_{\ell > j} 2^{(j-\ell)(\alpha+\frac{1}{2})} 2^{\frac{1}{2}(j-\ell)} = C \sum_{\ell > j} 2^{(j-\ell)(\alpha+1)} = C.$$

As we have already observed in Remark 3, $\left\| \varphi_{2^{-j}} * f \right\|_{L^\infty} \leq 2^{-j\alpha} \left\| f \right\|_{L^\infty}$ when $j \leq 0$ and $\alpha > 0$. Thus,

$$\sup_{j \in \mathbb{Z}} 2^{j\alpha} \left\| \varphi_{2^{-j}} * f \right\|_{L^\infty} \leq C,$$

and, hence, $f \in \mathbb{B}^\alpha$.   ∎

Combining Theorem 7.16 with Theorem 7.7 and Theorem 7.11 we obtain the following corollary.

THEOREM 7.18 *Let ψ be an orthonormal wavelet.*

(a) *If $\psi \in \mathfrak{R}^0$ and $0 < \alpha < 1$, then $f \in \Lambda_\alpha$ if and only if $f \in L^\infty(\mathbb{R})$ and*

$$\left| \langle f, \psi_{j,k} \rangle \right| \leq C\, 2^{-j(\alpha+\frac{1}{2})} \qquad \text{for all } k \in \mathbb{Z},\ j = 1, 2, \cdots.$$

(b) *If $\psi \in \mathfrak{R}^1$, then $f \in \Lambda_*$ if and only if $f \in L^\infty(\mathbb{R})$ and*

$$\left| \langle f, \psi_{j,k} \rangle \right| \leq C\, 2^{-\frac{3}{2}j} \qquad \text{for all } k \in \mathbb{Z},\ j = 1, 2, \cdots.$$

If $\alpha > 1$ we could have defined the spaces $\Lambda_\alpha(\mathbb{R})$ for $\alpha \notin \mathbb{Z}^+$ and $\Lambda_{*,\alpha}(\mathbb{R})$ for $\alpha \in \mathbb{Z}^+$ inductively. That is, if $\alpha \notin \mathbb{Z}^+$, we say that $f \in \Lambda_\alpha$ if and only if $f \in L^\infty(\mathbb{R})$ and $Df \in \Lambda_{\alpha-1}$. If $\alpha \in \mathbb{Z}^+$ we define $f \in \Lambda_{*,\alpha}$ if and only if $f \in L^\infty(\mathbb{R})$ and $Df \in \Lambda_{*,\alpha-1}$ (here $\Lambda_{*,1} \equiv \Lambda_*$ is the Zygmund class). One can show that $\mathbb{B}^\alpha = \Lambda_\alpha$ if $\alpha \notin \mathbb{Z}^+$ and $\mathbb{B}^\alpha = \Lambda_{*,\alpha}$ if $\alpha \in \mathbb{Z}^+$, but we shall not pursue this matter here.

We finish this section with a simple application of the characterization of the Sobolev spaces $L^{p,s}(\mathbb{R})$ and the spaces $\mathbb{B}^\alpha(\mathbb{R})$. If $f \in L^{p,s}(\mathbb{R})$, $1 < p < \infty$ and $s \in \mathbb{Z}^+$, we use Theorem 6.27 to deduce that, for $j, k \in \mathbb{Z}$,

$$C\|f\|_{L^{p,s}} \geq \|\mathcal{W}_\psi^s f\|_{L^p} \geq \left|\langle f, \psi_{j,k}\rangle\right| 2^{js} 2^{\frac{j}{2}} \left(\int_\mathbb{R} \chi_{I_{j,k}}(x)\, dx\right)^{\frac{1}{p}}$$

$$= \left|\langle f, \psi_{j,k}\rangle\right| 2^{j(s-\frac{1}{p})} 2^{\frac{j}{2}}.$$

This shows that $L^{p,s}(\mathbb{R})$ is continuously embedded in $\mathbb{B}^{s-\frac{1}{p}}(\mathbb{R})$, which is the Sobolev embedding theorem in one dimension.

6.8 Notes and references

1. The Shannon sampling theorem (Theorem 1.4) can be found in [Sha]. A generalization of this theorem in the setting of multiresolution analysis has been presented in [Wal2]. This result is as follows. Suppose that the subspaces $\{V_j : j \in \mathbb{Z}\}$ form an MRA for $L^2(\mathbb{R})$ with a scaling function φ satisfying

$$|\varphi(x)| \leq \frac{1}{(1+|x|)^{1+\varepsilon}} \qquad \text{for some } \varepsilon > 0,$$

and

$$\Phi(\xi) \equiv \sum_{k\in\mathbb{Z}} \hat\varphi(\xi + 2k\pi) \neq 0 \qquad \text{for all } \xi \in \mathbb{R}.$$

Then, for any $f \in V_0$,

$$f(x) = \sum_{k\in\mathbb{Z}} f(k) S(x - k), \qquad x \in \mathbb{R}, \tag{8.1}$$

where

$$\widehat{S}(\xi) = \frac{\hat{\varphi}(\xi)}{\Phi(\xi)}, \qquad \xi \in \mathbb{R}.$$

When (8.1) is applied to the scaling function, $\varphi = \chi_{[0,1]}$, of the Haar wavelet, one obtains the easily established equality

$$f(x) = \sum_{k \in \mathbb{Z}} f(k)\varphi(x - k)$$

for all functions f which is constant on intervals of the form $[k, k+1)$, $k \in \mathbb{Z}$. Moreover, if (8.1) is written for the scaling function, $\varphi(x) = \mathrm{sinc}\,(\pi x)$, of the Shannon wavelet the result obtained is the Shannon sampling theorem for every $f \in L^2(\mathbb{R})$ such that $\mathrm{supp}\,(\hat{f}) \subseteq [-\pi, \pi]$.

2. Expressions of the form (1.7), and its continuous analogues, have appeared with great generality in Fourier Analysis. They are known as Calderón's identities (see [Cal] and [CT]). These general Calderón's identities are used in [FJ1], [FJ2] and [FJ3] to study large classes of function spaces. The monograph [FJW] contains a discussion of the convergence of formulas like (1.10) and (1.11).

3. There is much literature concerning Littlewood-Paley functions. The reader could look in [Ste1] (Chapter IV), [EG], [CW1] or [FJW] (Chapter 4) to find good survey articles and introductions to this theory. The original work on this subject by Littlewood and Paley can be found in [LP]. Theorem 2.9 is taken from [GR] (Theorem 5.3, pages 505-506).

4. The Hardy-Littlewood maximal function $\mathcal{M}$ has been studied by mathematicians for more than half a century. The original paper on the boundedness of $\mathcal{M}$ on $L^p(\mathbb{R})$, $1 < p < \infty$, is [HL]. For a modern description of this result, including the n-dimensional version, see [Ste1]. The result contained in Theorem 3.4, concerning vector-valued inequalities for the maximal function $\mathcal{M}$, was first published by [FS2]. It can also be proved as an application of the theory of vector-valued Calderón-Zygmund operators, precisely the same result that is used in the proof of Theorem 2.9. The interested reader can see this approach in [GR] (pages 497-498). The other kind of "maximal" function studied in section 6.3 is g_λ^* in (3.1). It has been introduced in [FS1] to characterize the Hardy spaces H^p, $0 < p \leq 1$. The approach we follow in section 6.3 to obtain results for g_λ^* is closely related to the work in [Pe2].

5. The Plancherel-Polya inequality (Theorem 3.8) is a classical result proved in [PP]. By now there are many proofs in the literature (see [FJ1] and the references given there). The opposite inequality in Theorem 3.8, that is,

$$\|g\|_{L^p(\mathbb{R})}^p \leq 2^{-j} \sum_{k \in \mathbb{Z}} \sup_{z \in I_{j,k}} |g(z)|^p$$

is trivial, since the right-hand side is the upper-Riemann sum of $|g|^p$. The idea behind the proof of Theorem 3.8 is that since g is band-limited, the Paley-Wiener theorem implies that g is of exponential type with greater smoothness, in some sense, as $j \to -\infty$; thus, the quantities

$$g(2^{-j}k), \qquad \sup_{z \in I_{j,k}} |g(z)| \qquad \text{and} \qquad \left(\int_{I_{j,k}} |g(x)|^p \, dz \right)^{\frac{1}{p}}$$

should be of the same size. The Plancherel-Polya inequality has been used in [Pe3] and [FJ1] in connection with the study of Besov spaces.

6. The inequalities contained in Lemma 3.16 are a variant of the following well-known inequality

$$\int_{\mathbb{R}} \frac{|f(y)|}{(1 + |x - y|)^{1+\varepsilon}} \, dy \leq C \, (Mf)(x), \qquad x \in \mathbb{R}, \ \varepsilon > 0,$$

for locally integrable f. The proof follows the same idea as the one of Lemma 3.16. Here is a sketch. Choose

$$A_0 = \{y \in \mathbb{R} : |x - y| \leq 2\} \qquad \text{and}$$

$$A_s = \{y \in \mathbb{R} : 2^s < |x - y| \leq 2^{s+1}\} \qquad \text{for} \ s = 1, 2, \cdots.$$

Since on A_s, $1 + |x - y| \geq 2^s$, we have, since $\varepsilon > 0$,

$$\int_{\mathbb{R}} \frac{|f(y)|}{(1 + |x - y|)^{1+\varepsilon}} \, dy = \sum_{s=0}^{\infty} \int_{A_s} \frac{|f(y)|}{(1 + |x - y|)^{1+\varepsilon}} \, dy$$

$$\leq \int_{A_0} |f(y)| \, dy + \sum_{s=1}^{\infty} 2^{-s(1+\varepsilon)} \int_{A_s} |f(y)| \, dy$$

$$\leq C \left\{ 1 + \sum_{s=1}^{\infty} 2^{-s\varepsilon} \right\} (Mf)(x) = C \, (Mf)(x).$$

7. Some readers might be aware of the existence of the Hardy spaces $H^p(\mathbb{R})$, $0 < p \leq 1$. Although there are several definitions of these spaces (see [SW], [Ste2] or [GR]), the one that is best suited to wavelet treatment is the atomic one: it establishes that the "elements" of $H^p(\mathbb{R})$, $0 < p \leq 1$, can be built using elementary particles called atoms. More precisely, a (p, ∞)-atom, a, is a measurable function defined on $\mathbb{R}$ that satisfies

$$\begin{cases} \text{(i)} & \text{supp}\,(a) \subseteq I, \text{ where } I \text{ is a finite interval on } \mathbb{R}, \\[2mm] \text{(ii)} & \|a\|_{L^\infty} \leq |I|^{-\frac{1}{p}}, \\[2mm] \text{(iii)} & \int_\mathbb{R} x^\ell a(x)\,dx = 0 \quad \text{for } \ell = 0, 1, \cdots, \left[\frac{1}{p}\right] - 1, \end{cases} \qquad (8.2)$$

where $[t]$ denotes the integer part of the positive real number t. The space $H^p(\mathbb{R})$, $0 < p \leq 1$, is the space of all distributions f that can be written as

$$f = \sum_{j=1}^\infty \lambda_j a_j \qquad \text{(convergence in } \mathcal{S}'), \qquad (8.3)$$

where the a_j's are p-atoms and $\sum_{j=1}^\infty |\lambda_j|^p < \infty$. Moreover,

$$\|f\|_{H^p}^p \equiv \inf \left\{ \sum_{j=1}^\infty |\lambda_j|^p \right\},$$

where the infimum is taken over all possible decompositions of f of the form (8.3), makes $H^p(\mathbb{R})$ a complete quasi-normed space (see Theorem 5.4 of Chapter 5 for the case $p = 1$). For applications of the atomic theory of these spaces see [CW2]. There is a characterization of $H^p(\mathbb{R})$, $0 < p \leq 1$, in terms of wavelet coefficients. Let ψ be a function defined on $\mathbb{R}$ for which there exist constants $\varepsilon > 0$, $\gamma > 0$, $C < \infty$ and $C_n < \infty$, $n = 0, 1, 2, \cdots, \left[\frac{1}{p}\right]$, such that

$$\begin{cases} \text{i)} & \int_\mathbb{R} x^\ell \psi(x)\,dx = 0 \qquad\qquad \text{for } \ell = 0, 1, \cdots, \left[\frac{1}{p}\right] - 1; \\[3mm] \text{ii)} & |\psi(x)| \leq \dfrac{C}{(1 + |x|)^{1 + [\frac{1}{p}] + \gamma}} \qquad \text{for } x \in \mathbb{R}; \\[3mm] \text{iii)} & |D^n \psi(x)| \leq \dfrac{C_n}{(1 + |x|)^{\frac{1}{p} + \varepsilon}} \qquad \text{for } x \in \mathbb{R},\ n = 0, 1, 2, \cdots, \left[\frac{1}{p}\right]. \end{cases} \qquad (8.4)$$

Then, there exist two constants A and B, $0 < A \leq B < \infty$, such that

$$A\|f\|_{H^p} \leq \left\|\sum_{j\in\mathbb{Z}}\sum_{k\in\mathbb{Z}}|\langle f,\psi_{j,k}\rangle|^2 2^j \chi_{I_{j,k}}(\cdot)\right\|_{L^p} \leq B\|f\|_{H^p}$$

for all $f \in H^p(\mathbb{R})$, provided ψ is an orthonormal wavelet satisfying (8.4). Notice that for orthonormal wavelets part i) of (8.4) is implied by part ii) and part iii) (see Theorem 3.4 of Chapter 2). Observe, also, that for the decomposition

$$f \sim \sum_{j\in\mathbb{Z}}\sum_{k\in\mathbb{Z}}\langle f,\psi_{j,k}\rangle\psi_{j,k}$$

to be expressed as an atomic decomposition, when ψ is a compactly supported orthonormal wavelet, we need ψ to satisfy part iii) of (8.2), which is precisely part i) of (8.4). This result can be proved using techniques similar to the ones developed in this chapter.

8. Consider

$$\mathbf{g}^{s,q}(f)(x) = \left\{\sum_{j\in\mathbb{Z}}(2^{js}|\varphi_{2^{-j}} * f(x)|)^q\right\}^{\frac{1}{q}} \tag{8.5}$$

for any real number s and any $0 < q \leq \infty$, where φ satisfies the same conditions as in the definition of (6.10) (observe that for $s \in \mathbb{N}$ and $q = 2$, $\mathbf{g}^{s,2}(f) = \mathbf{g}^s(f)$, where $\mathbf{g}^s(f)$ is given in (6.10)). For $s \in \mathbb{R}$, $0 < q \leq \infty$ and $0 < p < \infty$, we define $F_p^{s,q} \equiv F_p^{s,q}(\mathbb{R})$ as the space of all functions $f \in L^\infty(\mathbb{R})$ such that $\mathbf{g}^{s,q}(f) \in L^p(\mathbb{R})$, and the quasi-norm in $F_p^{s,q}(\mathbb{R})$ is given by

$$\|f\|_{F_p^{s,q}} \equiv \|f\|_{L^p} + \|\mathbf{g}^{s,q}(f)\|_{L^p}.$$

These spaces are called **inhomogeneous Triebel-Lizorkin spaces**. Obviously, $F_p^{0,2} = L^p$ if $1 < p < \infty$, and $F_p^{s,2} = L^{p,s}$ if $s \in \mathbb{N}$ and $1 < p < \infty$. The methods developed in this chapter can be used to show that, for $s \in \mathbb{R}$, $s \geq 0$, $1 \leq q \leq \infty$ and $1 < p < \infty$, there exist constants A and B, $0 < A \leq B < \infty$, such that

$$A\|f\|_{F_p^{s,q}} \leq \left\|\left\{\sum_{j\in\mathbb{Z}}\sum_{k\in\mathbb{Z}}|\langle f,\psi_{j,k}\rangle|^q (1+2^{qjs})2^{\frac{1}{2}qj}\chi_{I_{j,k}}(\cdot)\right\}^{\frac{1}{q}}\right\|_{L^p} \leq B\|f\|_{F_p^{s,q}}$$

for all $f \in F_p^{s,q}$, provided ψ is an orthonormal wavelet in $\mathfrak{R}^s$. The results for the general case, $0 < q \leq \infty$, $s \in \mathbb{R}$ and $0 < p < \infty$ are essentially the same but the regularity of the wavelets depends not only on s, but also on p and q, if they are smaller than 1. See also paragraph 9 of this section.

9. The **homogeneous Triebel-Lizorkin spaces** are the spaces $\dot{F}_p^{s,q} \equiv \dot{F}_p^{s,q}(\mathbb{R})$, $s \in \mathbb{R}$, $0 < q \leq \infty$ and $0 < p < \infty$, formed by the class of all distributions (modulo polynomials) f such that $\mathfrak{g}^{s,q}(f) \in L^p(\mathbb{R})$, where $\mathfrak{g}^{s,q}(f)$ is given in (8.5). The quasi-norm in $\dot{F}_p^{s,q}(\mathbb{R})$ is given by

$$\|f\|_{\dot{F}_p^{s,q}} \equiv \|\mathfrak{g}^{s,q}(f)\|_{L^p}.$$

These spaces were introduced by H. Triebel in [Tr1] (for a complete treatment and applications the reader can look at [Pe1], [Tr2], [Tr3] and [Tr4]). There is also a wavelet characterization of these spaces: for an appropriate orthonormal wavelet ψ, there exist constants A and B, $0 < A \leq B < \infty$, such that

$$A\|f\|_{\dot{F}_p^{s,q}} \leq \left\|\left\{\sum_{j\in\mathbb{Z}}\sum_{k\in\mathbb{Z}}(|\langle f,\psi_{j,k}\rangle|\,2^{j(s+\frac{1}{2})}\chi_{I_{j,k}}(\cdot))^q\right\}^{\frac{1}{q}}\right\|_{L^p} \leq B\|f\|_{\dot{F}_p^{s,q}}$$

for all $f \in \dot{F}_p^{s,q}$. The regularity of ψ depends on s, p and q. We invite the reader to look at [FJ3], where this is done in the context of the φ-transform.

10. The space BMO of functions with bounded mean oscillation (see section 5.5) can also fit into the scale of the Triebel-Lizorkin spaces $\dot{F}_p^{s,q}$. But the definition is not as simple. Since $H^1 = \dot{F}_1^{0,2}$, and the dual space of H^1 is BMO, the definition has to be made to preserve this duality. More precisely, the space $\dot{F}_\infty^{s,q}$ is the class of all distributions (modulo polynomials) f such that

$$\|f\|_{\dot{F}_\infty^{s,q}} = \sup_{\ell,m\in\mathbb{Z}}\left(2^\ell\int_{I_{\ell,m}}\sum_{j\geq\ell}2^{sq}|\varphi_{2^{-j}}*f(x)|^q\,dx\right)^{\frac{1}{q}} < \infty,$$

where $I_{\ell,m} = [2^{-\ell}m, 2^{-\ell}(m+1)]$. It can be shown that $\dot{F}_\infty^{0,2} = BMO$, and this characterization can be used to obtain a wavelet characterization of BMO. This characterization will involve the finiteness of the expression

$$\sup_{\ell,m\in\mathbb{Z}}\left(2^\ell\int_{I_{\ell,m}}\sum_{I_{j,k}\subseteq I_{\ell,m}}|\langle f,\psi_{j,k}\rangle|^2 2^j\chi_{I_{j,k}}(x)\,dx\right)^{\frac{1}{2}}.$$

This fact is equivalent to the existence of a finite constant C such that, for all dyadic intervals $I_{\ell,m}$,

$$\sum_{I_{j,k}\subseteq I_{\ell,m}}|\langle f,\psi_{j,k}\rangle|^2 \leq C\,2^{-\ell}.$$

The above characterizations are true for wavelets with a certain regularity. In [Me1] (Chapter V) this result is proved for wavelets with compact support which are differentiable. We invite the reader to apply the methods developed in this chapter to find a larger class of wavelets for which this characterization holds.

11. Many authors have considered Lipschitz spaces and its generalizations. The reader can look at [Ste1] (Chapter II) or [Tai] for these general definitions. The spaces $\mathbb{B}^\alpha$ are special cases of the so-called non-homogeneous Besov spaces $\mathbb{B}_p^{\alpha,q}$ which, for $\alpha \in \mathbb{R}$ and $0 < p, q \leq \infty$, are defined by the finiteness of the expression

$$\|f\|_{\mathbb{B}_p^{\alpha,q}} \equiv \|f\|_{L^p} + \left\{ \sum_{j \in \mathbb{Z}} \left(2^{j\alpha} \|\Phi_{2^{-j}} * f\|_{L^p} \right)^q \right\}^{\frac{1}{q}}$$

for appropriate functions Φ (see [Pe1] and [Pe3]). The characterization of the spaces $\mathbb{B}^\alpha$ in terms of wavelet coefficients, and also that of $\mathbb{B}_p^{\alpha,q}$ with $\alpha \in \mathbb{R}$ and $1 \leq p, q \leq \infty$, was first given in [LM], and an expanded account can be found in [Me1]. For the case $\alpha > 0$ and $p = q = \infty$ we have presented a different proof from the one contained in [Me1]; our proof follows the ideas developed in [FJ1] and [FJ3] (see also [FJW]). The ideas we have developed can be used to characterize the spaces $\mathbb{B}_p^{\alpha,q}$ in terms of wavelets. To give a result, we mention that if ψ is an orthonormal wavelet such that $\psi \in \mathfrak{R}^{[\alpha]}$ with $\alpha > 0$ and $1 \leq p, q \leq \infty$, then $f \in \mathbb{B}_p^{\alpha,q}$ if and only if

$$\|f\|_{L^p} + \left\{ \sum_{j \in \mathbb{Z}} \left(\sum_{k \in \mathbb{Z}} \left(|\langle f, \psi_{j,k} \rangle| \, 2^{j(\alpha + \frac{1}{2} - \frac{1}{p})} \right)^p \right)^{\frac{q}{p}} \right\}^{\frac{1}{q}} < \infty$$

(see [FJ1] or [FJ3], where this is done in the context of the so-called φ-transform).

12. The homogeneous Besov spaces $\dot{\mathbb{B}}_p^{\alpha,q}$, $\alpha \in \mathbb{R}$ and $0 < p, q \leq \infty$, are defined by using the finiteness of the expression

$$\|f\|_{\dot{\mathbb{B}}_p^{\alpha,q}} \equiv \left\{ \sum_{j \in \mathbb{Z}} \left(2^{j\alpha} \|\Phi_{2^{-j}} * f\|_{L^p} \right)^q \right\}^{\frac{1}{q}}$$

for appropriate functions Φ (see [Pe1] and [Pe3]). For $\|\cdot\|_{\dot{\mathbb{B}}_p^{\alpha,q}}$ to be a norm we have to consider the elements of $\dot{\mathbb{B}}_p^{\alpha,q}$ as equivalent classes of distributions modulo polynomials. A wavelet characterization for these homogeneous Besov spaces $\dot{\mathbb{B}}_p^{\alpha,q}$ can be given. For example, if $\alpha > 0$ and $1 \le p, q \le \infty$, then $f \in \dot{\mathbb{B}}_p^{\alpha,q}$ if and only if

$$\left\{ \sum_{j \in \mathbb{Z}} \left(\sum_{k \in \mathbb{Z}} \left(|\langle f, \psi_{j,k} \rangle| \, 2^{j(\alpha + \frac{1}{2} - \frac{1}{p})} \right)^p \right)^{\frac{q}{p}} \right\}^{\frac{1}{q}} < \infty$$

(see, again, [FJ1] or [FJ3]).

13. The remark that follows the proof of Lemma 3.14 allows us to provide wavelet characterizations using a larger class of wavelets than the corresponding one employed for L^p, H^1, $L^{p,s}$ and $\mathbb{B}^\alpha$. The condition on the $(N+1)^{\text{th}}$ derivative can be replaced by a Lipschitz condition on the Nth derivative, as in the aforementioned remark. For the spaces L^p, $1 < p < \infty$, and H^1, this means that the Franklin wavelet can be used to characterize these spaces, while this wavelet is not allowed in Theorem 4.15 and Theorem 5.18 (since the Franklin wavelet does not belong to $\mathfrak{R}^0$).

7

Characterizations in the theory of wavelets

This chapter is dedicated to the characterization of several properties that appear in the theory of wavelets and the concept of multiresolution analysis (MRA).

There are two basic equations that characterize all orthonormal wavelets in $L^2(\mathbb{R})$. These basic equations have already appeared in section 3.2 in the setting of band-limited wavelets. In section 7.1 we present a detailed proof of this characterization of all wavelets.

A single equation characterizes the orthonormal wavelets that arise from an MRA. This result, as well as characterizations of low-pass filters and scaling functions, are developed in sections 7.3, 7.4 and 7.5.

We apply these results to obtain several examples of wavelets (some new and some known, that we present throughout all the sections in this chapter), and to study (in section 7.2) the minimally supported frequency (MSF) wavelets (that is, the ones whose set of frequencies is as small as possible, as dictated by Corollary 2.4 of Chapter 2).

In section 7.6 we show that an orthonormal wavelet in $H^2(\mathbb{R})$ (the space of L^2-functions whose Fourier transform is 0 at the negative reals) cannot exist if we impose some regularity on its Fourier transform.

7.1   The basic equations

As we have indicated in the preface, there are two simple equations that characterize **all** orthonormal wavelets. We have introduced and used these equations on several occasions (see equations (2.1) and (2.2) in Chapter 3). Recall that an orthonormal wavelet is a function $\psi \in L^2(\mathbb{R})$ such that the system $\{\psi_{j,k} : j, k \in \mathbb{Z}\}$ is an orthonormal basis of $L^2(\mathbb{R})$, where

$$\psi_{j,k}(x) = 2^{\frac{j}{2}} \psi(2^j x - k), \qquad j, k \in \mathbb{Z}.$$

We shall establish the following basic result:

THEOREM 1.1 *A function $\psi \in L^2(\mathbb{R})$, with $\|\psi\|_2 = 1$, is an orthonormal wavelet if and only if*

$$\sum_{j \in \mathbb{Z}} |\hat{\psi}(2^j \xi)|^2 = 1 \quad \text{for a.e. } \xi \in \mathbb{R} \tag{1.2}$$

and

$$\sum_{j=0}^{\infty} \hat{\psi}(2^j \xi) \, \overline{\hat{\psi}(2^j(\xi + 2m\pi))} = 0 \quad \text{for a.e. } \xi \in \mathbb{R}, \quad m \in 2\mathbb{Z} + 1. \tag{1.3}$$

We have seen in section 3.1 that the system $\{\psi_{j,k} : j, k \in \mathbb{Z}\}$ is orthonormal if and only if

$$\sum_{k \in \mathbb{Z}} |\hat{\psi}(\xi + 2k\pi)|^2 = 1 \quad \text{for a.e. } \xi \in \mathbb{R} \tag{1.4}$$

and

$$\sum_{k \in \mathbb{Z}} \hat{\psi}(2^j(\xi + 2k\pi)) \, \overline{\hat{\psi}(\xi + 2k\pi)} = 0 \quad \text{for a.e. } \xi \in \mathbb{R}, \quad j \geq 1. \tag{1.5}$$

As we have seen, this is (relatively) easy to prove and follows from a periodization argument. The almost everywhere convergence of the series in (1.4), (1.5) and (1.3) are simple applications of Beppo-Levi's theorem.

Beppo-Levi's theorem states that if $\sum_n g_n(x)$ is a series of integrable functions such that

$$\sum_n \int_{\mathbb{X}} |g_n(x)|\, dx$$

converges, then $\sum_n g_n(x)$ converges almost everywhere to an integrable function and the series can be integrated term by term.

The convergence of the series in (1.4) follows by integrating the absolute values of its terms over the interval $[0, 2\pi]$. (Observe that the left-hand side of (1.4) is 2π-periodic.) In fact,

$$\sum_{k \in \mathbb{Z}} \int_0^{2\pi} |\hat{\psi}(\xi + 2k\pi)|^2\, d\xi = \sum_{k \in \mathbb{Z}} \int_{2k\pi}^{2(k+1)\pi} |\hat{\psi}(\xi)|^2\, d\xi = \|\hat{\psi}\|_2^2 < \infty.$$

The almost everywhere convergence now follows by Beppo-Levi's theorem. For the series in (1.5) we have

$$\sum_{k \in \mathbb{Z}} |\hat{\psi}(2^j(\xi + 2k\pi))\hat{\psi}(\xi + 2k\pi)|$$

$$\leq \left(\sum_{k \in \mathbb{Z}} |\hat{\psi}(2^j(\xi + 2k\pi))|^2\right)^{\frac{1}{2}} \left(\sum_{k \in \mathbb{Z}} |\hat{\psi}(\xi + 2k\pi)|^2\right)^{\frac{1}{2}}.$$

Thus the a.e. convergence of the series in (1.5) follows from the same result we just proved for (1.4). For the convergence of the series in (1.3) we make use of the fact that the sum ranges over the non-negative integers j: by a change of variables and Schwarz's inequality

$$\sum_{j=0}^{\infty} \int_{-\infty}^{\infty} |\hat{\psi}(2^j\xi)\hat{\psi}(2^j(\xi + 2m\pi))|\, d\xi$$

$$= \sum_{j=0}^{\infty} \frac{1}{2^j} \int_{\mathbb{R}} |\hat{\psi}(\eta)\hat{\psi}(\eta + 2^{j+1}m\pi)|\, d\eta \leq \sum_{j=0}^{\infty} \frac{1}{2^j} \|\hat{\psi}\|_2^2 < \infty.$$

Again, the almost everywhere convergence follows by Beppo-Levi's theorem. One of the difficulties with the series that appears in (1.2) is that the sum is extended over all the integers and, as a consequence, the almost everywhere convergence is not established easily. It will follow, however, from the arguments we shall present below.

An immediate consequence of Theorem 1.1 is that (1.4) and (1.5) follow from (1.2) and (1.3), provided $\|\psi\|_2 = 1$. Recall that, for band-limited wavelets we have shown that (1.2) and (1.3) characterize the completeness of the system $\{\psi_{j,k} : j, k \in \mathbb{Z}\}$ (see section 3.2).

The hypothesis $\|\psi\|_2 = 1$ in Theorem 1.1 is very natural; the system $\{\psi_{j,k} : j, k \in \mathbb{Z}\}$ cannot be normal without this assumption ($\|\psi_{j,k}\|_2 = \|\psi\|_2 = 1$ for all $j, k \in \mathbb{Z}$). It is important to point out, however, that the two equations (1.2) and (1.3), without this hypothesis $\|\psi\|_2 = 1$, characterize systems that are not necessarily orthonormal bases for $L^2(\mathbb{R})$, but they do satisfy the basic "analyzing" and "reconstruction" properties of these bases. We shall show, in fact, that the following result is true:

THEOREM 1.6 *If* $\psi \in L^2(\mathbb{R})$, *the following statements are equivalent:*

(A) $\displaystyle\sum_{j,k\in\mathbb{Z}} |\langle f, \psi_{j,k}\rangle|^2 = \|f\|_2^2$ *for all* $f \in L^2(\mathbb{R})$;

(B) $\displaystyle f = \sum_{j,k\in\mathbb{Z}} \langle f, \psi_{j,k}\rangle\, \psi_{j,k}$, *with convergence in* $L^2(\mathbb{R})$ *for all* $f \in L^2(\mathbb{R})$;

(C) ψ *satisfies equations* (1.2) *and* (1.3).

Before embarking into the proof of this theorem let us make several remarks and observations.

Some of the results we shall encounter are abstract statements about a general Hilbert space $\mathbb{H}$, while others involve measure-theoretic properties of Lebesgue measure on $\mathbb{R}$. For example, the equivalence between (A) and (B) is a simple observation involving a general Hilbert space $\mathbb{H}$. For simplicity let $\{e_j : j = 1, 2, \cdots\}$ be a family of vectors in $\mathbb{H}$. (We let the indexing set be the natural numbers; it will be clear that more general indexing sets, like $\mathbb{Z} \times \mathbb{Z}$, can equally well be used.) Then the equivalence of (A) and (B) can be formulated as follows:

THEOREM 1.7 *Let* $\mathbb{H}$ *be a Hilbert space and* $\{e_j : j = 1, 2, \cdots\}$ *be a family of elements of* $\mathbb{H}$. *Then,*

(i) $\displaystyle \|f\|^2 = \sum_{j=1}^{\infty} |\langle f, e_j\rangle|^2$ *holds for all* $f \in \mathbb{H}$

if and only if

(ii) $\displaystyle f = \sum_{j=1}^{\infty} \langle f, e_j\rangle\, e_j$, *with convergence in* $\mathbb{H}$, *for all* $f \in \mathbb{H}$.

PROOF : It is immediate that (ii) implies (i): the convergence in $\mathbb{H}$ in the equality in (ii) implies

$$\|f\|^2 = \langle f, f \rangle = \lim_{M \to \infty} \sum_{j=1}^{M} \langle f, e_j \rangle \langle e_j, f \rangle = \sum_{j=1}^{\infty} |\langle f, e_j \rangle|^2.$$

In order to show that (i) implies (ii) we first observe that

$$\{S_N\} = \left\{ \sum_{j=1}^{N} \langle f, e_j \rangle e_j \right\}, \qquad N = 1, 2, \cdots,$$

is a Cauchy sequence: if $1 \leq M \leq N$, then

$$\left\| \sum_{j=M}^{N} \langle f, e_j \rangle e_j \right\| = \sup_{\|g\| \leq 1} \left| \left\langle \sum_{j=M}^{N} \langle f, e_j \rangle e_j, g \right\rangle \right|$$

$$\leq \sup_{\|g\| \leq 1} \sum_{j=M}^{N} |\langle f, e_j \rangle| \cdot |\langle g, e_j \rangle| \leq \left(\sum_{j=M}^{N} |\langle f, e_j \rangle|^2 \right)^{\frac{1}{2}},$$

since, by (i),

$$\sum_{j=M}^{N} |\langle g, e_j \rangle|^2 \leq \|g\|^2 \leq 1.$$

Using (i) again, we see that

$$\lim_{\substack{M \to \infty \\ M \leq N}} \sum_{j=M}^{N} |\langle f, e_j \rangle|^2 = 0.$$

Thus, the series $\displaystyle\sum_{j=1}^{\infty} \langle f, e_j \rangle e_j$ converges to an element

$$h = \sum_{j=1}^{\infty} \langle f, e_j \rangle e_j \in \mathbb{H}.$$

By polarization (in $\mathbb{H}$ and in $\ell^2(\mathbb{Z})$) applied to equality (i) we have

$$\langle f, g \rangle = \sum_{j=1}^{\infty} \langle f, e_j \rangle \langle e_j, g \rangle$$

for all $f, g \in \mathbb{H}$. Hence,

$$\langle h, g \rangle = \left\langle \sum_{j=1}^{\infty} \langle f, e_j \rangle e_j, g \right\rangle = \sum_{j=1}^{\infty} \langle f, e_j \rangle \langle e_j, g \rangle = \langle f, g \rangle$$

for all $g \in \mathbb{H}$. This shows $h = f$ and (ii) is established. ∎

Equality (A) of Theorem 1.6 (or, in general, equality (i) in Theorem 1.7) tells us that the system $\{\psi_{j,k} : j, k \in \mathbb{Z}\}$ is a special "frame." In the next chapter we shall study the properties of frames. The following result shows that Theorem 1.1 is an immediate consequence of Theorem 1.6.

THEOREM 1.8 *Suppose that $\{e_j : j = 1, 2, \cdots\}$ is a system of vectors in a Hilbert space $\mathbb{H}$ satisfying condition* (i) *in Theorem 1.7. If*

$$\|e_j\| \geq 1 \qquad for \ \ j = 1, 2, \cdots,$$

then $\{e_j : j = 1, 2, \cdots\}$ is an orthonormal basis for $\mathbb{H}$.

PROOF : Applying equality (i) to $f = e_{j_o}$ we obtain

$$\|e_{j_o}\|^2 = |\langle e_{j_o}, e_{j_o} \rangle|^2 + \sum_{j \neq j_o} |\langle e_{j_o}, e_j \rangle|^2.$$

That is,

$$\|e_{j_o}\|^2 (1 - \|e_{j_o}\|^2) = \sum_{j \neq j_o} |\langle e_{j_o}, e_j \rangle|^2.$$

Since $\|e_{j_o}\| \geq 1$ the left-hand side of the last equality does not exceed 0. But the right-hand side is non-negative and, thus, equals 0. Thus, $\langle e_{j_o}, e_j \rangle = 0$ for all $j \neq j_o$. Since the left-hand side must also be 0 and $\|e_{j_o}\| \geq 1$, we must have $\|e_{j_o}\| = 1$. The fact that j_o is arbitrary implies that the system is orthonormal. Condition (i) (or (ii)) of Theorem 1.7 for all $f \in \mathbb{H}$ gives us completeness. ∎

Let us now give an example of a function $\psi \in L^2(\mathbb{R})$ satisfying (1.2) and (1.3) such that $\{\psi_{j,k} : j, k \in \mathbb{Z}\}$ is **not** an orthonormal basis for $L^2(\mathbb{R})$. Let

b be a non-negative even function such that $\{t \geq 0 : b(t) > 0\} \subseteq [\frac{1}{4}\pi, \pi]$ and

$$b^2(t) + b^2(\tfrac{1}{2}t) = 1 \qquad \text{for } t \in [\tfrac{1}{2}\pi, \pi]. \tag{1.9}$$

For example, b can be chosen to be an arbitrarily smooth function on $\mathbb{R}^+ = \{t \in \mathbb{R} : t \geq 0\}$ satisfying (3.9) of Chapter 1 with $\alpha = \frac{3}{8}\pi$, $\beta = \frac{3}{4}\pi$, $\varepsilon = \frac{1}{8}\pi$ and $\varepsilon' = 2\varepsilon = \frac{1}{4}\pi$. The support of b is contained in $E = [-\pi, -\frac{1}{4}\pi] \cup [\frac{1}{4}\pi, \pi]$. The translate of E by $2^j 2q\pi$ lies totally to the right (left) of E if q is a positive (negative) odd integer and $j \geq 0$. It follows that if ψ is chosen so that $|\hat{\psi}| = b$, then ψ satisfies (1.3). Equality (1.9) shows that (1.2) is satisfied by ψ. On the other hand, Corollary 2.4 in Chapter 2 implies that $\{\psi(\cdot - k) : k \in \mathbb{Z}\}$ cannot be an orthonormal system since the measure of the support of $\hat{\psi}$ does not exceed $|E| = \frac{3}{2}\pi < 2\pi$. Yet, by Theorem 1.6, ψ does enjoy the basic properties (A) and (B) of an orthonormal wavelet. This is a special case of the "φ **and** ψ **transform**" theory developed by Frazier and Jawerth at the same time modern wavelets were introduced.

Let us now turn to the proof of Theorem 1.6. In view of Theorem 1.7 it suffices to show the equivalence of (A) and (C). It turns out that we can weaken property (A):

LEMMA 1.10 *Suppose* $\{e_j : j = 1, 2, \cdots\}$ *is a family of elements in a Hilbert space* $\mathbb{H}$ *such that equality* (i) *in* Theorem 1.7 *holds for all* f *belonging to a dense subset* $\mathcal{D}$ *of* $\mathbb{H}$, *then this equality is valid for all* $f \in \mathbb{H}$.

PROOF: Suppose $f \in \mathbb{H}$ and $\{f_n\}$ is a sequence in $\mathcal{D}$ converging to f:

$$\lim_{n \to \infty} \|f - f_n\| = 0.$$

Then, for a fixed positive integer N, we have

$$\sum_{j=1}^{N} |\langle f, e_j \rangle|^2 = \lim_{n \to \infty} \sum_{j=1}^{N} |\langle f_n, e_j \rangle|^2 \leq \lim_{n \to \infty} \sum_{j=1}^{\infty} |\langle f_n, e_j \rangle|^2$$

$$= \lim_{n \to \infty} \|f_n\|^2 = \|f\|^2.$$

Since N is arbitrary we see that

$$\sum_{j=1}^{\infty} |\langle f, e_j \rangle|^2 \leq \|f\|^2. \tag{1.11}$$

We now establish the reverse inequality: choose $\varepsilon > 0$ and $g \in \mathcal{D}$ such that $\|f - g\| < \varepsilon$. Then, by Minkowski's inequality in ℓ^2, (1.11) and the fact that $\|f\| \le \|g\| + \varepsilon$ we have

$$\|f\| - 2\varepsilon \le \|g\| - \varepsilon \le \|g\| - \|g - f\| = \left(\sum_{j=1}^{\infty} |\langle g, e_j \rangle|^2\right)^{\frac{1}{2}} - \|g - f\|$$

$$\le \left(\sum_{j=1}^{\infty} |\langle g, e_j \rangle|^2\right)^{\frac{1}{2}} - \left(\sum_{j=1}^{\infty} |\langle g - f, e_j \rangle|^2\right)^{\frac{1}{2}} \le \left(\sum_{j=1}^{\infty} |\langle f, e_j \rangle|^2\right)^{\frac{1}{2}}.$$

Since ε is arbitrary we see that the reverse inequality to (1.11) must also be true. $\blacksquare$

In the case $\mathbb{H} = L^2(\mathbb{R})$ and the system we are studying is $\{\psi_{j,k} : j, k \in \mathbb{Z}\}$, where

$$\psi_{j,k}(x) = 2^{\frac{j}{2}} \psi(2^j x - k), \qquad j, k \in \mathbb{Z},$$

we let $\mathcal{D}$ be the class of all those functions $f \in L^2(\mathbb{R})$ such that $\hat{f} \in L^\infty(\mathbb{R})$ and $\hat{f}$ is compactly supported in $\mathbb{R} \setminus \{0\}$. We begin by examining the expression

$$I = \sum_{j,k \in \mathbb{Z}} |\langle f, \psi_{j,k} \rangle|^2, \qquad f \in \mathcal{D}.$$

Since

$$(\psi_{j,k})^\wedge(\xi) = 2^{-\frac{j}{2}} \hat{\psi}(2^{-j}\xi) e^{-i 2^{-j} k \xi},$$

an application of Plancherel's theorem gives us

$$I = \frac{1}{(2\pi)^2} \sum_{j \in \mathbb{Z}} \sum_{k \in \mathbb{Z}} 2^{-j} \left| \int_{\mathbb{R}} \hat{f}(2^{-j}\xi) \overline{\hat{\psi}(\xi)} e^{ik\xi} d\xi \right|^2. \tag{1.12}$$

Let $F_j(\xi) = \hat{f}(2^{-j}\xi) \overline{\hat{\psi}(\xi)}$ for $j \in \mathbb{Z}$. Each F_j is compactly supported in $\mathbb{R} \setminus \{0\}$ and belongs to $L^2(\mathbb{R})$. If F is such a function,

$$\sum_{k \in \mathbb{Z}} F(\xi + 2k\pi)$$

is a 2π-periodic function in $L^2(\mathbb{T})$ whose Fourier coefficients are $\frac{1}{2\pi} \widehat{F}(k)$, $k \in \mathbb{Z}$. Thus,

$$\sum_{k \in \mathbb{Z}} F(\xi + 2k\pi) = \frac{1}{2\pi} \sum_{k \in \mathbb{Z}} \widehat{F}(k) e^{ik\xi},$$

with convergence in $L^2(\mathbb{T})$. Hence,

$$\frac{1}{2\pi} \int_{\mathbb{R}} \overline{F(\xi)} \sum_{k\in\mathbb{Z}} \widehat{F}(k)\, e^{ik\xi}\, d\xi = \int_{\mathbb{R}} \overline{F(\xi)} \sum_{k\in\mathbb{Z}} F(\xi + 2k\pi)\, d\xi,$$

with each series converging in $L^2(\mathrm{supp}\,(F))$. But the left side of the last equality equals

$$\frac{1}{2\pi} \sum_{k\in\mathbb{Z}} \widehat{F}(k) \overline{\int_{\mathbb{R}} F(\xi)\, e^{-ik\xi}\, d\xi} = \frac{1}{2\pi} \sum_{k\in\mathbb{Z}} |\widehat{F}(k)|^2.$$

It follows that

$$\frac{1}{2\pi} \sum_{k\in\mathbb{Z}} |\widehat{F}(k)|^2 = \int_{\mathbb{R}} \overline{F(\xi)} \sum_{k\in\mathbb{Z}} F(\xi + 2k\pi)\, d\xi. \tag{1.13}$$

Applying (1.13) when $F = F_j$ in (1.12) we obtain

$$I = \frac{1}{2\pi} \sum_{j\in\mathbb{Z}} 2^{-j} \int_{\mathbb{R}} \overline{\hat{f}(2^{-j}\xi)\,\hat{\psi}(\xi)} \sum_{k\in\mathbb{Z}} \hat{f}(2^{-j}(\xi + 2k\pi))\, \overline{\hat{\psi}(\xi + 2k\pi)}\, d\xi.$$

In the sum with respect to k we separate the term $k = 0$ from the others and write $I = I_0 + I_1$, where

$$I_0 = \frac{1}{2\pi} \sum_{j\in\mathbb{Z}} 2^{-j} \int_{\mathbb{R}} |\hat{f}(2^{-j}\xi)|^2 |\hat{\psi}(\xi)|^2\, d\xi = \frac{1}{2\pi} \int_{\mathbb{R}} |\hat{f}(\xi)|^2 \sum_{j\in\mathbb{Z}} |\hat{\psi}(2^j\xi)|^2\, d\xi$$

and

$$I_1 = \frac{1}{2\pi} \sum_{j\in\mathbb{Z}} 2^{-j} \int_{\mathbb{R}} \overline{\hat{f}(2^{-j}\xi)\,\hat{\psi}(\xi)} \sum_{k\neq 0} \hat{f}(2^{-j}(\xi + 2k\pi))\, \overline{\hat{\psi}(\xi + 2k\pi)}\, d\xi.$$

Because of the positivity of the summands we can commute integration and summation in I_0. The following lemma shows that this can also be done in I_1:

LEMMA 1.14

$$\sum_{j\in\mathbb{Z}} 2^{-j} \int_{\mathbb{R}} |\hat{f}(2^{-j}\xi)\hat{\psi}(\xi)| \sum_{k\neq 0} |\hat{f}(2^{-j}(\xi + 2k\pi))\hat{\psi}(\xi + 2k\pi)|\, d\xi < \infty$$

whenever $f \in \mathcal{D}$.

REMARK: A consequence of this lemma is that I is finite for all $f \in \mathcal{D}$ if and only if I_0 is finite and the latter is true if and only if

$$\sum_{j \in \mathbb{Z}} |\hat{\psi}(2^j \xi)|^2$$

is locally integrable in $\mathbb{R} \setminus \{0\}$.

In order to prove Lemma 1.14 we observe that, since

$$2|\hat{\psi}(\xi)| \cdot |\hat{\psi}(\xi + 2k\pi)| \leq |\hat{\psi}(\xi)|^2 + |\hat{\psi}(\xi + 2k\pi)|^2,$$

it suffices to show

$$\int_{\mathbb{R}} \sum_{j \in \mathbb{Z}} \sum_{k \neq 0} 2^{-j} |\hat{f}(2^{-j}\xi)| \cdot |\hat{f}(2^{-j}(\xi + 2k\pi)| \cdot |\hat{\psi}(\xi)|^2 \, d\xi < \infty, \qquad (1.15)$$

since the term involving $|\hat{\psi}(\xi + 2k\pi)|^2$ equals the integral in (1.15) (after the change of variables $\eta = \xi + 2k\pi$ and replacing $-k$ by k in the sum over k). But, inequality (1.15) (and, thus, Lemma 1.14) is an immediate consequence of

LEMMA 1.16 *Suppose* $0 < a < b < \infty$, $f \in \mathcal{D}$ *with*

$$supp\,(\hat{f}) \subseteq \{\xi : a < |\xi| < b\}$$

and $\delta = \mathrm{diam}\,(supp\,(\hat{f}))$. *Then*

$$\sigma(\xi) = \sum_{j \in \mathbb{Z}} \sum_{k \neq 0} 2^{-j} |\hat{f}(2^{-j}\xi)| \cdot |\hat{f}(2^{-j}(\xi + 2k\pi))| \leq \frac{\delta}{\pi} \left(1 + \log_2 \tfrac{b}{a}\right) \|\hat{f}\|_{L^\infty}^2$$

for all $\xi \in \mathbb{R}$.

PROOF: If $2^{-j}2\pi > \delta$ then at most one of the points $2^{-j}\xi + 2^{-j}2k\pi$ and $2^{-j}\xi$ can lie in the support of $\hat{f}$ (since $k \neq 0$). Thus, in the sum over j we need only consider $j \geq j_o$, where j_o is the least integer such that $2^{-j_o}2\pi \leq \delta$. For each such j the summand $2^{-j}|\hat{f}(2^{-j}\xi)| \cdot |\hat{f}(2^{-j}(\xi + 2k\pi))|$ does not exceed $2^{-j}\|\hat{f}\|_\infty^2$. Also, for each such j the number of k for which $\hat{f}(2^{-j}(\xi + 2k\pi))$ differs from 0 cannot exceed $1 + (2\pi)^{-1}2^j\delta$. We see, therefore, that each such j contributes at most

$$\left(1 + \frac{2^j \delta}{2\pi}\right) 2^{-j} \|\hat{f}\|_{L^\infty}^2$$

to the size of $\sigma(\xi)$. Since $j \geq j_o$ this last expression is dominated by

$$\left(2^{-j} + \frac{\delta}{2\pi}\right)\|\hat{f}\|_{L^\infty}^2 \leq \left(2^{-j_o} + \frac{\delta}{2\pi}\right)\|\hat{f}\|_{L^\infty}^2 \leq \frac{\delta'}{\pi}\|\hat{f}\|_{L^\infty}^2.$$

Finally, we observe that the fact that $\hat{f}(2^{-j}\xi)$ differs from 0 only if $a \leq 2^{-j}|\xi| \leq b$ implies that j must lie in the interval

$$\left[\log_2 \frac{|\xi|}{b}, \log_2 \frac{|\xi|}{a}\right]$$

to produce a non-zero summand in the series defining $\sigma(\xi)$. But the number of integers j in this interval is, at most, $1 + \log_2(\frac{b}{a})$. This shows that

$$\sigma(\xi) \leq \left(1 + \log_2 \frac{b}{a}\right)\frac{\delta}{\pi}\|\hat{f}\|_{L^\infty}^2.$$

∎

Let

$$t_q(\xi) = \sum_{\ell=0}^{\infty} \hat{\psi}(2^\ell \xi)\,\overline{\hat{\psi}(2^\ell(\xi + 2q\pi))} \qquad \text{for } q \in \mathbb{Z}. \tag{1.17}$$

The function t_q belongs to $L^1(\mathbb{R})$ since

$$\int_{\mathbb{R}} |t_q(\xi)|\,d\xi \leq \left(\int_{\mathbb{R}} \sum_{\ell=0}^{\infty} |\hat{\psi}(2^\ell \xi)|^2\,d\xi\right)^{\frac{1}{2}} \left(\int_{\mathbb{R}} \sum_{\ell=0}^{\infty} |\hat{\psi}(2^\ell(\xi + 2q\pi))|^2\,d\xi\right)^{\frac{1}{2}}$$

$$= \left(\sum_{\ell=0}^{\infty} \int_{\mathbb{R}} 2^{-\ell}|\hat{\psi}(\xi)|^2\,d\xi\right)^{\frac{1}{2}} \left(\sum_{\ell=0}^{\infty} \int_{\mathbb{R}} 2^{-\ell}|\hat{\psi}(\xi)|^2\,d\xi\right)^{\frac{1}{2}} = 2\|\hat{\psi}\|_2^2.$$

It will be useful to express I_1 as a sum involving these functions t_q with q ranging throughout the odd integers $2\mathbb{Z} + 1$.

LEMMA 1.18 *If* $f \in \mathcal{D}$, *then*

$$I_1 = \frac{1}{2\pi}\int_{\mathbb{R}}\sum_{p\in\mathbb{Z}}\sum_{q\in 2\mathbb{Z}+1} \overline{\hat{f}(\xi)}\,\hat{f}(\xi + 2^p 2q\pi)\,t_q(2^{-p}\xi)\,d\xi.$$

PROOF : In the expression for I_1, we presented immediately preceding Lemma 1.14 the parameter k is a non-zero integer. For each such k there exists a unique non-negative integer ℓ and a unique odd integer q such that $k = 2^\ell q$. Since, by Lemma 1.14, the sum defining I_1, is absolutely convergent, the following equalities, that lead to an expression involving t_q, are valid:

$$2\pi I_1 = \sum_{j \in \mathbb{Z}} \int_{\mathbb{R}} \overline{\hat{f}(\xi)}\, \hat{\psi}(2^j \xi) \sum_{k \neq 0} \hat{f}(\xi + 2^{-j}2k\pi)\, \overline{\hat{\psi}(2^j \xi + 2k\pi)}\, d\xi$$

$$= \sum_{j \in \mathbb{Z}} \int_{\mathbb{R}} \overline{\hat{f}(\xi)}\hat{\psi}(2^j \xi) \sum_{\ell=0}^{\infty} \sum_{q \in 2\mathbb{Z}+1} \hat{f}(\xi + 2^{\ell-j}2q\pi)\, \overline{\hat{\psi}(2^\ell(2^{j-\ell}\xi + 2q\pi))}\, d\xi$$

$$= \int_{\mathbb{R}} \overline{\hat{f}(\xi)} \sum_{q \in 2\mathbb{Z}+1} \sum_{\ell=0}^{\infty} \sum_{p \in \mathbb{Z}} \hat{f}(\xi + 2^p 2q\pi)\hat{\psi}(2^\ell 2^{-p}\xi)\, \overline{\hat{\psi}(2^\ell(2^{-p}\xi + 2q\pi))}\, d\xi$$

$$= \int_{\mathbb{R}} \overline{\hat{f}(\xi)} \sum_{q \in 2\mathbb{Z}+1} \sum_{p \in \mathbb{Z}} \hat{f}(\xi + 2^p 2q\pi)\, t_q(2^{-p}\xi)\, d\xi.$$

∎

Let us collect the results we have obtained about the sum of squares I:

PROPOSITION 1.19 *Suppose $\psi \in L^2(\mathbb{R})$ and $f \in \mathcal{D}$, then*

$$\sum_{j,k \in \mathbb{Z}} |\langle f, \psi_{j,k} \rangle|^2 = \frac{1}{2\pi} \int_{\mathbb{R}} |\hat{f}(\xi)|^2 \sum_{j \in \mathbb{Z}} |\hat{\psi}(2^j \xi)|^2\, d\xi$$

$$+ \frac{1}{2\pi} \int_{\mathbb{R}} \overline{\hat{f}(\xi)} \sum_{p \in \mathbb{Z}} \sum_{q \in 2\mathbb{Z}+1} \hat{f}(\xi + 2^p 2q\pi)\, t_q(2^{-p}\xi)\, d\xi. \quad (1.20)$$

The last integrand is integrable and so is the first when $\sum_{j \in \mathbb{Z}} |\hat{\psi}(2^j \xi)|^2$ is locally integrable in $\mathbb{R} \setminus \{0\}$.

Since equation (1.3) asserts that $t_q(\xi) = 0$ for all $q \in 2\mathbb{Z}+1$, Proposition 1.19 immediately implies that condition (C) guarantees condition (A) (in Theorem 1.6) for all $f \in \mathcal{D}$. Lemma 1.10, in turn, shows that (A) is valid for all $f \in L^2(\mathbb{R})$. Theorem 1.7 tells us that this is equivalent to (B). In order to complete the proof of Theorem 1.6, therefore, all we need to do is to show that if (A) is satisfied for all $f \in \mathcal{D}$ then the two equations (1.2) and (1.3) are satisfied.

If condition (A) is satisfied for all $f \in \mathcal{D}$, then, by the remark that follows Lemma 1.14,

$$\sum_{j \in \mathbb{Z}} |\hat{\psi}(2^j \xi)|^2$$

is locally integrable in $\mathbb{R} \setminus \{0\}$. Let $\xi_o \neq 0$ be a Lebesgue point for this function; then,

$$\lim_{\delta \to 0+} \frac{1}{2\delta} \int_{\xi_o - \delta}^{\xi_o + \delta} \sum_{j \in \mathbb{Z}} |\hat{\psi}(2^j \xi)|^2 \, d\xi = \sum_{j \in \mathbb{Z}} |\hat{\psi}(2^j \xi_o)|^2,$$

where $[\xi_o - \delta, \xi_o + \delta] \subset \mathbb{R} \setminus \{0\}$. Let I^δ, I_0^δ and I_1^δ be the quantities I, I_0 and I_1 when $f = f_\delta$, where

$$\hat{f}_\delta(\xi) = \frac{1}{\sqrt{2\delta}} \, \chi_{[\xi_o - \delta, \xi_o + \delta]}(\xi),$$

and use Proposition 1.19 together with (A) to obtain

$$I^\delta = \|f_\delta\|_2^2 = \frac{1}{2\pi} \|\hat{f}_\delta\|_2^2 = \frac{1}{2\pi} \int_{\xi_o - \delta}^{\xi_o + \delta} \frac{1}{2\delta} \sum_{j \in \mathbb{Z}} |\hat{\psi}(2^j \xi)|^2 \, d\xi + I_1^\delta.$$

Since $\|\hat{f}_\delta\|_2 = 1$, we must have

$$1 = \sum_{j \in \mathbb{Z}} |\hat{\psi}(2^j \xi_o)|^2 + 2\pi \lim_{\delta \to 0+} I_1^\delta. \tag{1.21}$$

If we can show that the last limit is 0 we then have

$$\sum_{j \in \mathbb{Z}} |\hat{\psi}(2^j \xi_o)|^2 = 1$$

at each such Lebesgue point ξ_o and equality (1.2) must hold. In order to estimate I_1^δ we argue as we did in the proof of Lemma 1.14 (see, in particular, (1.15)), to observe that

$$|I_1^\delta| \leq \int_{\mathbb{R}} \sum_{j \in \mathbb{Z}} \sum_{k \neq 0} 2^{-j} |\hat{f}_\delta(2^{-j} \xi)| \cdot |\hat{f}_\delta(2^{-j}(\xi + 2k\pi))| \cdot |\hat{\psi}(\xi)|^2 \, d\xi. \tag{1.22}$$

Let us apply Lemma 1.16 to this choice of f. The diameter of the support of $\hat{f} = \hat{f}_\delta$ is 2δ. Hence, since $k \neq 0$, $|\hat{f}_\delta(2^{-j} \xi)| \cdot |\hat{f}(2^{-j}(\xi + 2k\pi))| = 0$ if

$2^{-j}2\pi > 2\delta$. Thus, if j_o is the least integer such that $2^{j_o}\delta \geq \pi$ we need only consider $j \geq j_o$ in the sum in (1.22). Also, if $\widehat{f_\delta}(2^{-j}\xi) \neq 0$ we must have $\xi_o - \delta < 2^{-j}\xi$ and, without loss of generality, we can assume $0 < \xi_o - \delta$. Thus, the integral in (1.22) can be assumed to be over the region

$$\left\{\xi : 2^{j_o}(\xi_o - \delta) < \xi\right\} \subseteq \left\{\xi : \frac{\pi}{\delta}(\xi_o - \delta) < \xi\right\}.$$

Therefore, using the notation of Lemma 1.16,

$$|I_1^\delta| \leq \int_{\frac{\pi}{\delta}(\xi_o - \delta)}^{\infty} \sigma_\delta(\xi)|\hat{\psi}(\xi)|^2 \, d\xi$$

$$\leq \frac{\delta}{\pi}\left(1 + \log_2 \frac{\xi_o + \delta}{\xi_o - \delta}\right)\frac{1}{2\delta}\int_{\frac{\pi}{\delta}(\xi_o - \delta)}^{\infty} |\hat{\psi}(\xi)|^2 \, d\xi,$$

since $\left\|\widehat{f_\delta}\right\|_{L^\infty}^2 = \frac{1}{2\delta}$. It is now clear that

$$\lim_{\delta \to 0+} I_1^\delta = 0.$$

Finally, we must show that if (A) holds for all $f \in \mathcal{D}$, then equation (1.3) is true. But it follows from equality (1.20) and the just established fact that (1.2) holds, that (A) implies

$$\int_{\mathbb{R}} \sum_{p \in \mathbb{Z}} \sum_{q \in 2\mathbb{Z}+1} \overline{\hat{f}(\xi)}\, \hat{f}(\xi + 2^p 2q\pi)\, t_p(2^{-p}\xi)\, d\xi = 0$$

for all $f \in \mathcal{D}$. By polarization we then have

$$\int_{\mathbb{R}} \sum_{p \in \mathbb{Z}} \sum_{q \in 2\mathbb{Z}+1} \overline{\hat{f}(\xi)}\, \hat{g}(\xi + 2^p 2q\pi)\, t_q(2^{-p}\xi)\, d\xi = 0 \qquad (1.23)$$

for all f and $g \in \mathcal{D}$.

Let us fix an odd integer q_o and let ξ_o be a Lebesgue point of t_{q_o} such that ξ_o and $\xi_o + 2q_o\pi$ are not 0. We can also assume δ to be sufficiently small and positive so that neither $[\xi_o - \delta, \xi_o + \delta]$ nor $[\xi_o + 2q_o\pi - \delta, \xi_o + 2q_o\pi + \delta]$ contain 0. In fact, without loss of generality, we can assume $\xi_o > 0$ and $0 < \delta < \frac{1}{3}\xi_o$. Let $f = f_\delta$ be the function satisfying

$$\widehat{f_\delta} = \frac{1}{\sqrt{2\delta}}\, \chi_{[\xi_o - \delta, \xi_o + \delta]}$$

and $g = g_\delta$ such that $\widehat{g_\delta}(\xi) = \widehat{f_\delta}(\xi - 2q_o\pi)$. Then

$$\widehat{f_\delta}(\xi)\,\widehat{g_\delta}(\xi + 2q_o\pi) = \frac{1}{2\delta}\,\chi_{[\xi_o-\delta,\xi_o+\delta]}(\xi).$$

Equality (1.23) then can be written in the form

$$0 = \frac{1}{2\delta}\int_{\xi_o-\delta}^{\xi_o+\delta} t_{q_o}(\xi)\,d\xi + \sum_{\substack{p\in\mathbb{Z},\,q\in 2\mathbb{Z}+1 \\ (p,q)\neq(0,q_o)}}\int_\mathbb{R}\overline{\widehat{f_\delta}(\xi)}\,\widehat{g_\delta}(\xi + 2^p 2q\pi)\,t_q(2^{-p}\xi)\,d\xi$$

$$\equiv \frac{1}{2\delta}\int_{\xi_o-\delta}^{\xi_o+\delta} t_{q_o}(\xi)\,d\xi + J_\delta\,.$$

Since the first summand tends to $t_{q_o}(\xi_o)$ as $\delta \to 0+$, it suffices to prove that

$$\lim_{\delta\to 0+} J_\delta = 0.$$

If $\overline{\widehat{f_\delta}(\xi)}\,\widehat{g_\delta}(\xi + 2^p 2q\pi) \neq 0$ we must have

$$|\xi - \xi_o| \leq \delta \qquad\text{and}\qquad |\xi - 2q_o\pi + 2^p 2q\pi - \xi_o| \leq \delta.$$

Thus,

$$|q_o - 2^p q| \leq \frac{1}{2\pi}\left\{|\xi - 2q_o\pi + 2^p 2q\pi - \xi_o| + |\xi - \xi_o|\right\} \leq \frac{\delta}{\pi}. \qquad (1.24)$$

Since we are interested in the behavior of these quantities as $\delta \to 0+$, we can assume that $\delta < \pi$. Now if $p > 0$, since q_o is odd, $|q_o - 2^p q|$ must be odd and, therefore, cannot be less than 1. If $p = 0$, in which case $q \neq q_o$, we cannot have $|q_o - 2^p q| = |q_o - q| < 1$. Finally, if $p < 0$, then $|q_o - 2^p q| = 2^p|2^{-p}q_o - q| \geq 2^p$ since q is odd; inequality (1.24), therefore, implies $2^p \leq \delta/\pi$. This shows that, if $\delta < \pi$ and j_o is the largest integer satisfying $2^{j_o} \leq \delta/\pi$,

$$J_\delta = \sum_{p\leq j_o}\sum_{q\in 2\mathbb{Z}+1}\int_\mathbb{R}\overline{\widehat{f_\delta}(\xi)}\,\widehat{g_\delta}(\xi + 2^p 2q\pi)\,t_q(2^{-p}\xi)\,d\xi$$

$$= \sum_{p\leq j_o}\sum_{q\in 2\mathbb{Z}+1} 2^p\int_\mathbb{R}\overline{\widehat{f_\delta}(2^p\xi)}\,\widehat{g_\delta}(2^p(\xi + 2q\pi))\,t_q(\xi)\,d\xi.$$

Since

$$2|t_q(\xi)| \leq \sum_{\ell=0}^\infty |\hat\psi(2^\ell\xi)|^2 + \sum_{\ell=0}^\infty |\hat\psi(2^\ell(\xi + 2q\pi))|^2,$$

we have

$$|J_\delta| \leq J_\delta^{(1)} + J_\delta^{(2)},$$

where

$$J_\delta^{(1)} = \sum_{p \leq j_o} \sum_{q \in 2\mathbb{Z}+1} 2^p \int_{\mathbb{R}} |\widehat{f_\delta}(2^p \xi)| \cdot |\widehat{g_\delta}(2^p(\xi + 2q\pi))| \, [\tau(\xi)]^2 \, d\xi$$

with

$$\int_{\mathbb{R}} [\tau(\xi)]^2 \, d\xi = \sum_{\ell=0}^{\infty} \int_{\mathbb{R}} |\hat{\psi}(2^\ell \xi)|^2 \, d\xi = 2\|\hat{\psi}\|_2^2 < \infty,$$

and

$$J_\delta^{(2)} = \sum_{p \leq j_o} \sum_{q \in 2\mathbb{Z}+1} 2^p \int_{\mathbb{R}} |\widehat{f_\delta}(2^p \xi)| \cdot |\widehat{g_\delta}(2^p(\xi + 2q\pi))| \, [\tau(\xi + 2q\pi)]^2 \, d\xi$$

$$= \sum_{p \leq j_o} \sum_{q \in 2\mathbb{Z}+1} 2^p \int_{\mathbb{R}} |\widehat{f_\delta}(2^p(\eta - 2q\pi))| \cdot |\widehat{g_\delta}(2^p \eta)| \, [\tau(\eta)]^2 \, d\eta.$$

Thus, $J_\delta^{(2)}$ has the same form as $J_\delta^{(1)}$ with the rôles of $\widehat{f_\delta}$ and $\widehat{g_\delta}$ interchanged. Hence, since $\widehat{f_\delta} = \frac{1}{\sqrt{2\delta}} \chi_{[\xi_o - \delta, \xi_o + \delta]}$, we deduce that

$$J_\delta^{(1)} = \sum_{p \leq j_o} \sum_{q \in 2\mathbb{Z}+1} \frac{2^p}{\sqrt{2\delta}} \int_{2^{-p}(\xi_o - \delta)}^{2^{-p}(\xi_o + \delta)} |\hat{g}(2^p(\xi + 2q\pi))| \, [\tau(\xi)]^2 \, d\xi. \qquad (1.25)$$

Fix p. For $\hat{g}(2^p(\xi + 2q\pi))$ not to be 0 we must have

$$2^p(\xi + 2q\pi) \in \left[\xi_o + 2q_o\pi - \delta, \xi_o + 2q_o\pi + \delta\right].$$

Since $2^p \xi \in [\xi_o - \delta, \xi_o + \delta]$ a simple calculation shows that

$$-2^{-p}\frac{\delta}{\pi} \leq q - 2^{-p}q_o \leq 2^{-p}\frac{\delta}{\pi} \qquad (1.26)$$

if $\hat{g}(2^p(\xi + 2q\pi)) \neq 0$. But q is odd and $2^{-p}q_o$ is even. Thus, $q - 2^{-p}q_o$ is odd. We claim that there are at most $2^{-p}2\delta/\pi$ odd integers in the interval $[-2^{-p}\delta/\pi, 2^{-p}\delta/\pi]$ and this gives us an upper bound for the number of integers of the form $q - 2^{-p}q_o$ that satisfy (1.26).

Thus, by (1.25),

$$J_\delta^{(1)} = \sum_{p \leq j_o} \frac{2^p}{\sqrt{2\delta}} \int_{2^{-p}(\xi_o-\delta)}^{2^{-p}(\xi_o+\delta)} [\tau(\xi)]^2 \sum_{q \in 2\mathbb{Z}+1} |\hat{g}(2^p(\xi + 2q\pi))| \, d\xi$$

$$\leq \sum_{p \leq j_o} \frac{2^p}{\sqrt{2\delta}} \int_{2^{-p}(\xi_o-\delta)}^{2^{-p}(\xi_o+\delta)} 2^{-p} \frac{2\delta}{\pi} \frac{1}{\sqrt{2\delta}} [\tau(\xi)]^2 \, d\xi$$

$$= \frac{1}{\pi} \sum_{p \leq j_o} \int_{2^{-p}(\xi_o-\delta)}^{2^{-p}(\xi_o+\delta)} [\tau(\xi)]^2 \, d\xi.$$

Recall that we are assuming $0 < \delta < \frac{1}{3}\xi_o$. This implies that the intervals

$$[2^{-p}(\xi_o - \delta), 2^{-p}(\xi_o + \delta)], \qquad p = j_o, j_o - 1, j_o - 2, \cdots,$$

are disjoint. Therefore,

$$J_\delta^{(1)} \leq \frac{1}{\pi} \sum_{p \leq j_o} \int_{2^{-p}(\xi_o-\delta)}^{2^{-p}(\xi_o+\delta)} [\tau(\xi)]^2 \, d\xi$$

$$\leq \frac{1}{\pi} \int_{2^{-j_o}(\xi_o-\delta)}^{\infty} [\tau(\xi)]^2 \, d\xi \leq \frac{1}{\pi} \int_{\frac{\pi}{3}(\xi_o-\delta)}^{\infty} [\tau(\xi)]^2 \, d\xi.$$

The last expression tends to 0 as $\delta \to 0+$ since $\tau^2 \in L^1(\mathbb{R})$. In the case $\xi_o < 0$ this argument can be modified in an obvious way to obtain this same result:

$$\lim_{\delta \to 0+} J_\delta^{(1)} = 0.$$

The case involving $J_\delta^{(2)}$ is similar, with the roles of $\hat{f}_\delta$ and $\hat{g}_\delta$ interchanged and the point $\xi_o + 2q_o\pi$ replacing ξ_o in the above argument. This finishes the proof of Theorem 1.6.

∎

7.2   Some applications of the basic equations

In Proposition 2.6 of Chapter 3 it has been proved that any band-limited wavelet ψ for which $|\hat{\psi}|$ is continuous at zero must satisfy $\hat{\psi}(0) = 0$. We can now prove this result for any wavelet, not necessarily band-limited.

PROPOSITION 2.1 *If ψ is an orthonormal wavelet and $|\hat{\psi}|$ is continuous at zero, then $\hat{\psi}(0) = 0$.*

PROOF : By Theorem 1.1, the wavelet ψ satisfies

$$\sum_{j \in \mathbb{Z}} |\hat{\psi}(2^j \xi)|^2 = 1 \quad \text{for} \quad a.e. \ \xi \in \mathbb{R}.$$

Dividing the above equality by ξ and integrating over the interval $[1, 2]$ we obtain

$$\int_0^\infty \frac{|\hat{\psi}(\xi)|^2}{\xi} \, d\xi = \log 2.$$

If we suppose that $\hat{\psi}(0) = C \neq 0$, a real number $\varepsilon > 0$ can be chosen such that $|\hat{\psi}(\xi)| \geq \frac{1}{2}|C|$ for every $\xi \in (0, \varepsilon)$. Thus,

$$\int_0^\infty \frac{|\hat{\psi}(\xi)|^2}{\xi} \, d\xi \geq \int_0^\varepsilon \frac{|C|^2}{4\xi} \, d\xi = \infty,$$

contradicting the above equality.   ∎

Observe that Proposition 2.6 in Chapter 3 is valid not only when ψ is an orthonormal wavelet, but also when $\{\psi_{j,k} : j, k \in \mathbb{Z}\}$ is an orthonormal system in $L^2(\mathbb{R})$.

We have seen, when we studied the Lemarie-Meyer wavelets in Chapter 3, that if ψ is such a wavelet and we define $\tilde{\psi}$ by

$$\widehat{\tilde{\psi}}(\xi) = e^{i\alpha(\xi)} \hat{\psi}(\xi)$$

with α being any real-valued, 2π-periodic, measurable function on $\mathbb{R}$, then $\tilde{\psi}$ is also a wavelet. This is an immediate consequence of the phase equation in property (vi) of Theorem 3.1 in Chapter 3. We can now see that this fact is completely general. It follows immediately from Theorem 1.1 since (1.2) and (1.3) are also satisfied when $\hat{\psi}$ is replaced by $\widehat{\tilde{\psi}}$. We have obtained:

THEOREM 2.2 *Suppose that ψ is an orthonormal wavelet, and α is a 2π-periodic, real-valued, measurable function on $\mathbb{R}$. If $\tilde{\psi}$ is defined by*

$$\widehat{\tilde{\psi}}(\xi) = e^{i\alpha(\xi)}\hat{\psi}(\xi) \quad for \ a.e. \ \xi \in \mathbb{R},$$

then $\tilde{\psi}$ is also a wavelet.

Equations (1.2) and (1.3) are particularly useful in determining whether a function $\psi \in L^2(\mathbb{R})$ is a wavelet, if we have a "nice" expression for $\hat{\psi}$. In particular, this is the case if $|\hat{\psi}|$ is the characteristic function of a measurable set $K \subset \mathbb{R}$. We have already seen (see Corollary 2.4 of Chapter 2) that for these wavelets $|K| = 2\pi$. The minimality of the measure of supp $(\hat{\psi})$ in these cases motivated us to call such wavelets **Minimally Supported Frequency** (MSF) wavelets. For MSF wavelets we have the following characterization

THEOREM 2.3 *Let $\psi \in L^2(\mathbb{R})$ be such that $|\hat{\psi}| = \chi_K$ for some measurable $K \subset \mathbb{R}$. Then, ψ is an MSF wavelet if and only if (1.2) and (1.4) are satisfied.*

PROOF: We know that for any wavelet (1.2) and (1.4) are satisfied; in particular they are satisfied by any MSF wavelet. Thus, we only need to show that (1.2) and (1.4) imply that ψ is a wavelet. But it is immediate that (1.4) implies (1.3) if $|\hat{\psi}| = \chi_K$. In fact, if $2^j\xi \in$ supp $(\hat{\psi})$, then $|\hat{\psi}(2^j\xi)| = 1$. By (1.4) it then follows that

$$|\hat{\psi}(2^j\xi + 2k\pi)| = 0 \qquad \text{for all } k \in \mathbb{Z}, \ k \neq 0.$$

In particular, $\hat{\psi}(2^j(\xi + 2m\pi)) = 0$ for every $m \in 2\mathbb{Z}+1$. Hence, (1.3) is satisfied. By Theorem 1.1, this, together with (1.2), tells us that ψ is a wavelet.

∎

REMARK: We already know that (1.5) follows from (1.2) and (1.3). When $|\hat{\psi}| = \chi_K$ it is particularly easy to see that (1.2) implies (1.5). We just repeat the last argument with (1.2) playing the role of (1.4).

An easy consequence of Theorem (2.3) is that if ψ is an MSF wavelet and β is any real-valued, measurable function defined on $\mathbb{R}$ and $\tilde{\psi}$ is defined by $\hat{\tilde{\psi}}(\xi) = e^{i\beta(\xi)}\hat{\psi}(\xi)$, then $\tilde{\psi}$ is also an MSF wavelet. Observe that for these wavelets we do not need β to be 2π-periodic.

A set theoretic version of Theorem (2.3) is:

COROLLARY 2.4 *Let K be a measurable set in $\mathbb{R}$ and let ψ be a function such that $|\hat{\psi}| = \chi_K$. Then, ψ is an orthonormal wavelet if and only if*

(a) $\{K + 2k\pi : k \in \mathbb{Z}\}$ *is a partition of* $\mathbb{R}$ *(a.e.), and*

(b) $\{2^j K : j \in \mathbb{Z}\}$ *is a partition of* $\mathbb{R}$ *(a.e.).*

EXAMPLE A: Corollary (2.4) allows us to conclude immediately when certain functions are MSF wavelets. One of them is the Shannon wavelet, ψ_s, for which $|\hat{\psi_s}| = \chi_{K^s}$ with

$$K^s = [-2\pi, -\pi] \cup [\pi, 2\pi]$$

(see Example C of Chapter 2). It can now be easily proved (this was mentioned without proof after the proof of Proposition 2.17 in Chapter 2) that if ψ_J is given by $\hat{\psi_J} = \chi_{K^J}$, where

$$K^J = [-\tfrac{32}{7}\pi, -4\pi] \cup [-\pi, -\tfrac{4}{7}\pi] \cup [\tfrac{4}{7}\pi, \pi] \cup [4\pi, \tfrac{32}{7}\pi],$$

then ψ_J is an orthonormal wavelet, called the **Journé wavelet** (see [Mal1]). Another example, due to Lemarié ([Le1]), is $\psi_{\mathcal{L}}$ given by $\hat{\psi_{\mathcal{L}}} = \chi_{K^{\mathcal{L}}}$, where

$$K^{\mathcal{L}} = [-\tfrac{8}{7}\pi, -\tfrac{4}{7}\pi] \cup [\tfrac{4}{7}\pi, \tfrac{6}{7}\pi] \cup [\tfrac{24}{7}\pi, \tfrac{32}{7}\pi].$$

Corollary (2.4) gives, immediately, that $\psi_{\mathcal{L}}$ is an orthonormal wavelet.

EXAMPLE B: For $\ell \in \mathbb{Z}, \ell \geq 0$, let $K_\ell = K_\ell^+ \cup K_\ell^-$, where $K_\ell^- = -K_\ell^+$ and

$$K_\ell^+ = [\frac{2^\ell}{2^{\ell+1}-1}\pi, \pi] \cup [2^\ell\pi, 2^\ell\pi + \frac{2^\ell}{2^{\ell+1}-1}\pi] \equiv K_\ell^{(1)} \cup K_\ell^{(2)}.$$

We claim that ψ_ℓ is an MSF wavelet if $|\widehat{\psi_\ell}| = \chi_{K_\ell}$. To see this observe that

$$\{2^j K_\ell^{(2)} : j \in \mathbb{Z}\} = \{2^j[\pi, (1 + \theta)\pi] : j \in \mathbb{Z}\},$$

where $\theta = \dfrac{1}{2^{\ell+1} - 1}$, since $[\pi, (1 + \theta)\pi] = 2^{-\ell} K_\ell^{(2)}$; moreover,

$$\{2^j K_\ell^{(1)} : j \in \mathbb{Z}\} = \{2^j[(1 + \theta)\pi, 2\pi] : j \in \mathbb{Z}\}$$

since $2K_\ell^{(1)} = [(1+\theta)\pi, 2\pi]$. Since $[\pi, 2\pi] = [\pi, (1+\theta)\pi] \cup [(1+\theta)\pi, 2\pi]$ and $\{2^j[\pi, 2\pi] : j \in \mathbb{Z}\}$ is a partition of $(0, \infty)$ we see, therefore, that K_ℓ satisfies condition (b) of Corollary (2.4). To see that condition (a) of Corollary (2.4) is satisfied by K_ℓ it suffices to observe that $[-\pi, \pi]$ is the almost everywhere disjoint union of the intervals

$$K_\ell^{(2)} - 2^\ell\pi, K_\ell^{(1)}, -K_\ell^{(1)}, \text{ and } - K_\ell^{(2)} + 2^\ell\pi.$$

Notice that when $\ell = 0$ we obtain the Shannon wavelet; when $\ell = 2$ we obtain the Journé wavelet (see Example A).

EXAMPLE C: The functions ψ_a such that

$$\widehat{\psi_a} = \chi_{[-2a,-a]\cup[2\pi-a,4\pi-2a]}, \quad 0 < a < 2\pi,$$

also form a family of MSF wavelets. This follows easily from Corollary (2.4). Letting $a \to 0$ we obtain (as an L^2-limit) the function ψ_H such that

$$\widehat{\psi_H} = \chi_{[2\pi,4\pi]},$$

which **cannot** be a wavelet for $L^2(\mathbb{R})$ since $\widehat{\psi_H}(\xi) = 0$ for $\xi < 0$. But ψ_H is a wavelet for the classical Hardy space $H^2(\mathbb{R})$ defined as the set of all $f \in L^2(\mathbb{R})$ such that $\hat{f}(\xi) = 0$ for all $\xi < 0$. This example shows that wavelets (as elements of the unit sphere in $L^2(\mathbb{R})$) can converge (in $L^2(\mathbb{R})$) to a function that is not a wavelet in $L^2(\mathbb{R})$.

If $K \subset \mathbb{R}$ is the support of the Fourier transform of an MSF wavelet we say that K is an **MSF set**. MSF sets can be constructed from the "Shannon set," $I = [-2\pi, -\pi] \cup [\pi, 2\pi]$, as shown by the following result:

THEOREM 2.5 *Let $\psi \in L^2(\mathbb{R})$ be such that $|\hat{\psi}| = \chi_K$ for a measurable set $K \subset \mathbb{R}$. Then, ψ is a wavelet if and only if there exist a partition $\{I_\ell : \ell \in \mathbb{Z}\}$ of I, a partition $\{K_\ell : \ell \in \mathbb{Z}\}$ of K, and two integer-valued sequences $\{j_\ell : \ell \in \mathbb{Z}\}, \{k_\ell : \ell \in \mathbb{Z}\}$ such that*

$\quad$ (1) $\quad K_\ell = 2^{j_\ell} I_\ell, \ell \in \mathbb{Z},$ *and*

$\quad$ (2) $\quad \{K_\ell + 2k_\ell \pi : \ell \in \mathbb{Z}\}$ *is a partition of* I.

PROOF : To prove the sufficiency we need to show that (1.2) and (1.4) are satisfied. The second one of these conditions follows from (2) and the fact that the Shannon wavelet satisfies (1.4):

$$\sum_{k \in \mathbb{Z}} |\hat{\psi}(\xi + 2k\pi)|^2 = \sum_{n \in \mathbb{Z}} \chi_{K + 2n\pi}(\xi) = \sum_{n \in \mathbb{Z}} \sum_{\ell \in \mathbb{Z}} \chi_{K_\ell + 2n\pi}(\xi)$$

$$= \sum_{\ell \in \mathbb{Z}} \sum_{n \in \mathbb{Z}} \chi_{K_\ell + 2k_\ell \pi + 2(n - k_\ell)\pi}(\xi) = \sum_{\ell \in \mathbb{Z}} \sum_{m \in \mathbb{Z}} \chi_{K_\ell + 2k_\ell \pi + 2m\pi}(\xi)$$

$$= \sum_{m \in \mathbb{Z}} \sum_{\ell \in \mathbb{Z}} \chi_{K_\ell + 2k_\ell \pi + 2m\pi}(\xi) = \sum_{m \in \mathbb{Z}} \chi_{I + 2m\pi}(\xi) = 1 \qquad \text{for a.e. } \xi \in \mathbb{T},$$

where, in the second equality we used the fact that $\{K_\ell : \ell \in \mathbb{Z}\}$ forms a partition of K. Using (1) we deduce that (1.2) is true:

$$\sum_{j \in \mathbb{Z}} |\hat{\psi}(2^j \xi)|^2 = \sum_{j \in \mathbb{Z}} \chi_{2^j K}(\xi) = \sum_{j \in \mathbb{Z}} \sum_{\ell \in \mathbb{Z}} \chi_{2^j K_\ell}(\xi) = \sum_{\ell \in \mathbb{Z}} \sum_{j \in \mathbb{Z}} \chi_{2^j 2^{j_\ell} I_\ell}(\xi)$$

$$= \sum_{\ell \in \mathbb{Z}} \sum_{m \in \mathbb{Z}} \chi_{2^m I_\ell}(\xi) = \sum_{m \in \mathbb{Z}} \sum_{\ell \in \mathbb{Z}} \chi_{2^m I_\ell}(\xi) = \sum_{m \in \mathbb{Z}} \chi_{2^m I}(\xi) = 1, \quad \text{a.e. } \xi \in \mathbb{R}.$$

Suppose now that ψ is a wavelet. Equations (1.2) and (1.4) show that both $\{K + 2k\pi : k \in \mathbb{Z}\}$ and $\{2^j K : j \in \mathbb{Z}\}$ are partitions of $\mathbb{R}$. Let

$$E_n = K \cap (2^n I), \quad J_n = 2^{-n} E_n = (2^{-n} K) \cap I, \quad n \in \mathbb{Z}.$$

Since $\{2^n I : n \in \mathbb{Z}\}$ and $\{2^{-n} K : n \in \mathbb{Z}\}$ are partitions of $\mathbb{R}$, $\{E_n : n \in \mathbb{Z}\}$ and $\{J_n : n \in \mathbb{Z}\}$ are partitions of K and I, respectively. For each $n \in \mathbb{Z}$ we define

$$K_{n,j} = E_n \cap (I + 2j\pi), \quad I_{n,j} = J_n \cap (2^{-n}(I + 2j\pi)), \quad j \in \mathbb{Z}.$$

Since $\{I + 2j\pi : j \in \mathbb{Z}\}$ is a partition of $\mathbb{R}$, $\{K_{n,j} : j \in \mathbb{Z}\}$ is a partition of E_n and $\{I_{n,j} : j \in \mathbb{Z}\}$ is a partition of J_n. It follows that $\{K_{n,j} : n, j \in \mathbb{Z}\}$ is a partition of K and $\{I_{n,j} : n, j \in \mathbb{Z}\}$ is a partition of I.

Since $E_n = 2^n J_n$, we have $K_{n,j} = 2^n I_{n,j}$, which is condition (1) with (n, j) corresponding to ℓ and n corresponding to j_ℓ. Finally, we observe that $K_{n,j} - 2j\pi = (E_n - 2j\pi) \cap I$. This shows that, for each fixed $j \in \mathbb{Z}$, $\{K_{n,j} - 2j\pi : n \in \mathbb{Z}\}$ is a partition of $(K - 2j\pi) \cap I$. Therefore,

$$\{K_{n,j} - 2j\pi : n, j \in \mathbb{Z}\}$$

is a partition of I, which gives us (2) with (n, j) corresponding to ℓ and j corresponding to $-k_\ell$.

∎

The following result follows immediately from the above theorem. We will show that it is particularly useful for constructing MSF wavelets.

COROLLARY 2.6 *Suppose that $\{I_\ell : \ell \in \mathbb{Z}\}$ is a partition of*

$$I = [-2\pi, -\pi] \cap [\pi, 2\pi],$$

and there exist two integer-valued sequences $\{j_\ell : \ell \in \mathbb{Z}\}$ and $\{k_\ell : \ell \in \mathbb{Z}\}$ such that $\{2^{j_\ell} I_\ell + 2k_\ell \pi : \ell \in \mathbb{Z}\}$ forms a partition of I. Then all ψ such that $|\hat{\psi}| = \chi_K$, with

$$K = \bigcup_{\ell \in \mathbb{Z}} 2^{j_\ell} I_\ell,$$

are MSF wavelets.

EXAMPLE D : To illustrate how this corollary can be applied we consider the following partition of $I = [-2\pi, -\pi] \cup [\pi, 2\pi]$:

$$I = I_1 \cup I_2 \cup I_3 \cup I_4$$

with

$$I_1 = [-2\pi, -\tfrac{8}{7}\pi], \ I_2 = [-\tfrac{8}{7}\pi, -\pi], \ I_3 = [\pi, \tfrac{8}{7}\pi], \ I_4 = [\tfrac{8}{7}\pi, 2\pi]$$

$(I_\ell = \emptyset$ if $\ell \in \mathbb{Z} \setminus \{1, 2, 3, 4\})$. If we choose

$$\{j_1, j_2, j_3, j_4\} = \{-1, 2, 2, -1\},$$

then, by Corollary 2.6,

$$K = 2^{-1}I_1 \cup 2^2 I_2 \cup 2^2 I_3 \cup 2^{-1}I_4$$

$$= [-\pi, -\tfrac{4}{7}\pi] \cup [-\tfrac{32}{7}\pi, -4\pi] \cup [4\pi, \tfrac{32}{7}\pi] \cup [\tfrac{4}{7}\pi, \pi]$$

yields the Journé wavelet (see Example A). Observe that if

$$\{k_1, k_2, k_3, k_4\} = \{1, 3, -3, 1\},$$

then $\{K_j + 2k_j\pi : j = 1, 2, 3, 4\}$ is the partition of I

$$[\pi, \tfrac{10}{7}\pi], \ [\tfrac{10}{7}\pi, 2\pi], \ [-2\pi, -\tfrac{10}{7}\pi], \ [-\tfrac{10}{7}\pi, -\pi].$$

Corollary 2.6 is used in [FW] to obtain several new families of MSF orthonormal wavelets. In particular, they have constructed MSF wavelets that are not band-limited.

7.3   The characterization of MRA wavelets

We have seen in section 3.4 that if ψ is a wavelet obtained from an MRA then $D_\psi(\xi) = 1$ a.e. (see equation (4.10) of Chapter 3). We shall show that this is a necessary and sufficient condition for an orthonormal wavelet to be "associated" with an MRA. We start by explaining what we mean by saying that a wavelet is "associated" with an MRA.

Suppose that ψ is an orthonormal wavelet. For $j \in \mathbb{Z}$ let W_j be the closure in $L^2(\mathbb{R})$ of the span of $\{\psi_{j,k} : k \in \mathbb{Z}\}$. Since ψ is an orthonormal wavelet,

$$L^2(\mathbb{R}) = \bigoplus_{j \in \mathbb{Z}} W_j.$$

Define

$$V_j = \bigoplus_{\ell=-\infty}^{j-1} W_\ell, \qquad j \in \mathbb{Z}.$$

The sequence of subspaces $\{V_j : j \in \mathbb{Z}\}$ satisfies properties (1.1), (1.2), (1.3) and (1.4) of the definition of MRA (see section 2.1). Thus, $\{V_j : j \in \mathbb{Z}\}$ will generate an MRA if there exists a function $\varphi \in L^2(\mathbb{R})$ such that the system $\{\varphi(\cdot - k) : k \in \mathbb{Z}\}$ is an orthonormal basis for V_0. In this case, we say that the wavelet ψ is **associated with an MRA** or, more simply, that ψ is an **MRA wavelet**.

Recall that we constructed a wavelet ψ from such an MRA by using equation (2.12) in Chapter 2. We also have shown that if $\widehat{\tilde{\psi}}(\xi) = e^{i\beta(\xi)}\hat{\psi}(\xi)$, where ψ is a wavelet, and β is a real-valued, 2π-periodic measurable function, then $\tilde{\psi}$ is also a wavelet (see Theorem 2.2 in this chapter). The wavelet $\tilde{\psi}$ does not necessarily satisfy equation (2.12) in Chapter 2 with the same scaling function φ. We shall show later, however, that an appropriate change of the phase of $\hat{\varphi}$ will produce a scaling function $\tilde{\varphi}$ for which (2.12) in Chapter 2 is satisfied by $\tilde{\psi}$. It will then follow that if ψ is an MRA wavelet, as defined above, we can then find a scaling function φ such that (2.12) of Chapter 2 holds.

Thus, given a wavelet ψ, we shall study the properties of

$$D(\xi) \equiv D_\psi(\xi) \equiv \sum_{j=1}^{\infty} \sum_{k \in \mathbb{Z}} |\hat{\psi}(2^j(\xi + 2k\pi))|^2. \tag{3.1}$$

As mentioned above, this is the function occurring in (4.10) of Chapter 3. We already observed that, since

$$\int_0^{2\pi} D_\psi(\xi)\, d\xi = \|\hat{\psi}\|_2^2,$$

(see Lemma 4.16 of Chapter 3), $D_\psi(\xi)$ is well defined for almost every $\xi \in \mathbb{R}$ whenever $\psi \in L^2(\mathbb{R})$.

The main result of this section is the following theorem:

THEOREM 3.2 *A wavelet $\psi \in L^2(\mathbb{R})$ is an MRA wavelet if and only if $D_\psi(\xi) = 1$ for almost every $\xi \in \mathbb{R}$.*

As stated above we have already shown that $D_\psi(\xi) = 1$ a.e. when ψ is an MRA wavelet. Thus, to finish the proof of Theorem 3.2 we must show that if $D_\psi(\xi) = 1$ a.e., then the wavelet is an MRA wavelet. We shall break up the proof of this into several lemmas.

LEMMA 3.3 *If ψ is an orthonormal wavelet, then*

$$\hat{\psi}(2^n\xi) = \sum_{j=1}^{\infty}\sum_{k\in\mathbb{Z}} \hat{\psi}(2^n(\xi+2k\pi))\,\overline{\hat{\psi}(2^j(\xi+2k\pi))}\,\hat{\psi}(2^j\xi) \quad a.e. \qquad (3.4)$$

for all $n \geq 1$.

PROOF : We start by showing that the series in (3.4) is well defined. We use Schwarz's inequality and (1.4) to obtain

$$\sum_{k\in\mathbb{Z}} |\hat{\psi}(2^n(\xi+2k\pi))\,\overline{\hat{\psi}(2^j(\xi+2k\pi))}\,|$$

$$\leq \left(\sum_{k\in\mathbb{Z}} |\hat{\psi}(2^n(\xi+2k\pi))|^2\right)^{\frac{1}{2}} \left(\sum_{k\in\mathbb{Z}} |\hat{\psi}(2^j(\xi+2k\pi))|^2\right)^{\frac{1}{2}}$$

$$\leq \left(\sum_{\ell\in\mathbb{Z}} |\hat{\psi}(2^n\xi+2\ell\pi)|^2\right)^{\frac{1}{2}} \left(\sum_{k\in\mathbb{Z}} |\hat{\psi}(2^j(\xi+2k\pi))|^2\right)^{\frac{1}{2}}$$

$$= 1\left(\sum_{k\in\mathbb{Z}} |\hat{\psi}(2^j(\xi+2k\pi))|^2\right)^{\frac{1}{2}}.$$

Summing over all $j \geq 1$, using Schwarz's inequality and (1.2), we deduce

$$\sum_{j=1}^{\infty}\sum_{k\in\mathbb{Z}} |\hat{\psi}(2^n(\xi+2k\pi))\,\overline{\hat{\psi}(2^j(\xi+2k\pi))}\,\hat{\psi}(2^j\xi)|$$

$$\leq \sum_{j=1}^{\infty}\left(\sum_{k\in\mathbb{Z}} |\hat{\psi}(2^j(\xi+2k\pi))|^2\right)^{\frac{1}{2}} |\hat{\psi}(2^j\xi)|$$

$$\leq \left(\sum_{j=1}^{\infty}\sum_{k\in\mathbb{Z}} |\hat{\psi}(2^j(\xi+2k\pi))|^2\right)^{\frac{1}{2}} \left(\sum_{j=1}^{\infty} |\hat{\psi}(2^j\xi)|^2\right)^{\frac{1}{2}}$$

$$\leq \left(\sum_{j=1}^{\infty}\sum_{k\in\mathbb{Z}} |\hat{\psi}(2^j(\xi+2k\pi))|^2\right)^{\frac{1}{2}} \cdot 1 = \sqrt{D_\psi(\xi)}\,.$$

All the above inequalities are true for a.e. $\xi \in \mathbb{R}$. This shows that the series in (3.4) is well defined almost everywhere. Let $G_n(\xi)$ be the right-hand side of (3.4). We have to show that $G_n(\xi) = \hat{\psi}(2^n\xi)$ for a.e. $\xi \in \mathbb{R}$. We first show that $G_n(\xi) = G_{n-1}(2\xi)$, and, then, that $G_1(\xi) = \hat{\psi}(2\xi)$; this will clearly give us (3.4).

Using (1.5) with j replaced by n we obtain

$$G_n(\xi) = \sum_{k \in \mathbb{Z}} \hat{\psi}(2^n(\xi + 2k\pi)) \sum_{j=1}^{\infty} \overline{\hat{\psi}(2^j(\xi + 2k\pi))} \, \hat{\psi}(2^j \xi)$$

$$= \sum_{k \in \mathbb{Z}} \hat{\psi}(2^n(\xi + 2k\pi)) \, \overline{\hat{\psi}(\xi + 2k\pi)} \, \hat{\psi}(\xi)$$

$$+ \sum_{k \in \mathbb{Z}} \hat{\psi}(2^n(\xi + 2k\pi)) \sum_{j=1}^{\infty} \overline{\hat{\psi}(2^j(\xi + 2k\pi))} \, \hat{\psi}(2^j \xi)$$

$$= \sum_{k \in \mathbb{Z}} \hat{\psi}(2^n(\xi + 2k\pi)) \sum_{j=0}^{\infty} \overline{\hat{\psi}(2^j(\xi + 2k\pi))} \, \hat{\psi}(2^j \xi).$$

In the sum over j the terms where k is odd are zero a.e. by (1.3). Hence, replacing k by 2ℓ we deduce

$$G_n(\xi) = \sum_{\ell \in \mathbb{Z}} \hat{\psi}(2^n(\xi + 4\ell\pi)) \sum_{j=0}^{\infty} \overline{\hat{\psi}(2^j(\xi + 4\ell\pi))} \, \hat{\psi}(2^j \xi)$$

$$= \sum_{\ell \in \mathbb{Z}} \hat{\psi}(2^{n+1}(\tfrac{\xi}{2} + 2\ell\pi)) \sum_{j=0}^{\infty} \overline{\hat{\psi}(2^{j+1}(\tfrac{\xi}{2} + 2\ell\pi))} \, \hat{\psi}(2^{j+1} \tfrac{\xi}{2})$$

$$= \sum_{\ell \in \mathbb{Z}} \hat{\psi}(2^{n+1}(\tfrac{\xi}{2} + 2\ell\pi)) \sum_{j=1}^{\infty} \overline{\hat{\psi}(2^j(\tfrac{\xi}{2} + 2\ell\pi))} \, \hat{\psi}(2^j \tfrac{\xi}{2}) = G_{n+1}(\tfrac{\xi}{2}).$$

This shows that $G_n(\xi) = G_{n-1}(2\xi)$ almost everywhere.

We now calculate $G_1(\xi)$. Changing variables in the sum over j we obtain

$$G_1(\xi) = \sum_{k \in \mathbb{Z}} \hat{\psi}(2(\xi + 2k\pi)) \sum_{j=1}^{\infty} \overline{\hat{\psi}(2^j(\xi + 2k\pi))} \, \hat{\psi}(2^j \xi)$$

$$= \sum_{k \in \mathbb{Z}} \hat{\psi}(2\xi + 4k\pi) \sum_{j=0}^{\infty} \overline{\hat{\psi}(2^j(2\xi + 4k\pi))} \, \hat{\psi}(2^j 2\xi).$$

To the last sum over k we add all the corresponding terms with $2k$ replaced by $2k + 1$, which are zero by (1.3). This gives us

$$G_1(\xi) = \sum_{k \in \mathbb{Z}} \hat{\psi}(2\xi + 2k\pi) \sum_{j=0}^{\infty} \overline{\hat{\psi}(2^j(2\xi + 2k\pi))} \, \hat{\psi}(2^j 2\xi).$$

Interchanging the order of summation, using (1.5) when $j \geq 1$ and (1.4) when $j = 0$ we obtain $G_1(\xi) = \hat{\psi}(2\xi)$.

∎

Let us now consider $\ell^2(\mathbb{Z})$, and denote its (usual) norm by $\| \cdot \|_{\ell^2}$ and the inner product by $\langle \cdot, \cdot \rangle_{\ell^2}$. If ψ is an orthonormal wavelet, we define the vector

$$\Psi_j(\xi) = \{\hat{\psi}(2^j(\xi + 2k\pi)) : k \in \mathbb{Z}\}, \quad j \geq 1.$$

For almost every ξ this vector belongs to $\ell^2(\mathbb{Z})$, since (1.4) implies

$$\|\Psi_j(\xi)\|_{\ell^2} = \left(\sum_{k \in \mathbb{Z}} |\hat{\psi}(2^j(\xi + 2k\pi))|^2\right)^{\frac{1}{2}}$$

$$\leq \left(\sum_{\ell \in \mathbb{Z}} |\hat{\psi}(2^j\xi + 2\ell\pi)|^2\right)^{\frac{1}{2}} = 1 \quad \text{for a.e. } \xi \in \mathbb{R}.$$

Let $\mathbb{F}_\psi(\xi)$ be the closure of the span of the set of vectors $\{\Psi_j(\xi) : j \geq 1\}$; this is a well-defined subspace of $\ell^2(\mathbb{Z})$ for almost every $\xi \in \mathbb{R}$. We can rewrite (3.4) in terms of this notation:

$$\hat{\psi}(2^n\xi) = \sum_{j=1}^{\infty} \langle \Psi_n(\xi), \Psi_j(\xi) \rangle_{\ell^2} \hat{\psi}(2^j\xi) \quad \text{for a.e. } \xi \in \mathbb{R}.$$

Replacing ξ by $\xi + 2\ell\pi$ we obtain, for $n \geq 1$,

$$\hat{\psi}(2^n(\xi + 2\ell\pi)) = \sum_{j=1}^{\infty} \langle \Psi_n(\xi), \Psi_j(\xi) \rangle_{\ell^2} \hat{\psi}(2^j(\xi + 2\ell\pi)) \quad \text{a.e.}$$

since $\langle \Psi_n(\xi), \Psi_j(\xi) \rangle_{\ell^2}$ is 2π-periodic. We can write this equality vectorially as

$$\Psi_n(\xi) = \sum_{j=1}^{\infty} \langle \Psi_n(\xi), \Psi_j(\xi) \rangle_{\ell^2} \Psi_j(\xi). \tag{3.5}$$

Observe that $D_\psi(\xi)$ can also be expressed as follows:

$$D_\psi(\xi) = \sum_{j=1}^{\infty} \|\Psi_j(\xi)\|_{\ell^2}^2. \tag{3.6}$$

We shall use:

LEMMA 3.7 *Let* $\{v_j : j \geq 1\}$ *be a family of vectors in a Hilbert space* $\mathbb{H}$ *such that*

(1) $\displaystyle\sum_{n=1}^{\infty} \|v_n\|^2 = C < \infty$, *and*

(2) $\displaystyle v_n = \sum_{m=1}^{\infty} \langle v_n, v_m \rangle v_m$ *for all* $n \geq 1$.

Let $\mathbb{F} = \overline{\text{span}\{v_j : j \geq 1\}}$. *Then* $\dim \mathbb{F} = \displaystyle\sum_{j=1}^{\infty} \|v_j\|^2 = C$.

PROOF: Let us first observe that the series $\displaystyle\sum_{m=1}^{\infty} \langle v, v_m \rangle v_m$ converges in $\mathbb{H}$. In fact,

$$\left\| \sum_{m=M}^{N} \langle v, v_m \rangle v_m \right\| \leq \sum_{m=M}^{N} \|\langle v, v_m \rangle v_m\| \leq \|v\| \sum_{m=M}^{N} \|v_m\|^2 \longrightarrow 0$$

as $M, N \to \infty$ because of (1). In particular, the series in (2) gives us a well-defined element of $\mathbb{H}$ and

$$Tv = \sum_{n=1}^{\infty} \langle v, v_n \rangle v_n, \quad \text{for } v \in \mathbb{H},$$

is a well-defined linear operator on $\mathbb{H}$; moreover, it is bounded since

$$\|Tv\| \leq \sum_{n=1}^{\infty} |\langle v, v_n \rangle| \cdot \|v_n\| \leq \sum_{n=1}^{\infty} \|v\| \cdot \|v_n\|^2 = C\|v\|.$$

We claim that $\mathbb{F} = \text{Ker}\,(T - I) = \text{Range}\,(T)$. In fact, $\mathbb{F} \subseteq \text{Ker}\,(T - I)$ follows immediately from (2) and the continuity of T, $\text{Ker}\,(T - I) \subseteq \text{Range}\,(T)$ is true for any linear operator, and $\text{Range}\,(T) \subseteq \mathbb{F}$ follows from the definitions of T and $\mathbb{F}$.

Let $\{e_k\}$ be an orthonormal basis of $\mathbb{F}$; then, since each $e_k \in \text{Ker}\,(T - I)$, we obtain

$$C = \sum_{n} \|v_n\|^2 = \sum_{n} \sum_{k} |\langle v_n, e_k \rangle|^2 = \sum_{n} \sum_{k} \langle v_n, e_k \rangle \langle e_k, v_n \rangle$$

$$= \sum_{k} \left\langle \sum_{n} \langle e_k, v_n \rangle v_n, e_k \right\rangle = \sum_{k} \langle T e_k, e_k \rangle = \sum_{k} \langle e_k, e_k \rangle.$$

The absolute convergence of the second and third series above follows from Schwarz's inequality and (1). This shows that the number of basis elements of $\mathbb{F}$ is C, as we wanted to prove.

∎

Let S be the subset of $\mathbb{T}$ on which $D_\psi(\xi) < \infty$ and (3.5) is satisfied. Then the vectors $\Psi_j(\xi)$, $j \geq 1$, are well defined on S (observe that $|S| = 2\pi$). For $\xi \in S$, let $\mathbb{F}_\psi(\xi)$ be the closure, in $\ell^2(\mathbb{Z})$, of the span of $\{\Psi_j(\xi) : j \geq 1\}$. Then the hypotheses of Lemma 3.7 are satisfied if $v_j = \Psi_j(\xi)$. This gives us

$$\dim \mathbb{F}_\psi(\xi) = D_\psi(\xi) \quad \text{on} \quad S. \tag{3.8}$$

We are now ready to prove the sufficiency part of Theorem 3.2. That is, we assume that $\psi \in L^2(\mathbb{R})$ is a wavelet for which $D_\psi(\xi) = 1$ for a.e. $\xi \in \mathbb{R}$. Then, by (3.8), $\dim \mathbb{F}_\psi(\xi) = 1$ for almost every $\xi \in \mathbb{T}$. This shows that, for each $\xi \in S$, $\mathbb{F}_\psi(\xi)$ is generated by a single unit vector $U(\xi)$. We now choose a particular one. For $j \geq 1$, let

$$E_j = \{\xi \in S : \Psi_j(\xi) \neq \vec{0} \quad \text{and} \quad \Psi_m(\xi) = \vec{0} \quad \text{for all} \quad m < j\}.$$

The sets E_j, for $j \geq 1$, are mutually disjoint and, together with

$$E_0 = \{\xi \in \mathbb{T} : D_\psi(\xi) = 0\},$$

form a partition of S. Hence, for $\xi \in S \setminus E_0$, there exists a unique $j \geq 1$ such that $\xi \in E_j$. Thus, since E_0 has measure 0,

$$U(\xi) = \frac{1}{\|\Psi_j(\xi)\|_{\ell^2}} \Psi_j(\xi), \quad \xi \in E_j \text{ for some } j \geq 1,$$

is well defined and $\|U(\xi)\|_{\ell^2} = 1$ for almost every $\xi \in \mathbb{T}$. Write

$$U(\xi) = \{u_k(\xi) : k \in \mathbb{Z}\}.$$

It is natural, in view of the first equality in (4.10) of Chapter 3, to hope that $u_k(\xi) = \hat{\varphi}(\xi + 2k\pi)$ if we find the scaling function φ we are seeking. Thus, we let

$$\hat{\varphi}(\xi) = u_k(\xi - 2k\pi) \quad \text{if} \quad \xi \in \mathbb{T} + 2k\pi \text{ for some } k \in \mathbb{Z}.$$

This defines $\hat{\varphi}$ on $\mathbb{R}$. We claim $\hat{\varphi} \in L^2(\mathbb{R})$:

$$\|\hat{\varphi}\|_2^2 = \sum_{k \in \mathbb{Z}} \int_{\mathbb{T}} |\hat{\varphi}(\xi + 2k\pi)|^2 \, d\xi = \sum_{k \in \mathbb{Z}} \int_{\mathbb{T}} |u_k(\xi)|^2 \, d\xi$$

$$= \int_{\mathbb{T}} \|U(\xi)\|_{\ell^2}^2 \, d\xi = 2\pi$$

since $U(\xi)$ is a unit vector. We also have

$$\sum_{k \in \mathbb{Z}} |\hat{\varphi}(\xi + 2k\pi)|^2 = \sum_{k \in \mathbb{Z}} |u_k(\xi)|^2 = \|U(\xi)\|_{\ell^2}^2 = 1 \quad \text{for} \quad a.e. \ \xi \in \mathbb{R}, \quad (3.9)$$

which, as we have pointed out several times, is equivalent to the fact that $\{\varphi(\cdot - k) : k \in \mathbb{Z}\}$ is an orthonormal system in $L^2(\mathbb{R})$. Define $V_0^\sharp$ as the closed subspace of $L^2(\mathbb{R})$ generated by $\{\varphi(\cdot - k) : k \in \mathbb{Z}\}$. We claim

$$V_0^\sharp = V_0 = \bigoplus_{j<0} W_j. \tag{3.10}$$

From this it follows that $\{V_j : j \in \mathbb{Z}\}$ is the desired MRA (see the discussion at the beginning of this section).

For each $j \geq 1$, there exists a measurable function ν_j, defined on $\mathbb{T}$, such that $\Psi_j(\xi) = \nu_j(\xi)U(\xi)$ for a.e. $\xi \in \mathbb{T}$. Componentwise,

$$\hat{\psi}(2^j(\xi + 2k\pi)) = \nu_j(\xi)\hat{\varphi}(\xi + 2k\pi) \quad \text{for} \quad a.e. \ \xi \in \mathbb{T}, \quad k \in \mathbb{Z}.$$

Hence, by (3.9), for a.e. $\xi \in \mathbb{T}$,

$$\sum_{k \in \mathbb{Z}} |\hat{\psi}(2^j(\xi + 2k\pi))|^2 = \sum_{k \in \mathbb{Z}} |\nu_j(\xi)|^2 |\hat{\varphi}(\xi + 2k\pi)|^2 = |\nu_j(\xi)|^2, \tag{3.11}$$

which shows that $\nu_j \in L^2(\mathbb{T})$ with $\|\nu_j\|_{L^2(\mathbb{T})}^2 = 2^{-j}(2\pi)$. Write the Fourier series of ν_j, $j \geq 1$, as

$$\nu_j(\xi) = \sum_{k \in \mathbb{Z}} a_k^j e^{-ik\xi} \quad \text{for} \quad a.e. \ \xi \in \mathbb{T},$$

with convergence in the $L^2(\mathbb{T})$-norm, and $\{a_k^j\}_{k \in \mathbb{Z}} \in \ell^2(\mathbb{Z})$. Extending ν_j 2π-periodically we obtain

$$\hat{\psi}(2^j\xi) = \nu_j(\xi)\hat{\varphi}(\xi) \quad \text{for} \quad a.e. \ \xi \in \mathbb{R}, \quad j \geq 1. \tag{3.12}$$

Taking inverse Fourier transform on both sides, we have

$$\psi_{-j,0}(x) = 2^{-\frac{j}{2}}\psi(2^{-j}x) = 2^{\frac{j}{2}}\sum_{k\in\mathbb{Z}}a_k^j\varphi(x-k), \quad j \geq 1.$$

Hence, $\psi_{-j,0} \in V_0^\sharp$ for $j \geq 1$. Since $V_0^\sharp$ is invariant under integral translations and $\psi_{-j,k}(x) = 2^{-\frac{j}{2}}\psi(2^{-j}(x-2^jk))$, we have $\psi_{-j,k} \in V_0^\sharp$ for all $k \in \mathbb{Z}$ and $j \geq 1$. Thus, $W_{-j} \subseteq V_0^\sharp$ for all $j \geq 1$ and, hence, $V_0 \subseteq V_0^\sharp$.

We need to show that $V_0^\sharp \subseteq V_0$. We do this by showing that φ is perpendicular to W_j for all $j \geq 0$. For $j \geq 0$ and $\ell \in \mathbb{Z}$, the Plancherel theorem, a change of variables and a periodization argument allow us to write

$$2\pi\langle\varphi,\psi_{j,\ell}\rangle = \langle\hat{\varphi},(\psi_{j,\ell})^\wedge\rangle = 2^{-\frac{j}{2}}\int_{\mathbb{R}}\hat{\varphi}(\xi)\,\overline{\hat{\psi}(2^{-j}\xi)}\,e^{i2^{-j}\ell\xi}\,d\xi$$

$$= 2^{\frac{j}{2}}\int_{\mathbb{R}}\hat{\varphi}(2^j\xi)\,\overline{\hat{\psi}(\xi)}\,e^{i\ell\xi}\,d\xi$$

$$= 2^{\frac{j}{2}}\int_{\mathbb{T}}\left(\sum_{k\in\mathbb{Z}}\hat{\varphi}(2^j(\xi+2k\pi))\,\overline{\hat{\psi}(\xi+2k\pi)}\right)e^{i\ell\xi}\,d\xi. \qquad (3.13)$$

The convergence of the last series in $L^2(\mathbb{T})$ is guaranteed by the fact that $\varphi, \psi \in L^2(\mathbb{R})$. From (3.11) and our assumption $D_\psi(\xi) = 1$ a.e. we obtain

$$\sum_{j=1}^{\infty}|\nu_j(\xi)|^2 = \sum_{j=1}^{\infty}\sum_{k\in\mathbb{Z}}|\hat{\psi}(2^j(\xi+2k\pi))|^2 = 1 \quad \text{for a.e. } \xi \in \mathbb{R}.$$

Hence, for such ξ and for each $j \geq 0$, there exists $j_0 \equiv j_0(2^j\xi) \geq 1$ such that $\nu_{j_0}(2^j\xi) \neq 0$. This and (3.12) imply, for such ξ,

$$\hat{\varphi}(2^j(\xi+2k\pi)) = \frac{1}{\nu_{j_0}(2^j\xi)}\hat{\psi}(2^{j+j_0}(\xi+2k\pi)), \quad k \in \mathbb{Z}.$$

We now use (1.5) to obtain (observe that $j + j_0 \geq 1$)

$$\sum_{k\in\mathbb{Z}}\hat{\varphi}(2^j(\xi+2k\pi))\,\overline{\hat{\psi}(\xi+2k\pi)}$$

$$= \frac{1}{\nu_{j_0}(2^j\xi)}\sum_{k\in\mathbb{Z}}\hat{\psi}(2^{j+j_0}(\xi+2k\pi))\,\overline{\hat{\psi}(\xi+2k\pi)} = 0$$

for a.e. $\xi \in \mathbb{T}$ and all $j \geq 0$. Therefore, from this result and (3.13), we obtain $\langle\varphi,\psi_{j,\ell}\rangle = 0$ for all $\ell \in \mathbb{Z}$ and $j \geq 0$. This shows that φ is orthogonal to W_j for all $j \geq 0$. Since W_j is invariant under integral translation, we deduce that $V_0^\sharp \perp W_j$ for all $j \geq 0$. Hence $V_0^\sharp \subseteq V_0$, and the proof of Theorem 3.2 is finished.

∎

PROPOSITION 3.14 *For any orthonormal wavelet* $\psi \in L^2(\mathbb{R})$, *the following statements are equivalent:*

   (1) ψ *is an MRA wavelet;*

   (2) $D_\psi(\xi) = 1$ *for a.e.* $\xi \in \mathbb{T}$;

   (3) $D_\psi(\xi) > 0$ *for a.e.* $\xi \in \mathbb{T}$;

   (4) $\dim \mathbb{F}_\psi(\xi) = 1$ *for a.e.* $\xi \in \mathbb{T}$.

where $\mathbb{F}_\psi(\xi)$ *is the closure, in* $\ell^2(\mathbb{Z})$, *of the span of* $\{\Psi_j(\xi) : j \geq 1\}$ *and* $\Psi_j(\xi)$ *is the vector* $\{\hat{\psi}(2^j(\xi + 2k\pi)) : k \in \mathbb{Z}\}$.

PROOF : The equivalence between (1), (2) and (4) has already been proved (see Theorem 3.2 and equality (3.8)). Since, obviously, (2) implies (3), it is enough to show that (3) implies (2). If $D_\psi(\xi) > 0$ for a.e. $\xi \in \mathbb{T}$, the fact that $D_\psi(\xi)$ is an integer a.e. implies that $D_\psi(\xi) \geq 1$ almost everywhere. But this and the equality $\int_\mathbb{T} D_\psi(\xi) \, d\xi = 2\pi$ (see Lemma 4.16 of Chapter 3) clearly implies (2).

∎

Several corollaries can be immediately deduced from these results.

COROLLARY 3.15 *Any compactly supported orthonormal wavelet must be an MRA wavelet.*

PROOF : If ψ is a compactly supported wavelet, then $\hat{\psi}$ is the restriction to $\mathbb{R}$ of an entire function. Thus, $\hat{\psi}(\xi) = 0$ only on a denumerable set, which shows that condition (3) of Proposition 3.14 is true.

∎

COROLLARY 3.16 *If* ψ *is an orthonormal wavelet such that* $|\hat{\psi}|$ *is continuous and* $|\hat{\psi}(\xi)| = O(|\xi|^{-\frac{1}{2}-\alpha})$ *at* ∞ *for some* $\alpha > 0$, *then* ψ *is an MRA wavelet.*

PROOF : The behavior at infinity of $|\hat{\psi}|$ tells us that the series

$$s(\xi) = \sum_{j=1}^{\infty} |\hat{\psi}(2^j \xi)|^2$$

converges uniformly on compact subsets of $\mathbb{R} \setminus \{0\}$. Moreover, an easy calculation shows that $s(\xi) = O(|\xi|^{-1-2\alpha})$ at ∞. It follows that

$$\sum_{k \in \mathbb{Z}} s(\xi + 2k\pi) = D_\psi(\xi)$$

converges uniformly on compact subsets of $\mathbb{T}$. Thus, $D_\psi(\xi)$ is continuous on $(0, 2\pi)$. Since $D_\psi(\xi)$ is integer-valued and $\int_{\mathbb{T}} D_\psi(\xi) \, d\xi = 2\pi$ (see Lemma 4.16 of Chapter 3), we must have $D_\psi(\xi) = 1$ a.e. on $\mathbb{T}$. ∎

Every band-limited wavelet obviously exhibits the behavior at infinity assumed for $|\hat{\psi}|$ in Corollary 3.16. Thus, we have the following result:

COROLLARY 3.17 *If ψ is a band-limited wavelet such that $|\hat{\psi}|$ is continuous, then ψ is an MRA wavelet.*

The characterization of MRA wavelets described in Theorem 3.2 can be used to obtain some results about the set of all wavelets as a subset of the unit sphere of $L^2(\mathbb{R})$. For example,

THEOREM 3.18 *Suppose that $\{\psi^{(n)} : n = 1, 2, \cdots\}$ is a sequence of MRA wavelets converging to ψ in $L^2(\mathbb{R})$. If ψ is also a wavelet, then ψ must be an MRA wavelet.*

PROOF : A periodization argument as in the proof of Lemma 4.16 of Chapter 3 shows that

$$\int_{\mathbb{T}} D_{\psi^{(n)} - \psi}(\xi) \, d\xi = 2\pi \|\psi^{(n)} - \psi\|_2^2.$$

Since, by hypothesis, the right-hand side of the above equality tends to 0 as $n \to \infty$, by Fatou's lemma we have

$$\lim_{n \to \infty} D_{\psi^{(n)} - \psi}(\xi) = 0 \quad \text{for} \quad a.e. \ \xi \in \mathbb{T}.$$

From (3.6) we see that $\sqrt{D_\psi(\xi)}$ defines a norm for ψ and we deduce that

$$\sqrt{D_\psi(\xi)} \leq \sqrt{D_{\psi^{(n)}-\psi}(\xi)} + \sqrt{D_{\psi^{(n)}}(\xi)}\,.$$

Since each $\psi^{(n)}$ is an MRA wavelet, we have

$$\sqrt{D_\psi(\xi)} \leq \sqrt{D_{\psi^{(n)}-\psi}(\xi)} + 1.$$

Taking $\underline{\lim}$ (when $n \to \infty$) of both sides, we see that $\sqrt{D_\psi(\xi)} \leq 1$. But, since $\int_{\mathbb{T}} D_\psi(\xi)\, d\xi = 2\pi$, we must have $D_\psi(\xi) = 1$ for almost every $\xi \in \mathbb{T}$. The conclusion of our theorem follows from Proposition 3.14.

∎

This theorem shows that a non-MRA wavelet cannot be approximated in $L^2(\mathbb{R})$ by MRA wavelets. Because of Corollary 3.16, therefore, we see that a non-MRA wavelet cannot be arbitrarily close, in the L^2-norm, to a smooth wavelet that decays "mildly" at ∞; in particular, it cannot be arbitrarily close to wavelets that belong to the Schwartz class.

REMARK: Not every L^2-converging sequence of MRA wavelets converges to a wavelet in $L^2(\mathbb{R})$. The family of wavelets ψ_a defined by

$$\widehat{\psi_a} = \chi_{[-2a,-a]\cup[2\pi-a,4\pi-2a]}, \quad 0 < a < 2\pi,$$

(see Example C in section 7.2) converges in $L^2(\mathbb{R})$, as $a \to 0$, to the function $\psi_H = (\chi_{[2\pi,4\pi]})^\vee$. This function ψ_H, as we have already observed, is not a wavelet for $L^2(\mathbb{R})$ since it does not satisfy (1.2).

7.4   A characterization of low-pass filters

In section 2.3, in which we constructed compactly supported wavelets, we found sufficient conditions for a trigonometric polynomial m_0 to be the low-pass filter of an MRA. This was done in Theorem 3.13 and the paragraphs that follow its proof; we required m_0 to satisfy (3.1) of Chapter 2 and $m_0(\xi) \neq 0$ for all $\xi \in [-\frac{1}{2}\pi, \frac{1}{2}\pi]$.

In this section we shall characterize the functions $m_0 \in C^1$ satisfying $m_0(0) = 1$ that are low-pass filters for MRAs. This result will allow us to exhibit new wavelets. In addition, we shall give a characterization of low-pass filters that produces scaling functions with **polynomial decay** in $L^2(\mathbb{R})$ (see the definition of "polynomial decay" before Proposition 2.24 in Chapter 2). An MRA whose scaling function has this polynomial decay is sometimes called **localized** in the literature.

Let us recall how an MRA scaling function φ produces a low-pass filter m_0. We first observe that, since $\varphi(\frac{\cdot}{2}) \in V_{-1} \subset V_0$, we must have

$$\tfrac{1}{2}\varphi(\tfrac{1}{2}x) = \sum_{k\in\mathbb{Z}} \alpha_k \varphi(x - k) \tag{4.1}$$

with $\sum_{k\in\mathbb{Z}} |\alpha_k|^2 = \frac{1}{2} < \infty$. By taking the Fourier transform of both sides of equation (4.1) we obtain

$$\hat{\varphi}(2\xi) = m_0(\xi)\hat{\varphi}(\xi) \tag{4.2}$$

where

$$m_0(\xi) = \sum_{k\in\mathbb{Z}} \alpha_k e^{-ik\xi}$$

is a 2π-periodic function in $L^2([0, 2\pi])$. This function m_0 is the low-pass filter in question (see section 2.2). We shall assume in this section that $\hat{\varphi}$ is continuous at 0 — a very mild condition. By Theorem 1.7 of Chapter 2 we can assume, and do, in this section, that $\hat{\varphi}(0) = 1$. Thus, from (4.2), we deduce

$$m_0(0) = 1. \tag{4.3}$$

Using the orthonormality of the system $\{\varphi(\cdot - k) : k \in \mathbb{Z}\}$ and (4.2), we have shown (see (2.5) of Chapter 2)

$$|m_0(\xi)|^2 + |m_0(\xi + \pi)|^2 = 1 \quad \text{for} \ \ a.e. \ \xi \in \mathbb{R}, \tag{4.4}$$

which is a very important property of the low-pass filter.

Iterating (4.2) we obtain

$$\hat{\varphi}(\xi) = \lim_{N\to\infty} \hat{\varphi}(2^{-N}\xi) \prod_{j=1}^{N} m_0(2^{-j}\xi).$$

Since $\hat{\varphi}$ is continuous at 0 and $\hat{\varphi}(0) = 1$,

$$\lim_{N\to\infty} \hat{\varphi}(2^{-N}\xi) = 1.$$

As for the convergence of the product that appears in the above equality we have the following result:

LEMMA 4.5 *Suppose that m_0 is a 2π-periodic function in the class C^1 and satisfies (4.3) and (4.4). Then, the sequence*

$$\left\{ \prod_{j=1}^{N} m_0(2^{-j}\xi) : N = 1, 2, \cdots \right\}$$

converges uniformly on bounded sets of $\mathbb{R}$. Moreover,

$$\hat{\varphi}(\xi) = \prod_{j=1}^{\infty} m_0(2^{-j}\xi) \tag{4.6}$$

is continuous.

PROOF :   The fact that $\{\prod_{j=1}^{N} m_0(2^{-j}\xi) : N = 1, 2, \cdots\}$ converges uniformly on bounded sets of $\mathbb{R}$ has been proved at the beginning of section 2.3 (see the argument that follows (3.1) in Chapter 2). The continuity of $\hat{\varphi}$ follows since it is the limit of a sequence converging uniformly on bounded sets.

∎

We know that if m_0 is the low-pass filter for an MRA for which $\hat{\varphi}$ is continuous at 0 we must have (4.3) and (4.4). We now state and prove another condition that m_0 must satisfy.

PROPOSITION 4.7 *Suppose that the low-pass filter m_0 of an MRA is a C^1 function and the scaling function φ satisfies $\hat{\varphi}(0) = 1$ and $|\hat{\varphi}(\xi)| = O(|\xi|^{-\frac{1}{2}-\alpha})$ at ∞ for some $\alpha > 0$. Then m_0 must satisfy the following property:*

(C) *There exists a set $K \subset \mathbb{R}$ which is a finite union of closed bounded intervals such that 0 is in the interior of K,*

$$\sum_{k \in \mathbb{Z}} \chi_K(\xi + 2k\pi) = 1 \quad for \quad a.e. \ \xi \in \mathbb{R},$$

and $m_0(2^{-j}\xi) \neq 0$ for all $j = 1, 2, \cdots$ and all $\xi \in K$.

PROOF: We can apply Lemma 4.5 to obtain that $\hat{\varphi}$ is continuous. The decay of $\hat{\varphi}$ at ∞ and the orthonormality of the system $\{\varphi(\cdot - k) : k \in \mathbb{Z}\}$ then imply

$$\sum_{k \in \mathbb{Z}} |\hat{\varphi}(\xi + 2k\pi)|^2 = 1 \quad \text{for all} \ \ \xi \in \mathbb{T}.$$

Therefore, for each $\xi \in [-\pi, \pi]$, there exists a $k = k(\xi) \in \mathbb{Z}$ such that $\hat{\varphi}(\xi + 2k(\xi)\pi) \neq 0$. Observe that $k(0) = 0$ because $\hat{\varphi}(0) = 1$. Since $\hat{\varphi}$ is continuous, for every $\xi \in [-\pi, \pi]$ there exists a real number $C_\xi > 0$ and an open interval V_ξ such that

$$|\hat{\varphi}(\mu + 2k(\xi)\pi)|^2 > C_\xi \quad \text{for all} \ \ \mu \in V_\xi.$$

Since $[-\pi, \pi]$ is compact, there exist $\xi_0, \xi_1, \cdots, \xi_n$ such that $\bigcup_{j=0}^{n} V_{\xi_j}$ covers $[-\pi, \pi]$ and we can assume $0 \in V_{\xi_o}$. Let

$$R_0 = V_{\xi_o} \cap [-\pi, \pi]$$

and, for $j = 1, 2, \cdots, n$, define, inductively,

$$R_j = (V_{\xi_j} \cap [-\pi, \pi]) \setminus \bigcup_{\ell=0}^{j-1} R_\ell.$$

Finally, take

$$K = \bigcup_{j=0}^{n} \overline{(R_j + 2k(\xi_j)\pi)}.$$

The set K is clearly a finite union of closed bounded intervals; moreover, since $0 \in V_{\xi_o}$, K contains the origin in its interior. The collection

$$\{R_j : j = 0, 1, \cdots, n\}$$

is a partition of $[-\pi, \pi]$; hence, $\sum_{j=0}^{n} \chi_{R_j}(\xi) = 1$ for a.e. $\xi \in [-\pi, \pi]$. Thus, if $k \neq k(\xi_j)$, $j = 0, 1, \cdots, n$, $\chi_K(\xi + 2k\pi) = 0$ for a.e. $\xi \in [-\pi, \pi]$, and

$$\sum_{k \in \mathbb{Z}} \chi_K(\xi + 2k\pi) = \sum_{j=0}^{n} \chi_K(\xi + 2k(\xi_j)\pi)$$

$$= \sum_{j=0}^{n} \chi_{R_j + 2k(\xi_j)\pi}(\xi + 2k(\xi_j)\pi)$$

$$= \sum_{j=0}^{n} \chi_{R_j}(\xi) = 1 \qquad \text{for} \quad a.e. \ \xi \in [-\pi, \pi].$$

Since $\sum_{k \in \mathbb{Z}} \chi_K(\xi + 2k\pi)$ is 2π-periodic we have proved that this sum equals 1 for a.e. $\xi \in \mathbb{R}$. Moreover, for all $\xi \in K$,

$$|\hat{\varphi}(\xi)|^2 \geq C = \min\{C_{\xi_o}, C_{\xi_1}, \cdots, C_{\xi_n}\} > 0.$$

By (4.6) we must have $m_0(2^{-j}\xi) \neq 0$ for all $\xi \in K$ and all $j = 1, 2, \cdots$. ∎

EXAMPLE E: For the Lemarié-Meyer wavelets given in Example D of Chapter 2 we can take $K = [-\pi, \pi]$. Observe that K is not unique, since for these wavelets we can also take $K = [-\pi - \frac{1}{2}\varepsilon, \pi - \frac{1}{2}\varepsilon]$, with ε small enough.

EXAMPLE F: For the Haar wavelet (see Example B of Chapter 2) $m_0(\xi) = \frac{1}{2}(1 + e^{i\xi})$; we can take $K = [-\pi, \pi]$. The same K can be used for any of the compactly supported wavelets developed in section 2.3. Observe that in these cases the filter is a trigonometric polynomial and, hence, a C^∞, 2π-periodic function.

REMARK: In some of the wavelet literature (see [Da1] and [Co2]) the condition

$$\sum_{k \in \mathbb{Z}} \chi_K(\xi + 2k\pi) = 1$$

is stated by saying that K is congruent to $[-\pi, \pi]$ modulo 2π; that is, for almost every $\xi \in [-\pi, \pi]$ there exists a unique $k(\xi) \in \mathbb{Z}$ such that $\xi + 2k(\xi)\pi \in K$. Obviously, the two conditions are equivalent.

The following theorem allows us to decide when a C^1 2π-periodic function m_0 is a filter for an MRA.

THEOREM 4.8 *Let $m_0 \in C^1(\mathbb{R})$ be a 2π-periodic function which satisfies (4.3), (4.4) and condition (C) of Proposition 4.7. Then, m_0 is the low-pass filter for an MRA.*

PROOF: We first claim that if g is a 2π-periodic function on $\mathbb{R}$ and K is the set described in condition (C) of Proposition 4.7 we have

$$\int_K g(\xi) \, d\xi = \int_{-\pi}^{\pi} g(\xi) \, d\xi. \tag{4.9}$$

This follows easily from the hypothesis $\sum\limits_{k\in\mathbb{Z}} \chi_K(\xi + 2k\pi) = 1$ a.e. on $[-\pi, \pi]$ and the periodization of g: the left side of (4.9) equals

$$\int_{\mathbb{R}} g(\xi)\chi_K(\xi)\, d\xi = \sum_{k\in\mathbb{Z}} \int_{(k-1)\pi}^{(k+1)\pi} g(\xi)\chi_K(\xi)\, d\xi$$

$$= \int_{-\pi}^{\pi} g(\xi) \sum_{k\in\mathbb{Z}} \chi_K(\xi + 2k\pi)\, d\xi = \int_{-\pi}^{\pi} g(\xi)\, d\xi.$$

Also helpful for the proof is that any 2π-periodic function h defined on $\mathbb{R}$ is easily seen to satisfy

$$\int_{-\pi}^{\pi} h(\xi)\, d\xi = \int_{-\frac{\pi}{2}}^{\frac{\pi}{2}} \left[h(\xi) + h(\xi + \pi) \right] d\xi \qquad (4.10)$$

To prove the theorem we define φ by (4.6), Let us first show that the system $\{\varphi(\cdot - k) : k \in \mathbb{Z}\}$ is orthonormal. Let $\{f_n : n = 1, 2, \cdots\}$ be defined by

$$\widehat{f_n}(\xi) = \chi_{2^n K}(\xi) \prod_{j=1}^{n} m_0(2^{-j}\xi).$$

Since 0 belongs to the interior of K the sets $2^n K$ cover $\mathbb{R}$ as $n \to \infty$; thus $\lim\limits_{n\to\infty} \widehat{f_n}(\xi) = \hat{\varphi}(\xi)$ for all $\xi \in \mathbb{R}$. We start by showing that

$$\langle f_n, f_n(\cdot - \ell)\rangle = \delta_{0,\ell} \qquad \text{for all } n \in \mathbb{N}, \ \ell \in \mathbb{Z}. \qquad (4.11)$$

By the Plancherel theorem

$$\langle f_n, f_n(\cdot - \ell)\rangle = \frac{1}{2\pi} \int_{\mathbb{R}} |\widehat{f_n}(\xi)|^2 e^{i\ell\xi}\, d\xi.$$

The change of variables $\xi = 2^n \mu$ gives us

$$\int_{\mathbb{R}} |\widehat{f_n}(\xi)|^2 e^{i\ell\xi}\, d\xi = \int_{2^n K} \left[\prod_{j=1}^{n} |m_0(2^{-j}\xi)|^2 \right] e^{i\ell\xi}\, d\xi$$

$$= 2^n \int_K \left[\prod_{j=1}^{n} |m_0(2^{n-j}\mu)|^2 \right] e^{i\ell 2^n \mu}\, d\mu = 2^n \int_K \left[\prod_{k=0}^{n-1} |m_0(2^k\mu)|^2 \right] e^{i\ell\mu 2^n}\, d\mu.$$

We use (4.9) and (4.10) and the 2π-periodicity of the functions involved to obtain

$$\int_{\mathbb{R}} |\widehat{f_n}(\xi)|^2 e^{i\ell\xi} d\xi = 2^n \int_{-\pi}^{\pi} \left[\prod_{k=0}^{n-1} |m_0(2^k\mu)|^2 \right] e^{i\ell\mu 2^n} d\mu$$

$$= 2^n \int_{-\frac{\pi}{2}}^{\frac{\pi}{2}} \prod_{k=1}^{n-1} |m_0(2^k\mu)|^2 \left[|m_0(\mu)|^2 + |m_0(\mu+\pi)|^2 \right] e^{i\ell\mu 2^n} d\mu. \qquad (4.12)$$

Using (4.4) and the change of variables $\mu = \frac{1}{2}\xi$ we deduce

$$\int_{\mathbb{R}} |\widehat{f_n}(\xi)|^2 e^{i\ell\xi} d\xi = 2^n \int_{-\frac{\pi}{2}}^{\frac{\pi}{2}} \left[\prod_{k=1}^{n-1} |m_0(2^k\mu)|^2 \right] e^{i\ell\mu 2^n} d\mu$$

$$= 2^{n-1} \int_{-\pi}^{\pi} \left[\prod_{j=0}^{n-2} |m_0(2^j\xi)|^2 \right] e^{i\ell\xi 2^{n-1}} d\xi.$$

Comparing with (4.12) we see that

$$\int_{\mathbb{R}} |\widehat{f_n}(\xi)|^2 e^{i\ell\xi} d\xi = \int_{\mathbb{R}} |(f_{n-1})^{\wedge}(\xi)|^2 e^{i\ell\xi} d\xi.$$

Repeating the argument involving (4.9) and (4.10) we obtain

$$\int_{\mathbb{R}} |\widehat{f_n}(\xi)|^2 e^{i\ell\xi} d\xi = \int_{\mathbb{R}} |\widehat{f_1}(\xi)|^2 e^{i\ell\xi} d\xi = \int_{2K} |m_0(\tfrac{1}{2}\xi)|^2 e^{i\ell\xi} d\xi$$

$$= 2 \int_{K} |m_0(\mu)|^2 e^{i\ell 2\mu} d\mu$$

$$= 2 \int_{-\frac{\pi}{2}}^{\frac{\pi}{2}} \left[|m_0(\mu)|^2 + |m_0(\mu+\pi)|^2 \right] e^{i\ell 2\mu} d\mu$$

$$= 2 \int_{-\frac{\pi}{2}}^{\frac{\pi}{2}} e^{i\ell 2\mu} d\mu = 2\pi\delta_{0,\ell}.$$

This proves (4.11).

Using Fatou's lemma we have

$$\int_{\mathbb{R}} |\varphi(x)|^2 dx = \frac{1}{2\pi} \int_{\mathbb{R}} |\hat{\varphi}(\xi)|^2 d\xi = \frac{1}{2\pi} \int_{\mathbb{R}} \lim_{n\to\infty} |\widehat{f_n}(\xi)|^2 d\xi$$

$$\leq \frac{1}{2\pi} \lim_{n\to\infty} \int_{\mathbb{R}} |\widehat{f_n}(\xi)|^2 d\xi = 1.$$

This shows that $\varphi \in L^2(\mathbb{R})$ and $\|\varphi\|_2 \leq 1$. We claim that

$$\lim_{n \to \infty} f_n = \varphi \quad \text{in the norm of } L^2(\mathbb{R}).$$

To see this observe that from condition (C) of Proposition 4.7 we have $m_0(2^{-j}\xi) \neq 0$ for $j = 1, 2, \cdots$ and $\xi \in K$. Since $\hat{\varphi}(0) = 1$ there exists $j_1 \in \mathbb{N}$ such that if $j > j_1$, $|m_0(2^{-j}\xi)| \geq C_1 > 0$ (C_1 can be chosen to be arbitrarily close to 1) for all $\xi \in K$. For $1 \leq j \leq j_1$, $m_0(2^{-j}\xi)$ represent only a finite number of non-zero continuous functions on the compact set K. Hence, there exists C_2 such that

$$|m_0(2^{-j}\xi)| \geq C_2 > 0 \quad \text{for} \quad j = 1, 2, \cdots \quad \text{and all } \xi \in K. \qquad (4.13)$$

Since $|m_0(\xi) - m_0(0)| \leq C_3|\xi|$, by the mean value theorem and the fact $|m_0(0)| = 1$, we have $|m_0(\xi)| \geq 1 - C_3|\xi|$. Since K is bounded we can find $j_2 \in \mathbb{N}$ such that $2^{-j}C_3|\xi| < \frac{1}{2}$ if $\xi \in K$ for all $j \geq j_2$. Using $1 - x \geq e^{-2x}$ for $0 \leq x \leq \frac{1}{2}$ we have, for $\xi \in K$,

$$|\hat{\varphi}(\xi)| = \left[\prod_{j=1}^{j_2} |m_0(2^{-j}\xi)|\right] \cdot \left[\prod_{j=j_2+1}^{\infty} |m_0(2^{-j}\xi)|\right]$$

$$\geq C_2^{j_2} \prod_{j=j_2+1}^{\infty} e^{-2C_3 2^{-j}|\xi|}$$

$$\geq C_2^{j_2} e^{-2C_3 2^{-j_2} \max_{\{\xi \in K\}} |\xi|} = B > 0.$$

From this estimate it follows that

$$|\widehat{f_n}(\xi)| = \chi_{2^n K}(\xi) \prod_{j=1}^{n} |m_0(2^{-j}\xi)|$$

$$\leq \frac{|\hat{\varphi}(2^{-n}\xi)|}{B} \prod_{j=1}^{n} |m_0(2^{-j}\xi)| = \frac{1}{B}|\hat{\varphi}(\xi)|.$$

We can now apply the Lebesgue dominated convergence theorem to obtain

$$\lim_{n \to \infty} \int_{\mathbb{R}} |\widehat{f_n}(\xi) - \hat{\varphi}(\xi)|^2 d\xi = 0.$$

From (4.11) we deduce that

$$\langle \varphi, \varphi(\cdot - \ell) \rangle = \lim_{n \to \infty} \langle f_n, f_n(\cdot - \ell) \rangle = \delta_{0,\ell}.$$

This shows that $\{\varphi(\cdot - \ell) : \ell \in \mathbb{Z}\}$ is an orthonormal system. Once we have this result it is easy to show that φ defines an MRA with the low-pass filter m_0 (see the construction that follows Theorem 3.13 of Chapter 2). In particular the function ψ such that $\hat{\psi}(2\xi) = e^{i\xi} \overline{m_0(\xi + \pi)} \, \hat{\varphi}(\xi)$ is an orthonormal wavelet associated with this MRA.

∎

This theorem, therefore, can be used to construct orthonormal wavelets. If $m_0 \in C^1(\mathbb{R})$ is a 2π-periodic function which satisfies (4.3) and (4.4) for all $\xi \in [-\pi, \pi]$, and we suppose that $m_0(\xi) \neq 0$ on $[-\frac{1}{2}\pi, \frac{1}{2}\pi]$, then condition (C) of Proposition 4.7 is satisfied with $K = [-\pi, \pi]$. Hence, Theorem 3.13 of Chapter 2 is a simple corollary of Theorem 4.8. Recall that it was Theorem 3.13 of Chapter 2 that allowed us to construct orthonormal wavelets with compact support. Let us present some new examples below.

EXAMPLE G: This example can be considered to be a continuation and elaboration of Example C in section 7.2. Let $0 < a \le \pi$ and $0 < \varepsilon \le \frac{1}{6}a$. Let b be a bell function (see Chapter 1) associated with the interval $[-\frac{1}{2}a, \pi - \frac{1}{2}a]$ and the extension by ε at both endpoints (see Figure 1.2 in section 1.3). In particular, $b(\xi)^2 + b(\xi + \pi)^2 = 1$ for all $\xi \in [-\pi - \frac{1}{2}a + \varepsilon, \pi - \frac{1}{2}a - \varepsilon]$. Let m_0 be the 2π-periodic function that coincides with b on $[-\pi, \pi]$. Condition (C) of Proposition 4.7 is satisfied with $K = [-a, 2\pi - a]$ and so are (4.3) and (4.4). By Theorem 4.8, m_0 produces a scaling function φ such that

$$\hat{\varphi}(\xi) = \prod_{j=1}^{\infty} m_0(2^{-j}\xi).$$

We claim that the function $\hat{\varphi}$ has support contained in the interval

$$[-a - 2\varepsilon, 2\pi - a + 2\varepsilon].$$

For $\xi \in [2^j(2\pi - a + 2\varepsilon), 2^{j+1}(2\pi - a + 2\varepsilon)]$ with $j \ge 0$, we have $t \equiv 2^{-(j+1)}\xi \in [\pi - \frac{1}{2}a + \varepsilon, 2\pi - a + 2\varepsilon]$; if $t \in [\pi - \frac{1}{2}a + \varepsilon, \pi]$, $m_0(t) = 0$ and if $t \in [\pi, 2\pi - a + 2\varepsilon]$, $t - 2\pi \in [-\pi, -a + 2\varepsilon]$, and $m_0(t) = m_0(t - 2\pi) = 0$ since $-a + 2\varepsilon \le -\frac{1}{2}a - \varepsilon$ (use $\varepsilon \le \frac{1}{6}a$). Thus

$$\hat{\varphi}(\xi) = \hat{\varphi}(2^{j+1}t) = m_0(2^j t) \cdots m_0(t) \prod_{\ell=1}^{\infty} m_0(2^{-\ell}t) = 0.$$

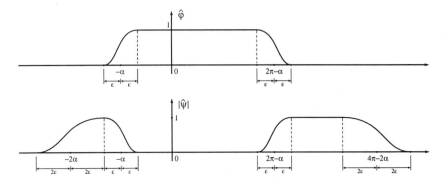

Figure 7.1: Graphs of $\hat{\varphi}$ and $|\hat{\psi}|$ for the wavelet in Example G.

This shows that $\hat{\varphi}(\xi) = 0$ for all $\xi \in [2\pi - a + 2\varepsilon, \infty)$. In the case $\xi \in [-2^{j+1}(a + 2\varepsilon), -2^j(a + 2\varepsilon)]$ with $j \geq 0$, we have

$$t \equiv 2^{-(j+1)}\xi \in [-a - 2\varepsilon, -\tfrac{1}{2}a - \varepsilon].$$

Since $-a - 2\varepsilon \geq -\pi - \tfrac{1}{2}a + \varepsilon$ (use $3\varepsilon \leq \tfrac{1}{2}a$ and $a \leq \pi$), we have $m_0(t) = 0$. As before, $\hat{\varphi}(\xi) = 0$ for all $\xi \in (-\infty, -a - 2\varepsilon]$. This proves our claim.

Since $m_0(2^{-j}\xi) = 1$ for all $\xi \in [-a - 2\varepsilon, 2\pi - a + 2\varepsilon]$ and all $j \geq 2$ we have

$$\hat{\varphi}(\xi) = m_0(\tfrac{1}{2}\xi)\chi_{[-a-2\varepsilon, 2\pi - a + 2\varepsilon]}(\xi) = b(\tfrac{1}{2}\xi).$$

Figure 7.1 contains graphs of the Fourier transform of φ and the absolute value of the Fourier transform of the wavelet ψ. This last one is obtained from the equality

$$\hat{\psi}(\xi) = e^{i\frac{\xi}{2}} \overline{m_0(\tfrac{\xi}{2} + \pi)} \, \hat{\varphi}(\tfrac{\xi}{2}).$$

Using Theorem 4.8 we can prove a result that generalizes Theorem 3.13 of Chapter 2.

COROLLARY 4.14 *Let $m_0 \in C^1(\mathbb{R})$ be a 2π-periodic function which satisfies (4.3), (4.4) and $m_0(\xi) \neq 0$ for all $\xi \in [-\tfrac{1}{3}\pi, \tfrac{1}{3}\pi]$. Then m_0 is the low-pass filter for an MRA.*

PROOF: We have to find a set K that satisfies Condition (C) of Proposition 4.7. Let A^+ be the set of zeros of m_0 in $[\tfrac{1}{3}\pi, \tfrac{1}{2}\pi]$ and let A^- be the set of zeros of m_0 in $[-\tfrac{1}{2}\pi, -\tfrac{1}{3}\pi]$.

If $\alpha \in A^-$ there exists an open interval $V(\alpha)$ such that $|m_0(\xi)| < \frac{1}{\sqrt{2}}$ for all $\xi \in V(\alpha)$; using (4.4) we obtain $|m_0(\xi + \pi)| > \frac{1}{\sqrt{2}}$ for such ξ. Since m_0 is continuous, the set $A^- = \{\xi \in [-\frac{1}{2}\pi, -\frac{1}{3}\pi] : m_0(\xi) = 0\}$ is compact, and, hence, a finite union of the $V(\alpha)$'s covers A^-. Suppose that $\bigcup_{j=1}^n V(\alpha_j)$ covers A^-. Let us define

$$A_j^- = V(\alpha_j) \cap [-\tfrac{1}{2}\pi, -\tfrac{1}{3}\pi] \qquad \text{for} \;\; j = 1, 2, \cdots, n.$$

This gives us a finite collection of intervals $A_j^- \subseteq [-\frac{1}{2}\pi, -\frac{1}{3}\pi]$, $1 \le j \le n$, such that $\bigcup_{j=1}^n A_j^-$ covers A^- and $|m_0(\xi + \pi)| > \frac{1}{\sqrt{2}}$ when $\xi \in A_j^-$, $j = 1, 2, \cdots, n$.

The same procedure can be aplied to A^+ to obtain a finite collection of intervals $A_j^+ \subseteq [\frac{1}{3}\pi, \frac{1}{2}\pi]$, $j = 1, 2, \cdots, k$, such that $\bigcup_{j=1}^k A_j^+$ covers A^+ and $|m_0(\xi - \pi)| > \frac{1}{\sqrt{2}}$ if $\xi \in A_j^+$, $j = 1, 2, \cdots, k$.

We now define the compact set K which will satisfy condition (C) of Proposition 4.7: Let $E^- = \bigcup_{j=1}^n 2A_j^-$, $E^+ = \bigcup_{j=1}^k 2A_j^+$ and define K by

$$K = \overline{\{[-\pi, \pi] \setminus (E^- \cup E^+)\}} \cup \{(E^- + 2\pi) \cup (E^+ - 2\pi)\}.$$

It is clear that K is compact; $0 \in \overset{\circ}{K}$ since $K \supset [-\frac{2}{3}\pi, \frac{2}{3}\pi]$. The set K is congruent to $[-\pi, \pi]$ modulo 2π by construction.

For $\xi \in \overline{[-\pi, \pi] \setminus (E^- \cup E^+)}$ we have $|m_0(\frac{1}{2}\xi)| > 0$ by our construction. If $\xi \in \overline{2A_j^- + 2\pi}$, $\eta = \frac{1}{2}\xi - \pi \in \overline{A_j^-}$ and $|m_0(\frac{1}{2}\xi)| = |m_0(\eta + \pi)| \ge \frac{1}{\sqrt{2}}$; similarly, $|m_0(\frac{1}{2}\xi)| \ge \frac{1}{\sqrt{2}}$ on $\overline{2A_j^+ - 2\pi}$. For $j \ge 2$ and $\xi \in K$ we have $2^{-j}\xi \in [-\frac{1}{3}\pi, \frac{1}{3}\pi]$, since $K \subset [-\frac{4}{3}\pi, \frac{4}{3}\pi]$. Our hypothesis implies that $m_0(2^{-j}\xi) \ne 0$ for $\xi \in K$ and $j \ge 2$. Thus, $m_0(2^{-j}\xi) \ne 0$ if $j \ge 1$ when $\xi \in K$. This shows that K satisfies condition (C) of Proposition 4.7 and this finishes the proof of the corollary. ∎

This result is the best possible in the following sense: it is impossible to find $\beta < \frac{1}{3}$ such that if m_0 has no zeros on $[-\beta\pi, \beta\pi]$, then $\{\varphi(\cdot - k) : k \in \mathbb{Z}\}$ is an orthonormal system. This is a consequence of Example F in Chapter 2 (recall that $m_0(\xi) = \frac{1}{2}(1 + e^{3i\xi})$).

We now show how to construct a wavelet using Corollary 4.14.

EXAMPLE H: Let b be a C^∞ function, whose graph is drawn in Fig-

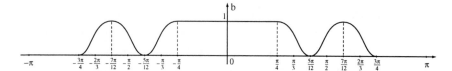

Figure 7.2: A function that gives rise to a low-pass filter.

ure 7.2, chosen so that

$$b^2(\xi) + b^2(\xi + \pi) = 1 \qquad \text{for all } \xi \in [-\pi, 0].$$

We leave it as an exercise for the reader to check that this is possible. Let m_0 be the 2π-periodic function that coincides with b on $[-\pi, \pi]$; this function m_0 satisfies (4.3) and (4.4) and, clearly, $m_0(\xi) \neq 0$ for all $\xi \in [-\frac{1}{3}\pi, \frac{1}{3}\pi] \subset [-\frac{5}{12}\pi, \frac{5}{12}\pi]$. Corollary 4.14 allows us to conclude that m_0 produces an MRA and we obtain a wavelet ψ such that

$$\hat{\psi}(2\xi) = e^{i\xi}\,\overline{m_0(\xi + \pi)}\,\hat{\varphi}(\xi) \quad \text{and} \quad \hat{\varphi}(\xi) = \prod_{j=1}^{\infty} m_0(2^{-j}\xi).$$

Unlike the wavelet constructed in Example G, this one is not band-limited. It is enough to show that $\hat{\varphi}$ does not have compact support (from (2.16) of Chapter 2, we see that if ψ is band-limited, so is φ).

Observe that the set $\{-\frac{2}{3}\pi, \frac{2}{3}\pi\}$ is invariant under the transformation $x \mapsto 2x$ (modulo 2π); hence, for all $j \geq 0$, $2^j(\frac{2}{3}\pi) \equiv \pm\frac{2}{3}\pi$ (mod 2π). Let $\xi_k = 2^k(\frac{2}{3}\pi)$ for $k \geq 1$; then

$$\hat{\varphi}(\xi_k) = \hat{\varphi}(2^k\tfrac{2}{3}\pi) = \prod_{j=1}^{\infty} m_0(2^k 2^{-j}\tfrac{2}{3}\pi)$$

$$= m_0(2^{k-1}\tfrac{2}{3}\pi)\cdots m_0(\tfrac{2}{3}\pi)\prod_{\ell=1}^{\infty} m_0(2^{-\ell}\tfrac{2}{3}\pi)$$

$$= m_0(\pm\tfrac{2}{3}\pi)\cdots m_0(\pm\tfrac{2}{3}\pi)\prod_{\ell=1}^{\infty} m_0(2^{-\ell}\tfrac{2}{3}\pi).$$

Clearly, $m_0(\frac{2}{3}\pi) = m_0(-\frac{2}{3}\pi) = \frac{1}{\sqrt{2}} \neq 0$ and since, when $\ell \geq 1$, $2^{-\ell}\frac{2}{3}\pi \in [-\frac{1}{3}\pi, \frac{1}{3}\pi]$, we have

$$m_0(\tfrac{1}{3}\pi) = \tfrac{1}{\sqrt{2}} \qquad \text{and} \qquad m_0(2^{-\ell}\tfrac{2}{3}\pi) = 1 \quad \text{for all } \ell \geq 2.$$

Hence, $\hat{\varphi}(\xi_k) \neq 0$. Since $\hat{\varphi}$ is in C^∞, supp $(\hat{\varphi})$ contains a non-trivial closed interval around $2^k(\frac{2}{3}\pi)$ for all $k = 1, 2, \cdots, \infty$. This establishes our claim.

We now turn our attention to the characterization of low-pass filters associated with localized multiresolution analysis. Recall that an MRA is called **localized** if the scaling function φ has polynomial decay in $L^2(\mathbb{R})$; that is

$$\int_\mathbb{R} (1 + |x|^2)^n |\varphi(x)|^2 \, dx < \infty \quad \text{for all } n \in \mathbb{N}$$

(see (2.22) in section 2.2, where this concept was introduced).

Our main result is Theorem 4.23 below, which is due to A. Cohen ([Co1] and [Co2]). We present its proof in several steps.

THEOREM 4.15 *Let m_0 be a 2π-periodic function in the class C^n, $n \in \mathbb{N}$, which satisfies (4.3) and (4.4). Let*

$$\hat{\varphi}(\xi) = \prod_{j=1}^\infty m_0(2^{-j}\xi). \tag{4.16}$$

Then, $\hat{\varphi}$ and all its derivatives up to order n, in the sense of distributions, belong to $L^2(\mathbb{R})$.

PROOF : Notice that Lemma 4.5 allows us to write (4.16), and to assert that $\hat{\varphi}$ is continuous. We have already shown that $\varphi \in L^2(\mathbb{R})$ (see the proof of Theorem 4.8). We now show that $\frac{d}{d\xi}\hat{\varphi}$ belongs to $L^2(\mathbb{R})$. To do this we approximate $\hat{\varphi}$ by using regular "cut off" functions. Let γ be a C^∞ function such that

$$\left.\begin{array}{lll} \text{(i)} & 0 \leq \gamma(\xi) \leq 1, & \\ \text{(ii)} & \gamma(\xi) = 1 & \text{if } \xi \in [-\frac{1}{2}\pi, \frac{1}{2}\pi], \\ \text{(iii)} & \gamma(\xi) = 0 & \text{if } \xi \notin [-\pi, \pi]. \end{array}\right\} \tag{4.17}$$

We define

$$g_n(\xi) = \gamma(2^{-n}\xi) \prod_{j=1}^n m_0(2^{-j}\xi)$$

so that $\lim_{n\to\infty} g_n(\xi) = \hat{\varphi}(\xi)$ for every $\xi \in \mathbb{R}$. We want to obtain a bound for the L^2-norm of $\frac{d}{d\xi}g_n$. We have

$$\frac{dg_n}{d\xi} = 2^{-n}\frac{d\gamma}{d\xi}(2^{-n}\xi)\prod_{j=1}^{n}m_0(2^{-j}\xi)$$

$$+\gamma(2^{-n}\xi)\sum_{\ell=1}^{n}2^{-\ell}\frac{dm_0}{d\xi}(2^{-\ell}\xi)\prod_{\substack{j=1\\j\neq\ell}}^{n}m_0(2^{-j}\xi). \qquad (4.18)$$

We shall use the following lemma:

LEMMA 4.19 *Let $\mu_1, \cdots, \mu_n$ be 2π-periodic functions and set*

$$M_j = \sup_{\xi\in\mathbb{T}}\left(|\mu_j(\xi)|^2 + |\mu_j(\xi+\pi)|^2\right).$$

Then,

$$\int_{-2^n\pi}^{2^n\pi}\prod_{j=1}^{n}|\mu_j(2^{-j}\xi)|^2\,d\xi \leq 2\pi M_1\cdots M_n.$$

PROOF: (The argument is similar to the one that appears in Proposition 3.9 of Chapter 2.) Let

$$\Pi_n(\xi) = \prod_{j=1}^{n}\mu_j(2^{-j}\xi),$$

which is a $2^{n+1}\pi$-periodic function. Hence,

$$I_n \equiv \int_{-2^n\pi}^{2^n\pi}\left|\Pi_n(\xi)\right|^2\,d\xi = \int_{0}^{2^{n+1}\pi}\left|\Pi_n(\xi)\right|^2\,d\xi$$

$$= \int_{0}^{2^n\pi}\left|\Pi_n(\xi)\right|^2\,d\xi + \int_{2^n\pi}^{2^{n+1}\pi}\left|\Pi_n(\xi)\right|^2\,d\xi$$

$$= \int_{0}^{2^n\pi}\left\{\left|\Pi_n(\xi)\right|^2 + \left|\Pi_n(\xi+2^n\pi)\right|^2\right\}d\xi$$

$$= \int_{0}^{2^n\pi}\left|\Pi_{n-1}(\xi)\right|^2\left\{|\mu_n(2^{-n}\xi)|^2 + |\mu_n(2^{-n}\xi+\pi)|^2\right\}d\xi$$

$$\leq M_n I_{n-1}.$$

Repeating this argument we obtain

$$I_n \leq M_n \cdots M_2 I_1 = M_n \cdots M_2 \int_0^{4\pi} |\mu_1(\tfrac{1}{2}\xi)|^2 \, d\xi$$

$$= 2M_n \cdots M_2 \int_0^{2\pi} |\mu_1(\xi)|^2 \, d\xi$$

$$= 2M_n \cdots M_2 \int_0^{\pi} \left\{ |\mu_1(\xi)|^2 + |\mu_1(\xi + \pi)|^2 \right\} d\xi \leq 2\pi M_n \cdots M_2 M_1.$$

∎

We continue with the proof of Theorem 4.15. Using (4.18) we obtain

$$\left\| \frac{dg_n}{d\xi} \right\|_{L^2} \leq 2^{-n} \sup_{\xi \in \mathbb{R}} \left| \frac{d\gamma}{d\xi} \right| \left(\int_{-2^n \pi}^{2^n \pi} \left(\prod_{j=1}^n |m_0(2^{-j}\xi)|^2 \right) d\xi \right)^{\frac{1}{2}}$$

$$+ \sum_{\ell=1}^n 2^{-\ell} \left(\int_{-2^n \pi}^{2^n \pi} \left| \frac{dm_0}{d\xi} \right| \prod_{\substack{j=1 \\ j \neq \ell}}^n |m_0(2^{-j}\xi)|^2 \, d\xi \right)^{\frac{1}{2}}. \qquad (4.20)$$

The integrals in (4.20) are of the type that appears in Lemma 4.19. In the first one, each μ_j is equal to m_0. (By Proposition 3.9 of Chapter 2, the integral equals 2π in this case.) In the other cases, all the μ_j but one are equal to m_0, the remaining one being $\frac{dm_0}{d\xi}$. By Lemma 4.19 each one of these integrals is bounded by $4\pi \left\| \frac{dm_0}{d\xi} \right\|_\infty^2$. Thus, we deduce

$$\left\| \frac{dg_n}{d\xi} \right\|_{L^2} \leq C 2^{-n} \sqrt{2\pi} + C \sum_{\ell=1}^n 2^{-\ell} (2\sqrt{\pi}) \leq C'. \qquad (4.21)$$

Since C' does not depend on n, there exists $h \in L^2(\mathbb{R})$ such that $\frac{d}{d\xi} g_n \to h$, as $n \to \infty$, in the weak sense (by the Banach-Alaoglu theorem) and, thus, in $\mathcal{S}'(\mathbb{R})$. Hence, the distributional derivative $\hat{\varphi}'$ of $\hat{\varphi}$ satisfies

$$(\hat{\varphi}', \eta) = -\int_{\mathbb{R}} \hat{\varphi} \, \eta' = -\lim_{n \to \infty} \int_{\mathbb{R}} g_n \eta' = \lim_{n \to \infty} \int_{\mathbb{R}} g_n' \eta = (h, \eta)$$

for all $\eta \in \mathcal{S}(\mathbb{R})$. Thus, $\frac{d}{d\xi} \hat{\varphi} = \hat{\varphi}' = h \in L^2(\mathbb{R})$.

For higher derivatives, of order $k \leq n$, the proof is similar. We shall encounter products of the type

$$\prod_{j=1}^{n} \mu_j(2^{-j}\xi),$$

where the μ_j are chosen from the collection of m_0 and its derivatives up to order k. Moreover, at least $n - k$ of the functions μ_j equal m_0, and in this case, $M_j = 1$. Thus, the products $2\pi M_1 M_2 \cdots M_n$ are uniformly bounded with respect to n. Combining this observation with the geometric series that we obtain when computing $\frac{d^k g_n}{d\xi^n}$, we deduce that the L^2-norm of $\frac{d^k g_n}{d\xi^n}$ is bounded independently of n. ∎

COROLLARY 4.22 *Suppose m_0 is as in Theorem 4.15, then $\hat\varphi \in C^{n-1}$. If, in addition, $m_0 \in C^\infty$, then $\hat\varphi \in C^\infty$.*

PROOF : Recall that if $n \in \mathbb{N}$, the Sobolev space $L^{2,n}(\mathbb{R})$ is the space of all $f \in L^2(\mathbb{R})$ such that

$$\frac{d^k}{dx^k} f \in L^2(\mathbb{R}) \qquad \text{for all } k \le n,$$

where the derivatives are taken in the sense of distributions. Theorem 4.15 shows that $\hat\varphi \in L^{2,n}(\mathbb{R})$. By the Sobolev imbedding theorem (see [Ste1] or the last paragraph in section 6.7), $L^{2,n}(\mathbb{R}) \subset C^{n-1}$. The last assertion of the corollary is a consequence of this and the hypothesis $m_0 \in C^\infty$. ∎

THEOREM 4.23 *A function m_0 is the low-pass filter for a localized MRA if and only if m_0 is a C^∞ 2π-periodic function which satisfies (4.3), (4.4) and condition (C) of Proposition 4.7.*

PROOF : Suppose that m_0 is the low-pass filter for a localized MRA. Then φ has polynomial decay and we have shown in Proposition 2.24 of Chapter 2 that $m_0 \in C^\infty$. The rest of the properties of m_0 have been already proved (see the derivation of (4.3) and (4.4) and the proof of Proposition 4.7).

Assume that m_0 is a C^∞ 2π-periodic function which satisfies (4.3), (4.4) and condition (C) of Proposition 4.7. We have already proved (see Theorem 4.8) that m_0 is the low-pass filter for an MRA. It remains to show that this MRA is localized.

Theorem 4.15 shows that $\frac{d^n}{d\xi^n}\hat{\varphi} \in L^2(\mathbb{R})$ for all $n = 0, 1, 2, \cdots$. Since $\frac{d^n}{d\xi^n}\hat{\varphi}(\xi) = (-i)^n((\cdot)^n\varphi)^\wedge(\xi)$ we have

$$\int_{\mathbb{R}} |x|^{2n}|\varphi(x)|^2 \, dx = C \int_{\mathbb{R}} \left|\frac{d^n}{d\xi^n}\hat{\varphi}(\xi)\right|^2 d\xi < \infty.$$

Thus,

$$\int_{\mathbb{R}} (1 + |x|^2)^n |\varphi(x)|^2 \, dx \leq \int_{|x|\leq 1} 2^n |\varphi(x)|^2 \, dx + \int_{|x|>1} 2|x|^{2n}|\varphi(x)|^2 \, dx$$

$$\leq C(n)(\|\varphi\|_2^2 + \|(\cdot)^n\varphi\|_2^2) < \infty.$$

This shows that φ has polynomial decay, and, hence, the MRA is localized.

$\blacksquare$

7.5   A characterization of scaling functions

We have characterized all orthonormal wavelets by the two basic equations (1.2) and (1.3), under the L^2-normalization, in section 7.1. We presented a characterization of all orthonormal wavelets that arise from an MRA in terms of a single equality in section 7.3 (see Theorem 3.2). Necessary and sufficient conditions for certain 2π-periodic functions to be low-pass filters have been described in section 7.4. It is then natural to pose the question of how one can characterize those functions that are scaling functions for an MRA. This question, in fact, was posed by R. Strichartz in the survey article [St].

We should state carefully what we mean when we say that a function is a scaling function for an MRA. Given $\varphi \in L^2(\mathbb{R})$, we define the closed subspaces V_j, $j \in \mathbb{Z}$, by

$$V_j = \begin{cases} \overline{\text{span}\{\varphi(\cdot - k) : k \in \mathbb{Z}\}}, & \text{if } j = 0; \\ \{f : f(2^{-j}\cdot) \in V_o\}, & \text{if } j \neq 0. \end{cases} \tag{5.1}$$

We say that $\varphi \in L^2(\mathbb{R})$ is a scaling function for an MRA if the sequence of closed subspaces $\{V_j : j \in \mathbb{Z}\}$ given by (5.1) constitutes an MRA for $L^2(\mathbb{R})$ with $\{\varphi(\cdot - k) : k \in \mathbb{Z}\}$ being an orthonormal basis for V_0.

THEOREM 5.2 *A function $\varphi \in L^2(\mathbb{R})$ is a scaling function for an MRA if and only if*

$$\sum_{k\in\mathbb{Z}} |\hat{\varphi}(\xi + 2k\pi)|^2 = 1 \qquad \text{for a.e. } \xi \in \mathbb{T}; \tag{5.3}$$

$$\lim_{j\to\infty} |\hat{\varphi}(2^{-j}\xi)| = 1 \qquad \text{for a.e. } \xi \in \mathbb{R}; \tag{5.4}$$

$$\left.\begin{array}{l} \text{There exists a } 2\pi\text{-periodic function } m_0 \text{ such that} \\[4pt] \hat{\varphi}(2\xi) = m_0(\xi)\hat{\varphi}(\xi) \qquad \text{for a.e. } \xi \in \mathbb{R}. \end{array}\right\} \tag{5.5}$$

PROOF : Suppose that φ is a scaling function for an MRA as defined in section 2.1. Then $\{\varphi(\cdot - k) : k \in \mathbb{Z}\}$ is an orthonormal system in $L^2(\mathbb{R})$, which implies (5.3) by Proposition 1.11 of Chapter 2. Since $\{V_j : j \in \mathbb{Z}\}$ is an MRA for $L^2(\mathbb{R})$, the closure of $\cup_{j\in\mathbb{Z}}V_j$ is $L^2(\mathbb{R})$, and according to the paragraph that follows the proof of Theorem 1.7 in Chapter 2, we must have

$$\lim_{j\to\infty} \frac{1}{2^{-j+1}} \int_{-2^{-j}}^{2^{-j}} |\hat{\varphi}(\mu)|\, d\mu = 1.$$

The change of variables $\mu = 2^{-j}\xi$ shows that

$$\lim_{j\to\infty} \frac{1}{2} \int_{-1}^{1} |\hat{\varphi}(2^{-j}\xi)|\, d\xi = 1.$$

Since $|m_0(\xi)| \le 1$ for a.e. $\xi \in \mathbb{R}$ (see (2.5) of Chapter 2), the equality $\hat{\varphi}(2\xi) = m_0(\xi)\hat{\varphi}(\xi)$ (see (2.3) of Chapter 2) shows that $|\hat{\varphi}(2^{-j}\xi)|$ is non-decreasing for almost every $\xi \in \mathbb{R}$ as $j \to \infty$. Let

$$g(\xi) = \lim_{j\to\infty} |\hat{\varphi}(2^{-j}\xi)|.$$

The fact that $|\hat{\varphi}(\xi)| \le 1$ a.e. (as a consequence of (5.3)) and the Lebesgue dominated convergence theorem give us

$$\frac{1}{2} \int_{-1}^{1} g(\xi)\, d\xi = 1,$$

and (5.4) then follows since $0 \le g(\xi) \le 1$ for a.e. $\xi \in \mathbb{R}$. Finally, (5.5) is equality (2.3) of Chapter 2.

Assume now that (5.3), (5.4) and (5.5) are satisfied. The orthonormality of $\{\varphi(\cdot - k) : k \in \mathbb{Z}\}$ follows immediately from (5.3) and Proposition 1.11 of

Chapter 2. This, and the definition of V_0 by (5.1), give us (1.5) of Chapter 2 in the definition of an MRA.

The definition of the subspaces V_j given in (5.1) also shows that $f \in V_j$ if and only if $f(2(\cdot)) \in V_{j+1}$, which is (1.2) of the definition of an MRA. Moreover, for each $j \in \mathbb{Z}$, we claim that

$$V_j = \{f : \hat{f}(2^j \xi) = \mu_j(\xi)\hat{\varphi}(\xi) \text{ for some } 2\pi\text{-periodic } \mu_j \in L^2(\mathbb{T})\}. \quad (5.6)$$

This claim is established by expressing $f(2^{-j}\cdot) \in V_0$ as a linear combination of $\varphi(\cdot - k)$, $k \in \mathbb{Z}$, and then taking Fourier transforms. By (5.5), the 2π-periodicity of m_0 and (5.3) we have

$$1 = \sum_{k \in \mathbb{Z}} |\hat{\varphi}(2\xi + 2k\pi)|^2 = \sum_{k \in \mathbb{Z}} |m_0(\xi + k\pi)|^2 |\hat{\varphi}(\xi + k\pi)|^2$$

$$= |m_0(\xi)|^2 \sum_{\ell \in \mathbb{Z}} |\hat{\varphi}(\xi + 2\ell\pi)|^2 + |m_0(\xi + \pi)|^2 \sum_{\ell \in \mathbb{Z}} |\hat{\varphi}(\xi + (2\ell+1)\pi)|^2$$

$$= |m_0(\xi)|^2 + |m_0(\xi + \pi)|^2 \qquad\qquad \text{for } a.e.\ \xi \in \mathbb{T}.$$

In particular, this shows $|m_0(\xi)| \leq 1$ for a.e. $\xi \in \mathbb{T}$. Property (1.1) of the definition of an MRA follows if we show $V_0 \subset V_1$. By (5.6), given $f \in V_0$ there is a 2π-periodic function $\mu_o \in L^2(\mathbb{T})$, such that $\hat{f}(\xi) = \mu_o(\xi)\hat{\varphi}(\xi)$. Thus, using (5.5),

$$\hat{f}(2\xi) = \mu_o(2\xi)\hat{\varphi}(2\xi) = \mu_o(2\xi)m_0(\xi)\hat{\varphi}(\xi).$$

Clearly, $\mu_o(2\xi)m_0(\xi)$ is 2π-periodic, and is in $L^2(\mathbb{T})$ since $|m_0(\xi)| \leq 1$ for a.e. $\xi \in \mathbb{T}$. Using (5.6) again we see that $f \in V_1$.

We have already seen that property (1.3) in the definition of an MRA is a consequence of (1.1), (1.2) and (1.5) (see Theorem 1.6 of Chapter 2); thus, it remains to show property (1.4) in the definition of an MRA. That is,

$$L^2(\mathbb{R}) = \overline{\bigcup_{j \in \mathbb{Z}} V_j}. \qquad\qquad (5.7)$$

Let P_j be the projection onto V_j. It suffices to show that

$$\|P_j f - f\|_2^2 = \|f\|_2^2 - \|P_j f\|_2^2 \longrightarrow 0 \qquad \text{as } j \to \infty.$$

Furthermore, we can assume that our $f \in L^2(\mathbb{R})$ is such that $\hat{f}$ has compact support. Since $\{2^{\frac{j}{2}}\varphi(2^j \cdot -k) : k \in \mathbb{Z}\}$ is an orthonormal basis for V_j and, for large positive j, $\hat{f}(2^j)$ has support in $[-\pi, \pi]$,

$$\|P_j f\|_2^2 = 2^{-j} \sum_{k \in \mathbb{Z}} \left| \int_{\mathbb{R}} f(t) \, \overline{\varphi(2^j t - k)} \, 2^j \, dt \right|^2$$

$$= 2^{-j} \sum_{k \in \mathbb{Z}} \left| \int_{\mathbb{R}} f(2^{-j}t) \, \overline{\varphi(t - k)} \, dt \right|^2$$

$$= 2^j \sum_{k \in \mathbb{Z}} \left| \frac{1}{2\pi} \int_{-\pi}^{\pi} \hat{f}(2^j \xi) \, \overline{\hat{\varphi}(\xi)} \, e^{ik\xi} d\xi \right|^2$$

$$= \frac{2^j}{2\pi} \int_{-\pi}^{\pi} \left| \hat{f}(2^j \xi) \, \hat{\varphi}(\xi) \right|^2 d\xi = \frac{1}{2\pi} \int_{-2^j \pi}^{2^j \pi} \left| \hat{f}(\eta) \, \hat{\varphi}(2^{-j}\eta) \right|^2 d\eta.$$

But,

$$\frac{1}{2\pi} \int_{-2^j \pi}^{2^j \pi} |\hat{f}(\eta)|^2 \, |\hat{\varphi}(2^{-j}\eta)|^2 \, d\eta \longrightarrow \frac{1}{2\pi} \int_{-\infty}^{\infty} |\hat{f}(\eta)|^2 \, d\eta = \|f\|_2^2 \qquad (5.8)$$

as $j \to \infty$ because of the dominated convergence theorem (since $|\hat{\varphi}(\xi)| \leq 1$ by (5.3)) and (5.4).

∎

REMARK: The characterization given in Theorem 5.2 shows that if φ is a scaling function for an MRA, then the function $\varphi^{\natural}$, defined by $\widehat{\varphi^{\natural}}(\xi) = |\hat{\varphi}(\xi)|$, is also a scaling function for an MRA of $L^2(\mathbb{R})$.

The reader can easily check that if $\varphi = \chi_{[-1,0]}$, the function φ is a scaling function for an MRA. The only thing that needs to be done is to observe that

$$\hat{\varphi}(\xi) = e^{i\frac{\xi}{2}} \frac{\sin(\xi/2)}{\xi/2}$$

and show that (5.3), (5.4) and (5.5) are satisfied (by using equalities based on the first formula in the proof of Lemma 1.9 in Chapter 4). As we have seen in Chapter 2 this is the scaling function for the Haar wavelet (see Example B of Chapter 2).

The Shannon wavelet can also be obtained in this way (see Example C of Chapter 2). In this case we take

$$\varphi(\xi) = \frac{\sin(\pi x)}{\pi x}$$

so that $\hat{\varphi}(\xi) = \chi_{[-\pi,\pi]}(\xi)$ and apply Theorem 5.2.

A more interesting example arises when we consider spline wavelets. Let Δ^n, $n = 1, 2, 3, \cdots$, be the functions defined inductively by

$$\Delta^1 = \chi * \chi,$$
$$\Delta^n = \Delta^{n-1} * \chi, \quad n = 2, 3, \cdots \tag{5.9}$$

where $\chi = \chi_{[0,1]}$ (see section 4.2). By part (i) of Lemma 2.3 in Chapter 4, we have

$$\widehat{\Delta^n}(\xi) = e^{i\frac{n+1}{2}\xi} \left(\frac{\sin(\xi/2)}{\xi/2} \right)^{n+1}, \quad n = 1, 2, \cdots$$

The function Δ^n does not satisfy (5.3), but

$$\sum_{k \in \mathbb{Z}} |\widehat{\Delta^n}(\xi + 2k\pi)|^2 = \left(2 \sin \tfrac{\xi}{2} \right)^{2(n+1)} \sum_{k \in \mathbb{Z}} \frac{1}{(\xi + 2k\pi)^{2(n+1)}} = P_{2n+1}(\xi),$$

where P_N is the trigonometric polynomial that appears in Proposition 2.7 of Chapter 4. Since $P_{2n+1}(\xi) > 0$ we can consider the function φ^n given by

$$\widehat{\varphi^n}(\xi) = \frac{\widehat{\Delta^n}(\xi)}{\sqrt{P_{2n+1}(\xi)}}, \quad n = 1, 2, \cdots, \tag{5.10}$$

which clearly satisfies (5.3). Property (5.4) is satisfied since

$$\widehat{\varphi^n}(0) = \frac{\widehat{\Delta^n}(0)}{\sqrt{P_{2n+1}(0)}} = \frac{1}{\sqrt{1}} = 1, \quad n = 1, 2, \cdots,$$

and both $\widehat{\Delta^n}$ and P_{2n+1} are continuous. To show (5.5) observe that

$$\frac{\widehat{\varphi^n}(2\xi)}{\widehat{\varphi^n}(\xi)} = e^{i\frac{n+1}{2}\xi} \left(\cos \tfrac{\xi}{2} \right)^{n+1} \frac{\sqrt{P_{2n+1}(\xi)}}{\sqrt{P_{2n+1}(2\xi)}}$$

is a 2π-periodic function (consider separately the cases n odd and n even).

7.6   Non-existence of smooth wavelets in $H^2(\mathbb{R})$

So far in this book we have only worked with wavelets for $L^2(\mathbb{R})$. But some of the techniques developed in the previous sections of this chapter can be used to obtain results for wavelets in $H^2(\mathbb{R})$, which is a closed subspace of $L^2(\mathbb{R})$. The classical **Hardy space** $H^2(\mathbb{R})$ is defined by

$$H^2(\mathbb{R}) = \{f \in L^2(\mathbb{R}) : \hat{f}(\xi) = 0 \quad for \quad a.e. \ \xi \leq 0\}.$$

We say that a function $\psi \in H^2(\mathbb{R})$ is an **orthonormal wavelet for** $H^2(\mathbb{R})$ if the set $\{\psi_{j,k} : j, k \in \mathbb{Z}\}$ is an orthonormal basis for $H^2(\mathbb{R})$. For simplicity, we shall call such ψ an H^2-**wavelet**.

The first thing that should be noticed is that, by the Paley-Wiener theorem, the Hardy space $H^2(\mathbb{R})$ does not contain any compactly supported function other than the null function; hence, we can never find a compactly supported H^2-wavelet. This shows that nothing similar to the Haar wavelet or the compactly supported wavelets for $L^2(\mathbb{R})$ we constructed in Chapter 2 can be found for $H^2(\mathbb{R})$.

On the other hand, there exists an H^2-wavelet which is very similar to the Shannon wavelet for $L^2(\mathbb{R})$. This is given in the following example.

EXAMPLE I: Let ψ be defined by $\hat{\psi} = \chi_{[2\pi,4\pi]}$. We claim that ψ is an H^2-wavelet. The orthonormality of the system $\{\psi_{j,k} : j, k \in \mathbb{Z}\}$ follows easily by using the Plancherel theorem, the definition of $\hat{\psi}$ and the orthonormality of the system $\{e^{ikx} : k \in \mathbb{Z}\}$ on the interval $[2\pi, 4\pi]$. Since $H^2(\mathbb{R})$ is a Hilbert space (with the same inner product as $L^2(\mathbb{R})$) it is enough to show that we have the equality in Bessel's inequality (see (1.2) of Chapter 1) to establish the completeness of the system $\{\psi_{j,k} : j, k \in \mathbb{Z}\}$. This can be done as follows.

By the Plancherel theorem and the definition of $\hat{\psi}$, we obtain, for all $f \in H^2(\mathbb{R})$,

$$\sum_{j,k\in\mathbb{Z}} |<f, \psi_{j,k}>|^2 = \sum_{j,k\in\mathbb{Z}} \left| \frac{1}{2\pi} \int_{\mathbb{R}} \hat{f}(\xi) \overline{(\psi_{j,k})^\wedge(\xi)} \, d\xi \right|^2$$

$$= \sum_{j,k\in\mathbb{Z}} \frac{2^{-j}}{4\pi^2} \left| \int_{2^j 2\pi}^{2^j 4\pi} \hat{f}(\xi) e^{i2^{-j}k\xi} \, d\xi \right|^2$$

$$= \sum_{j \in \mathbb{Z}} 2^j \sum_{k \in \mathbb{Z}} \left| \frac{1}{2\pi} \int_{2\pi}^{4\pi} \hat{f}(2^j \mu) e^{ik\mu} \, d\mu \right|^2.$$

The sum over $k \in \mathbb{Z}$ in this very last expression is the sum of the sequence of the squares of the absolute values of the Fourier coefficients of the function $\chi_{[2\pi, 4\pi]}(\cdot) \hat{f}(2^j(\cdot))$; thus,

$$\sum_{j,k \in \mathbb{Z}} |<f, \psi_{j,k}>|^2 = \sum_{j \in \mathbb{Z}} 2^j \frac{1}{2\pi} \int_{2\pi}^{4\pi} |\hat{f}(2^j \mu)|^2 \, d\mu$$

$$= \sum_{j \in \mathbb{Z}} \frac{1}{2\pi} \int_{2^{-j}2\pi}^{2^{-j}4\pi} |\hat{f}(\xi)|^2 \, d\xi = \frac{1}{2\pi} \int_0^\infty |\hat{f}(\xi)|^2 \, d\xi = \|f\|_{H^2(\mathbb{R})}^2.$$

We have presented a direct proof of the fact that ψ is an orthonormal wavelet for $H^2(\mathbb{R})$. The most elementary argument follows from the observation that the dyadic dilations of $[2\pi, 4\pi]$ form a partition of $(0, \infty)$ and for each dilation by 2^j the functions $e^{i2^{-j}k\xi}$ provide us with the Fourier series expansions of the functions in $L^2(2^{j+1}\pi, 2^{j+2}\pi)$. This result also provides a good example of an application of the results stated in this section. We remark that the function ψ of this example has already appeared as the $L^2(\mathbb{R})$-limit of a family of wavelets for $L^2(\mathbb{R})$ (see the remark that follows the proof of Theorem 3.18 in this chapter).

The Fourier transform of the wavelet for $H^2(\mathbb{R})$ given in the above example is not continuous. The main purpose of this section is to show that $H^2(\mathbb{R})$ does not contain orthonormal wavelets ψ for which $|\hat{\psi}|$ is continuous and with a certain mild decrease at ∞. More precisely, we say that a function $\psi \in L^2(\mathbb{R})$ (or $H^2(\mathbb{R})$) satisfies the **regularity** condition $(\mathfrak{R}^0)$ if

$$\left. \begin{array}{l} |\hat{\psi}| \text{ is continuous on } \mathbb{R}, \quad \text{and} \\[2mm] |\hat{\psi}(\xi)| = O\big((1 + |\xi|)^{-\alpha - \frac{1}{2}}\big) \text{ at } \infty, \text{ for some } \alpha > 0. \end{array} \right\} \quad (6.1)$$

Notice that the regularity (or smoothness) condition is expressed in terms of the Fourier transform, rather than the function itself, which is natural since the space $H^2(\mathbb{R})$ is defined in terms of the Fourier transform. The decay at ∞ is expressed by the second condition in (6.1). We add a word of caution: in Chapter 6 (see (2.12)) we used the symbol $\mathfrak{R}^0$ to denote a different class.

We shall show that there is no wavelet for $H^2(\mathbb{R})$ that satisfies the regularity condition $(\mathfrak{R}^0)$. To prove this we need to extend some of the results proved in section 7.1 and section 7.3 to the setting of $H^2(\mathbb{R})$. Some of the simple modifications needed to do this will be omitted.

Let ψ be an orthonormal wavelet for $H^2(\mathbb{R})$. We have seen in section 3.1 that the system $\{\psi_{j,k} : j, k \in \mathbb{Z}\}$ is orthonormal if and only if

$$\sum_{k \in \mathbb{Z}} |\hat{\psi}(\xi + 2k\pi)|^2 = 1 \qquad \text{for } a.e. \ \xi \in \mathbb{R}; \tag{6.2}$$

$$\sum_{k \in \mathbb{Z}} \hat{\psi}(2^j(\xi + 2k\pi)) \overline{\hat{\psi}(\xi + 2k\pi)} = 0 \quad \text{for } a.e. \ \xi \in \mathbb{R}, \ j \geq 1. \tag{6.3}$$

These two equations, of course, are exactly the same as the corresponding ones for wavelets in $L^2(\mathbb{R})$. But the basic equations that characterize the wavelets for $H^2(\mathbb{R})$ are a little different (see Theorem 1.1 for the case of $L^2(\mathbb{R})$).

THEOREM 6.4 *A function $\psi \in H^2(\mathbb{R})$, with $\|\psi\|_2 = 1$, is an orthonormal wavelet for $H^2(\mathbb{R})$ if and only if*

$$\sum_{j \in \mathbb{Z}} |\hat{\psi}(2^j\xi)|^2 = \chi_{\mathbb{R}^+}(\xi) \qquad \text{for } a.e. \ \xi \in \mathbb{R}; \tag{6.5}$$

$$\sum_{j=0}^{\infty} \hat{\psi}(2^j\xi) \overline{\hat{\psi}(2^j(\xi + 2k\pi))} = 0 \quad \text{for } a.e. \ \xi \in \mathbb{R}, \ k \in 2\mathbb{Z}+1. \tag{6.6}$$

Notice that the only difference with Theorem 1.1 is that in the $H^2(\mathbb{R})$ case, the left-hand side of equality (6.5) cannot be equal to 1 for almost all $\xi \in \mathbb{R}$ since $\hat{\psi}(\xi) = 0$ when $\xi < 0$. The proof of Theorem 6.4 is similar to that of Theorem 1.1, and we leave it to the reader to make the necessary modifications in the proof of Theorem 1.1 to provide a proof of Theorem 6.4.

One can obtain several corollaries of this theorem, much in the same spirit of the applications given in section 7.2. We will not pursue this matter now. Instead we shall exhibit a new family of H^2-wavelets.

EXAMPLE J : For each $\ell \geq 0$, $\ell \in \mathbb{Z}$, define the set

$$K_\ell = \left[\frac{2^{\ell+1}}{2^{\ell+1} - 1}\pi, 2\pi\right] \cup \left[2^{\ell+1}\pi, \frac{2^{2\ell+2}}{2^{\ell+1} - 1}\pi\right],$$

and consider ψ_ℓ defined by $\widehat{\psi_\ell} = \chi_{K_\ell}$. It is easy to see that ψ_ℓ satisfies (6.5) and (6.6). Hence, ψ_ℓ, $\ell \geq 0$, is a family of H^2-wavelets. Notice that for $\ell = 0$ we recover the wavelet in Example I. When $\ell = 2$, we obtain $K_2 = \left[\frac{8}{7}\pi, 2\pi\right] \cup \left[8\pi, \frac{64}{7}\pi\right]$, which gives us the Journé-type H^2-wavelet ψ_2 (see the Journé wavelet for $L^2(\mathbb{R})$ after the proof of Proposition 2.17 in Chapter 2).

The concept of multiresolution analysis can also be defined for $H^2(\mathbb{R})$. A sequence $\{V_j : j \in \mathbb{Z}\}$ of closed subspaces of $H^2(\mathbb{R})$ is a **multiresolution analysis for** $H^2(\mathbb{R})$ (H^2-MRA) if the following properties are satisfied:

$$V_j \subset V_{j+1} \qquad \text{for all } j \in \mathbb{Z}; \tag{6.7}$$

$$f \in V_j \text{ if and only if } f(2(\cdot)) \in V_{j+1} \qquad \text{for all } j \in \mathbb{Z}; \tag{6.8}$$

$$\bigcap_{j \in \mathbb{Z}} V_j = \{0\}; \tag{6.9}$$

$$\overline{\bigcup_{j \in \mathbb{Z}} V_j} = H^2(\mathbb{R}); \tag{6.10}$$

$$\left.\begin{array}{l} \text{There exists a } \varphi \in V_0, \text{ such that} \\ \{\varphi(\cdot - k) : k \in \mathbb{Z}\} \text{ is an orthonormal basis for } V_0. \end{array}\right\} \tag{6.11}$$

The function φ whose existence is asserted in (6.11) is called a **scaling function** of the given MRA.

As in Chapter 2 for the case of $L^2(\mathbb{R})$, when we have an H^2-MRA with the scaling function φ, the low-pass filter can be defined as the 2π-periodic function m_0 such that

$$\hat{\varphi}(2\xi) = m_0(\xi)\hat{\varphi}(\xi) \qquad \text{for a.e. } \xi \in \mathbb{R}. \tag{6.12}$$

From this and the orthonormality of $\{\varphi(\cdot - k) : k \in \mathbb{Z}\}$ we easily obtain

$$|m_0(\xi)|^2 + |m_0(\xi + \pi)|^2 = 1 \qquad \text{for a.e. } \xi \in \mathbb{R}. \tag{6.13}$$

The function ψ defined by

$$\hat{\psi}(2\xi) = e^{i\xi}\, \overline{m_0(\xi + \pi)}\, \hat{\varphi}(\xi) \tag{6.14}$$

is an orthonormal wavelet for $H^2(\mathbb{R})$ (see (2.12) of Chapter 2 for the case of $L^2(\mathbb{R})$). The above three equations allow us to prove

$$|\hat{\varphi}(\xi)|^2 = \sum_{j=1}^{\infty} |\hat{\psi}(2^j \xi)|^2 \qquad \text{for} \ \ a.e. \ \xi \in \mathbb{R} \tag{6.15}$$

(see (2.16) of Chapter 2 for the case of $L^2(\mathbb{R})$).

Here is a simple example of an H^2-MRA.

EXAMPLE K: Let us choose $\hat{\varphi} = \chi_{[0,2\pi]}$, and define

$$V_0 = \overline{\text{span}\,\{\varphi(\cdot - k) : k \in \mathbb{Z}\}}, \quad \text{and}$$

$$V_j = \{f : f(2^{-j}(\cdot)) \in V_0\} \quad \text{for} \ j \in \mathbb{Z}.$$

It is not hard to verify that $\{V_j : j \in \mathbb{Z}\}$ forms an H^2-MRA. The corresponding low-pass filter is $m_0(\xi) = \chi_{[0,\pi]}$ on $[0, 2\pi)$, and extended 2π-periodically. This H^2-MRA gives us, via equation (6.14), an H^2-wavelet ψ such that $|\hat{\psi}| = \chi_{[2\pi,4\pi]}$.

For an H^2-wavelet ψ, we can define an associated 2π-periodic function D_ψ, as in (3.1), by

$$D_\psi(\xi) \equiv \sum_{j=1}^{\infty} \sum_{k \in \mathbb{Z}} |\hat{\psi}(2^j(\xi + 2k\pi))|^2 \qquad \text{for} \ \ a.e. \ \xi \in \mathbb{R}. \tag{6.16}$$

The same argument in the proof of Lemma 3.3 shows that ψ satisfies (3.4). This and Lemma 3.7 lead us to the following result:

LEMMA 6.17 *For an H^2-wavelet ψ, $D_\psi(\xi)$ is a.e. integer-valued.*

As in the case for $L^2(\mathbb{R})$, there is a single equation that determines when an H^2-wavelet arises from an H^2-MRA. If ψ is an H^2-wavelet associated with an H^2-MRA, then the orthonormality of the system $\{\varphi(\cdot - k) : k \in \mathbb{Z}\}$ and (6.15) imply

$$D_\psi(\xi) = 1 \qquad \text{for} \ \ a.e. \ \xi \in \mathbb{R}. \tag{6.18}$$

This condition is also sufficient, that is,

THEOREM 6.19 *An H^2-wavelet ψ is associated with an H^2-MRA if and only if $D_\psi(\xi) = 1$ for almost every $\xi \in \mathbb{R}$.*

The proof of this theorem is similar to the proof of Theorem 3.2.

Applying this theorem to the H^2-wavelet examples in Example J, we easily see that, when $\ell = 0$, ψ_0 is associated with an H^2-MRA. But when $\ell \geq 1$, ψ_ℓ is not associated with any H^2-MRA, since an easy calculation gives us, for a.e. $\xi \in [0, 2\pi]$,

$$D_{\psi_\ell}(\xi) = (\ell+1)\chi_{[0, \frac{2}{2^{\ell+1}-1}\pi]}(\xi) + \sum_{j=1}^{\ell}(\ell+1-j)\chi_{[\frac{2^j}{2^{\ell+1}-1}\pi, \frac{2^{j+1}}{2^{\ell+1}-1}\pi]}(\xi),$$

which is not equal to 1 on $[0, \frac{2}{2^{\ell+1}-1}\pi]$ when $\ell \geq 1$.

Now we are ready to show the non-existence of "regular" H^2-wavelets.

THEOREM 6.20 *There is no orthonormal wavelet for $H^2(\mathbb{R})$ satisfying the regularity condition $(\mathfrak{R}^0)$ given in (6.1).*

PROOF : We argue by contradiction. Suppose that there is an H^2-wavelet ψ that satisfies the regularity condition $(\mathfrak{R}^0)$, that is, $|\hat{\psi}|$ is continuous on $\mathbb{R}$ and $|\hat{\psi}(\xi)| = O(|\xi|^{-\alpha-\frac{1}{2}})$ at ∞ for some $\alpha > 0$. As observed in the proof of Corollary 3.16, these conditions tell us that the series

$$s(\xi) = \sum_{j=1}^{\infty}|\hat{\psi}(2^j\xi)|^2$$

converges uniformly on compact subsets of $\mathbb{R} \setminus \{0\}$ and, thus, represents a continuous function on this domain; moreover, $s(\xi) = O(|\xi|^{-1-2\alpha})$ at ∞. It follows that

$$\sum_{k\in\mathbb{Z}} s(\xi + 2k\pi) = D_\psi(\xi)$$

converges uniformly on compact subsets of $[-\pi, 0) \cup (0, \pi]$, which implies $D_\psi(\xi)$ is continuous on $[-\pi, 0) \cup (0, \pi]$. We know that

$$\int_{\mathbb{T}} D_\psi(\xi)\, d\xi = \|\hat{\psi}\|_2^2 = 2\pi$$

(see Lemma 4.16 of Chapter 3), and Lemma 6.17 tells us $D_\psi(\xi)$ is a.e. integer-valued; therefore, we must have $D_\psi(\xi) = 1$ on $[-\pi, 0) \cup (0, \pi]$.

The continuity of $s(\cdot - 2k\pi)$ on $\mathbb{R} \setminus \{2k\pi\}$, for each $k \in \mathbb{Z}$, and the fact that $s(\xi) = O(|\xi|^{-1-2\alpha})$ at ∞ also give us that the series

$$t(\xi) = \sum_{k\in\mathbb{Z}, k\neq 0} s(\xi + 2k\pi)$$

converges uniformly on $[-\pi, \pi]$ and, hence, is continuous on this interval.

Now on the set $[-\pi, 0) \cup (0, \pi]$, $D_\psi(\xi) = s(\xi) + t(\xi)$, and each of them is continuous. It follows that (since $\hat{\psi}(\xi) = 0$ for a.e. $\xi < 0$),

$$1 = \lim_{\xi \to 0^-} D_\psi(\xi) = \lim_{\xi \to 0^-} \left(s(\xi) + t(\xi)\right) = 0 + t(0), \qquad \text{and}$$

$$1 = \lim_{\xi \to 0^+} D_\psi(\xi) = \lim_{\xi \to 0^+} \left(s(\xi) + t(\xi)\right) = \lim_{\xi \to 0^+} s(\xi) + t(0).$$

Thus, we must have $\lim_{\xi \to 0} s(\xi) = 0$.

On the other hand, we can always find a $\xi_o > 0$ such that $|\hat{\psi}(\xi_o)| > 0$. From the definition of $s(\xi)$ we then have

$$s(2^{-j}\xi_o) = \sum_{\ell=1-j}^{\infty} |\hat{\psi}(2^\ell \xi_o)|^2$$

increases as $j \to \infty$ and is bounded below by $|\hat{\psi}(\xi_o)|^2 > 0$ for all $j \geq 1$. This contradicts $\lim_{\xi \to 0} s(\xi) = 0$, and this contradiction proves our theorem.

∎

As an easy corollary, we immediately obtain:

COROLLARY 6.21 *There is no band-limited orthonormal wavelet ψ for $H^2(\mathbb{R})$ such that $|\hat{\psi}|$ is continuous.*

REMARK: Most of the results in this section can be carried over to the closed subspace $L_E^2(\mathbb{R})$ defined, as before, by

$$L_E^2(\mathbb{R}) \equiv \{f \in L^2(\mathbb{R}) : \text{supp}\,(\hat{f}) \subseteq E \text{ a.e. }\},$$

where E is a measurable subset of $\mathbb{R}$ of positive measure such that

$$E = 2E \quad \text{and} \quad |\mathbb{R} \setminus E| \neq 0.$$

The classical Hardy space $H^2(\mathbb{R})$ is a special example of this type of subspace when $E = \mathbb{R}^+$.

In this setting, an **orthonormal wavelet for $L_E^2(\mathbb{R})$** can be defined as a function $\psi \in L_E^2(\mathbb{R})$ such that $\{\psi_{j,k} : j, k \in \mathbb{Z}\}$ forms an orthonormal basis for $L_E^2(\mathbb{R})$. All orthonormal wavelets ψ for $L_E^2(\mathbb{R})$ can be characterized by the condition $\|\psi\|_2 = 1$, the equation (6.6) and

$$\sum_{j \in \mathbb{Z}} |\hat{\psi}(2^j \xi)|^2 = \chi_E(\xi) \qquad \text{for a.e. } \xi \in \mathbb{R}. \tag{6.22}$$

For example, when

$$E = (-\infty, 0) \cup \left\{ \bigcup_{j \in \mathbb{Z}} 2^j [3\pi, 4\pi] \right\},$$

the function ψ defined by $\hat{\psi} = \chi_{[-2\pi, -\pi] \cup [3\pi, 4\pi]}$ is an orthonormal wavelet for $L_E^2(\mathbb{R})$. In fact, this ψ is the L^2-limit of the sequence $\{\psi^j\}_{j=3}^\infty$ of orthonormal wavelets for $L^2(\mathbb{R})$ defined by $\widehat{\psi^j} = \chi_{F_j}$, where

$$F_j = \left[-2\left(1 - \frac{3}{2^j - 1}\right)\pi, -\left(1 - \frac{3}{2^j - 1}\right)\pi \right]$$

$$\cup \left[\frac{4}{2^j - 1}\pi, \frac{6}{2^j - 1}\pi \right] \cup \left[3\pi + \frac{3}{2^j - 1}\pi, 4\pi + \frac{4}{2^j - 1}\pi \right].$$

This, by the way, also shows that the set of orthonormal wavelets for $L^2(\mathbb{R})$ is not a closed subset on the unit sphere of $L^2(\mathbb{R})$.

Similarly, a **multiresolution analysis for** $L_E^2(\mathbb{R})$ can be defined as a sequence $\{V_j : j \in \mathbb{Z}\}$ of closed subspaces of $L_E^2(\mathbb{R})$ satisfying conditions (6.7), (6.8), (6.9), (6.11) and

$$\overline{\bigcup_{j \in \mathbb{Z}} V_j} = L_E^2(\mathbb{R}).$$

It is left to the reader to extend the corresponding results for $H^2(\mathbb{R})$ to the space $L_E^2(\mathbb{R})$.

7.7 Notes and references

1. Yves Meyer told us that when he began his construction of the wavelets we called the "Lemarié-Meyer" wavelets he was aware that, if such wavelets exist, they must satisfy the system of equations (1.2), (1.3), (1.4) and (1.5). He tried to construct wavelets with smooth Fourier transforms by using these four equations. When doing this he did encounter the example we constructed involving (1.9) and, at first, he thought that there could not exist such wavelets. He was, of course, successful in his search of such wavelets based on these four equations (as explained in Chapter 3). He attributes the idea of introducing the hypothesis $\|e_j\| \geq 1$ (see Theorem 1.8) to J.L. Journé, and this implies, in particular, that equations (1.2) and (1.3) imply the orthonormality equations (1.4) and (1.5). This observation by Journé was made a few years later. In this exposition we are exploiting the fact that all we need are the two equations in order to construct wavelets and more general systems that represent functions.

Thus, equations (1.2) and (1.3) have been known since the beginning of the development of the modern theory of wavelets (see, also, [Le1] and [Le4]). However, the proof that these two equations characterize orthonormal wavelets in full generality did not appear in the published literature until recently. It can be found in [Gri1], [Wan] and [HWW3]. The proof presented in section 7.1 combines the ideas presented in these references. We show elsewhere that the argument presented in this book extends rather easily to higher dimensions.

2. Minimally supported frequency (MSF) wavelets are treated in [FW], [HKLS] and [Wan]. (In [HKLS] they are called **unimodular wavelets**.) All the examples and results given in section 7.2 are (essentially) contained in one or more of the aforementioned articles. In addition, an example of an MSF wavelet which is not band-limited is given in [FW]. In spite of the characterizations given in Corollary 2.4 and Theorem 2.5 for the supports of the Fourier transforms of MSF wavelets, we still lack a classification of all of them. (In particular cases, this is done in [HKLS].)

3. The articles [HWW1] and [HWW2] also deal with MSF wavelets. Although a new family of MSF wavelets is exhibited in the second one of these articles, the main purpose of them is to obtain wavelets $\psi \in \mathcal{S}(\mathbb{R})$, the Schwartz class, such that they are "close" in the L^2-norm to a given MSF wavelet. (This is, in fact, the relation between the Shannon wavelet and the Lemarié-Meyer wavelets.) By Theorem 3.18 this is impossible for MSF wavelets that do not arise from an MRA. For a large family of those MSF wavelets that do arise from an MRA a procedure has been described in [HWW2]. In some cases, this procedure gives surprising results: for some band-limited wavelets, their approximations obtained by this procedure are not band-limited. This phenomenon is related to the concept of **invariant cycle** under the transformation $x \mapsto 2x \pmod{2\pi}$ (see [HWW2]).

4. The question of whether an orthonormal wavelet is associated with an MRA is treated in [Au2], [Au5], [Le2] and [Le3]. It is the circle of ideas presented in [Au2] that we have exploited to obtain Theorem 3.2 and Proposition 3.14. In particular, the proof that $D_\psi(\xi)$ is the dimension of a certain vector space is contained in this article. A different proof of Theorem 3.2 can be found in [Gri1]. A generalization of these results to **biorthogonal wavelets** arising from MRA's is given in [Han] and [Wan].

5. The characterization of low-pass filters (Theorem 4.8) is essentially due to A. Cohen. The proof presented here is an adapted and expanded version of the ideas presented in [Co1] and [Co2], where only the case $m_0 \in C^\infty(\mathbb{R})$ is treated. Another approach to this problem is presented in [Law], where the characterization is given in terms of the multiplicity of the eigenvalue 1 of an operator associated with the low-pass filter (see also section 6.3 of [Da1]). The question of characterizing low-pass filters that are not necessarily $C^1(\mathbb{R})$ remains open. A step in this direction has been taken in [HWW2], where the case $m_0 = \chi_F$ (F is a measurable set in $\mathbb{R}$) is studied.

Example G (a similar one also appears in [Co1]) exhibits a so-called "quasi-analytic" wavelet (that is, a wavelet $\psi \in \mathcal{S}(\mathbb{R})$ such that the set $\text{supp}\,(\hat{\psi}) \cap \mathbb{R}^-$ is very small). This type of wavelet is important to the study of the analytic part of acoustic signals.

6. The question of characterizing scaling functions that produce MRAs was posed by R. Strichartz in the survey article [St]. The answer provided in Theorem 5.2 can be found in [HWW3] and [Wan], where it is used to study if the convolution of two scaling functions can be used to generate an MRA. The elegant argument, in the proof of Theorem 5.2, that shows that

$$L^2(\mathbb{R}) = \overline{\bigcup_{j \in \mathbb{Z}} V_j}$$

is due to Paolo M. Soardi. It is a considerable improvement of the argument we originally introduced and appears in [HWW3].

7. Y. Meyer (section 4.3 of [Me1]) posed the question of finding a wavelet ψ for $H^2(\mathbb{R})$ such that $\psi \in \mathcal{S}(\mathbb{R})$, the Schwartz class. A negative answer (Theorem 6.20) was announced in [Au4] and a complete proof was given in [Au2]. Our presentation follows this last article.

8

Frames

The main purpose of this book is to study orthonormal bases generated by dilations and translations of a single function. This theory has been presented in the preceding chapters. In some cases, the orthonormality condition can be relaxed, without losing the reconstruction property (see (1.1) in Chapter 1 and the paragraph that follows this formula), which is vital for applications. In fact, we showed immediately after Lemma 1.8 in Chapter 2 that property (1.5) of the definition of an MRA (see section 2.1) can be relaxed by assuming that the integer translates of the scaling function form a Riesz basis for V_0.

In this chapter we introduce the notion of a "frame" which allows us to replace the orthonormality condition by a weaker one. In applications, frames are often used to encode and reconstruct sounds and other signals. The precise mathematical description of the concept of "frame" is given in section 8.1. Although frames are more general than orthonormal wavelet bases, the "uncertainty" principle expressed in the Balian-Low theorem (see Theorem 2.1 in Chapter 1) is still true; this will be explained in the second section. The third section is dedicated to finding sufficient conditions for sets of dilates and translates of a single function to form a frame. This will provide us with new examples of frames. Finally we shall exhibit examples of frames in $H^2(\mathbb{R})$ with smooth Fourier transforms; a surprising result, since we have already proved that there are no orthonormal wavelet bases in $H^2(\mathbb{R})$ with smooth Fourier transforms. We call an L^2-function ψ a **frame wavelet** when $\{\psi_{j,k} : j, k \in \mathbb{Z}\}$ is a frame. Frame wavelets, therefore, are more general than orthonormal wavelets. We shall see that the former enjoy important properties of the latter. Nevertheless, there are basic differences. We shall exhibit in section 8.3 a frame, whose elements are in the Schwartz class, that cannot be used to characterize the spaces $L^p(\mathbb{R})$ when $1 < p < 2$. Thus, important results we obtained in Chapter 6 cannot be extended to frames.

8.1   The reconstruction formula for frames

It is well known that a collection of unitary elements $\{\varphi_j : j \in \mathbb{J}\}$ in a Hilbert space $\mathbb{H}$ is an orthonormal basis of $\mathbb{H}$ if and only if

$$\sum_{j \in \mathbb{J}} |\langle f, \varphi_j \rangle|^2 = \|f\|_{\mathbb{H}}^2 \qquad \text{for all} \ \ f \in \mathbb{H}. \qquad (1.1)$$

We have written $\langle \cdot, \cdot \rangle$ for the inner product in $\mathbb{H}$ and $\|\cdot\| = \|\cdot\|_{\mathbb{H}}$ for its associated norm. Moreover, in this situation,

$$f = \sum_{j \in \mathbb{J}} \langle f, \varphi_j \rangle \, \varphi_j \qquad (1.2)$$

with convergence in $\mathbb{H}$. This shows that the element f can be recovered from the numbers $\langle f, \varphi_j \rangle$ and the elements of the basis. This last property is particularly important for applications and need not involve orthonormality.

In the previous chapters we have constructed several orthonormal bases of $L^2(\mathbb{R})$. The ones described in section 1.4 do not have a closed form in the space variable. It is particularly useful for applications to relax condition (1.1), and yet still be able to recover f, as in (1.2), and in addition have a "nice" expression for the elements of the basis. These considerations lead us to the notion of a "frame."

A collection of elements $\{\varphi_j : j \in \mathbb{J}\}$ in a Hilbert space $\mathbb{H}$ is called a **frame** if there exist constants A and B, $0 < A \le B < \infty$, such that

$$A\|f\|^2 \le \sum_{j \in \mathbb{J}} |\langle f, \varphi_j \rangle|^2 \le B\|f\|^2 \qquad \text{for all} \ \ f \in \mathbb{H}. \qquad (1.3)$$

The constants A and B are called **frame bounds**. A frame is not necessarily a basis; for example, adding the zero element to the collection $\{\varphi_j : j \in \mathbb{J}\}$ does not alter the inequalities (1.3). The elements of a frame do not have to be normal, but they must satisfy $\|\varphi_k\| \le \sqrt{B}$ for all $k \in \mathbb{J}$. To see this we can assume $\varphi_k \ne 0$; then, with $f = \varphi_k$ on (1.3) we have

$$\|\varphi_k\|^4 = |\langle \varphi_k, \varphi_k \rangle|^2 \le \sum_{j \in \mathbb{J}} |\langle \varphi_k, \varphi_j \rangle|^2 \le B\|\varphi_k\|^2,$$

and it follows that $\|\varphi_k\| \leq \sqrt{B}$.

REMARK: When $\{\varphi_j : j \in \mathbb{J}\}$ is a frame, the closure of the span of $\{\varphi_j : j \in \mathbb{J}\}$ must be $\mathbb{H}$. This is easily seen, since if we assume that f is perpendicular to φ_j for all $j \in \mathbb{J}$, the first inequality in (1.3) implies that f must be the zero element of $\mathbb{H}$.

EXAMPLE A: Even when all the zero elements are removed from a frame, the new frame is not necessarily a basis. Here is an example. Let $\mathbb{H} = \mathbb{C}^2$ and take

$$\varphi_1 = (0, 1), \qquad \varphi_2 = \left(\tfrac{\sqrt{3}}{2}, \tfrac{1}{2}\right), \qquad \varphi_3 = \left(\tfrac{\sqrt{3}}{2}, -\tfrac{1}{2}\right).$$

Then, for $f = (f_1, f_2) \in \mathbb{C}^2$ we have

$$\sum_{j=1}^{3} |\langle f, \varphi_j \rangle|^2 = |f_2|^2 + \left|\tfrac{\sqrt{3}}{2} f_1 + \tfrac{1}{2} f_2\right|^2 + \left|\tfrac{\sqrt{3}}{2} f_1 - \tfrac{1}{2} f_2\right|^2$$

$$= \tfrac{3}{2} \|f\|^2.$$

Therefore, $A = B = \tfrac{3}{2}$ in (1.3). The bound $\tfrac{3}{2}$ is a measure of the "redundancy" of the system: three vectors in two dimensions.

When $A = B$ in (1.3) we say that the frame is **tight**. Any orthonormal basis in a Hilbert space is a tight frame with $A = B = 1$; the frame of Example A is also tight.

If $\{\varphi_j : j \in \mathbb{J}\}$ is a tight frame, and $\|\varphi_{j_o}\| \geq 1$ for some $j_o \in \mathbb{J}$, then

$$A = \|\varphi_{j_o}\|^2 + \sum_{j \neq j_o} \frac{1}{\|\varphi_{j_o}\|^2} |\langle \varphi_{j_o}, \varphi_j \rangle|^2 \geq 1.$$

In this situation, if the frame is tight and $A = 1$ we see that $\varphi_j \perp \varphi_{j_o}$ for all $j \in \mathbb{J}$ and $\|\varphi_{j_o}\| = 1$. Therefore, if the elements of a tight frame all have norm 1 and $A = 1$, the frame is an orthonormal basis for $\mathbb{H}$ (see, also, Theorem 1.8 in Chapter 7).

As was the case in (1.2) for orthonormal bases, when $\{\varphi_j : j \in \mathbb{J}\}$ is a frame we can always reconstruct f from the coefficients $\{\langle f, \varphi_j \rangle : j \in \mathbb{J}\}$. Let us first assume that the frame is tight. Then, we have

$$\sum_{j \in \mathbb{J}} |\langle f, \varphi_j \rangle|^2 = A \|f\|^2 \qquad \text{for all } f \in \mathbb{H}, \tag{1.4}$$

and the "same" argument that establishes (ii) from (i) in Theorem 1.7 of Chapter 7 shows that

$$f = \frac{1}{A} \sum_{j \in \mathbb{J}} \langle f, \varphi_j \rangle \, \varphi_j \qquad \text{for all } f \in \mathbb{H}, \tag{1.5}$$

with convergence in $\mathbb{H}$. This shows that f can be reconstructed from the coefficients $\langle f, \varphi_j \rangle$.

We claim that, even if the frame is not tight, we can find a reconstruction formula similar to (1.5). Consider a general frame $\{\varphi_j : j \in \mathbb{J}\}$ and denote by $\ell^2(\mathbb{J})$ the space of all sequences $\mathbf{c} = \{c_j\}_{j \in \mathbb{J}}$ such that

$$\|\mathbf{c}\|_2^2 = \sum_{j \in \mathbb{J}} |c_j|^2 < \infty.$$

We define the **frame operator** $\mathcal{F} : \mathbb{H} \longrightarrow \ell^2(\mathbb{J})$ by

$$\mathcal{F} f = \left\{ \langle f, \varphi_j \rangle \right\}_{j \in \mathbb{J}}. \tag{1.6}$$

Since $\mathcal{F}$ is linear, (1.3) implies the continuity of $\mathcal{F}$ with $\|\mathcal{F}\| \leq \sqrt{B}$. Hence, $\mathcal{F}$ has an adjoint, $\mathcal{F}^* : \ell^2(\mathbb{J}) \longrightarrow \mathbb{H}$, which satisfies, $\|\mathcal{F}^*\| = \|\mathcal{F}\| \leq \sqrt{B}$ and

$$\langle f, \mathcal{F}^*(\{c_j\}_{j \in \mathbb{J}}) \rangle = \langle \mathcal{F} f, \{c_j\}_{j \in \mathbb{J}} \rangle = \sum_{j \in \mathbb{J}} \langle f, \varphi_j \rangle \, \overline{c_j}$$

$$= \left\langle f, \sum_{j \in \mathbb{J}} c_j \, \varphi_j \right\rangle.$$

Thus, the continuity of $\mathcal{F}$ implies

$$\mathcal{F}^*(\{c_j\}_{j \in \mathbb{J}}) = \sum_{j \in \mathbb{J}} c_j \, \varphi_j \tag{1.7}$$

with convergence in the norm of $\mathbb{H}$. If $\mathsf{S} = \mathcal{F}^* \mathcal{F}$, we have

$$\mathsf{S} f = \mathcal{F}^* \mathcal{F} f = \sum_{j \in \mathbb{J}} \langle f, \varphi_j \rangle \, \varphi_j, \tag{1.8}$$

for all $f \in \mathbb{H}$, and (1.3) can be rewritten as

$$A \langle f, f \rangle \leq \langle \mathsf{S} f, f \rangle \leq B \langle f, f \rangle \qquad \text{for all } f \in \mathbb{H}. \tag{1.9}$$

This shows that S is a positive bounded linear operator satisfying

$$AI \leq S \leq BI, \tag{1.10}$$

where I is the identity operator on the Hilbert space $\mathbb{H}$. From (1.10) it can be deduced that S has an inverse which satisfies

$$\frac{1}{B} I \leq S^{-1} \leq \frac{1}{A} I. \tag{1.11}$$

These inequalities allow us to define a frame associated with $\{\varphi_j : j \in \mathbb{J}\}$, the **dual frame**:

LEMMA 1.12 *If $\{\varphi_j : j \in \mathbb{J}\}$ is a frame on a Hilbert space $\mathbb{H}$ with frame bounds A and B, the collection $\{\widetilde{\varphi_j} \equiv S^{-1}(\varphi_j) : j \in \mathbb{J}\}$ is also a frame for $\mathbb{H}$, with frame bounds $\frac{1}{B}$ and $\frac{1}{A}$, which is called the* **dual frame** *to $\{\varphi_j : j \in \mathbb{J}\}$.*

PROOF : It suffices to show

$$\sum_{j \in \mathbb{J}} |\langle f, \widetilde{\varphi_j} \rangle|^2 = \langle S^{-1} f, f \rangle \qquad \text{for all } f \in \mathbb{H}, \tag{1.13}$$

since (1.10) implies

$$\frac{1}{B} \|f\|^2 \leq \langle S^{-1} f, f \rangle \leq \frac{1}{A} \|f\|^2$$

and, hence, (1.3) is satisfied with φ_j replaced by $\widetilde{\varphi_j}$. To prove (1.13) we use that S^{-1} is self-adjoint (which follows from the self-adjointness of S). Then, we have

$$\sum_{j \in \mathbb{J}} |\langle f, \widetilde{\varphi_j} \rangle|^2 = \sum_{j \in \mathbb{J}} |\langle f, S^{-1} \varphi_j \rangle|^2 = \sum_{j \in \mathbb{J}} |\langle S^{-1} f, \varphi_j \rangle|^2$$

$$= \|\mathcal{F}(S^{-1} f)\|_2^2 = \langle \mathcal{F}(\mathcal{F}^* \mathcal{F})^{-1} f, \mathcal{F}(\mathcal{F}^* \mathcal{F})^{-1} f \rangle_2$$

$$= \langle (\mathcal{F}^* \mathcal{F})^{-1} f, (\mathcal{F}^* \mathcal{F})(\mathcal{F}^* \mathcal{F})^{-1} f \rangle = \langle S^{-1} f, f \rangle.$$

∎

If we define $\langle g, f \rangle_\sharp = \langle S^{-1} g, f \rangle$ for $g, f \in \mathbb{H}$, we obtain an inner product on $\mathbb{H}$ equivalent to the original one $\langle \cdot, \cdot \rangle$. Then, (1.13) can be rewritten as

$$\sum_{j \in \mathbb{J}} |\langle f, \varphi_j \rangle_\sharp|^2 = \|f\|_\sharp^2.$$

Consequently, $\{\varphi_j : j \in \mathbb{J}\}$ is a tight frame with frame constant 1 if we use the inner product $\langle \cdot, \cdot \rangle_\sharp$ instead of $\langle \cdot, \cdot \rangle$. Thus, by (1.5),

$$f = \sum_{j \in \mathbb{J}} \langle f, \varphi_j \rangle_\sharp \varphi_j = \sum_{j \in \mathbb{J}} \langle f, \widetilde{\varphi_j} \rangle \varphi_j \qquad \text{for all } f \in \mathbb{H} \qquad (1.14)$$

with convergence in $\mathbb{H}$. We also have the dual formula

$$f = \sum_{j \in \mathbb{J}} \langle f, \varphi_j \rangle \widetilde{\varphi_j} \qquad \text{for all } f \in \mathbb{H} \qquad (1.15)$$

in the norm of $\mathbb{H}$. This follows from (1.14) since

$$\langle S^{-1} f, g \rangle = \langle f, g \rangle_\sharp = \sum_{j \in \mathbb{J}} \langle f, \varphi_j \rangle_\sharp \langle \varphi_j, g \rangle_\sharp$$

$$= \sum_{j \in \mathbb{J}} \langle S^{-1} f, \varphi_j \rangle \langle \widetilde{\varphi_j}, g \rangle,$$

which implies

$$S^{-1} f = \sum_{j \in \mathbb{J}} \langle S^{-1} f, \varphi_j \rangle \widetilde{\varphi_j}.$$

Since S is invertible, $S^{-1} f$ is a general element of $\mathbb{H}$, which proves (1.15) in the weak sense. To prove the convergence in the strong sense we take $d_1 \subset d_2 \subset \mathbb{J}$; then

$$\left\| \sum_{j \in d_2 \setminus d_1} \langle S^{-1} f, \varphi_j \rangle \widetilde{\varphi_j} \right\|_{\mathbb{H}}$$

$$= \sup_{\|g\|=1} \left| \left\langle \sum_{j \in d_2 \setminus d_1} \langle S^{-1} f, \varphi_j \rangle \widetilde{\varphi_j}, g \right\rangle \right| = \sup_{\|g\|=1} \left| \sum_{j \in d_2 \setminus d_1} \langle S^{-1} f, \varphi_j \rangle \langle \widetilde{\varphi_j}, g \rangle \right|$$

$$\le \sup_{\|g\|=1} \left(\sum_{j \in d_2 \setminus d_1} |\langle S^{-1} f, \varphi_j \rangle|^2 \right)^{\frac{1}{2}} \left(\sum_{j \in d_2 \setminus d_1} |\langle \widetilde{\varphi_j}, g \rangle|^2 \right)^{\frac{1}{2}}$$

$$= \sup_{\|g\|=1} \left(\sum_{j \in d_2 \setminus d_1} |\langle f, \varphi_j \rangle_\sharp|^2 \right)^{\frac{1}{2}} \left(\sum_{j \in \mathbb{J}} |\langle \varphi_j, g \rangle_\sharp|^2 \right)^{\frac{1}{2}}.$$

The first sum is small since $\{\varphi_j : j \in \mathbb{J}\}$ is a tight frame with bound 1 using the inner product $\langle \cdot, \cdot \rangle_\sharp$. The same reason shows that the last sum is bounded by $\|g\|_\sharp$.

We have shown that when $\{\varphi_j : j \in \mathbb{J}\}$ is a frame, f can be reconstructed from the coefficients $\langle f, \varphi_j \rangle$ using the dual frame $\{\widetilde{\varphi_j} : j \in \mathbb{J}\}$ and that f is also a superposition of φ_j's with coefficients $\langle f, \widetilde{\varphi_j} \rangle$ (see (1.14) and (1.15)).

8.2   The Balian-Low theorem for frames

We now examine when a system of the form

$$g_{m,n}(x) = e^{2\pi i m x} g(x - n), \qquad m, n \in \mathbb{Z},$$

generated by translations and modulations of a single function $g \in L^2(\mathbb{R})$, can be a frame for $L^2(\mathbb{R})$. We start by showing that, in this case, the dual frame is also of the same type; that is, it is generated by translations and modulations of a single function. It will be shown later that the same Balian-Low phenomenon described in section 1.2 for orthonormal bases holds in this, more general, situation.

PROPOSITION 2.1 *Suppose that $g \in L^2(\mathbb{R})$ and that*

$$\left\{ g_{m,n}(x) = e^{2\pi i m x} g(x - n) : \ m, n \in \mathbb{Z} \right\}$$

is a frame for $L^2(\mathbb{R})$. Let $\tilde{g} \equiv \mathsf{S}^{-1} g$, where $\mathsf{S} = \mathcal{F}^* \mathcal{F}$ and $\mathcal{F}$ is the frame operator defined in (1.6). Then, the dual frame $\{\widetilde{g_{m,n}}\}$, of $\{g_{m,n}\}$ (see Lemma 1.12) satisfies

$$\widetilde{g_{m,n}}(x) = e^{2\pi i m x}\, \tilde{g}(x - n) = \tilde{g}_{m,n}(x) \qquad \textit{for } m, n \in \mathbb{Z}.$$

PROOF : We start by showing that the operator $\mathcal{F}^* \mathcal{F}$ commutes with translations by integers, $(\tau_k f)(x) = f(x - k)$, and with integral modulations, $(\mathbf{e}_j f)(x) = e^{2\pi i j x} f(x)$; that is

$$
\begin{cases}
(\mathcal{F}^* \mathcal{F})(\tau_k f) = \tau_k((\mathcal{F}^* \mathcal{F})f), & \text{and} \\
(\mathcal{F}^* \mathcal{F})(\mathbf{e}_j f) = \mathbf{e}_j((\mathcal{F}^* \mathcal{F})f),
\end{cases}
\tag{2.2}
$$

for all $f \in L^2(\mathbb{R})$ and all $k, j \in \mathbb{Z}$. The first equality in (2.2) follows from the following computations (see (1.8)):

$$\left((\mathcal{F}^* \mathcal{F})(\tau_k f)\right)(x) = \sum_{m \in \mathbb{Z}} \sum_{n \in \mathbb{Z}} \langle \tau_k f, g_{m,n} \rangle\, g_{m,n}(x)$$

$$= \sum_{m\in\mathbb{Z}} \sum_{n\in\mathbb{Z}} \langle f, g_{m,n-k} \rangle\, g_{m,n}(x)$$

$$= \sum_{m\in\mathbb{Z}} \sum_{\ell\in\mathbb{Z}} \langle f, g_{m,\ell} \rangle\, g_{m,\ell}(x-k)$$

$$= \tau_k\big((\mathcal{F}^*\mathcal{F})f\big)(x).$$

The second equality follows by a similar argument:

$$\big((\mathcal{F}^*\mathcal{F})(\mathbf{e}_j f)\big)(x) = \sum_{m\in\mathbb{Z}} \sum_{n\in\mathbb{Z}} \langle \mathbf{e}_j f, g_{m,n} \rangle\, g_{m,n}(x)$$

$$= \sum_{m\in\mathbb{Z}} \sum_{n\in\mathbb{Z}} \langle f, g_{m-j,n} \rangle\, g_{m,n}(x)$$

$$= \sum_{\ell\in\mathbb{Z}} \sum_{n\in\mathbb{Z}} \langle f, g_{\ell,n} \rangle\, g_{\ell+j,n}(x)$$

$$= e^{2\pi ijx} \sum_{\ell\in\mathbb{Z}} \sum_{n\in\mathbb{Z}} \langle f, g_{\ell,n} \rangle\, g_{\ell,n}(x)$$

$$= \mathbf{e}_j\big((\mathcal{F}^*\mathcal{F})f\big)(x).$$

It follows that $\mathsf{S}^{-1} = (\mathcal{F}^*\mathcal{F})^{-1}$ also commutes with these translations and modulations. Thus,

$$\widetilde{g_{m,n}}(x) = \mathsf{S}^{-1} g_{m,n}(x) = (\mathsf{S}^{-1}\mathbf{e}_m\tau_n g)(x)$$

$$= \mathbf{e}_m\tau_n(\mathsf{S}^{-1}g)(x) = \mathbf{e}_m\tau_n\tilde{g}(x) = e^{2\pi imx}\tilde{g}(x-n).$$

∎

THEOREM 2.3 **(Balian-Low theorem for frames)**

   Suppose that $g \in L^2(\mathbb{R})$ and

$$g_{m,n}(x) = e^{2\pi imx}g(x-n), \qquad m, n \in \mathbb{Z}.$$

If $\{g_{m,n} : m, n \in \mathbb{Z}\}$ is a frame for $L^2(\mathbb{R})$, then either

$$\int_{\mathbb{R}} x^2|g(x)|^2\, dx = \infty \qquad or \qquad \int_{\mathbb{R}} \xi^2|\hat{g}(\xi)|^2\, d\xi = \infty.$$

The proof of this theorem shares some of the features of the proof of Theorem 2.1 in Chapter 1. As before, we examine the operators

$$(Qf)(x) = xf(x) \quad \text{and} \quad (Pf)(x) = -if'(x).$$

As in the proof of Theorem 2.1 in Chapter 1, we show that both Qf and Pf **cannot** belong to $L^2(\mathbb{R})$. This leads us to replace equalities (2.2), (2.3) and (2.4) in Chapter 1 by

$$\langle Qg, P\tilde{g} \rangle = \sum_{m \in \mathbb{Z}} \sum_{n \in \mathbb{Z}} \langle Qg, \tilde{g}_{m,n} \rangle \langle g_{m,n}, P\tilde{g} \rangle, \tag{2.4}$$

$$\langle Pg, Q\tilde{g} \rangle = \sum_{m \in \mathbb{Z}} \sum_{n \in \mathbb{Z}} \langle Pg, \tilde{g}_{m,n} \rangle \langle g_{m,n}, Q\tilde{g} \rangle, \tag{2.5}$$

$$\langle Qg, \tilde{g}_{m,n} \rangle = \langle g_{-m,-n}, Q\tilde{g} \rangle \quad \text{for all } m, n \in \mathbb{Z}, \tag{2.6}$$

$$\langle Pg, \tilde{g}_{m,n} \rangle = \langle g_{-m,-n}, P\tilde{g} \rangle \quad \text{for all } m, n \in \mathbb{Z}. \tag{2.7}$$

For these equalities to make sense we need to prove that $Q\tilde{g}$ and $P\tilde{g}$ belong to $L^2(\mathbb{R})$ when Qg and Pg belong to $L^2(\mathbb{R})$. Assuming this fact, which we shall prove later, (2.4) and (2.5) are easy consequences of (1.14) with $f = Qg$ and $f = Pg$, respectively, and Proposition 2.1. Continuing with the proof, let us assume that (2.6) and (2.7) hold. The above four equalities readily imply

$$\langle Qg, P\tilde{g} \rangle = \langle Pg, Q\tilde{g} \rangle. \tag{2.8}$$

Using the integration by parts formula (see (1.5) in Chapter 1) we obtain

$$\langle Qg, P\tilde{g} \rangle = \int_{\mathbb{R}} xg(x) \overline{\{-i\tilde{g}'(x)\}} \, dx$$

$$= -i \int_{\mathbb{R}} \{g(x) + xg'(x)\} \overline{\tilde{g}(x)} \, dx$$

$$= -i \langle g, \tilde{g} \rangle + \langle Pg, Q\tilde{g} \rangle.$$

Equality (2.8) implies $i \langle g, \tilde{g} \rangle = 0$. Our final step in the proof will be to show $\langle g, \tilde{g} \rangle = 1$, so that the contradiction $i = 0$ proves Theorem 2.3.

We still have to show the technical results we assumed:

$$\begin{cases} \text{(a)} & \text{if } Qg, Pg \in L^2(\mathbb{R}), \text{ then } Q\tilde{g}, P\tilde{g} \in L^2(\mathbb{R}); \\ \text{(b)} & \text{equalities (2.6) and (2.7);} \\ \text{(c)} & \langle g, \tilde{g} \rangle = 1 \text{ where } \tilde{g} = \mathsf{S}^{-1}g. \end{cases} \tag{2.9}$$

In the proof of these facts, the frame (in fact, the orthonormal basis) of Example C in Chapter 1, for which $g = \chi = \chi_{[0,1]}$, will play a major role. That is

$$\left\{ E_{j,k}(x) = e^{2\pi ijx} \chi(x - k) : j, k \in \mathbb{Z} \right\}$$

is an orthonormal basis of $L^2(\mathbb{R})$. Clearly, the system

$$\left\{ e_{j,k}(s,t) = e^{2\pi ijs} e^{2\pi ikt} : j, k \in \mathbb{Z} \right\},$$

is an orthonormal basis of $L^2([0,1) \times [0,1)) \equiv L^2(\mathbb{T}^2)$.

Let $\mathcal{R} : L^2(\mathbb{R}) \longrightarrow L^2(\mathbb{T}^2)$ be defined by letting

$$\mathcal{R}(E_{j,k}) = e_{j,k} \qquad \text{for all } j, k \in \mathbb{Z}.$$

The operator $\mathcal{R}$ is called the **Zak transform** (see note 7 in section 8.5). The map $\mathcal{R}$ is unitary since it maps an orthonormal basis of $L^2(\mathbb{R})$ onto an orthonormal basis of $L^2(\mathbb{T}^2)$. Hence,

$$\langle f, g \rangle \equiv \langle f, g \rangle_{L^2(\mathbb{R})} = \langle \mathcal{R}f, \mathcal{R}g \rangle_{L^2(\mathbb{T}^2)} \equiv [\mathcal{R}f, \mathcal{R}g].$$

LEMMA 2.10 *Suppose that $g \in L^2(\mathbb{R})$ and that*

$$\left\{ g_{m,n}(x) = e^{2\pi imx} g(x - n) : m, n \in \mathbb{Z} \right\}$$

is a frame with frame bounds A and B. Then, we have

$$0 < A \leq \left| \mathcal{R}g(s,t) \right|^2 \leq B < \infty \qquad \textit{a.e. on } \mathbb{T}^2, \tag{2.11}$$

and

$$(\mathcal{R}\tilde{g}_{m,n})(s,t) = \frac{e^{2\pi ims} e^{2\pi int}}{\overline{\mathcal{R}g(s,t)}}. \tag{2.12}$$

PROOF : We begin by computing $\mathcal{R}g_{m,n}$:

$$(\mathcal{R}g_{m,n})(s,t) = \sum_{j\in\mathbb{Z}} \sum_{k\in\mathbb{Z}} \langle g_{m,n}, E_{j,k} \rangle \, e_{j,k}(s,t)$$

$$= \sum_{j\in\mathbb{Z}} \sum_{k\in\mathbb{Z}} \langle g, E_{j-m,k-n} \rangle \, e_{j,k}(s,t)$$

$$= e^{2\pi ims} e^{2\pi int} \sum_{j\in\mathbb{Z}} \sum_{k\in\mathbb{Z}} \langle g, E_{j,k} \rangle \, e_{j,k}(s,t)$$

$$= e^{2\pi ims} e^{2\pi int} (\mathcal{R}g)(s,t). \tag{2.13}$$

It follows that, if $f \in L^2(\mathbb{R})$,

$$\sum_{m \in \mathbb{Z}} \sum_{n \in \mathbb{Z}} |\langle f, g_{m,n} \rangle|^2 = \sum_{m \in \mathbb{Z}} \sum_{n \in \mathbb{Z}} |[\mathcal{R}f, \mathcal{R}g_{m,n}]|^2$$

$$= \sum_{m \in \mathbb{Z}} \sum_{n \in \mathbb{Z}} \left| \int_0^1 \int_0^1 \mathcal{R}f(s,t) e^{-2\pi i m s} e^{-2\pi i n t} \overline{\mathcal{R}g(s,t)} \, ds \, dt \right|^2$$

$$= \int_0^1 \int_0^1 |\mathcal{R}f(s,t)|^2 |\mathcal{R}g(s,t)|^2 \, ds \, dt. \tag{2.14}$$

Since $\{g_{m,n} : m, n \in \mathbb{Z}\}$ is a frame, with frame bounds A and B, the unitary property of $\mathcal{R}$ and this last equality give us

$$A[\mathcal{R}f, \mathcal{R}f] = A\|f\|^2 \le \int_0^1 \int_0^1 |\mathcal{R}f(s,t)|^2 |\mathcal{R}g(s,t)|^2 \, ds \, dt$$

$$\le B\|f\|^2 = B[\mathcal{R}f, \mathcal{R}f]$$

for all $f \in L^2(\mathbb{R})$. As f ranges throughout $L^2(\mathbb{R})$, $|\mathcal{R}f(s,t)|$ ranges throughout all positive functions on $L^2(\mathbb{T}^2)$. Thus,

$$0 < A \le |\mathcal{R}g(s,t)|^2 \le B < \infty \qquad \text{a.e. on } \mathbb{T}^2, \tag{2.15}$$

which is (2.11).

If we let $\varphi = |\mathcal{R}g|^2$ and $\mathcal{F}$ be the frame operator (see (1.6)), then (2.14) gives us, for $f \in L^2(\mathbb{R})$,

$$\langle Sf, f \rangle = \langle \mathcal{F}f, \mathcal{F}f \rangle = \sum_{m \in \mathbb{Z}} \sum_{n \in \mathbb{Z}} |\langle f, g_{m,n} \rangle|^2$$

$$= \int_0^1 \int_0^1 \varphi(s,t) \mathcal{R}f(s,t) \overline{\mathcal{R}f(s,t)} \, ds \, dt.$$

Polarizing both sides of this equality we obtain

$$[\mathcal{R}Sf, \mathcal{R}h] = \langle Sf, h \rangle = [\varphi \mathcal{R}f, \mathcal{R}h] \qquad \text{for all } f, h \in L^2(\mathbb{R}).$$

Thus $\mathcal{R}S = \varphi \mathcal{R}$, or equivalently, $\varphi^{-1} \mathcal{R} = \mathcal{R}S^{-1}$ (observe that φ^{-1} is well defined by (2.11)). Then, by Proposition 2.1 and equality (2.13) we obtain

$$(\mathcal{R}\tilde{g}_{m,n})(s,t) = (\mathcal{R}\widetilde{g_{m,n}})(s,t) = (\mathcal{R}S^{-1}g_{m,n})(s,t) = \frac{1}{\varphi(s,t)} \mathcal{R}g_{m,n}(s,t)$$

$$= \frac{1}{|\mathcal{R}g(s,t)|^2} e^{2\pi i m s} e^{2\pi i n t} (\mathcal{R}g)(s,t) = \frac{e^{2\pi i m s} e^{2\pi i n t}}{\overline{\mathcal{R}g(s,t)}}.$$

This finishes the proof of Lemma 2.10.

∎

It is now easy to prove part (c) of (2.9). When $m = n = 0$, (2.12) implies $\mathcal{R}\tilde{g} = (\overline{\mathcal{R}g})^{-1}$. Thus,

$$\langle g, \tilde{g} \rangle = [\mathcal{R}g, \mathcal{R}\tilde{g}] = \int_0^1 \int_0^1 \mathcal{R}g(s,t) \frac{1}{\mathcal{R}g(s,t)} \, ds \, dt = 1.$$

A more general result, which we shall use below, is true:

$$\langle g, \tilde{g}_{m,n} \rangle = \delta_{m,0} \delta_{n,0}, \qquad m, n \in \mathbb{Z}. \tag{2.16}$$

This follows from (2.13):

$$\langle g, \tilde{g}_{m,n} \rangle = \langle g_{-m,-n}, \tilde{g} \rangle = [\mathcal{R}g_{-m,-n}, \mathcal{R}\tilde{g}]$$

$$= \int_0^1 \int_0^1 e^{-2\pi i m s} e^{-2\pi i n t} \mathcal{R}g(s,t) \frac{1}{\mathcal{R}g(s,t)} \, ds \, dt = \delta_{m,0} \delta_{n,0}.$$

The proof of (2.9) (a) requires the following lemma which relates the operators $\mathcal{R}$, Q and P.

LEMMA 2.17 *For any* $h \in L^2(\mathbb{R})$, *if* $Qh \in L^2(\mathbb{R})$ *and* $Ph \in L^2(\mathbb{R})$, *we have*

(i) $\quad \mathcal{R}(Qh)(s,t) = s(\mathcal{R}h)(s,t) + \dfrac{1}{2\pi i} \dfrac{\partial}{\partial t}(\mathcal{R}h)(s,t) \qquad$ *and*

(ii) $\quad \mathcal{R}(Ph)(s,t) = -i \dfrac{\partial}{\partial s}(\mathcal{R}h)(s,t).$

PROOF : Let $f \in L^2(\mathbb{R})$. The "double" Fourier coefficients of $(\mathcal{R}f)(s,t)$ are

$$(\mathcal{R}f)^{\wedge}(j,k) = [\mathcal{R}f, e_{j,k}] = [\mathcal{R}f, \mathcal{R}E_{j,k}] = \langle f, E_{j,k} \rangle$$

$$= \int_{\mathbb{R}} f(x) e^{-2\pi i j x} \chi(x - k) \, dx$$

$$= \int_0^1 f(u + k) e^{-2\pi i j u} \, du.$$

Writing $\mathcal{R}f$ in terms of its double Fourier series we obtain

$$(\mathcal{R}f)(s,t) = \sum_{k \in \mathbb{Z}} \sum_{j \in \mathbb{Z}} \left(\int_0^1 f(u+k)\, e^{-2\pi i j u}\, du \right) e^{2\pi i j s} e^{2\pi i k t}$$

$$= \sum_{k \in \mathbb{Z}} f(s+k)\, e^{2\pi i k t}.$$

Consequently, if $f = Qh$,

$$(\mathcal{R}f)(s,t) = (\mathcal{R}\,Qh)(s,t) = \sum_{k \in \mathbb{Z}} (s+k)h(s+k)\, e^{2\pi i k t}$$

$$= s \sum_{k \in \mathbb{Z}} h(s+k)\, e^{2\pi i k t} + \frac{1}{2\pi i} \sum_{k \in \mathbb{Z}} 2\pi i k\, h(s+k)\, e^{2\pi i k t}$$

$$= s(\mathcal{R}h)(s,t) + \frac{1}{2\pi i} \frac{\partial}{\partial t}(\mathcal{R}h)(s,t),$$

which shows part (i) of Lemma 2.17. To prove (ii) in the same lemma, let $f = Ph$ so that

$$(\mathcal{R}f)(s,t) = (\mathcal{R}\,Ph)(s,t)$$

$$= -i \sum_{k \in \mathbb{Z}} h'(s+k)\, e^{2\pi i k t} = -i \frac{\partial}{\partial s}(\mathcal{R}h)(s,t).$$

∎

We now show part (a) of (2.9). If $Qg \in L^2(\mathbb{R})$, part (i) of Lemma 2.17 shows that $\frac{\partial}{\partial t}(\mathcal{R}g) \in L^2(\mathbb{T}^2)$. Since $\mathcal{R}\tilde{g} = (\overline{\mathcal{R}g})^{-1}$, it follows that

$$\frac{\partial}{\partial t}(\mathcal{R}\tilde{g}) = -(\overline{\mathcal{R}g})^{-2} \frac{\partial}{\partial t}(\overline{\mathcal{R}g}).$$

The inequalities (2.15) show that $|\mathcal{R}g|$ is bounded away from zero, so that $\frac{\partial}{\partial t}(\mathcal{R}\tilde{g}) \in L^2(\mathbb{T}^2)$. Applying now part (i) of Lemma 2.17 to $h = \tilde{g}$ we deduce that $Q\tilde{g} \in L^2(\mathbb{R})$. A similar argument, using Lemma 2.17 (ii) shows the result for Pg and $P\tilde{g}$ in (2.9) (a).

It remains to show (2.9) (b), that is, inequalities (2.6) and (2.7). To do this we use (2.16) to obtain:

$$\langle Qg, \tilde{g}_{m,n} \rangle = \langle Qg, \tilde{g}_{m,n} \rangle - n \langle g, \tilde{g}_{m,n} \rangle$$

$$= \int_{\mathbb{R}} g(x)(x-n)\,\overline{\tilde{g}(x-n)}\,e^{-2\pi imx}\,dx$$

$$= \int_{\mathbb{R}} g(y+n)\,y\,\overline{\tilde{g}(y)}\,e^{-2\pi imy}\,dy = \langle g_{-m,-n}, Q\tilde{g}\rangle.$$

This shows (2.6). For (2.7) we have, using integration by parts and (2.16),

$$\langle Pg, \tilde{g}_{m,n}\rangle = -i \int_{\mathbb{R}} g'(x)\,e^{-2\pi imx}\,\overline{\tilde{g}(x-n)}\,dx$$

$$= \int_{\mathbb{R}} g(x)i\,\overline{\tilde{g}'(x-n)}\,e^{-2\pi imx}\,dx$$

$$+ \int_{\mathbb{R}} g(x)2\pi m\,\overline{\tilde{g}(x-n)}\,e^{-2\pi imx}\,dx$$

$$= \int_{\mathbb{R}} g(y+n)\,i\,\overline{\tilde{g}'(y)}\,e^{-2\pi imy}\,dy + 2\pi m\,\langle g, \tilde{g}_{m,n}\rangle$$

$$= \langle g_{-m,-n}, P\tilde{g}\rangle + 0 = \langle g_{-m,-n}, P\tilde{g}\rangle.$$

This finishes the proof of the Balian-Low theorem for frames (Theorem 2.3).

8.3   Frames from translations and dilations

In this section we consider the question of finding conditions on a system of the form

$$\psi_{j,k}(x) = 2^{\frac{j}{2}}\psi(2^j x - k), \qquad j, k \in \mathbb{Z}, \tag{3.1}$$

generated by translations and dilations of a single function $\psi \in L^2(\mathbb{R})$, so that it becomes a frame in $L^2(\mathbb{R})$. Obviously, every orthonormal wavelet is a frame of this type, but we shall show that the converse is not true. Nevertheless, they still have perfect reconstruction (as shown in section 8.1) and they have been used in several applications.

If (3.1) is a frame, the general theory provides us with the dual frame $\widetilde{\psi}_{j,k} = S^{-1}\psi_{j,k}$, where $S = \mathcal{F}^*\mathcal{F}$ and $\mathcal{F}$ is the frame operator. The operator $S = \mathcal{F}^*\mathcal{F}$ commutes with the dilations $(\delta^m f)(x) = 2^{\frac{m}{2}} f(2^m x)$, $m \in \mathbb{Z}$:

$$((\mathcal{F}^*\mathcal{F})(\delta^m f))(x) = \sum_{j\in\mathbb{Z}}\sum_{k\in\mathbb{Z}} \langle \delta^m f, \psi_{j,k}\rangle\,\psi_{j,k}(x)$$

$$= \sum_{j \in \mathbb{Z}} \sum_{k \in \mathbb{Z}} \langle f, \psi_{j-m,k} \rangle \, \psi_{j,k}(x)$$

$$= 2^{\frac{m}{2}} \sum_{j \in \mathbb{Z}} \sum_{k \in \mathbb{Z}} \langle f, \psi_{j-m,k} \rangle \, \psi_{j-m,k}(2^m x)$$

$$= 2^{\frac{m}{2}} (\mathcal{F}^* \mathcal{F} f)(2^m x) = \delta^m (\mathcal{F}^* \mathcal{F} f)(x).$$

Thus, S^{-1} also commutes with these dilations, and we have

$$\widetilde{\psi_{j,k}}(x) = (S^{-1}\psi_{j,k})(x) = (S^{-1}\delta^j \psi_{0,k})(x)$$

$$= \delta^j S^{-1}\psi_{0,k}(x) = \delta^j \widetilde{\psi_{0,k}}(x) = 2^{\frac{j}{2}} \widetilde{\psi_{0,k}}(2^j x).$$

Thus, for k fixed, the functions $\widetilde{\psi_{j,k}}$ are all dilations of a single function $\widetilde{\psi_{0,k}}$. Unfortunately, this is not the case for translations $(\tau_k f)(x) = f(x-k)$. The computation

$$\big((\mathcal{F}^* \mathcal{F})(\tau_\ell f)\big)(x) = \sum_{k \in \mathbb{Z}} \sum_{j \in \mathbb{Z}} \langle \tau_\ell f, \psi_{j,k} \rangle \, \psi_{j,k}(x)$$

$$= \sum_{k \in \mathbb{Z}} \sum_{j \in \mathbb{Z}} \langle f, \psi_{j,k-2^j \ell} \rangle \, \psi_{j,k}(x)$$

$$= \sum_{n \in \mathbb{Z}} \sum_{j \in \mathbb{Z}} \langle f, \psi_{j,n} \rangle \, \psi_{j,n+2^j \ell}(x) = \tau_\ell(\mathcal{F}^* \mathcal{F} f)(x)$$

is valid if $2^j \ell$ is an integer, which is true for every ℓ only when $j \geq 0$. Thus, in general, one cannot expect the dual frame to be generated by dilations and translations of a single function (see the end of section 8.3 and note 4 in section 8.5).

We now address the problem of finding sufficient conditions on ψ for (3.1) to be a frame. It turns out that this problem shares some features with the one that led to the basic equations that characterize wavelets (see section 7.1).

Define

$$t_m(\xi) = \sum_{j=0}^{\infty} \hat{\psi}(2^j \xi) \, \overline{\hat{\psi}(2^j(\xi + 2m\pi))}, \qquad \xi \in \mathbb{R}, m \in \mathbb{Z},$$

as we did in section 7.1 (see (1.17) in Chapter 7), and

$$S(\xi) = \sum_{j \in \mathbb{Z}} |\hat{\psi}(2^j \xi)|^2, \qquad \xi \in \mathbb{R}.$$

Consider

$$\underline{S}_\psi = \text{ess} \inf_{\xi \in \mathbb{R}} S(\xi), \qquad \overline{S}_\psi = \text{ess} \sup_{\xi \in \mathbb{R}} S(\xi)$$

and

$$\beta_\psi(m) = \text{ess} \sup_{\xi \in \mathbb{R}} \sum_{k \in \mathbb{Z}} |t_m(2^k \xi)|.$$

Observe that all the expressions inside the infimum and suprema in the above definitions are invariant under the usual dilations (scaling) by 2, so that these infimum and suprema need only be computed over $1 \leq |\xi| \leq 2$ (a dilation "period").

THEOREM 3.2 *Let $\psi \in L^2(\mathbb{R})$ be such that*

$$A_\psi = \underline{S}_\psi - \sum_{q \in 2\mathbb{Z}+1} \left[\beta_\psi(q) \beta_\psi(-q) \right]^{\frac{1}{2}} > 0,$$

and

$$B_\psi = \overline{S}_\psi + \sum_{q \in 2\mathbb{Z}+1} \left[\beta_\psi(q) \beta_\psi(-q) \right]^{\frac{1}{2}} < \infty.$$

Then, $\{\psi_{j,k} : j, k \in \mathbb{Z}\}$ is a frame with frame bounds A_ψ and B_ψ.

REMARK: If $S(\xi) = 1$ for a.e. $\xi \in \mathbb{R}$ and $t_m(\xi) = 0$ for a.e. $\xi \in \mathbb{R}$ and all $m \in 2\mathbb{Z}+1$, $A_\psi = B_\psi = 1$. If, in addition, $\|\psi\|_2 = 1$, the frame $\{\psi_{j,k} : j, k \in \mathbb{Z}\}$ is an orthonormal basis of $L^2(\mathbb{R})$, and thus a wavelet (see section 8.1). This result is the sufficiency of Theorem 1.1 in Chapter 7.

To prove Theorem 3.2 we need to generalize Lemma 1.10 of Chapter 7 to the setting of frames. The proof in this more general case is a simple modification of the one given for the aforementioned lemma.

LEMMA 3.3 *Suppose that $\{e_j : j = 1, 2, \cdots\}$ is a family of elements in a Hilbert space $\mathbb{H}$ such that there exist constants $0 < A \leq B < \infty$ satisfying*

$$A\|f\|^2 \leq \sum_{j=1}^{\infty} |\langle f, e_j \rangle|^2 \leq B\|f\|^2$$

for all f belonging to a dense subset $\mathcal{D}$ of $\mathbb{H}$. Then, the same inequalities are true for all $f \in \mathbb{H}$; that is, $\{e_j : j = 1, 2, \cdots\}$ is a frame for $\mathbb{H}$.

PROOF: As in the proof of Lemma 1.10 of Chapter 7, a density argument establishes the inequality

$$\sum_{j=1}^{\infty} |\langle f, e_j \rangle|^2 \leq B \|f\|^2$$

for all $f \in \mathbb{H}$. To show the other inequality, we choose $\varepsilon > 0$ and $g \in \mathcal{D}$ such that $\|g - f\| < \varepsilon$. Then, by Minkowski's inequality in ℓ^2, the above inequality, and $\|f\| \leq \|g\| + \varepsilon \leq \|g\| + \sqrt{\frac{B}{A}}\,\varepsilon$, we obtain

$$\|f\| - 2\frac{\sqrt{B}}{\sqrt{A}}\varepsilon \leq \|g\| - \frac{\sqrt{B}}{\sqrt{A}}\varepsilon \leq \|g\| - \frac{\sqrt{B}}{\sqrt{A}}\|g - f\|$$

$$\leq \left(\frac{1}{A}\sum_{j=1}^{\infty}|\langle g, e_j\rangle|^2\right)^{\frac{1}{2}} - \frac{\sqrt{B}}{\sqrt{A}}\left(\frac{1}{B}\sum_{j=1}^{\infty}|\langle g - f, e_j\rangle|^2\right)^{\frac{1}{2}}$$

$$\leq \left(\frac{1}{A}\sum_{j=1}^{\infty}|\langle f, e_j\rangle|^2\right)^{\frac{1}{2}}.$$

This finishes the proof since ε is arbitrary. ∎

We now prove Theorem 3.2. Let $\mathcal{D}$ be the class of all $f \in L^2(\mathbb{R})$ such that $\hat{f} \in L^{\infty}(\mathbb{R})$ and $\hat{f}$ is compactly supported in $\mathbb{R} \setminus \{0\}$. By Proposition 1.19 of Chapter 7 we have

$$\sum_{j \in \mathbb{Z}} \sum_{k \in \mathbb{Z}} |\langle f, \psi_{j,k}\rangle|^2$$

$$= \frac{1}{2\pi}\int_{\mathbb{R}} |\hat{f}(\xi)|^2 S(\xi)\,d\xi + \frac{1}{2\pi}\int_{\mathbb{R}} \overline{\hat{f}(\xi)} \sum_{p \in \mathbb{Z}} \sum_{q \in 2\mathbb{Z}+1} \hat{f}(\xi + 2^p 2q\pi)\, t_q(2^{-p}\xi)\,d\xi$$

$$\equiv \frac{1}{2\pi}\int_{\mathbb{R}} |\hat{f}(\xi)|^2 S(\xi)\,d\xi + \frac{1}{2\pi} R_\psi(f) \tag{3.4}$$

for all $f \in \mathcal{D}$. The Schwarz inequality gives us

$$|R_\psi(f)| \leq \sum_{q \in 2\mathbb{Z}+1} \sum_{p \in \mathbb{Z}} \left(\int_{\mathbb{R}} |\hat{f}(\eta)|^2\, |t_q(2^{-p}\eta)|\,d\eta\right)^{\frac{1}{2}}$$

$$\cdot \left(\int_{\mathbb{R}} |\hat{f}(\eta + 2^p 2q\pi)|^2 \, |t_q(2^{-p}\eta)| \, d\eta \right)^{\frac{1}{2}}.$$

In the second integral we change variables to obtain

$$\left(\int_{\mathbb{R}} |\hat{f}(\eta)|^2 \, |t_q(2^{-p}\eta - 2q\pi)| \, d\eta \right)^{\frac{1}{2}}.$$

Since $t_q(\xi - 2q\pi) = \overline{t_{-q}(\xi)}$, we deduce, after applying Schwarz's inequality for series,

$$|R_\psi(f)| \leq \sum_{q \in 2\mathbb{Z}+1} \left(\sum_{p \in \mathbb{Z}} \int_{\mathbb{R}} |\hat{f}(\eta)|^2 \, |t_q(2^{-p}\eta)| \, d\eta \right)^{\frac{1}{2}}$$

$$\cdot \left(\sum_{p \in \mathbb{Z}} \int_{\mathbb{R}} |\hat{f}(\eta)|^2 \, |t_{-q}(2^{-p}\eta)| \, d\eta \right)^{\frac{1}{2}}$$

$$\leq \sum_{q \in 2\mathbb{Z}+1} [\beta_\psi(q)\beta_\psi(-q)]^{\frac{1}{2}} \|\hat{f}\|_2^2.$$

Hence,

$$- \sum_{q \in 2\mathbb{Z}+1} [\beta_\psi(q)\beta_\psi(-q)]^{\frac{1}{2}} \|\hat{f}\|_2^2 \leq R_\psi(f) \leq \sum_{q \in 2\mathbb{Z}+1} [\beta_\psi(q)\beta_\psi(-q)]^{\frac{1}{2}} \|\hat{f}\|_2^2.$$

These inequalities, together with (3.4), give us

$$A_\psi \|f\|_2^2 \leq \sum_{j \in \mathbb{Z}} \sum_{k \in \mathbb{Z}} |\langle f, \psi_{j,k} \rangle|^2 \leq B_\psi \|f\|_2^2$$

for all $f \in \mathcal{D}$. Since $\mathcal{D}$ is dense in $L^2(\mathbb{R})$, the same inequalities hold for all $f \in L^2(\mathbb{R})$ by Lemma 3.3. This finishes the proof of Theorem 3.2. ∎

EXAMPLE B: Let $\psi \in \mathcal{S}$ be such that $\operatorname{supp}(\hat{\psi})$ is contained in the set $\{\xi \in \mathbb{R} : \frac{1}{2} \leq |\xi| \leq 2\}$ and

$$\sum_{j \in \mathbb{Z}} |\hat{\psi}(2^j \xi)|^2 = 1 \qquad \text{for all} \ \xi \neq 0.$$

Then $t_q(\xi) = 0$ for all $\xi \in \mathbb{R}$ and, consequently, $A_\psi = B_\psi = 1$. Thus, $\{\psi_{j,k} : j, k \in \mathbb{Z}\}$ is a tight frame. These are the type of frames considered by M. Frazier and B. Jawerth (see section 8.5).

EXAMPLE C: An example of a frame of the type discussed in this section is the one generated by the **Mexican hat** function. This is the function

$$\psi(x) = \frac{2}{\sqrt{3}}\,\pi^{-\frac{1}{4}}(1 - x^2)e^{-\frac{1}{2}x^2},$$

which coincides with $-\frac{d^2}{dx^2}\left(e^{-\frac{1}{2}x^2}\right)$, when normalized in $L^2(\mathbb{R})$. The graph of the function ψ is given in Figure 8.1 (a). The Fourier transform of ψ is

$$\hat{\psi}(\xi) = \frac{2}{\sqrt{3}}\,\pi^{-\frac{1}{4}}\,\sqrt{2\pi}\,\xi^2\,e^{-\frac{1}{2}\xi^2},$$

whose graph is given in Figure 8.1 (b). I. Daubechies has reported frame bounds of 3.223 and 3.596 for the frame obtained by translations and dilations of the Mexican hat function (see [Da3], pages 986–987). An approximate quotient of these frame bounds is 1.116, which indicates that this frame is "close" to a tight frame.

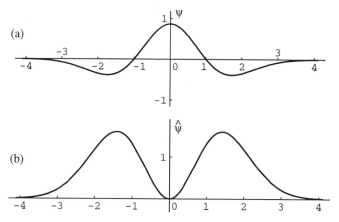

Figure 8.1: Graphs of the Mexican hat ψ and its Fourier transform $\hat{\psi}$.

Let ψ be a Lemarié-Meyer wavelet (see Corollary 4.7 in Chapter 1 or Example D in Chapter 2) such that ψ is in the Schwartz class $\mathcal{S}$. For $-1 < r < 1$ define

$$\eta(x) = \psi(x) - r\sqrt{2}\,\psi(2x). \tag{3.5}$$

Since $\eta_{j,k}(x) = \psi_{j,k}(x) - r\psi_{j+1,2k}(x)$ and ψ is an orthonormal wavelet, it follows immediately that the set $\{\eta_{j,k} : j, k \in \mathbb{Z}\}$ is a frame for $L^2(\mathbb{R})$ with frame bounds $(1 - |r|)^2$ and $(1 + |r|)^2$. Obviously, $\eta \in \mathcal{S}$ and has infinitely many vanishing moments since $\hat{\eta}(\xi) \equiv 0$ in a neighborhood of the origin. We shall show that given $p \in (1, 2)$ we can choose $r \in (-1, 1)$ such that $\tilde{\eta} = S^{-1}\eta$ does not belong to $L^p(\mathbb{R})$. Hence, there is no hope to have a representation like (1.15) for functions $f \in L^p(\mathbb{R})$ with the frame $\{\eta_{j,k} : j, k \in \mathbb{Z}\}$.

We start by computing the elements of the dual frame. Let

$$U : L^2(\mathbb{R}) \longrightarrow L^2(\mathbb{R})$$

be the operator given by

$$U(\psi_{j,k}) = \psi_{j+1,2k} \qquad \text{for all } j, k \in \mathbb{Z}.$$

Its adjoint, $U^* : L^2(\mathbb{R}) \longrightarrow L^2(\mathbb{R})$ is given by

$$\begin{cases} U^*(\psi_{j,2k}) = \psi_{j-1,k}, \\ U^*(\psi_{j,2k+1}) = 0, \end{cases} \qquad j, k \in \mathbb{Z}.$$

Observe that $\eta_{j,k} = (I - rU)(\psi_{j,k})$. Thus, from (1.8) we obtain

$$S(\psi_{j,k}) = \sum_{\ell \in \mathbb{Z}} \sum_{m \in \mathbb{Z}} \langle \psi_{j,k}, \eta_{\ell,m} \rangle \, \eta_{\ell,m}$$

$$= \psi_{j,k} - rU^*(\psi_{j,k}) - rU(\psi_{j,k}) + r^2 UU^*(\psi_{j,k})$$

$$= (I - rU)(I - rU^*)(\psi_{j,k}).$$

Thus, for the dual frame we have

$$\widetilde{\eta_{j,k}} = \left[(I - rU)(I - rU^*) \right]^{-1} (\eta_{j,k})$$

$$= (I - rU^*)(\psi_{j,k}) = \sum_{\ell=0}^{\infty} r^\ell (U^*)^\ell (\psi_{j,k}), \qquad (3.6)$$

where we have used that $(I - rU^*)$ is invertible when $r \in (-1, 1)$. For $j = k = 0$ we obtain

$$\tilde{\eta} \equiv \widetilde{\eta_{0,0}} = \sum_{\ell=0}^{\infty} r^\ell \psi_{-\ell,0} \, . \qquad (3.7)$$

To compute the L^p-norm of $\tilde{\eta}$, $1 < p < \infty$, we use the criterion proved in Theorem 4.5 of Chapter 6. Hence, the L^p-norm of $\tilde{\eta}$ is equivalent to

$$\left(\int_{\mathbb{R}} \left| \sum_{\ell=0}^{\infty} r^{2\ell} 2^{-\ell} \chi_{[0,2^\ell)}(x) \right|^{\frac{p}{2}} dx \right)^{\frac{1}{p}}.$$

Writing

$$\mathbb{R}^+ = [0,1) \cup [1,2) \cup [2,2^2) \cup \cdots \cup [2^m, 2^{m+1}) \cup \cdots$$

it can be shown that this expression is finite if and only if

$$\sum_{m=0}^{\infty} 2^m \left(\frac{r^2}{2}\right)^{m\frac{p}{2}} < \infty.$$

This holds if and only if $2(\frac{|r|}{\sqrt{2}})^p < 1$, or equivalently,

$$|r| < \sqrt{2}\,(\tfrac{1}{2})^{\frac{1}{p}}.$$

Hence, given $p \in (1,2)$, for any $r \in (2^{\frac{1}{2}-\frac{1}{p}}, 1)$ the function $\tilde{\eta}$ does not belong to $L^p(\mathbb{R})$, as we wanted to show.

We can now show that the dual frame of $\{\eta_{j,k} : j, k \in \mathbb{Z}\}$ cannot be obtained by dilations and translations of a single function $\tilde{\eta}$. If this were the case we would have

$$\widetilde{\eta_{0,1}}(x) = \tilde{\eta}(x-1).$$

On the other hand, it follows from (3.6) and the definition of U^* that

$$\widetilde{\eta_{0,1}}(x) = \psi_{0,1}(x).$$

Hence, by (3.7),

$$\psi(x) = \widetilde{\eta_{0,1}}(x+1) = \tilde{\eta}(x) = \psi(x) + \sum_{\ell=1}^{\infty} r^\ell \psi_{-\ell,0}(x).$$

Thus,

$$\sum_{\ell=1}^{\infty} r^\ell \psi_{-\ell,0}(x) = 0, \qquad x \in \mathbb{R},$$

for all $r \in (-1,1)$. It follows that $\psi(x) = 0$ for all $x \in \mathbb{R}$, which is impossible. This construction was communicated to us by Y. Meyer.

8.4   Smooth frames for $H^2(\mathbb{R})$

We have shown in section 7.6 that there are no orthonormal wavelets in $H^2(\mathbb{R})$ whose Fourier transforms satisfy the regularity condition given by (6.1) in Chapter 7. In this section we shall show that it is possible to construct frames in $H^2(\mathbb{R})$ which are given by translations and dilations of a single function ψ, where ψ, as well as $\hat\psi$, belong to the Schwartz class. These examples share some similarities with the Lemarié-Meyer wavelets presented in Corollary 4.7 of Chapter 1. Recall that $H^2(\mathbb{R})$ is the set of all $f \in L^2(\mathbb{R})$ such that $\hat f(\xi) = 0$ for a. e. $\xi < 0$ (see section 7.6).

For $\varepsilon > 0$, let s_ε be a smooth function (say C^r, $r = 0, 1, 2, \cdots$, or C^∞) such that $s_\varepsilon(x) = 0$ if $x < \varepsilon$ and

$$s_\varepsilon^2(x) + c_\varepsilon^2(x) = 1, \qquad (4.1)$$

where $c_\varepsilon(x) = s_\varepsilon(-x)$ (see Figure 1.1 in Chapter 1). For $0 < \varepsilon \le \frac{1}{3}\pi$ let

$$b_\varepsilon(x) = s_\varepsilon(x - \pi)\, c_{2\varepsilon}(x - 2\pi) \qquad (4.2)$$

be a **bell function** associated with the interval $[\pi, 2\pi]$ (see Figure 1.2 in Chapter 1). Define ψ^ε by

$$\widehat{\psi^\varepsilon}(\xi) = b_\varepsilon(\xi), \qquad \xi \in \mathbb{R}. \qquad (4.3)$$

It is clear that ψ^ε belongs to $H^2(\mathbb{R})$. Moreover, using (4.1) we deduce

$$\sum_{j\in\mathbb{Z}} \left|\widehat{\psi^\varepsilon}(2^j\xi)\right|^2 = \chi_{(0,\infty)}(\xi), \qquad \xi \in \mathbb{R}. \qquad (4.4)$$

We now show that if $0 < \varepsilon \le \frac{1}{3}\pi$, the system $\{(\psi^\varepsilon)_{j,k} : j, k \in \mathbb{Z}\}$ is a frame for $H^2(\mathbb{R})$. For any $f \in H^2(\mathbb{R})$ we use Plancherel's theorem to obtain

$$\sum_{j\in\mathbb{Z}}\sum_{k\in\mathbb{Z}} \left|\langle f, (\psi^\varepsilon)_{j,k}\rangle\right|^2 = \frac{1}{4\pi^2}\sum_{j\in\mathbb{Z}}\sum_{k\in\mathbb{Z}} \left|\int_0^\infty \hat f(\xi) 2^{\frac{j}{2}} \overline{(\psi^\varepsilon)^\wedge(2^j\xi)}\, e^{-ik2^j\xi}\, d\xi\right|^2$$

$$= \sum_{j\in\mathbb{Z}} 2^j \sum_{k\in\mathbb{Z}} \left|\frac{1}{2\pi}\int_{2^{-j}(\pi-\varepsilon)}^{2^{-j}(2\pi+2\varepsilon)} \hat f(\xi) b_\varepsilon(2^j\xi) e^{-ik2^j\xi}\, d\xi\right|^2.$$

The sum over k in this last expression is the sum of the squares of the Fourier coefficients of the function $2^{-j}\hat{f}(2^{-j}\cdot)b_\varepsilon(\cdot)$ over the interval $\left[\frac{2}{3}\pi, \frac{8}{3}\pi\right]$. Thus, we have

$$\sum_{j\in\mathbb{Z}}\sum_{k\in\mathbb{Z}}\left|\langle f, (\psi^\varepsilon)_{j,k}\rangle\right|^2 = \frac{1}{2\pi}\sum_{j\in\mathbb{Z}}\int_{2^{-j}(\pi-\varepsilon)}^{2^{-j}(2\pi+2\varepsilon)}\left|\hat{f}(\xi)\right|^2\left|b_\varepsilon(2^j\xi)\right|^2 d\xi$$

$$= \frac{1}{2\pi}\sum_{j\in\mathbb{Z}}\int_0^\infty\left|\hat{f}(\xi)\right|^2\left|(\psi^\varepsilon)^\wedge(2^j\xi)\right|^2 d\xi.$$

Using (4.4) we obtain

$$\sum_{j\in\mathbb{Z}}\sum_{k\in\mathbb{Z}}\left|\langle f, (\psi^\varepsilon)_{j,k}\rangle\right|^2 = \frac{1}{2\pi}\int_0^\infty\left|\hat{f}(\xi)\right|^2 d\xi = \|f\|_{H^2(\mathbb{R})}^2.$$

We have proved the following result:

PROPOSITION 4.5 *For every ε such that $0 < \varepsilon \leq \frac{1}{3}\pi$, the system of functions $\{(\psi^\varepsilon)_{j,k} : j, k \in \mathbb{Z}\}$, where ψ^ε is given by (4.3), is a tight frame for $H^2(\mathbb{R})$ with frame bound 1. Moreover, ψ^ε is in the Schwartz class if we choose $s_\varepsilon \in C^\infty(\mathbb{R})$.*

A "folding" argument similar to the one presented at the end of section 1.2 shows

$$\left\|\psi^\varepsilon\right\|_{H^2(\mathbb{R})}^2 = \frac{1}{2\pi}\left\|(\psi^\varepsilon)^\wedge\right\|_2^2 = \frac{\pi}{2\pi} = \frac{1}{2}.$$

Thus, if $0 < \varepsilon \leq \frac{1}{3}\pi$, $\|\psi^\varepsilon\|_{H^2(\mathbb{R})} = \frac{1}{\sqrt{2}} < 1$, and we have a family of frames in $H^2(\mathbb{R})$ that **do not** form an orthonormal basis.

8.5   Notes and references

1. Frames appear in [DS] (1952) in the context of nonharmonic Fourier series (expansion of a function $f \in L^2([0,1])$ with complex exponentials $e^{i\lambda_n x}$, where $\lambda_n \neq 2n\pi$). They also appeared in [You]. An earlier version of this notion was introduced in 1947 by B. Sz-Nagy [Sz-N]. The modern development of frames is due to I. Daubechies; the proof of the reconstruction property for frames, several examples, the Balian-Low theorem in this setting and many other properties of frames are contained in [Da3]. More results for frames can be found in [DGM], [Da1] and [Me1]. A recent survey article concerning the Balian-Low theorem is [BHW], where more references are given. The sufficient condition given in Theorem 3.2 can be found in [Da3] (page 984), where it is attributed to Ph. Tchamitchian. A weaker result (sometimes better for computations), cited in [Me1] as "Daubechies' sufficient condition," can be found in [Me1] and [Da1].

2. We have studied (Schauder) bases for Banach spaces in Chapter 5 (see, in particular, section 5.1 and the first note in section 5.7). When the Banach space is a Hilbert space $\mathbb{H}$ the most "natural" bases are those that are orthonormal. A class that is somewhat more general is the collection of Riesz bases. We considered such bases in Chapter 2 (see, for example, Lemma 1.8 in that chapter). In this chapter we studied yet another family of vectors, the frames, that can be used to analyze and reconstruct the general element of a Hilbert space. We have seen, however, that a frame need not be a basis of the Hilbert space. But when it is a basis, the property of being a frame is equivalent to being a Riesz basis. Let us examine this equivalence in more detail.

We remind the reader that a set $E = \{e_1, e_2, \cdots, e_k, \cdots\}$ in the Hilbert space $\mathbb{H}$ is a **Riesz basis** if it is a basis for which there exist two positive constants A and B such that

$$A \sum_k |\alpha_k|^2 \le \left\| \sum_k \alpha_k e_k \right\|^2 \le B \sum_k |\alpha_k|^2 \tag{5.1}$$

for all finitely non-zero sequences $\alpha = \{\alpha_1, \alpha_2, \cdots, \alpha_k, \cdots\} \in \ell^2$. It is not hard to see that Riesz bases are precisely those that are the images, under invertible bounded linear operators on $\mathbb{H}$, of orthonormal bases. The two inequalities in (5.1) imply that the operator $T : \ell^2 \longrightarrow \mathbb{H}$, defined by

$$T\alpha = \sum_{k=1}^{\infty} \alpha_k e_k$$

is well defined for each $\alpha = \{\alpha_k\}_k \in \ell^2$, has operator norm $\|T\| \le \sqrt{B}$, is invertible and T^{-1} has operator norm $\|T^{-1}\| \le 1/\sqrt{A}$.

Let $T^* : \mathbb{H} \longrightarrow \ell^2$ be the **adjoint** operator to T, defined by the equality

$$(\alpha, T^* h) = \langle T\alpha, h \rangle$$

for all $\alpha \in \ell^2$ and $h \in \mathbb{H}$, where

$$(\alpha, \beta) = \sum_{k=1}^{\infty} \alpha_k \overline{\beta_k}$$

denotes the inner product for ℓ^2 and $\langle \cdot, \cdot \rangle$ the inner product for $\mathbb{H}$. Then,

$$(\alpha, T^*h) = \langle T\alpha, h \rangle = \left\langle \sum_{k=1}^{\infty} \alpha_k e_k, h \right\rangle = \sum_{k=1}^{\infty} \alpha_k \langle e_k, h \rangle = (\alpha_k, \{\langle h, e_k \rangle\}).$$

This show that $T^*h = \{\langle h, e_k \rangle\}$. Since $\|T^*\| = \|T\| \leq \sqrt{B}$ and $\|(T^*)^{-1}\| = \|(T^{-1})^*\| = \|T^{-1}\| \leq 1/\sqrt{A}$ we then have

$$A\|h\|^2 \leq \sum_{k=1}^{\infty} |\langle h, e_k \rangle|^2 \leq B\|h\|^2, \tag{5.2}$$

which is precisely the defining property (1.3) for a frame.

Now let us assume that the set E is a frame and a basis. The frame operator $\mathcal{F}$ defined by (1.6) is well defined and bounded even when E is not a basis. It is easy to see that the adjoint operator $\mathcal{F}^* : \ell^2 \longrightarrow \mathbb{H}$ is given by

$$\mathcal{F}^*\alpha = \sum_{k=1}^{\infty} \alpha_k e_k .$$

The assumption that E is a basis is now used to see that $\mathcal{F}^*$ is one to one. This allows us to show that $(\mathcal{F}^*)^{-1} = (\mathcal{F}^{-1})^*$ exists and is a bounded operator. In fact, $\|\mathcal{F}^*\| \leq \sqrt{B}$ and $\|(\mathcal{F}^*)^{-1}\| \leq 1/\sqrt{A}$. These inequalities can then be written in the form given by (5.1) showing that E is a Riesz basis (see [You] for more detailed proofs of these facts).

3. A frame with frame bounds A and B obtained from translations and dilations of a single function $\psi \in L^2(\mathbb{R})$ (see (3.1) and (1.3)) must satisfy

$$A \leq \sum_{j \in \mathbb{Z}} |\hat{\psi}(2^j \xi)|^2 \leq B \qquad \text{for a.e. } \xi \in \mathbb{R}.$$

For the case of orthonormal wavelets for $L^2(\mathbb{R})$, these inequalities are proved in Theorem 1.1 of Chapter 7 (with $A = B = 1$). A complete proof in the general case can be found in [CS] and a sketch is given in [Chu]. Dividing the above inequalities by $|\xi|$ and integrating over $1 \le |\xi| \le 2$ we deduce

$$2A \log 2 \le \int_{\mathbb{R}} \frac{|\hat{\psi}(\xi)|^2}{|\xi|} \, d\xi \le 2B \log 2.$$

This is the so-called "**admissible condition**" for frames of this type that was first proved by I. Daubechies in [Da3] using techniques from trace-class operators.

4. At the end of section 8.3 we have shown that given $p \in (1, 2)$ we can find a frame of the form $\{\eta_{j,k} : j, k \in \mathbb{Z}\}$ for which (1.15) fails for $f \in L^p(\mathbb{R})$. One might think that this is due to the fact that the dual frame is not obtained from translations and dilations of a single function. But this is not the case, since there exist frames of the type $\{\psi_{j,k} : j, k \in \mathbb{Z}\}$, for which the dual frame is also obtained from translations and dilations of a single function $\tilde{\psi}$, and they do not enjoy the characterization properties given by wavelets in Chapter 6. See [Da3] (page 989) and the references given there for details.

5. How does the reconstruction formula for frames work?

The (exact) reconstruction formula for frames given in (1.14) and (1.15) requires the knowledge of the dual frame $\{\widetilde{\varphi}_j : j \in \mathbb{J}\}$. Sometimes it is difficult to find the elements of this dual frame. Nevertheless, the following argument shows that if the frame bounds are "close" to each other, there is a rapidly converging algorithm to reconstruct the function up to a small error, which does not need the elements of the dual frame. Assume that B is "close" to A (i.e. $r = \frac{B}{A} - 1$ is small). Since $AI \le \mathsf{S} \le BI$, we have $\mathsf{S} \sim \frac{1}{2}(A + B)I$. Thus, $\mathsf{S}^{-1} = \frac{2}{A+B} I$. Hence,

$$\widetilde{\varphi}_j = \mathsf{S}^{-1}\varphi_j \sim \frac{2}{A + B} \varphi_j.$$

Let us write

$$f = \frac{2}{A + B} \sum_{j \in \mathbb{J}} \langle f, \varphi_j \rangle \varphi_j + \mathcal{R}_1 f \tag{5.3}$$

where (see (1.8))

$$\mathcal{R}_1 = I - \frac{2}{A + B} \mathsf{S}.$$

It follows that

$$-\frac{B - A}{A + B} I \le \mathcal{R}_1 \le \frac{B - A}{A + B} I \quad \text{and} \quad \|\mathcal{R}_1\| \le \frac{B - A}{A + B} = \frac{r}{r + 2}.$$

Thus, (5.3) reconstructs f up to an error of $\dfrac{r}{r+2}\|f\|_2$ in $L^2(\mathbb{R})$. We can iterate (5.3) to obtain

$$\mathcal{R}_1 f = \frac{2}{A+B} \sum_{j \in \mathbb{J}} \langle \mathcal{R}_1 f, \varphi_j \rangle \, \varphi_j + \mathcal{R}_1(\mathcal{R}_1 f),$$

where $\|\mathcal{R}_1(\mathcal{R}_1 f)\| \leq \left(\frac{r}{r+2}\right)^2 \|f\|_2$. In general, after k iterations, the approximation error is smaller than $\left(\frac{r}{r+2}\right)^k \|f\|_2$ in $L^2(\mathbb{R})$.

Another approach is to observe that $\mathsf{S} = \frac{A+B}{2}(I - \mathcal{R}_1)$, so that

$$\mathsf{S}^{-1} = \frac{2}{A+B}(I - \mathcal{R}_1)^{-1} = \frac{2}{A+B} \sum_{k=0}^{\infty} \mathcal{R}_1^k$$

with convergence in norm, since $\|\mathcal{R}_1\| \leq \frac{B-A}{A+B} < 1$. Thus,

$$\widetilde{\varphi_j} = \mathsf{S}^{-1}\varphi_j = \frac{2}{A+B} \sum_{k=0}^{\infty} \mathcal{R}_1^k \varphi_j \qquad \text{for all } j \in \mathbb{J}.$$

Considering

$$\tilde{\varphi}_j^N = \frac{2}{A+B} \sum_{k=0}^{N} \mathcal{R}_1^k \varphi_j \qquad \text{for } N = 0, 1, 2, \cdots,$$

it can be shown that

$$\left\| f - \sum_{j \in \mathbb{J}} \langle f, \varphi_j \rangle \, \tilde{\varphi}_j^N \right\|_2 \leq \left(\frac{r}{2+r}\right)^N \|f\|_2,$$

so that a good approximation is obtained without too many iterations if r is small. Details of this approximation result can be found in [Da1] (pages 61-62).

6. Biorthogonal wavelets for $L^2(\mathbb{R})$.

Two functions $\psi, \tilde{\psi} \in L^2(\mathbb{R})$ are called biorthogonal wavelets if each one of the sets $\{\psi_{j,k} : j, k \in \mathbb{Z}\}$ and $\{\tilde{\psi}_{j,k} : j, k \in \mathbb{Z}\}$ is a Riesz basis of $L^2(\mathbb{R})$ and they are biorthogonal to each other in the following sense

$$\langle \psi_{j,k}, \tilde{\psi}_{\ell,m} \rangle = \delta_{j,\ell}\delta_{k,m} \qquad \text{for all } j, k, \ell, m \in \mathbb{Z}. \tag{5.4}$$

For the theory, applications, and examples of biorthogonal wavelets, we refer the reader to [CDF], [Co2], [Da1] and [CD]. (All these references treat only the case in which the two sets $\{\psi_{j,k} : j, k \in \mathbb{Z}\}$ and $\{\tilde{\psi}_{j,k} : j, k \in \mathbb{Z}\}$ arise from MRAs with Riesz bases; see the comment that follows (1.5) in Chapter 2.) In the spirit of the proof of Theorem 1.1 in Chapter 7 and Theorem 3.2 in this chapter, it is shown in [Wan] that a pair of biorthogonal wavelets must satisfy certain equations similar to the basic equations (1.2) and (1.3) in Chapter 7 and (2.1) and (2.2) in Chapter 3.

7. The Zak transform

We have defined the Zak transform as the unitary operator that maps $E_{j,k}$ to $e_{j,k}$ (see the proof of Theorem 2.3). We have shown in the proof of Lemma 2.17 that

$$(\mathcal{R}f)(s,t) = \sum_{k \in \mathbb{Z}} f(s+k)\, e^{2\pi i k t}.$$

The series on the right-hand side converges for every continuous function f such that

$$|f(x)| \le \frac{C}{(1+|x|)^{1+\varepsilon}}$$

for some $\varepsilon > 0$. For these functions, $(\mathcal{R}f)(s,t)$ is a continuous function. Moreover, it is immediate that $(\mathcal{R}f)(s,t)$ is a periodic function of t, with period 1, and

$$(\mathcal{R}f)(s+1,t) = e^{-2\pi i t}(\mathcal{R}f)(s,t). \tag{5.5}$$

The following is a result related to the Balian-Low theorem, which has been communicated to us by Y. Meyer.

THEOREM 5.6 *If g is a continuous function on $\mathbb{R}$ and there exists $\varepsilon > 0$ such that*

$$|g(x)| \le \frac{C}{(1+|x|)^{1+\varepsilon}},$$

the system

$$\{g_{m,n}(x) = e^{2\pi i m x} g(x-n) : \ m, n \in \mathbb{Z}\}$$

cannot be a frame for $L^2(\mathbb{R})$.

PROOF : We prove it by contradiction. Suppose that $\{g_{m,n} : m, n \in \mathbb{Z}\}$ is a frame. By (2.11), $(\mathcal{R}g)(s,t) \ne 0$ for all $s, t \in [0,1] \times [0,1] = \mathbb{T}^2$. This leads us to a contradiction. In fact, if $(\mathcal{R}g)(s,t) \neq 0$ for all $s, t \in \mathbb{T}^2$, for $s \in [0,1]$ we can define the complex curve $\mathcal{Z}_s : [0,1] \longrightarrow \mathbb{C}$ given by

$$\mathcal{Z}_s(t) = \frac{(\mathcal{R}g)(s,t)}{(\mathcal{R}g)(0,t)} \, .$$

These curves depend continuously on s and satisfy

$$\begin{cases} \mathcal{Z}_0(t) = 1, \\ \mathcal{Z}_1(t) = e^{-2\pi i t}, \end{cases} \qquad \text{for all } t \in [0,1]$$

where the last equality follows from (5.5). But it is impossible to have a continuous deformation from $\mathcal{Z}_1$ to $\mathcal{Z}_0$ without passing through the origin. Hence, the system $\{g_{m,n} : m, n \in \mathbb{Z}\}$ cannot be a frame. ∎

8. Frazier-Jawerth frames

Example B is a very special case of a much more general result, valid in $\mathbb{R}^n$, due to M. Frazier and B. Jawerth ([FJ1], [FJ2], [FJ3] and [FJW]). These results involve extensions of the notion of frames that produce unconditional bases for many of the spaces used in analysis.

9

Discrete transforms and algorithms

In the previous chapters we have studied functions defined on the real number continuum. But in the study of signals, or other applications, one is presented with data that are represented by a finite number of values; thus, it is important in such applications to study how the material we developed applies to such "discrete functions." In particular we shall discuss the discrete versions of the Fourier transform and discrete versions of the bases we have presented. Some of the constructions of bases, such as the ones obtained by the local sine and cosine bases, are carried out in this discrete setting.

We also describe the decomposition and reconstruction algorithms for wavelets that have proved to be quite useful in applications. Finally, new bases of $L^2(\mathbb{R})$ are obtained by making use of "wavelets packets," a concept that is explained in section 9.5.

9.1 The discrete Fourier transform (DFT) and the fast Fourier transform (FFT)

Let $\mathbb{T}$ be the group of complex numbers of absolute value 1. Let Γ_N be the subgroup of $\mathbb{T}$ that consists of the Nth roots of unity; that is,

$$\Gamma_N = \left\{ \gamma \in \mathbb{T} : \gamma = e^{2\pi i \frac{k}{N}}, \ k = 0, 1, \cdots, N-1 \right\}.$$

We denote by $\ell^2(\Gamma_N)$ the collection of all functions $f : \Gamma_N \longrightarrow \mathbb{C}$ endowed with the norm

$$\|f\|_{\ell^2(\Gamma_N)} = \Big(\sum_{\gamma \in \Gamma_N} |f(\gamma)|^2 \Big)^{\frac{1}{2}},$$

and its associated inner product

$$(f,g) = (f,g)_{\ell^2(\Gamma_N)} = \sum_{\gamma \in \Gamma_N} f(\gamma)\, \overline{g(\gamma)}.$$

The subgroup Γ_N tends to the circle group as N tends to ∞ in various ways; for example,

$$\lim_{N \to \infty} \frac{1}{N} \sum_{\gamma \in \Gamma_N} f(\gamma) = \frac{1}{2\pi} \int_0^{2\pi} f(e^{ix})\, dx \qquad (1.1)$$

when f is continuous on $\mathbb{T}$.

There is a Fourier analysis on Γ_N that is in close analogy with the harmonic analysis on $\mathbb{T}$. The functions E_k where $E_k(\gamma) = \gamma^k$ are the analogues of the exponentials $e^{2\pi i k \theta}$ in the case of classical Fourier series. It can be seen that

$$\Big\{ \tfrac{1}{\sqrt{N}} E_k : k = 0, 1, \cdots, N-1 \Big\}$$

is an orthonormal basis for $\ell^2(\Gamma_N)$. To see this we only need to check that this is an orthonormal system in the N-dimensional space $\ell^2(\Gamma_N)$. In fact,

$$\Big(\tfrac{1}{\sqrt{N}} E_k, \tfrac{1}{\sqrt{N}} E_\ell \Big) = \tfrac{1}{N} \sum_{\gamma \in \Gamma_N} \gamma^{k-\ell}.$$

If $k = \ell$, the right-hand side of the above formula is 1; if $k \ne \ell$, $z = e^{2\pi i \frac{k-\ell}{N}}$ is an Nth root of unity different from 1. Since

$$0 = (1 - z^N) = (1-z)(1 + z + z^2 + \cdots + z^{N-1}),$$

we must have $0 = \sum_{n=0}^{N-1} z^n = \sum_{\gamma \in \Gamma_N} \gamma^{k-l}$. Therefore,

$$\Big(\tfrac{1}{\sqrt{N}} E_k, \tfrac{1}{\sqrt{N}} E_\ell \Big) = \delta_{k,\ell}.$$

Given $f \in \ell^2(\Gamma_N)$ we define the Nth **discrete Fourier transform** (DFT) of f as the function $\hat{f} : \{0, 1, \cdots, N-1\} \longrightarrow \mathbb{C}$ whose values are

$$\hat{f}(k) = \frac{1}{N} \sum_{\gamma \in \Gamma_N} f(\gamma) \overline{E_k(\gamma)}; \qquad (1.2)$$

$\hat{f}(k)$ is called the kth **Fourier coefficient** of f. We then have the "inversion formula"

$$f(\gamma) = \sum_{k=0}^{N-1} \hat{f}(k) E_k(\gamma), \qquad (1.3)$$

which follows from (1.2) and the fact that $\{\frac{1}{\sqrt{N}} E_k : k = 0, \cdots, N-1\}$ is an orthonormal basis for $L^2(\Gamma_N)$. Another way of expressing (1.3) is to write (1.2) in matricial form: let

$$\zeta = e^{2\pi i \frac{1}{N}} \qquad \text{and} \qquad \zeta_j = \zeta^j \quad \text{for} \quad j = 0, 1, \cdots, N-1,$$

then

$$\begin{pmatrix} \hat{f}(0) \\ \hat{f}(1) \\ \vdots \\ \hat{f}(N-1) \end{pmatrix} = \frac{1}{N} \begin{pmatrix} 1 & 1 & \cdots & 1 \\ 1 & \overline{\zeta_1} & \cdots & \overline{\zeta_{N-1}} \\ \vdots & \vdots & \ddots & \vdots \\ 1 & \overline{\zeta_1^{N-1}} & \cdots & \overline{\zeta_{N-1}^{N-1}} \end{pmatrix} \begin{pmatrix} f(1) \\ f(\zeta_1) \\ \vdots \\ f(\zeta_{N-1}) \end{pmatrix}$$

$$\equiv \frac{1}{N} C_N \left(f(\zeta_j) \right)_{j=0}^{N-1}, \qquad (1.4)$$

where C_N is a matrix of order $N \times N$. Expression (1.4) implies (1.3) since the matrix $\frac{1}{\sqrt{N}} C_N$ is unitary. (This is an immediate consequence of the orthonormality of the system given by $\{\frac{1}{\sqrt{N}} E_k : k = 0, 1, \cdots, N-1\}$.) In fact, (1.3) is obtained by multiplying both sides of (1.4) by $\frac{1}{\sqrt{N}} C_N^* = \left(\frac{1}{\sqrt{N}} C_N\right)^{-1}$.

In applications, when using the computer to calculate the discrete Fourier transform, it is particularly useful to try to simplify and reduce the number of operations performed. In general, multiplication is a more complicated operation to program than addition. It is clear that N^2 products are involved in equality (1.4). We shall show that when $N = 2^q$, $q \in \mathbb{N}$, the number of multiplications needed to calculate the DFT can be reduced to $2N \log_2 N$. This is a consequence of the following result:

THEOREM 1.5 *If $N = 2^q$, $C_N = E_1 E_2 \cdots E_q$, where each E_j is an $N \times N$ matrix such that each row has precisely two non-zero entries.*

PROOF : We prove this result by induction on N. When $q = 1$,

$$C_2 = \begin{pmatrix} 1 & 1 \\ 1 & -1 \end{pmatrix}$$

and the assertion is true. For $q = 2$ a simple computation shows

$$C_4 = \begin{pmatrix} 1 & 1 & 1 & 1 \\ 1 & -i & -1 & i \\ 1 & -1 & 1 & -1 \\ 1 & i & -1 & -i \end{pmatrix} = \begin{pmatrix} 1 & 1 & 0 & 0 \\ 0 & 0 & 1 & 1 \\ 1 & -1 & 0 & 0 \\ 0 & 0 & 1 & -1 \end{pmatrix} \begin{pmatrix} 1 & 0 & 1 & 0 \\ 0 & 1 & 0 & 1 \\ 1 & 0 & -1 & 0 \\ 0 & -i & 0 & i \end{pmatrix}.$$

Suppose that the result is true for $N = 2^q$. We introduce two operations on $n \times n$ matrices as follows. If $A = \left(a_{ij}\right)_{i,j=1}^{n}$ is an $n \times n$ matrix, we define $A^\sharp$ as the $2n \times 2n$ matrix given by

$$A^\sharp = \begin{pmatrix} a_{11} & a_{12} & \cdots & a_{1n} & 0 & 0 & \cdots & 0 \\ 0 & 0 & \cdots & 0 & a_{11} & a_{12} & \cdots & a_{1n} \\ a_{21} & a_{22} & \cdots & a_{2n} & 0 & 0 & \cdots & 0 \\ 0 & 0 & \cdots & 0 & a_{21} & a_{22} & \cdots & a_{2n} \\ \vdots & \vdots & \ddots & \vdots & \vdots & \vdots & \ddots & \vdots \\ a_{n1} & a_{n2} & \cdots & a_{nn} & 0 & 0 & \cdots & 0 \\ 0 & 0 & \cdots & 0 & a_{n1} & a_{n2} & \cdots & a_{nn} \end{pmatrix}.$$

Also, if B is an $n \times n$ matrix, we let B^d be the $2n \times 2n$ matrix given by

$$B^d = \begin{pmatrix} B & O \\ O & B \end{pmatrix}.$$

We claim that

$$P^\sharp Q^d = T^\sharp \tag{1.6}$$

whenever P, Q and T are $n \times n$ matrices such that $PQ = T$.

To prove (1.6) we write $\vec{p}_i$, $i = 1, 2, \cdots, n$, for the row vectors of P and $\vec{q}_j$, $j = 1, 2, \cdots, n$, for the column vectors of Q. Then

$$T = PQ = \left((\vec{p}_i, \vec{q}_j)\right)_{i,j=1}^{n}.$$

On the other hand,

$$
P^\sharp Q^d = \begin{pmatrix} \vec{p}_1 & 0 \\ 0 & \vec{p}_1 \\ \vdots & \vdots \\ \vec{p}_n & 0 \\ 0 & \vec{p}_n \end{pmatrix} \begin{pmatrix} \vec{q}_1 & \cdots & \vec{q}_n & 0 & \cdots & 0 \\ 0 & \cdots & 0 & \vec{q}_1 & \cdots & \vec{q}_n \end{pmatrix}
$$

$$
= \begin{pmatrix} (\vec{p}_1,\vec{q}_1) & \cdots & (\vec{p}_1\vec{q}_n) & 0 & \cdots & 0 \\ 0 & \cdots & 0 & (\vec{p}_1,\vec{q}_1) & \cdots & (\vec{p}_1,\vec{q}_n) \\ \vdots & \ddots & \vdots & \vdots & \ddots & \vdots \\ (\vec{p}_n,\vec{q}_1) & \cdots & (\vec{p}_n,\vec{q}_n) & 0 & \cdots & 0 \\ 0 & \cdots & 0 & (\vec{p}_1,\vec{q}_1) & \cdots & (\vec{p}_n,\vec{q}_n) \end{pmatrix} = T^\sharp.
$$

Let $\mu = e^{2\pi i \frac{1}{2N}}$ be the principal $(2N)^{\text{th}}$ root of unity and $\mu_j = \mu^j$ for $j = 0, 1, \cdots, 2N - 1$. Let U_N be the $2N \times 2N$ matrix given by

$$
U_N = \begin{pmatrix} I & I \\ D_N^1 & D_N^2 \end{pmatrix},
$$

where

$$
D_N^1 = \begin{pmatrix} 1 & 0 & 0 & \cdots & 0 \\ 0 & \overline{\mu_1} & 0 & \cdots & 0 \\ 0 & 0 & \overline{\mu_2} & \cdots & 0 \\ \vdots & \vdots & \vdots & \ddots & \vdots \\ 0 & 0 & 0 & \cdots & \overline{\mu_{N-1}} \end{pmatrix},
$$

and

$$
D_N^2 = \begin{pmatrix} \overline{\mu_N} & 0 & 0 & \cdots & 0 \\ 0 & \overline{\mu_{N+1}} & 0 & \cdots & 0 \\ 0 & 0 & \overline{\mu_{N+2}} & \cdots & 0 \\ \vdots & \vdots & \vdots & \ddots & \vdots \\ 0 & 0 & 0 & \cdots & \overline{\mu_{2N-1}} \end{pmatrix}.
$$

Then, we have

$$
C_N^\sharp U_N = C_{2N}, \tag{1.7}
$$

which can easily be proved using the relation $\zeta_1 = \mu_1^2$.

Suppose now that $C_N = E_1 E_2 \cdots E_q = E_1(E_2 \cdots E_q)$, and apply (1.6); we obtain $C_N^\sharp = E_1^\sharp (E_2 \cdots E_q)^d$. Since $(AB)^d = A^d B^d$, we deduce that $C_N^\sharp = E_1^\sharp E_2^d \cdots E_q^d$. Using (1.7) we can write $C_{2N} = E_1^\sharp E_2^d \cdots E_q^d U_N$, and each of the last $q + 1$ factors has precisely two non-zero entries in each row. ∎

Theorem 1.5 gives us a way of reducing the "complexity" of the computation for the Discrete Fourier Transform. Since multiplication of a column vector by the matrix E_j on the left only requires $2N$ products (because only two entries in each row of E_j are non-zero), the number of essential computations required in Theorem 1.5 to find the DFT is of the order of $2Nq = 2N \log_2 N$ (recall that $N = 2^q$). This number is much smaller than N^2; for example, for $q = 10$, $N^2 = 2^{20} = 1024^2 > 10^6$ and $2N \log_2 N = 2 \cdot 2^{10} \cdot 10 = 20 \cdot 1024 \sim 2 \cdot 10^4$.

The method described in Theorem 1.5 is known as the **fast Fourier transform** (FFT). Thus, the FFT is not a transform, but an algorithm to compute the DFT when $N = 2^q$.

The estimate $2N \log_2 N$ found above can be improved if we take into account that the matrices E_j in the proof of Theorem 1.5 have several entries equal to one. Looking into the proof of the theorem it is easy to see that N of the $2N$ non-zero entries of E_j are ones. Hence, the essential computations required to compute the DFT is reduced to $N \log_2 N$ for $N = 2^q$.

9.2 The discrete cosine transform (DCT) and the fast cosine transform (FCT)

In Theorem 4.1 of Chapter 1 we presented four orthonormal bases for $L^2([0,1])$ involving sines and cosines. One of them was

$$\{1, \sqrt{2}\cos(k\pi x) : k = 1, 2, \cdots\}.$$

With respect to this orthonormal basis we can write the cosine series

$$F(x) = \tfrac{1}{2}a_0 + \sum_{k=1}^{\infty} a_k \cos(k\pi x) \tag{2.1}$$

for a "general function" F on $[0, 1]$, where

$$a_k = 2 \int_0^1 F(x) \cos(k\pi x)\, dx, \qquad k = 0, 1, 2, \cdots. \qquad (2.2)$$

The sequence $\{a_k = \widehat{F}(k) : k = 0, 1, 2, \cdots\}$ is called the **cosine transform** of F. Similar results can be obtained for each one of the other three bases described in Theorem 4.1 of Chapter 1.

A discrete version of the cosine transform is obtained by approximating the integral in (2.2) by Riemann sums. More precisely, we consider the points $\frac{1}{2N}(2m + 1)$, $m = 0, 1, \cdots, N - 1$, in $[0, 1]$ and write

$$a_k^{(N)} = \frac{2}{N} \sum_{m=0}^{N-1} F\left(\frac{2m + 1}{2N}\right) \cos\left(\frac{k\pi}{N} \cdot \frac{2m + 1}{2}\right) \qquad (2.3)$$

for $k = 0, 1, \cdots, N - 1$, so that $\lim_{N \to \infty} a_k^{(N)} = a_k$. Given $N \in \mathbb{N}$ we consider the N vectors in $\mathbb{R}^N$ defined by

$$\begin{cases} C_0^{(N)} = \left(\frac{1}{\sqrt{2}}, \cdots, \frac{1}{\sqrt{2}}\right), \\ C_k^{(N)} = \left(\cos\left(\frac{k\pi}{N} \cdot \frac{2m + 1}{2}\right)\right)_{m=0}^{N-1}, \quad k = 1, \cdots, N - 1. \end{cases} \qquad (2.4)$$

THEOREM 2.5 *The vectors*

$$\left\{ \sqrt{\frac{2}{N}}\, C_k^{(N)} \ : \ k = 0, 1, \cdots, N - 1 \right\}$$

form an orthonormal basis for $\mathbb{R}^n$.

PROOF : In order to show that $\sqrt{\frac{2}{N}}\, C_0^{(N)}$ is orthogonal to $\sqrt{\frac{2}{N}}\, C_k^{(N)}$ for $k = 1, 2, \cdots, N - 1$, we make use of the equality

$$\sum_{m=0}^{N-1} \cos\left(\frac{\pi k}{N}(m + \tfrac{1}{2})\right) = 0 \qquad \text{for } k = 1, 2, \cdots, 2N - 1. \qquad (2.6)$$

To show (2.6) we use the formula

$$2\cos\alpha = e^{i\alpha} + e^{-i\alpha}$$

and write the corresponding two sums that are obtained as a sum with $2N$ terms; more precisely,

$$2\sum_{m=0}^{N-1}\cos\left(\frac{\pi k}{N}(m+\tfrac{1}{2})\right) = \sum_{m=0}^{N-1} e^{\pi i \frac{k}{N}(m+\frac{1}{2})} + e^{-\pi i \frac{k}{N}(m+\frac{1}{2})}$$

$$= e^{\pi i \frac{k}{2N}}\left(\sum_{m=0}^{N-1} e^{\pi i \frac{k}{N}m} + \sum_{m=0}^{N-1} e^{-\pi i \frac{k}{N}(m+1)}\right)$$

$$= e^{\pi i \frac{k}{2N}} \sum_{m=-N}^{N-1} e^{\pi i \frac{k}{N}m} = 0,$$

where the last equality is obtained by adding the terms of the sequence

$$\{e^{\pi i \frac{k}{N}m} : m = -N, -N+1, \cdots, N-1\},$$

which is a geometric sequence with $2N$ terms and ratio $e^{\pi i \frac{k}{N}}$.

With $k = 2\ell$, $\ell = 1, 2, \cdots, N-1$, formula (2.6) implies

$$\sum_{m=0}^{N-1} \cos\left(\frac{\pi \ell}{N}(2m+1)\right) = 0, \qquad \ell = 1, 2, \cdots, N-1. \qquad (2.7)$$

Using the formula $2\cos^2\alpha = 1 + \cos(2\alpha)$, (2.7) implies that the vectors $\sqrt{\frac{2}{N}} C_\ell^{(N)}$, $\ell = 1, 2, \cdots, N-1$, have norm 1 on $\mathbb{R}^N$ (that $\sqrt{\frac{2}{N}} C_0^{(N)}$ has norm 1 is trivial).

Finally, the orthogonality of $C_k^{(N)}$ and $C_\ell^{(N)}$, $1 \le k, \ell \le N-1$ and $k \ne \ell$ follows from $2\cos\alpha\cos\beta = \cos(\alpha+\beta) + \cos(\alpha-\beta)$ and (2.6). ∎

Theorem 2.5 shows that the matrix C whose rows are given by $\sqrt{\frac{2}{N}} C_k^{(N)}$ is unitary. Writing (2.3) in matricial form and using the fact that C^{-1} equals the transpose of C, we obtain an inversion formula, corresponding to (2.3), that gives us the values $F\left(\frac{1}{2N}(2m+1)\right)$, $m = 0, 1, \cdots, N-1$, in terms of the numbers $a_k^{(N)}$:

$$F\left(\frac{2m+1}{2N}\right) = \tfrac{1}{2} a_0^{(N)} + \sum_{k=1}^{N-1} a_k^{(N)} \cos\left(\frac{k\pi}{N} \cdot \frac{2m+1}{2}\right), \qquad (2.8)$$

$m = 0, 1, \cdots, N - 1$. The mapping that sends F to

$$\{a_k^{(N)} \; : \; k = 0, 1, \cdots, N - 1\}$$

is called the **discrete cosine transform** (DCT) of F.

We now consider the computational complexity of the DCT when $N = 2^q$. Writing (2.3) in terms of the unitary matrix C whose rows are given by the vectors $\sqrt{\frac{2}{N}} \, C_k^{(N)}$, one sees that the complexity of this calculation of the DCT is about N^2. We shall show that an appropriate algorithm will reduce the complexity of the computation of the DCT to that of the DFT, and, as shown in section 9.1, the complexity will be of the order of $N \log_2 N$.

To do this we consider the sequence

$$x_m = F\left(\frac{2m + 1}{2N}\right), \qquad m = 0, 1, 2, \cdots, N - 1;$$

extend this sequence to the finite sequence indexed by

$$m = -N, -N + 1, \cdots, -1, 0, 1, \cdots, N - 1$$

so that the new sequence is even with respect to $-\frac{1}{2}$; that is, $x_{-m} = x_{m-1}$ for $m = 1, 2, \cdots, N$. Thus, we can consider the function $f \in \ell^2(\Gamma_{2N})$ such that $x_\ell = f(e^{-2\pi i \frac{1}{N}\ell})$, $\ell = -N, -N + 1, \cdots, N - 2, N - 1$.

The discrete Fourier transform for f (see (1.2)) is given by the formula

$$\tilde{y}_k = \frac{1}{2N} \sum_{\ell=-N}^{N-1} x_\ell \, e^{-2\pi i \frac{k\ell}{2N}}. \tag{2.9}$$

We shall show that, except for a factor, $\tilde{y}_k$ coincides with the coefficients $a_k^{(N)}$ given in (2.3). This is done in the following calculation, which uses the fact that the sequence x_ℓ, $\ell = -N, -N + 1, \cdots, N - 1$, is even with respect to $-\frac{1}{2}$:

$$\tilde{y}_k = \frac{1}{2N} \sum_{\ell=0}^{N-1} x_\ell \, e^{-2\pi i \frac{k\ell}{2N}} + \frac{1}{2N} \sum_{\ell=1}^{N} x_{-\ell} \, e^{2\pi i \frac{k\ell}{2N}}$$

$$= \frac{1}{2N} \sum_{\ell=0}^{N-1} x_\ell \left(e^{-2\pi i \frac{k}{2N}\ell} + e^{2\pi i \frac{k}{2N}(\ell+1)} \right)$$

$$= \frac{1}{2N} e^{\pi i \frac{k}{2N}} \sum_{\ell=0}^{N-1} x_\ell \left(e^{-\pi i \frac{k}{N}(\ell+\frac{1}{2})} + e^{\pi i \frac{k}{N}(\ell+\frac{1}{2})} \right)$$

$$= \frac{1}{N} e^{\pi i \frac{k}{2N}} \sum_{\ell=0}^{N-1} x_\ell \cos\left(\frac{\pi k}{N} \cdot \frac{2\ell+1}{2} \right).$$

This shows that $\tilde{y}_k$ equals $\frac{1}{2} e^{\pi i \frac{k}{2N}}$ times the DCT coefficients $a_k^{(N)}$ for the function f. Proceeding in this way the complexity of the calculation of the DCT when $N = 2^q$ does not exceed the complexity involved in the calculation of the $(2N)^{\text{th}}$ Discrete Fourier Transform. As was shown at the end of section 9.1, this is of the order of $4N \log_2(2N)$ when $N = 2^q$. Since $\log_2(2N) \leq 2 \log_2 N$, the complexity of the calculation of the DCT does not exceed $8N \log_2 N$.

Thus, to find the DCT, which consists in obtaining the coefficients $a_k^{(N)}$ in (2.3), it is sufficient to calculate the discrete Fourier coefficients of the sequence $x_m = F\left(\frac{2m+1}{2N}\right)$, appropriately extended, by using the method described in Theorem 1.5 (that is, the FFT). This way of finding the discrete cosine coefficients $a_k^{(N)}$ is called the **fast cosine transform** (FCT). Again, the FCT is not a "transform;" it is an algorithm for computing the DCT of a function defined on $[0, 1]$.

Observe that, contrary to the case of the DFT, the DCT does not produce complex numbers when F is real-valued. This is sometimes advantageous.

The transform described above uses one of the cosine bases found in Theorem 4.1 of Chapter 1. Other discrete transforms can be developed using any of the bases described by this theorem. In particular, the **discrete sine transform** (DST) and the **fast sine transform** (FST) could also have been developed.

9.3 The discrete version of the local sine and cosine bases

In this section we shall describe a discrete version of the local cosine and sine bases introduced in section 1.4.

Let $\{a_j : j \in \mathbb{Z}\}$ be an increasing sequence of real numbers such that

$$\lim_{j \to \infty} a_j = +\infty, \qquad \lim_{j \to -\infty} a_j = -\infty \qquad \text{and} \qquad a_j \in \mathbb{Z} + \tfrac{1}{2}.$$

If $[a_j, a_{j+1}]$ is the collection of integers between a_j and a_{j+1} we obtain a partition of $\mathbb{Z}$. Let $\ell_j = a_{j+1} - a_j \in \mathbb{Z}$ and let $\{\eta_j : j \in \mathbb{Z}\}$ be a sequence of positive real numbers such that $n_j + \eta_{j+1} \le \ell_j$.

We could consider the DCT associated with each interval $[a_j, a_{j+1}]$ after appropriate dilations have been performed and, then, let j vary to obtain an orthonormal basis of $\ell^2(\mathbb{Z})$. This approach would lead to "abrupt" cut-offs. Instead, we will use the analogues of the "bell" functions that appeared in Chapter 1. These will be called "**windows**" and are functions of the form

$$w_j : \mathbb{Z} \longrightarrow [0, 1]$$

satisfying

$$w_j(x) = 1 \qquad \text{for } x \in [a_j + \eta_j, a_{j+1} - \eta_{j+1}], \tag{3.1}$$

$$w_j(x) = 0 \qquad \text{if } x \le a_j - \eta_j \ \text{ or } \ x \ge a_{j+1} + \eta_{j+1}, \tag{3.2}$$

and

$$\text{for } |t| \le \eta_j, \quad \begin{cases} w_{j-1}(a_j - t) = w_j(a_j + t), \quad \text{and} \\ w_{j-1}^2(a_j + t) + w_j^2(a_j + t) = 1. \end{cases} \tag{3.3}$$

If the windows w_j were so defined for x in $\mathbb{R}$, we would have the compatible bell functions introduced in Chapter 1; the reader can look at Figure 1.3 of the above mentioned chapter to "understand" the relations given in (3.3) for these windows w_j.

EXAMPLE A: If $\eta_j < \tfrac{1}{2}$ for all $j \in \mathbb{Z}$, the interval $[a_j - \eta_j, a_j + \eta_j]$ does not contain any integer, since

$$a_j - \eta_j = z_j + \tfrac{1}{2} - \eta_j > z_j \qquad \text{and}$$

$$a_j + \eta_j = z_j + \tfrac{1}{2} + \eta_j < z_j + 1$$

with $z_j \in \mathbb{Z}$. In this case $w_j(x)$ is the characteristic function of $[a_j, a_{j+1}] \cap \mathbb{Z}$.

We define the **discrete cosine bases** associated with the windows given above as the system of "functions" $\{u_{j,k} : j \in \mathbb{Z}, \, 0 \le k \le \ell_j - 1\}$ defined by

$$u_{j,k}(x) = \sqrt{\tfrac{2}{\ell_j}} \, w_j(x) \cos\left(\pi(k + \tfrac{1}{2})\left(\frac{x - a_j}{\ell_j}\right)\right), \qquad x \in \mathbb{Z}. \qquad (3.4)$$

The reader should compare the orthonormal basis given in (iii) of Theorem 4.3 of Chapter 1 with the elements $u_{j,k}(x)$ just defined.

Our first goal is to show that $\{u_{j,k} : j \in \mathbb{Z}, \, 0 \le k \le \ell_j - 1\}$ is an orthogonal system of $\ell^2(\mathbb{Z})$; we shall then show that it is a basis for $\ell^2(\mathbb{Z})$. In the proof of the orthonormality we shall make use of the identity

$$\sum_{x=0}^{N-1} \cos\left(\pi(x + \tfrac{1}{2})\frac{m}{N}\right) = 0 \qquad \text{for } 1 \le m \le 2N - 1, \qquad (3.5)$$

which is (2.6).

THEOREM 3.6 *The sequence* $\{u_{j,k} : j \in \mathbb{Z}, \, 0 \le k \le \ell_j - 1\}$ *given by (3.4) is an orthonormal basis for* $\ell^2(\mathbb{Z})$.

PROOF : Let F_j be the subspace of $\ell^2(\mathbb{Z})$ containing all $g \in \ell^2(\mathbb{Z})$ supported in $[a_j - \eta_j, a_j + \eta_{j+1}]$ and satisfying

$$g(a_j + t) = g(a_j - t) \qquad\qquad \text{if } |t| \le \eta_j \qquad (3.7)$$

$$g(a_{j+1} + t) = -g(a_{j+1} - t) \qquad \text{if } |t| \le \eta_{j+1}. \qquad (3.8)$$

A consequence of these equalities is the fact that a function g in F_j is determined uniquely by its values on $[a_j, a_{j+1}] \cap \mathbb{Z}$, so that

$$\dim F_j = a_{j+1} - a_j = \ell_j.$$

Let E_j be the set of all functions f defined on $\mathbb{Z}$ of the form $f = w_j g$, with $g \in F_j$. We shall show that for each j, the set $\{u_{j,k} : 0 \le k \le \ell_j - 1\}$ is an orthonormal basis of E_j; then we shall show that $\ell^2(\mathbb{Z})$ is an orthonormal direct sum of the subspaces E_j.

Observe that the functions

$$g_{j,k}(x) = \sqrt{\tfrac{2}{\ell_j}} \, \cos\left[\pi(k + \tfrac{1}{2})\left(\frac{x - a_j}{\ell_j}\right)\right]\chi_{[a_j - \eta_j, a_{j+1} + \eta_{j+1}]}(x)$$

belong to F_j for all $j = 0, \cdots, \ell_j - 1$; this is due to the fact that, putting $t = (x - a_j)/\ell_j$, these cosines are even functions with respect to the origin and odd with respect to 1. The details are left to the reader.

This shows that $u_{j,k} \in E_j$ for all $0 \leq k \leq \ell_j - 1$. To see that they form an orthonormal basis of E_j it suffices to show the orthonormality, since E_j has dimension ℓ_j.

If $k \neq s$ we use the equality $2\cos\alpha\cos\beta = \cos(\alpha - \beta) + \cos(\alpha + \beta)$ to obtain

$$\langle u_{j,k}, u_{j,s} \rangle$$

$$= \frac{1}{\ell_j} \sum_{x \in \mathbb{Z}} (w_j(x))^2 \left\{ \cos\left[\pi(k-s)\left(\frac{x - a_j}{\ell_j}\right)\right] + \cos\left[\pi(k+s+1)\left(\frac{x - a_j}{\ell_j}\right)\right] \right\}$$

$$\equiv \sum_{x \in \mathbb{Z}} (w_j(x))^2 \left\{ C_{k,s}^{(1)}\left(\frac{x - a_j}{\ell_j}\right) + C_{k,s}^{(2)}\left(\frac{x - a_j}{\ell_j}\right) \right\}.$$

Recall that $w_j(x) = 0$ if $x \notin [a_j - \eta_j, a_{j+1} + \eta_{j+1}]$. Define

$$\mathcal{Z}_j = [a_j - \eta_j, a_j + \eta_j] \cap \mathbb{Z} \qquad \text{and}$$

$$\mathcal{A}_j = [a_j + \eta_j, a_{j+1} - \eta_{j+1}] \cap \mathbb{Z}$$

so that $[a_j - \eta_j, a_{j+1} + \eta_{j+1}] \cap \mathbb{Z}$ is the disjoint union of $\mathcal{Z}_j, \mathcal{A}_j$ and $\mathcal{Z}_{j+1}$. On $\mathcal{A}_j$, $w_j(x) = 1$ by (3.1). The set $\mathcal{Z}_j$ can be divided into two parts: one to the left of a_j and the other to the right. Using the fact that the cosine is an even function and property (3.3) of the window function, the above sum over $\mathcal{Z}_j$ can be reduced to a sum over $\mathcal{Z}_j^+$. More precisely,

$$\sum_{x \in \mathcal{Z}_j} (w_j(x))^2 \left[C_{k,s}^{(1)}\left(\frac{x - a_j}{\ell_j}\right) + C_{k,s}^{(2)}\left(\frac{x - a_j}{\ell_j}\right) \right]$$

$$= \left\{ \sum_{x \in \mathcal{Z}_j^-} + \sum_{x \in \mathcal{Z}_j^+} \right\} \left\{ (w_j(x))^2 \left[C_{k,s}^{(1)}\left(\frac{x - a_j}{\ell_j}\right) + C_{k,s}^{(2)}\left(\frac{x - a_j}{\ell_j}\right) \right] \right\}$$

$$= \sum_{x = a_j + t \in \mathcal{Z}_j^+} \left\{ w_j^2(a_j - t) + w_j^2(a_j + t) \right\} \left[C_{k,s}^{(1)}\left(\frac{x - a_j}{\ell_j}\right) + C_{k,s}^{(2)}\left(\frac{x - a_j}{\ell_j}\right) \right]$$

$$= \sum_{x \in \mathcal{Z}_j^+} \left[C_{k,s}^{(1)}\left(\frac{x - a_j}{\ell_j}\right) + C_{k,s}^{(2)}\left(\frac{x - a_j}{\ell_j}\right) \right].$$

A similar expression is obtained for the sum over $x \in \mathcal{Z}_{j+1}$ by writing it as a sum over $\mathcal{Z}_{j+1}^{-}$. This gives us

$$\langle u_{j,k}, u_{j,s} \rangle$$

$$= \frac{1}{\ell_j} \sum_{x \in [a_j, a_{j+1}] \cap \mathbb{Z}} \left\{ \cos\left[\pi(k - s)\left(\frac{x - a_j}{\ell_j}\right)\right] + \cos\left[\pi(k + s + 1)\left(\frac{x - a_j}{\ell_j}\right)\right] \right\},$$

which is zero by (3.5). This proves the orthogonality of $u_{j,k}$, $1 \leq k \leq \ell_j - 1$, for j fixed. To prove normality we use the relation $2\cos^2\alpha = 1 + \cos(2\alpha)$ and the properties of the windows to obtain

$$\|u_{j,k}\|_+ 2^2 = \frac{1}{\ell_j} \sum_{x \in [a_j, a_{j+1}] \cap \mathbb{Z}} \left\{ 1 + \cos\left[\pi(2k + 1)\left(\frac{x - a_j}{\ell_j}\right)\right] \right\}$$

$$= \frac{1}{\ell_j} \sum_{x \in [a_j, a_{j+1}] \cap \mathbb{Z}} 1 = 1,$$

where we have again used equality (3.5). This completes the proof that $\{u_{j,k} : 1 \leq k \leq \ell_j - 1\}$ is an orthonormal basis for E_j.

We still have to show that $\ell^2(\mathbb{Z})$ is an orthogonal direct sum of the subspaces E_j. We start by proving that E_j and E_k are orthogonal if $j \neq k$. If $|j - k| \geq 2$, this result follows from the fact that the support of the elements in E_j is disjoint from the support of the elements in E_k. We only need to consider $k = j \pm 1$; by changing the roles played by k and j, it is enough to show that E_{j-1} and E_j are orthonormal.

Let $f_j = w_j g_j \in E_j$ and $f_{j-1} = w_{j-1} g_{j-1} \in E_{j-1}$. With respect to the variable $t = x - a_j$, the function $w_j w_{j-1}$ is even by (3.3), the function g_j is also even because of (3.7) and the function g_{j-1} is odd by (3.8). Hence,

$$\sum_{x \in \mathbb{Z}} f_j(x) f_{j-1}(x) = \sum_{x \in [a_j - \eta_j, a_j + \eta_j] \cap \mathbb{Z}} w_j(x) w_{j-1}(x) g_j(x) g_{j-1}(x) = 0.$$

We still have to show the completeness. Suppose that $f \in \ell^2(\mathbb{Z})$ is orthogonal to all the functions $f_j \in E_j$ for all $j \in \mathbb{Z}$. If we consider $k \in [a_j + \eta_j, a_{j+1} - \eta_{j+1}]$ the Dirac function $\delta_k(x) = w(x)\delta_k(x)$ belongs to E_j and, hence,

$$0 = \langle f, \delta_k \rangle = f(x)\delta_k(x) = f(k).$$

This shows that $f \equiv 0$ on $[a_j + \eta_j, a_{j+1} - \eta_{j+1}]$ for all $j \in \mathbb{Z}$. We have to show that f is also zero on $[a_j - \eta_j, a_j + \eta_j]$. Let $k \in [a_j - \eta_j, a_j + \eta_j]$ and consider

k' to be the point symmetric to k with respect to a_j. Thus, $\delta_k + \delta_{k'} \in F_j$ and $\delta_k - \delta_{k'} \in F_{j-1}$ so that $w_j(\delta_k + \delta_{k'}) \in E_j$ and $w_{j-1}(\delta_k - \delta_{k'}) \in E_{j-1}$. Since f is orthogonal to E_j and E_{j+1} we deduce

$$\begin{cases} f(k)w_j(k) + f(k')w_j(k') = 0, \\ f(k)w_{j-1}(k) - f(k')w_{j-1}(k') = 0. \end{cases}$$

This homogeneous system of two equations with unknowns $f(k)$ and $f(k')$ has determinant -1 by (3.3) and the symmetry of k and k' with respect to a_j. Thus, $f(k) = f(k') = 0$, and this completes the proof of Theorem 3.6. ∎

We have already mentioned that the discrete basis given in (3.4) corresponds to the orthonormal basis given in (iii) of Theorem 4.3 of Chapter 1. It is important to notice that the spaces E_j defined in the above proof satisfy the compatibility of "polarities" that is required to obtain the direct sum (3.18) of Chapter 1 for $L^2(\mathbb{R})$; namely, the "polarities" chosen at the end points a_j are $\cdots, (+,-), (+,-), (+,-), \cdots$.

There are several variations of the above construction that can be done. For a sequence $\{a_j : j \in \mathbb{Z}\}$ as described at the beginning of this section, we consider the same windows w_j as before, but we can change the definition of the spaces E_j, $j \in \mathbb{Z}$, as long as the compatibility of polarities is observed. One of these variations mimics the idea of the Wilson basis described at the end of section 1.4. In this basis the polarities are $\cdots, (+,+), (-,-), (+,+), \cdots$, so that discrete sine and cosine bases corresponding to (ii) and (iv) of Theorem 4.3 of Chapter 1 are used.

More precisely, the spaces E_j should be defined in the following way: if $j \in 2\mathbb{Z}$, E_j is the set of all functions of the form $w_j(x)q(x)$ where q is even with respect to a_j and a_{j+1}; if $j \in 2\mathbb{Z}+1$, E_j is the set of all functions of the form $w_j(x)q(x)$ where q is odd with respect to a_j and a_{j+1}. If $j \in 2\mathbb{Z}$, we define

$$\begin{cases} u_{j,0}(x) = \sqrt{\frac{1}{\ell_j}}\, w_j(x), \\ u_{j,k}(x) = \sqrt{\frac{2}{\ell_j}}\, w_j(x) \cos\left[k\pi\left(\frac{x - a_j}{\ell_j}\right)\right], \quad 1 \le k \le \ell_j - 1; \end{cases} \tag{3.9}$$

and, if $j \in 2\mathbb{Z}+1$, we define

$$u_{j,k}(x) = \sqrt{\frac{2}{\ell_j}}\, w_j(x) \sin\left[k\pi\left(\frac{x - a_j}{\ell_j}\right)\right], \quad 1 \le k \le \ell_j. \tag{3.10}$$

Using appropriate trigonometric identities together with (3.5), it can be shown that the functions defined by (3.9) and (3.10) form an orthonormal basis of $\ell^2(\mathbb{Z})$. The proof is similar to that of Theorem 3.6.

9.4   Decomposition and reconstruction algorithms for wavelets

The **multiresolution analysis** (MRA) method is well adapted to image analysis. The spaces V_j that appeared in the definition of an MRA (see section 2.1) can be interpreted as spaces where an approximation to the image at the "jth level" is obtained; the approximation produces a blurred picture when j is small and a clearer one as j grows. In addition, the "detail" in the approximation occurring in V_j, that is not in V_{j-1}, is "stored" in the spaces W_{j-1} which satisfy $V_{j-1} \oplus W_{j-1} = V_j$ (see also (2.1) in Chapter 2). In addition, the spaces V_j have a dilation and translation invariance structure given by (1.2) and (1.5) of the definition of an MRA (see section 2.1); the same structure is inherited by the orthogonal complements W_j. This leads to efficient decomposition and reconstruction algorithms which we shall presently explain for the one-dimensional case.

Let f be a function defined on $\mathbb{R}$ (which we may regard as a sound signal, for example). Choose an MRA with scaling function φ and wavelet ψ given by Proposition 2.13 in Chapter 2 with $\nu \equiv 1$. The orthogonal projection operators P_j, $j \in \mathbb{Z}$, from $L^2(\mathbb{R})$ onto V_j are given by

$$P_j f(x) = \sum_{k \in \mathbb{Z}} \langle f, \varphi_{j,k} \rangle \, \varphi_{j,k}(x).$$

Since $\lim_{j \to \infty} P_j f = f$ in the $L^2(\mathbb{R})$-norm, we can choose $j \in \mathbb{Z}$ such that $P_j f$ is a good approximation of f. (The choice of j may vary with the particular application we have in mind and the MRA chosen.) Thus, we have the coefficients

$$c_{j,k} = \langle f, \varphi_{j,k} \rangle, \qquad j, k \in \mathbb{Z},$$

and what we want to do is to **decompose** the sequence

$$\mathbf{c}^j = \left\{ c_{j,k} = \langle f, \varphi_{j,k} \rangle : k \in \mathbb{Z} \right\} \tag{4.1}$$

which belongs to $\ell^2(\mathbb{Z})$.

Before continuing with the decomposition algorithm let us mention that the way in which the coefficients $c_{j,k}$, $k \in \mathbb{Z}$, are calculated is not a matter of concern for the algorithm. On the other hand, in numerical applications the function f is already given as a (finite) sequence of samples; in this case the samples can be interpreted as the coefficients of the projection onto a subspace V_j associated with an MRA suited for the application considered. An algorithm to choose the **best basis** will be sketched in section 9.5.

We now continue with the decomposition algorithm. This is achieved by using the filters $m_0(\xi)$ and $m_1(\xi) = e^{i\xi}\,\overline{m_0(\xi + \pi)}$ introduced in section 2.1. Recall that $\frac{1}{2}\varphi(\frac{\cdot}{2}) \in V_{-1} \subset V_0$, so that

$$\tfrac{1}{2}\varphi\left(\tfrac{1}{2}x\right) = \sum_{k\in\mathbb{Z}} \alpha_n \varphi(x + n), \tag{4.2}$$

where

$$\alpha_n = \tfrac{1}{2}\int_{\mathbb{R}} \varphi\left(\tfrac{1}{2}x\right)\overline{\varphi(x+n)}\,dx \tag{4.3}$$

and

$$m_0(\xi) = \sum_{n\in\mathbb{Z}} \alpha_n\, e^{in\xi} \tag{4.4}$$

(see (2.2) and (2.3) in Chapter 2). Formula (4.2) is valid for the scales V_{-1} and V_0. For other scales, V_{j-1} and V_j, we have

$$\varphi_{j-1,k}(x) = 2^{\frac{1}{2}(j-1)}\varphi(2^{j-1}x - k) = 2^{\frac{1}{2}(j+1)}\tfrac{1}{2}\varphi\left(\tfrac{1}{2}(2^j x - 2k)\right)$$

$$= 2^{\frac{1}{2}(j+1)} \sum_{n\in\mathbb{Z}} \alpha_n \varphi(2^j x - 2k + n).$$

That is,

$$\varphi_{j-1,k}(x) = \sqrt{2} \sum_{n\in\mathbb{Z}} \alpha_n \varphi_{j,2k-n}(x), \qquad j,k \in \mathbb{Z}. \tag{4.5}$$

Similarly, $\frac{1}{2}\psi(\frac{\cdot}{2}) \in W_{-1} \subset V_0$ and, hence,

$$\tfrac{1}{2}\psi\left(\tfrac{1}{2}x\right) = \sum_{n\in\mathbb{Z}} \beta_n \varphi(x + n), \tag{4.6}$$

and

$$\psi_{j-1,k}(x) = \sqrt{2} \sum_{n\in\mathbb{Z}} \beta_n \varphi_{j,2k-n}(x), \qquad j,k \in \mathbb{Z}. \tag{4.7}$$

Moreover, the coefficients β_n are related to the coefficients α_n. Observe that (4.6) implies $\hat{\psi}(2\xi) = m_1(\xi)\hat{\varphi}(\xi)$, where

$$m_1(\xi) = \sum_{n \in \mathbb{Z}} \beta_n e^{in\xi},$$

and the choice of ψ gives us

$$m_1(\xi) = e^{i\xi} \overline{m_0(\xi + \pi)}$$

(see Proposition 2.13 in Chapter 2). Thus,

$$\sum_{n \in \mathbb{Z}} \beta_n e^{in\xi} = m_1(\xi) = e^{i\xi} \overline{m_0(\xi + \pi)} = e^{i\xi} \sum_{n \in \mathbb{Z}} \overline{\alpha_n}\, e^{-in(\xi + \pi)}$$

$$= \sum_{n \in \mathbb{Z}} \overline{\alpha_n}\, (-1)^n e^{i(1-n)\xi} = \sum_{n \in \mathbb{Z}} (-1)^{n+1}\, \overline{\alpha_{1-n}}\, e^{in\xi}.$$

It follows that

$$\beta_n = (-1)^{n+1}\, \overline{\alpha_{1-n}}, \tag{4.8}$$

so that the coefficients β_n do not require further computations.

Now substitute (4.5) in the definition of the coefficients $c_{j-1,k}$ (see (4.1)) to obtain

$$c_{j-1,k} = \langle f, \varphi_{j-1,k} \rangle = \left\langle f, \sqrt{2} \sum_{n \in \mathbb{Z}} \alpha_n \varphi_{j,2k-n} \right\rangle$$

$$= \sqrt{2} \sum_{n \in \mathbb{Z}} \overline{\alpha_n}\, c_{j,2k-n}. \tag{4.9}$$

This shows that the coefficients $c_{j-1,k}$ of the lowest resolution V_{j-1} can be obtained from the coefficients $c_{j,\ell}$ of the V_j resolution and the low-pass filter coefficients. For j fixed, the right-hand side of (4.9) is the convolution of the sequences

$$\tilde{\alpha}_n = \sqrt{2}\,\overline{\alpha_n} \quad \text{and} \quad \tilde{c}_n = c_{j,n},$$

followed by retaining only the convolution entries that appear in the even places. Thus, if there were N significant terms on the sequence $\mathbf{c}^j$, (4.9) tells us how to compute a blurred resolution with only $\frac{1}{2}N$ terms (approximately).

The rest of the terms, which contain the "details" in passing from V_{j-1} to V_j, are contained in W_{j-1}. More precisely,

$$D_j f(x) \equiv P_j f(x) - P_{j-1} f(x) = \sum_{n \in \mathbb{Z}} d_{j-1,k} \psi_{j-1,k}(x),$$

where $d_{j-1,k} = \langle f, \psi_{j-1,k} \rangle$. Using (4.7) we obtain

$$d_{j-1,k} = \left\langle f, \sqrt{2} \sum_{n \in \mathbb{Z}} \beta_n \varphi_{j,2k-n} \right\rangle = \sqrt{2} \sum_{n \in \mathbb{Z}} \overline{\beta_n} \, c_{j,2k-n}. \qquad (4.10)$$

As before, this last expression is a convolution followed by a "decimation" by 2. Thus, we have decomposed $\mathbf{c}^j$ into the sequences $\mathbf{c}^{j-1}$ and $\mathbf{d}^{j-1}$. The process can be continued with $\mathbf{c}^{j-1}$ to obtain the decomposition algorithm given in Figure 9.1. The decision to stop after M steps could be taken when the values of the sequence $\mathbf{c}^{j-M}$ fall within a certain a priori fixed range.

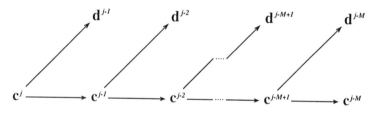

Figure 9.1: *Decomposition algorithm for wavelets. Each arrow represents a convolution followed by retaining only the even terms of the convolution (see (4.9) and (4.10)).*

If we consider only N terms of the original sequence $\mathbf{c}^j$, then we retain $\frac{1}{2}N$ terms (approximately) in $\mathbf{d}^{j-1}$, $\frac{1}{4}N$ terms (approximately) in $\mathbf{d}^{j-2}$, etc. Hence we keep $2^{-M}N$ terms (approximately) in $\mathbf{c}^{j-M}$ after M decompositions. Thus, the amount of data retained is of the same order as the original data

$$\left(\frac{N}{2} + \frac{N}{4} + \cdots + \frac{N}{2^M} \right) + \frac{N}{2^M} = N.$$

EXAMPLE B: Let us see how the Haar wavelet works for doing the decomposition algorithm. We take

$$\psi(x) = \chi_{[-1,-\frac{1}{2})}(x) - \chi_{[-\frac{1}{2},0)}(x) \qquad \text{and} \qquad \varphi(x) = \chi_{[-1,0)}(x),$$

to be consistent with Example B in Chapter 2. Then

$$\frac{1}{2}\varphi\left(\frac{1}{2}x\right) = \frac{1}{2}\varphi(x) + \frac{1}{2}\varphi(x+1),$$

and

$$\frac{1}{2}\psi\left(\frac{1}{2}x\right) = \frac{1}{2}\varphi(x+1) - \frac{1}{2}\varphi(x).$$

Thus,

$$\alpha_n = \begin{cases} \frac{1}{2} & \text{if } n = 0 \text{ or } 1, \\ 0 & \text{otherwise,} \end{cases}$$

and

$$\beta_n = \begin{cases} -\frac{1}{2} & \text{if } n = 0, \\ \frac{1}{2} & \text{if } n = 1, \\ 0 & \text{otherwise.} \end{cases}$$

From (4.9),

$$c_{j-1,k} = \sqrt{2}\left\{\frac{c_{j,2k} + c_{j,2k-1}}{2}\right\},$$

and, from (4.10),

$$d_{j-1,k} = \sqrt{2}\left\{\frac{c_{j,2k-1} - c_{j,2k}}{2}\right\}.$$

Hence, except for the factor $\sqrt{2}$, each coefficient $c_{j-1,k}$ is a moving average of two adjacent terms of the sequence $\mathbf{c}^j$, and each detail $d_{j-1,k}$ is half of the difference of two adjacent terms.

Let us now treat the problem of reconstructing $\mathbf{c}^j$ from the sequences $\mathbf{d}^{j-1}, \mathbf{d}^{j-2}, \cdots, \mathbf{d}^{j-M}$ and $\mathbf{c}^{j-M}$. By induction, it is enough to consider the reconstruction of $\mathbf{c}^j$ from $\mathbf{d}^{j-1}$ and $\mathbf{c}^{j-1}$. Since

$$P_j f = P_{j-1} f + D_{j-1} f,$$

we use (4.5) and (4.7) to obtain

$$\sum_{\ell \in \mathbb{Z}} c_{j,\ell} \varphi_{j,\ell}(x) = \sum_{k \in \mathbb{Z}} c_{j-1,k} \varphi_{j-1,k}(x) + \sum_{k \in \mathbb{Z}} d_{j-1,k} \psi_{j-1,k}(x)$$

$$= \sum_{k \in \mathbb{Z}} c_{j-1,k} \left(\sum_{n \in \mathbb{Z}} \sqrt{2}\, \alpha_n \varphi_{j,2k-n}(x)\right)$$

$$+ \sum_{k \in \mathbb{Z}} d_{j-1,k} \left(\sum_{n \in \mathbb{Z}} \sqrt{2}\, \beta_n \varphi_{j,2k-n}(x) \right)$$

$$= \sum_{\ell \in \mathbb{Z}} \left\{ \sum_{k \in \mathbb{Z}} [\sqrt{2}\, c_{j-1,k} \alpha_{2k-\ell} + \sqrt{2}\, d_{j-1,k} \beta_{2k-\ell}] \right\} \varphi_{j,\ell}(x).$$

Hence,

$$c_{j,\ell} = \sqrt{2} \sum_{k \in \mathbb{Z}} \left[c_{j-1,k} \alpha_{2k-\ell} + d_{j-1,k} \beta_{2k-\ell} \right]. \tag{4.11}$$

We let the reader check that in the case of the Haar wavelet, (4.11) gives

$$\begin{cases} c_{j,2k} = \frac{\sqrt{2}}{2} \left\{ c_{j-1,k} - d_{j-1,k} \right\}, & \text{and} \\[2mm] c_{j,2k-1} = \frac{\sqrt{2}}{2} \left\{ c_{j-1,k} + d_{j-1,k} \right\}; \end{cases}$$

this also follows easily from the formulas contained in Example B.

Formula (4.11) allows us to add the sequences $\mathbf{d}^{j-1}$ and $\mathbf{c}^{j-1}$ to obtain $\mathbf{c}^j$. If we start this process with $\mathbf{c}^{j-M}$ and $\mathbf{d}^{j-M}$ and we also know the "details" $\mathbf{d}^{j-M+1}, \cdots, \mathbf{d}^{j-1}$, we have the reconstruction algorithm given in Figure 9.2.

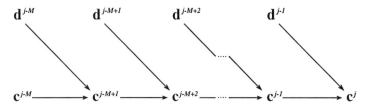

Figure 9.2: Reconstruction algorithm for wavelets. Each arrow
requires an "upsampling" by 2 and a convolution (see (4.11)).

Each of the summands in (4.11), represented by an arrow in Figure 9.2, is a convolution. To see this consider, for j fixed,

$$\tilde{c}_n = \begin{cases} c_{j-1,\frac{n}{2}} & \text{if } n \text{ is even,} \\[2mm] 0 & \text{if } n \text{ is odd,} \end{cases}$$

and $\tilde{\alpha}_n = \sqrt{2}\, \alpha_{-n}$. Then

$$(\tilde{c} * \tilde{\alpha})_\ell = \sum_{n \in \mathbb{Z}} \tilde{c}_n \tilde{\alpha}_{\ell-n} = \sum_{k \in \mathbb{Z}} \tilde{c}_{2k} \tilde{\alpha}_{\ell-2k} = \sum_{k \in \mathbb{Z}} c_{j-1,k} \sqrt{2}\, \alpha_{2k-\ell}.$$

Thus, each arrow in Figure 9.2 requires an "upsampling" by 2 (that is, a zero is inserted between each two adjacent entries in the original sequence) followed by a convolution.

Let us now study the complexity of the decomposition algorithm for wavelets. (The complexity of the reconstruction algorithm is obtained in a similar way.) Suppose that the sequence $\{\alpha_k : k \in \mathbb{Z}\}$ has only K non-zero terms; by (4.8) the same is true for the sequence $\{\beta_k : k \in \mathbb{Z}\}$. This kind of finite sequence gives rise to **finite impulse-response** (FIR) filters. If $\mathbf{c}^j$ has N terms, there are approximately $\frac{1}{2}N$ terms in $\mathbf{c}^{j-1}$, so that, by (4.9), we have to perform $\frac{1}{2}KN$ multiplications. Similarly, it follows from (4.10) and (4.8) that also $\frac{1}{2}KN$ multiplications have to be performed (approximately) to obtain the $\frac{1}{2}N$ terms of the sequence $\mathbf{d}^{j-1}$ from $\mathbf{c}^j$. Continuing the algorithm as described in Figure 9.1 we obtain

$$\left(K\frac{N}{2} + K\frac{N}{4} + K\frac{N}{8} + \cdots + K\frac{N}{2^M} \right) + K\frac{N}{2^M} = KN.$$

Thus, the complexity of the decomposition algorithm, in this case, is linear in N.

A typical example of a **finite impulse-response** (FIR) filter is the Haar low-pass filter, which has only two non-zero coefficients. The discontinuity of the Haar wavelet prompted I. Daubechies to find the compactly supported wavelets (see section 2.3); these are smoother than the Haar wavelet, but the filters are still FIR. In contrast to the Haar low-pass filter, the ones for the compactly supported wavelets could have irrational coefficients (see, for example, the filter given after (4.1) in Chapter 2, which corresponds to the first Daubechies wavelet). In this case the coefficients α_k have to be rounded-off (or "quantized") and the error needs to be estimated a priori.

If there is an infinite number of non-zero coefficients α_k, we have an **infinite impulse-response** (IIR) filter. The Shannon wavelet (see Example C of Chapter 2), the Lemarié-Meyer wavelets (see Example D of Chapter 2 and, also, section 3.3 and section 3.4), and the Franklin and spline wavelets (section 4.1 and section 4.2), are examples with IIR filters. In these cases the filters can be truncated to obtain FIR filters. The error introduced by this truncation needs to be controlled a priori. A detailed study of these algorithms and their implementations for spline wavelets can be found in [Chu].

At first sight, the decomposition algorithm (see Figure 9.1) does not seem to have good encoding properties: although the reconstruction is exact, it encodes N numbers with approximately N numbers, thus, achieving no apparent compression at all. But looking at it more carefully, one realizes that many of the "details"

$$d_{\ell,k}, \qquad \ell = j-1, \cdots, j-M, \quad k \in \mathbb{Z},$$

could be zero, or so small that they fall below a fixed threshold. Thus, we can omit these "small" details in the decomposition, still achieving good reconstruction.

To understand how this could be possible we describe a simplified example from image analysis. Suppose we have a picture as in Figure 9.3(a), where the colors have been numbered on each one of the 16 squares. The sequence

$$1, 1, 1, 1, 2, 3, 2, 3, 4, 5, 5, 4, 6, 6, 6, 6$$

completely encodes the picture. But one can have a "blurred" version and then **add** the details as in Figure 9.3(b); in this case only 8 numbers are needed to encode the picture (discard the zeroes), so that the compression is 50% and the reconstruction is exact.

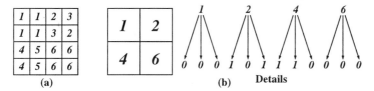

Figure 9.3: (a) *Color-coded picture* (b) *Blurred version and the details.*

9.5   Wavelet packets

The orthonormal wavelet bases we have developed in previous chapters have a frequency localization which is proportional to 2^j at the resolution level j. For example, for a band-limited wavelet ψ, the measure of supp $(\hat{\psi}_{j,k})$ is 2^j times the measure of supp $(\hat{\psi})$, since

$$(\psi_{j,k})^{\wedge}(\xi) = 2^{-\frac{j}{2}} \hat{\psi}(2^{-j}\xi) e^{-i2^{-j}k\xi}, \qquad j, k \in \mathbb{Z}.$$

Thus, the wavelet bases have poor frequency localization when j is large. For some applications, especially for speech signal processing, it is more convenient to have orthonormal bases with better frequency localization. This will be provided by the **wavelet packets**, which are obtained from wavelets associated with MRAs.

If ψ is an orthonormal wavelet generated by an MRA, we have used W_j, $j \in \mathbb{Z}$, to denote the closure in $L^2(\mathbb{R})$ of the linear space generated by the set $\{\psi_{j,k} : k \in \mathbb{Z}\}$. Thus, to reduce the frequency localization of the elements of W_j we need to write W_j as a direct sum of two subspaces generated by elements with better frequency localization than the original ones. Breaking up each one of the new subspaces, the frequency localization is further reduced. We shall describe this process below.

The process is not particular to subspaces of $L^2(\mathbb{R})$, but can be described for every infinite-dimensional separable Hilbert space $\mathbb{H}$. After working with this great generality, we shall obtain the **wavelet packets** by specifying the results to subspaces of $L^2(\mathbb{R})$.

Let $\{e_k : k \in \mathbb{Z}\}$ be an orthonormal system in an infinite-dimensional separable Hilbert space $\mathbb{H}$. A trivial way to separate this system into two orthonormal ones is to consider

$$\{e_{2k} : k \in \mathbb{Z}\} \quad \text{and} \quad \{e_{2k+1} : k \in \mathbb{Z}\}.$$

A more interesting one is to consider

$$\left\{\tfrac{1}{\sqrt{2}}\left[e_{2k-1} + e_{2k}\right] : k \in \mathbb{Z}\right\} \quad \text{and} \quad \left\{\tfrac{1}{\sqrt{2}}\left[e_{2k-1} - e_{2k}\right] : k \in \mathbb{Z}\right\}.$$

(We shall see later that this second example is related to the Haar wavelet.)

In general, consider two sequences $\{\alpha_k : k \in \mathbb{Z}\}$ and $\{\beta_k : k \in \mathbb{Z}\}$ in $\ell^2(\mathbb{Z})$. Define the elements $f_k \in \mathbb{H}$ by

$$\begin{cases} f_{2k} = \sqrt{2} \displaystyle\sum_{\ell \in \mathbb{Z}} \alpha_{2k-\ell} e_\ell \\ f_{2k+1} = \sqrt{2} \displaystyle\sum_{\ell \in \mathbb{Z}} \beta_{2k-\ell} e_\ell. \end{cases} \tag{5.1}$$

(We advise the reader to look at the decomposition formulas (4.9) and (4.10) to see the relation of (5.1) to wavelets.) We are interested in characterizing the sequences $\{\alpha_k\}$ and $\{\beta_k\}$ for which $\{f_k : k \in \mathbb{Z}\}$ is still an orthonormal sequence. To do this we consider the following **filters**:

$$
\begin{cases}
m_0(\xi) = \displaystyle\sum_{k \in \mathbb{Z}} \alpha_k e^{ik\xi}, \\[2mm]
m_1(\xi) = \displaystyle\sum_{k \in \mathbb{Z}} \beta_k e^{ik\xi},
\end{cases}
\qquad \xi \in \mathbb{R}. \tag{5.2}
$$

(These functions are 2π-periodic and related to the definition of a low-pass filter given in section 2.2.) Consider the matrix

$$
M(\xi) = \begin{pmatrix} m_0(\xi) & m_0(\xi + \pi) \\ m_1(\xi) & m_1(\xi + \pi) \end{pmatrix}, \qquad \xi \in \mathbb{R}. \tag{5.3}
$$

We shall show that $\{f_k : k \in \mathbb{Z}\}$ is an orthonormal sequence if and only if $M(\xi)$ is a unitary matrix. Before proving this result, let us find equivalent conditions for $M(\xi)$ to be unitary.

It is easy to see that $M(\xi)$ is unitary if and only if

$$
\begin{cases}
\text{a)} & |m_0(\xi)|^2 + |m_0(\xi + \pi)|^2 = 1, \\[1mm]
\text{b)} & |m_1(\xi)|^2 + |m_1(\xi + \pi)|^2 = 1, \\[1mm]
\text{c)} & m_0(\xi)\,\overline{m_1(\xi)} + m_0(\xi + \pi)\,\overline{m_1(\xi + \pi)} = 0.
\end{cases} \tag{5.4}
$$

By substituting (5.2) in part a) of (5.4) we obtain

$$
1 = \left(\sum_{k \in \mathbb{Z}} \alpha_k e^{ik\xi}\right)\left(\sum_{\ell \in \mathbb{Z}} \overline{\alpha_\ell}\, e^{-i\ell\xi}\right) + \left(\sum_{k \in \mathbb{Z}} \alpha_k (-1)^k e^{ik\xi}\right)\left(\sum_{\ell \in \mathbb{Z}} \overline{\alpha_\ell}\, (-1)^\ell e^{-i\ell\xi}\right)
$$

$$
= \sum_{k \in \mathbb{Z}} \left\{ \alpha_k \left(\sum_{\ell \in \mathbb{Z}} \overline{\alpha_\ell}\, e^{-i(\ell-k)\xi} + \sum_{\ell \in \mathbb{Z}} \overline{\alpha_\ell}\, (-1)^{\ell+k} e^{-i(\ell-k)\xi} \right) \right\}
$$

$$
= \sum_{m \in \mathbb{Z}} \left\{ \sum_{\ell \in \mathbb{Z}} \alpha_{\ell-m}\, \overline{\alpha_\ell}\, [1 + (-1)^m] \right\} e^{-im\xi}
$$

$$
= \sum_{k \in \mathbb{Z}} \left\{ 2 \sum_{\ell \in \mathbb{Z}} \alpha_{\ell-2k}\, \overline{\alpha_\ell} \right\} e^{-i2k\xi}.
$$

Hence, part a) of (5.4) is equivalent to

$$
2 \sum_{\ell \in \mathbb{Z}} \alpha_{\ell-2k}\, \overline{\alpha_\ell} = \delta_{k,0} \qquad \text{for all } k \in \mathbb{Z}.
$$

Similar arguments show the following result:

PROPOSITION 5.5 *The matrix* $M(\xi)$ *given by (5.3) and (5.2) is unitary if and only if*

a) $\quad \displaystyle\sum_{\ell \in \mathbb{Z}} \alpha_{\ell-2k}\, \overline{\alpha_\ell} = \tfrac{1}{2}\delta_{k,0} \qquad$ *for all* $k \in \mathbb{Z}$,

b) $\quad \displaystyle\sum_{\ell \in \mathbb{Z}} \beta_{\ell-2k}\, \overline{\beta_\ell} = \tfrac{1}{2}\delta_{k,0} \qquad$ *for all* $k \in \mathbb{Z}$, *and* $\qquad$ (5.6)

c) $\quad \displaystyle\sum_{\ell \in \mathbb{Z}} \alpha_{\ell-2k}\, \overline{\beta_\ell} = 0 \qquad$ *for all* $k \in \mathbb{Z}$.

THEOREM 5.7 *Let* $\{e_k : k \in \mathbb{Z}\}$ *be an orthonormal system in a Hilbert space* $\mathbb{H}$. *The sequence* $\{f_k : k \in \mathbb{Z}\}$ *given by (5.1) is an orthonormal system in* $\mathbb{H}$ *if and only if the matrix* $M(\xi)$, *given by (5.3) and (5.2), is unitary for every* $\xi \in \mathbb{R}$.

PROOF : Suppose that $M(\xi)$ is unitary. By part a) of (5.6),

$$\langle f_{2k}, f_{2s} \rangle = \left\langle \sqrt{2} \sum_{\ell \in \mathbb{Z}} \alpha_{2k-\ell} e_\ell,\ \sqrt{2} \sum_{\ell \in \mathbb{Z}} \alpha_{2s-\ell} e_\ell \right\rangle$$

$$= 2 \sum_{\ell \in \mathbb{Z}} \alpha_{2k-\ell}\, \overline{\alpha_{2s-\ell}} = 2 \sum_{p \in \mathbb{Z}} \alpha_{p-2(s-k)}\, \overline{\alpha_p} = \delta_{s-k,0} = \delta_{s,k}\,.$$

Similarly, $\langle f_{2k+1}, f_{2s+1} \rangle = \delta_{s,k}$ by part b) of (5.6). Finally, by (5.6) c),

$$\langle f_{2k}, f_{2s+1} \rangle = \left\langle \sqrt{2} \sum_{\ell \in \mathbb{Z}} \alpha_{2k-\ell} e_\ell,\ \sqrt{2} \sum_{\ell \in \mathbb{Z}} \beta_{2s-\ell} e_\ell \right\rangle$$

$$= 2 \sum_{\ell \in \mathbb{Z}} \alpha_{2k-\ell}\, \overline{\beta_{2s-\ell}} = 2 \sum_{p \in \mathbb{Z}} \alpha_{p-2(s-k)}\, \overline{\beta_p} = 0.$$

This shows that $\{f_k : k \in \mathbb{Z}\}$ is an orthonormal system. The equivalence is shown by reversing the above steps. $\blacksquare$

EXAMPLE C : With

$$\alpha_k = \tfrac{1}{\sqrt{2}}\delta_{k,0} \qquad \text{and} \qquad \beta_k = \tfrac{1}{\sqrt{2}}\delta_{k,-1}\,,$$

we obtain the trivial decomposition $f_{2k} = e_{2k}$ and $f_{2k+1} = e_{2k+1}$. With

$$\alpha_k = \tfrac{1}{2}\left[\delta_{k,1} + \delta_{k,0}\right] \qquad \text{and} \qquad \beta_k = \tfrac{1}{2}\left[\delta_{k,1} - \delta_{k,0}\right],$$

we obtain

$$f_{2k} = \tfrac{1}{\sqrt{2}}\left\{e_{2k-1} + e_{2k}\right\} \qquad \text{and} \qquad f_{2k+1} = \tfrac{1}{\sqrt{2}}\left\{e_{2k-1} - e_{2k}\right\}.$$

In this case

$$m_0(\xi) = \tfrac{1}{2}(e^{i\xi} + 1) \qquad \text{and} \qquad m_1(\xi) = \tfrac{1}{2}(e^{i\xi} - 1).$$

Observe that m_0 is the low-pass filter of the Haar wavelet (see Example B in section 2.2).

REMARK: The fact that $M(\xi)$, given by (5.3), is unitary implies that $m_1(\xi)$ and $m_0(\xi)$ are related. In fact, starting with part c) of (5.4) and arguing as in (2.7), (2.8) and (2.9) of Chapter 2 (through the use of (5.4) a) and b)) we deduce that

$$m_1(\xi) = e^{i\xi} s(2\xi) \overline{m_0(\xi + \pi)} \tag{5.8}$$

with $s \in L^2(\mathbb{T})$, 2π-periodic and $|s(\xi)| = 1$ for all $\xi \in \mathbb{R}$. This is necessary and sufficient since if m_1 is given by (5.8) it is easy to check that $M(\xi)$ is a unitary matrix.

We now show how to obtain a direct sum decomposition of a Hilbert space using (5.1).

THEOREM 5.9 *Let $\{e_k : k \in \mathbb{Z}\}$ be an orthonormal basis of a Hilbert space $\mathbb{H}$ and consider $\{f_k : k \in \mathbb{Z}\}$ and $M(\xi)$ as in (5.1), (5.2) and (5.3). The following properties are equivalent:*

   i)   *$\{f_k : k \in \mathbb{Z}\}$ is an orthonormal system in $\mathbb{H}$,*

   ii)   *$\{f_k : k \in \mathbb{Z}\}$ is an orthonormal basis of $\mathbb{H}$,*

   iii) *The matrix $M(\xi)$ is unitary.*

PROOF : By Theorem 5.7 it is enough to show that, if we assume i), the set $\{f_k : k \in \mathbb{Z}\}$ is complete in $\mathbb{H}$. This follows from the easily established equality

$$e_\ell = \sum_{k \in \mathbb{Z}} \sqrt{2} \, \overline{\alpha_{2k-\ell}} \, f_{2k} + \sum_{k \in \mathbb{Z}} \sqrt{2} \, \overline{\beta_{2k-\ell}} \, f_{2k+1}. \tag{5.10}$$

∎

It is now clear that given a Hilbert space $\mathbb{H}$ with orthonormal basis $\{e_k : k \in \mathbb{Z}\}$ and two sequences $\{\alpha_k : k \in \mathbb{Z}\}$ and $\{\beta_k : k \in \mathbb{Z}\}$ in $\ell^2(\mathbb{Z})$, we have

$$\mathbb{H} = \mathbb{H}_0 \oplus \mathbb{H}_1, \tag{5.11}$$

where $\mathbb{H}_0$ and $\mathbb{H}_1$ are the closures in $\mathbb{H}$ of the subspaces generated by $\{f_{2k} : k \in \mathbb{Z}\}$ and $\{f_{2k+1} : k \in \mathbb{Z}\}$, respectively (see (5.1)).

We now define the **basic wavelet packets** associated with a scaling function. Consider an MRA, as given in section 2.1, with a scaling function φ for which $\hat{\varphi}(0) = 1$. Let

$$m_0(\xi) = \sum_{k \in \mathbb{Z}} \alpha_k e^{ik\xi}$$

be the low-pass filter associated with φ so that

$$\hat{\varphi}(2\xi) = m_0(\xi)\hat{\varphi}(\xi) \tag{5.12}$$

(see (2.3) in Chapter 2). Also take $m_1(\xi) = e^{i\xi} \, \overline{m_0(\xi + \pi)}$ so that

$$m_1(\xi) = \sum_{k \in \mathbb{Z}} \beta_k e^{ik\xi}, \qquad \text{with} \quad \beta_k = (-1)^{k+1} \, \overline{\alpha_{1-k}}$$

(see (4.8)). With this choice of m_0 and m_1, the remark that preceeds Theorem 5.9 shows that the matrix $M(\xi)$ given by (5.3) is unitary.

Let $\omega_0 = \varphi$. The **basic wavelet packets** $\omega_n(x)$, $n = 0, 1, 2, \cdots$, associated with the scaling function φ are defined recursively by

$$\omega_{2n}(x) = 2 \sum_{k \in \mathbb{Z}} \alpha_k \omega_n(2x + k) \tag{5.13}$$

and

$$\omega_{2n+1}(x) = 2 \sum_{k \in \mathbb{Z}} \beta_k \omega_n(2x + k). \tag{5.14}$$

When $n = 0$ in (5.13) we obtain

$$\varphi(x) = 2 \sum_{k \in \mathbb{Z}} \alpha_k \varphi(2x + k)$$

which can be seen to be equivalent to (5.12) by taking Fourier transforms of both sides.

When $n = 0$ in (5.14) we deduce that

$$\omega_1(x) = 2 \sum_{k \in \mathbb{Z}} \beta_k \varphi(2x + k);$$

taking Fourier transforms of both sides we obtain

$$\widehat{\omega_1}(\xi) = m_1\left(\tfrac{1}{2}\xi\right) \hat{\varphi}\left(\tfrac{1}{2}\xi\right). \tag{5.15}$$

Since $m_1(\xi) = e^{i\xi}\,\overline{m_0(\xi + \pi)}$, Proposition 2.13 of Chapter 2 shows that $\omega_1 = \psi$ is an orthonormal wavelet associated with the given MRA.

From (5.12) and the assumption $\hat{\varphi}(0) = 1$ we deduce

$$\hat{\varphi}(\xi) = \prod_{j=1}^{\infty} m_0(2^{-j}\xi) = m_0\left(\tfrac{1}{2}\xi\right) \hat{\varphi}\left(\tfrac{1}{2}\xi\right). \tag{5.16}$$

Similarly, from (5.15) we obtain

$$\hat{\psi}(\xi) = \widehat{\omega_1}(\xi) = m_1\left(\tfrac{1}{2}\xi\right) \prod_{j=2}^{\infty} m_0(2^{-j}\xi) = m_1\left(\tfrac{1}{2}\xi\right) \hat{\varphi}\left(\tfrac{1}{2}\xi\right). \tag{5.17}$$

To generalize these results to the basic wavelet packets we need to consider the dyadic expansion of a positive integer n; that is,

$$n = \sum_{j=1}^{\infty} \varepsilon_j 2^{j-1}, \qquad \varepsilon_j \in \{0, 1\}. \tag{5.18}$$

Observe that the sum in (5.18) is always finite and the dyadic expansion of n is unique.

PROPOSITION 5.19 *Let n be a non-negative integer with dyadic expansion given by (5.18). Then the Fourier transform of the basic wavelet packets* *given by (5.13) and (5.14) satisfy*

$$\widehat{\omega_n}(\xi) = \prod_{j=1}^{\infty} m_{\varepsilon_j}(2^{-j}\xi) = \left\{ \prod_{j=1}^{k} m_{\varepsilon_j}(2^{-j}\xi) \right\} \hat{\varphi}(2^{-k}\xi). \tag{5.20}$$

Proof : Equalities (5.16) and (5.17) show that (5.20) is true for $n = 0$ and $n = 1$. We proceed by induction on n. Suppose $n = 2\ell$ is even and (5.20) is true for every non-negative integer $m < n$. By (5.13),

$$\widehat{\omega_{2\ell}}(\xi) = \sum_{k \in \mathbb{Z}} \alpha_k e^{ik\frac{\xi}{2}} \widehat{\omega_\ell}(\tfrac{1}{2}\xi) = m_0(\tfrac{1}{2}\xi)\widehat{\omega_\ell}(\tfrac{1}{2}\xi).$$

If $\ell = \sum_{j=1}^{\infty} \varepsilon_j 2^{j-1}$ is the dyadic decomposition of ℓ, the dyadic decomposition of $n = 2\ell$ is

$$n = 0 + 2\varepsilon_1 + 2^2\varepsilon_2 + \cdots + 2^j\varepsilon_j = \sum_{j=1}^{\infty} \varepsilon_{j-1} 2^{j-1}$$

with $\varepsilon_0 = 0$. Therefore,

$$\widehat{\omega_n}(\xi) = \widehat{\omega_{2\ell}}(\xi) = m_0(\tfrac{1}{2}\xi) \prod_{j=1}^{\infty} m_{\varepsilon_j}(2^{-j}\tfrac{1}{2}\xi) = m_0(\tfrac{1}{2}\xi) \prod_{j=2}^{\infty} m_{\varepsilon_{j-1}}(2^{-j}\xi),$$

which shows the desired result when n is even. For the odd case $n = 2\ell + 1$ use (5.14), the definition of $m_1(\xi)$ and the fact that, when $\ell = \sum_{j=1}^{\infty} \varepsilon_j 2^{j-1}$ is the dyadic decomposition of ℓ,

$$n = 2\ell + 1 = \sum_{j=1}^{\infty} \varepsilon_{j-1} 2^{j-1}$$

with $\varepsilon_0 = 1$. ∎

Using all the theory we have developed so far in this section it is easy to find a new orthonormal basis of $L^2(\mathbb{R})$ based on the basic wavelet packets.

Theorem 5.21 *Let $\{\omega_n : n \in \mathbb{Z}\}$ be the basic wavelet packets, defined by (5.13) and (5.14), associated with a scaling function φ of an MRA $\{V_j\}_{j \in \mathbb{Z}}$. Then,*

a) *For all $j = 0, 1, 2, \cdots$, the set $\{\omega_n(x - k) : k \in \mathbb{Z}, 0 \le n < 2^j\}$ is an orthonormal basis of V_j.*

b) *The set $\{\omega_n(x - k) : k \in \mathbb{Z}, n = 0, 1, 2, \cdots\}$ is an orthonormal basis of $L^2(\mathbb{R})$.*

PROOF : The case $j = 0$ in a) follows from the definition of an MRA (section 2.1) since $\omega_0 = \varphi$ and $\{\varphi(x - k) : k \in \mathbb{Z}\}$ is an orthonormal basis of V_0. We proceed by induction on j. Let us assume that

$$\{\omega_n(\cdot - k) : k \in \mathbb{Z}, 0 \le n < 2^{j-1}\}$$

is an orthonormal basis of V_{j-1}. Then

$$\{\sqrt{2}\,\omega_n(2(\cdot) - k) : k \in \mathbb{Z}, 0 \le n < 2^{j-1}\}$$

is an orthonormal basis of V_j by (1.2) of the definition of an MRA in section 2.1. But (5.13) and (5.14) can be written as

$$\omega_{2n}(x - m) = \sqrt{2} \sum_{k \in \mathbb{Z}} \alpha_k \left[\sqrt{2}\,\omega_n(2x - 2m + k)\right]$$

$$= \sqrt{2} \sum_{\ell \in \mathbb{Z}} \alpha_{2m-\ell} \left[\sqrt{2}\,\omega_n(2x - \ell)\right]$$

and

$$\omega_{2n+1}(x - m) = \sqrt{2} \sum_{\ell \in \mathbb{Z}} \beta_{2m-\ell} \left[\sqrt{2}\,\omega_n(2x - \ell)\right],$$

showing that we have a transformation of the form (5.1). Since φ is a scaling function of an MRA, $M(\xi)$ is unitary (with the choice $\beta_k = (-1)^{k+1} \overline{\alpha_{1-k}}$, which gives $m_1(\xi) = e^{i\xi} \overline{m_0(\xi + \pi)}$); hence, part ii) of Theorem 5.9 shows that $\{\omega_n(x - k) : k \in \mathbb{Z}, 0 \le n < 2^j\}$ is an orthonormal basis of V_j. This proves part a).

To prove part b) use the part just proved together with the fact that

$$L^2(\mathbb{R}) = \overline{\bigcup_{j=0}^{\infty} V_j}$$

and $V_j \subset V_{j+1}$ for all j (see the definition of an MRA in section 2.1).

∎

REMARK: The set $\{2^{\frac{j}{2}}\varphi(2^j x - k) : k \in \mathbb{Z}\}$ is an orthonormal basis of V_j, $j = 0, 1, 2, \cdots$. The band-width of the elements of this basis is proportional to 2^j. By part a) of Theorem 5.21, V_j has been decomposed into 2^j orthogonal channels $\{\omega_n(x - k) : k \in \mathbb{Z}\}, 0 \le n < 2^j$. Therefore, it is natural to expect that the band-width of the ω_n's be constant. This conjecture has been disproved in [CMW].

EXAMPLE D: Walsh series.

The wavelet packets associated with the scaling function of the Haar wavelet produce the Walsh series. Let $\varphi = \chi_{[-1,0]}$ be a scaling function for the Haar wavelet (see Example B in Chapter 2). Since

$$\tfrac{1}{2}\varphi(\tfrac{1}{2}x) = \tfrac{1}{2}\varphi(x) + \tfrac{1}{2}\varphi(x + 1),$$

we have

$$m_0(\xi) = \frac{1 + e^{i\xi}}{2}$$

and

$$m_1(\xi) = e^{i\xi}\,\overline{m_0(\xi + \pi)} = \frac{e^{i\xi} - 1}{2}.$$

Thus

$$\begin{cases} \alpha_k = \tfrac{1}{2}\delta_{k,0} + \tfrac{1}{2}\delta_{k,1}, \\ \beta_k = -\tfrac{1}{2}\delta_{k,0} + \tfrac{1}{2}\delta_{k,1}, \end{cases} \qquad k \in \mathbb{Z}.$$

Then, by (5.14) with $n = 0$ we obtain

$$\psi(x) = \omega_1(x) = 2\left(-\tfrac{1}{2}\varphi(2x) + \tfrac{1}{2}\varphi(2x - 1)\right) = \chi_{[-1,-\frac{1}{2})}(x) - \chi_{[-\frac{1}{2},0)}(x).$$

The rest of the elements of the wavelet packets associated with the Haar wavelet are obtained inductively using (5.13) and (5.14). The algorithm is

$$\begin{cases} \omega_{2n}(x) = \omega_n(2x) + \omega_n(2x + 1) \\ \omega_{2n+1}(x) = -\omega_n(2x) + \omega_n(2x + 1). \end{cases}$$

The first 8 elements of this wavelet packet are graphed in Figure 9.4. By Theorem 5.21 we have an orthonormal basis of $L^2(\mathbb{R})$.

In [Pal], R.E.A.C. Paley introduced a remarkable orthonormal basis. We give below Paley's description adapted to our situation (that is the

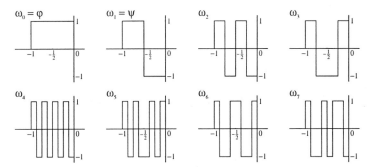

Figure 9.4: The first 8 elements of the wavelet packets associated with the scaling function of the Haar wavelet.

interval $[-1, 0]$, while the description in [Pal] is for the interval $[0, 1]$). Let $r(t)$ be the 1-periodic function whose value in $[-1, 0)$ is

$$r(t) = \chi_{[-1,-\frac{1}{2})}(t) - \chi_{[-\frac{1}{2},0)}(t).$$

For $k = 0, 1, 2, \cdots$, define the Rademacher functions by $r_k(t) = r(2^k t)$. For $n = 1, 2, 3, \cdots$, let $n = \varepsilon_0 + \varepsilon_1 2 + \cdots + \varepsilon_j 2^j$, $\varepsilon_i \in \{0, 1\}$, $i = 0, 1, \cdots, j$, be the dyadic expansion of n and define

$$\begin{cases} W_0(t) = \chi_{[-1,0)}(t), \\ W_n(t) = [r_0(t)]^{\varepsilon_0} [r_1(t)]^{\varepsilon_1} \cdots [r_j(t)]^{\varepsilon_j} & \text{for } n \geq 1. \end{cases}$$

The set $\{W_n : n = 0, 1, 2, \cdots\}$ is known as the **Walsh series** for the interval $[-1, 0)$. Considering all their integer translations we obtain an orthonormal basis of $L^2(\mathbb{R})$. Observe that $r^0 \equiv 1$ and $r^1 = r$. Then, using the 1-periodicity of r it is easy to see that

$$\begin{cases} W_{2n}(x) = W_n(2x) + W_n(2x + 1) \\ W_{2n+1}(x) = -W_n(2x) + W_n(2x + 1). \end{cases}$$

This shows that the elements W_n and ω_n coincide.

EXAMPLE E: Wavelet packets associated with the Shannon wavelet.

Example C in Chapter 2 shows that for the Shannon wavelet

$$m_0(\xi) = \sum_{k \in \mathbb{Z}} \chi_{[-\frac{\pi}{2},\frac{\pi}{2}]}(\xi + 2k\pi)$$

and

$$m_1(\xi) = e^{i\xi} \sum_{k \in \mathbb{Z}} \chi_{[\frac{\pi}{2}, \frac{3\pi}{2}]}(\xi + 2k\pi).$$

We know that $\hat{\varphi}(\xi) = \chi_{[-\pi,\pi]}(\xi)$, so that (5.20) allows us to find the Fourier transforms of the elements of the wavelet packets associated with φ. In Figure 9.5 we give computer graphs of the elements w_n, $0 \le n < 8$. Observe that for $n = \sum_{j=1}^{\infty} \varepsilon_j 2^{j-1}$, $\varepsilon_j \in \{0, 1\}$, the center of symmetry of w_n is the number in $(-1, 0]$ given by

$$c_n = -\sum_{j=1}^{\infty} \frac{\varepsilon_j}{2^j}.$$

For a scaling function φ with associated wavelet ψ we have constructed the corresponding wavelet packets given by (5.13) and (5.14). The set

$$\left\{ 2^{\frac{j}{2}} w_n(2^j x - k) : \ j \in \mathbb{Z}, \ n = 0, 1, 2, \cdots \right\}$$

is overcomplete in $L^2(\mathbb{R})$. In fact, this system contains the wavelet basis $\{ 2^{\frac{j}{2}} \psi(2^j x - k) : j, k \in \mathbb{Z} \}$ (choose $n = 1$) and the wavelet packets

$$\{ w_n(x - k) : k \in \mathbb{Z}, n = 0, 1, 2, \cdots \}$$

(choose $j = 0$). Many other subcollections of it give rise to orthonormal bases of $L^2(\mathbb{R})$.

For the family of wavelet packets $\{w_n\}$ corresponding to some orthonormal scaling function $\varphi = w_0$ define the family of subspaces of $L^2(\mathbb{R})$ given by

$$U_j^n \equiv \overline{\operatorname{span} \{w_n(2^j x - k) : \ k \in \mathbb{Z}\}}, \qquad j \in \mathbb{Z}, \ n = 0, 1, 2, \cdots. \qquad (5.22)$$

Observe that

$$\begin{cases} U_j^0 = V_j, \\ U_j^1 = W_j, \end{cases} \qquad j \in \mathbb{Z},$$

so that the orthogonal decomposition $V_{j+1} = V_j \oplus W_j$ can be written as

$$U_{j+1}^0 = U_j^0 \oplus U_j^1, \qquad j \in \mathbb{Z}.$$

This result can be generalized to other values of n.

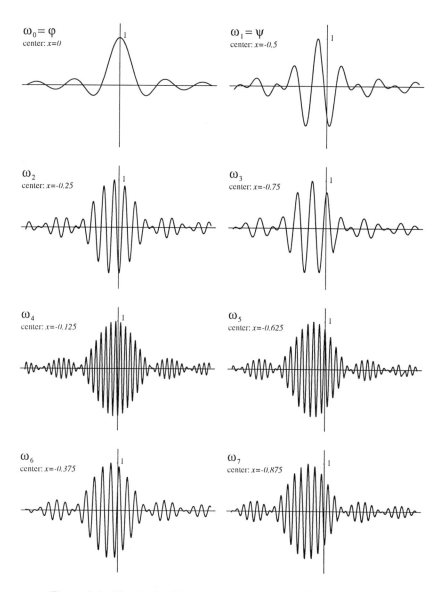

Figure 9.5: Wavelet packets associated with the Shannon wavelet.

PROPOSITION 5.23 *For* $n = 0, 1, 2, \cdots$ *we have*

$$U_{j+1}^n = U_j^{2n} \oplus U_j^{2n+1}, \qquad j \in \mathbb{Z},$$

where U_j^n *is defined by* (5.22).

PROOF : It follows from (5.13) that U_j^{2n} is a subspace of U_{j+1}^n; similarly, it follows from (5.14) that U_j^{2n+1} is a subspace of U_{j+1}^n. The set

$$\left\{ E_\ell(x) = 2^{\frac{j+1}{2}} \omega_n(2^{j+1} x - \ell) : \ell \in \mathbb{Z} \right\}$$

is an orthonormal basis of U_{j+1}^n. Thus, if $F_k(x)$, $k \in \mathbb{Z}$, are given by (5.1) with e_ℓ replaced by E_ℓ, Theorem 5.9 shows that

$$U_{j+1}^n = H_0 \oplus H_1,$$

where

$$\begin{cases} H_0 = \overline{\text{span}\,\{F_{2k}(\cdot) : \; k \in \mathbb{Z}\}}, \\ H_1 = \overline{\text{span}\,\{F_{2k+1}(\cdot) : \; k \in \mathbb{Z}\}}. \end{cases}$$

But

$$F_{2k}(x) = \sqrt{2} \sum_{\ell \in \mathbb{Z}} \alpha_{2k-\ell} E_\ell(x) = \sqrt{2} \sum_{\ell \in \mathbb{Z}} \alpha_{2k-\ell} 2^{\frac{j+1}{2}} \omega_n(2^{j+1} x - \ell)$$

$$= 2 \sum_{\ell \in \mathbb{Z}} \alpha_{2k-\ell} 2^{\frac{j}{2}} \omega_n(2^j 2x - \ell)$$

$$= 2 \sum_{m \in \mathbb{Z}} \alpha_m 2^{\frac{j}{2}} \omega_n(2(2^j x - k) + m)$$

$$= 2^{\frac{j}{2}} \omega_{2n}(2^j x - k),$$

where the last equality is due to (5.13). Thus, $H_0 = U_j^{2n}$. Similarly, with the aid of (5.14), one can show $H_1 = U_j^{2n+1}$. This finishes the proof of the proposition. ∎

Theorem 5.24 *For each $j = 1, 2, 3, \cdots$, we have*

$$
\begin{cases}
W_j = U_{j-1}^2 \oplus U_{j-1}^3 \, , \\[2mm]
W_j = U_{j-2}^4 \oplus U_{j-2}^5 \oplus U_{j-2}^6 \oplus U_{j-2}^7 \, , \\[2mm]
\qquad \vdots \\[2mm]
W_j = U_{j-k}^{2^k} \oplus U_{j-k}^{2^k+1} \oplus \cdots \oplus U_{j-k}^{2^{k+1}-1} \, , \\[2mm]
\qquad \vdots \\[2mm]
W_j = U_0^{2^j} \oplus U_0^{2^j+1} \oplus \cdots \oplus U_0^{2^{j+1}-1} \, ,
\end{cases}
\qquad (5.25)
$$

where U_j^n is defined in (5.22). Moreover, for each $j = 1, 2, \cdots$, $k = 1, 2, \cdots, j$, and $m = 0, 1, \cdots, 2^k - 1$, the set

$$
\left\{ 2^{\frac{j-k}{2}} \omega_{2^k+m}(2^{j-k}x - \ell) : \ \ell \in \mathbb{Z} \right\},
\qquad (5.26)
$$

is an orthonormal basis of $U_{j-k}^{2^k+m}$.

Proof : Since $U_j^1 = W_j$ we can apply Proposition 5.23 repeatedly to obtain (5.25). The definition of $U_{j-k}^{2^k+m}$ (see (5.22)) and the orthonormality of $\{\omega_{2^k+m}(x - k) : k \in \mathbb{Z}\}$ (see Theorem 5.21) show that (5.25) is an orthonormal basis of $U_{j-k}^{2^k+m}$. ∎

Theorem 5.24 allows us to obtain many orthonormal bases of $L^2(\mathbb{R})$. Since

$$
L^2(\mathbb{R}) = V_0 \oplus W_0 \oplus W_1 \oplus W_2 \oplus \cdots ,
$$

we can choose whether to decompose any W_j, $j = 0, 1, 2, \cdots$ further with any of the decompositions that appeared in (5.25). If we choose not to decompose any of the W_j we obtain the wavelet decomposition of $L^2(\mathbb{R})$. If we choose the last formula in (5.25) for each W_j we obtain the wavelet packets. In between these two decompositions, there are a denumerable number of ways to choose a decomposition of $L^2(\mathbb{R})$, and, hence, produce new orthonormal bases.

This flexibility of choosing the decomposition of $L^2(\mathbb{R})$ is well adapted for applications, where one can choose whether or not to decompose a W_j according to the available data. Also useful for applications is to label these families of wavelet packets using the dyadic intervals on $[0, 1)$ (see [CMW] and [Wi3]).

9.6   Notes and references

1. Several books describe the discrete transformations of sections 9.1 and 9.2 as well as related ones. We recommend [Walk] and [Wi2]. The continuous local cosine and sine bases were first described by R. Coifman and Y. Meyer ([CM2]) and the discrete version was introduced in some particular cases by H. Malvar ([Malv]). [Da2] contains a description of the decomposition and reconstruction algorithms for wavelets. The theory of wavelet packets was first published in [CMW].

2. A library of orthonormal bases is any finite collection of orthonormal bases. This finite collection may include the wavelet packets associated with orthonormal wavelets as well as local trigonometric bases developed in Chapter 1. One may wonder how to find from such a large collection of orthonormal bases one that best matches a given signal or family of signals. This can be done by defining an efficiency functional (the most attractive one is related to the Shannon entropy) and searching for a minimum of this functional. When the library is structured into a "tree" this minimum always exists and it is the "best" basis for the given problem. Details of these algorithms and their efficiency can be found in [CWi], [CMQW] and [Wi3].

3. The development of the theory of wavelets has closely followed its practical applications. Such is the case of the decomposition and reconstruction algorithms. Such algorithms first appeared with the **Laplacian pyramid algorithm** of P.J. Burt and E.H. Adelson ([BA]), in signal processing through the use of **quadrature mirror filters** (QMF) (see [SB1] and [SB2]), and were adapted to wavelets by S. Mallat ([Mal2]). Besides the references mentioned above, two recent books, [SN] and [VK], are recommended for the reader interested in knowing the theory of filters and their applications.

4. The theory of wavelets, and closely related ones such as the theory of frames presented in Chapter 8, have many applications, some of which are currently being developed. In 1992 a special issue on wavelet transforms and multiresolution signal analysis was published in the IEEE Transactions on Information Theory ([DMW]) which contains several articles describing some of these applications. The use of wavelets for fingerprinting is described in [Bri]. This theory can also be used to study purely mathematical objects, such as the Riemann function, whose fractal behavior is better analyzed by using wavelets. We strongly recommend [Me6] for those interested in pursuing a much more detailed account of applications.

References

[Au1] P. Auscher, *Remarks on local Fourier bases*, in Wavelets: Mathematics and Applications (J.J. Benedetto and M.W. Frazier, Ed.) CRC Press, (1994), 203-218.

[Au2] P. Auscher, *Solution of two problems on wavelets*, Journal of Geometric Analysis, Vol 5, No 2, (1995), 181-236.

[Au3] P. Auscher, *Wavelet bases for $L^2(\mathbb{R})$ with rational dilation factor*, in Wavelets And Their Applications, (M. Ruskai, et al., Eds.). Jones & Bartlett Publishers (1992), 439-451.

[Au4] P. Auscher, *Il n'existe pas de bases d'ondelettes régulières dans l'espace de Hardy H^2*, C. R. Acad. Sci. Paris, 315, Série I (1992), 769-772.

[Au5] P. Auscher, *Toute base d'ondelettes régulières de $L^2(\mathbb{R})$ est issue d'une analyse multirésolution régulière*, C. R. Acad. Sci. Paris, Série I, 315, (1992), 1227-1230.

[AWW] P. Auscher, G. Weiss, M.V. Wickerhauser, *Local sine and cosine basis of Coifman and Meyer and the construction of smooth wavelets*, in Wavelets: a tutorial in theory and applications (C.K. Chui, Ed.), Academic Press, (1992), 237-256.

[Bal] R. Balian, *Un principe d'incertitude fort en théorie du signal ou en mécanique quantique*, C. R. Acad. Sci. Paris, 292, Série II, (1981), 1357-1361.

[Bat1] G. Battle, *A block spin construction of ondelettes; Part I: Lemarié functions*, Comm. Math. Phys., 110, (1987), 601-615.

[Bat2] G. Battle, *Heisenberg proof of the Balian-Low theorem*, Lett. Math. Phys., 15, (1988), 175-177.

[BHW] J.J. Benedetto, C. Heil, D.F. Walnut, *Differentiation and the Balian-Low theorem*, The Journal of Fourier Analysis and Applications, Vol. 1, No. 4, (1995), 355-402.

[BL] J. Bergh, J. Löfström, Interpolation Spaces: An Introduction, Springer-Verlag, (1976).

[Ber] E. Berkson, *On the structure of the graph of the Franklin analyzing wavelet* in Analysis at Urbana 1, London Math. Soc., Lecture Note Ser. (E. Berkson, N.T. Peck, J. Uhl, Eds.), Cambridge U. Press, (1989), 366-394.

[BPW] E. Berkson, M. Paluzyński, G. Weiss, *Transference couples and their applications to convolution operators and maximal operators*, in Lecture Notes in Pure and Applied Math. (N. Kalton, E. Saab, S. Montgomery-Smith, Eds.), Vol. 175, Marcel Dekker, (1996), 69-84.

[BSW] A. Bonami, F. Soria, G. Weiss, *Band-limited wavelets*, J. of Geometric Analysis, 3(6), (1993), 544-578.

[Boo] C. de Boor, A practical guide to splines, Applied Mathematical Sciences, Vol. 27, Springer-Verlag, (1978).

[Bri] C.M. Brislaw, *Fingerprints go digital*, Notices of the AMS, Vol 42, November, (1995), 1272-1282.

[BGS] D.L. Burkholder, R.F. Gundy, M.L. Silverstein, *A maximal function characterization of the class H^p*, Trans. Amer. Math. Soc., 157, (1971), 137-153.

[BA] P.J. Burt, E.H. Adelson, *The Laplacian pyramid as a compact image code*, IEEE Trans. Comm., 31, (1983), 532-540.

[Cal] A.P. Calderón, *An atomic decomposition of distributions in parabolic H^p spaces*, Advances in Math., 25, (1978), 85-96.

[CT] A.P. Calderón, A. Torchinsky, *Parabolic maximal functions associated with a distribution I, II*, Advances in Math., 16, (1975), 1-64, and 24, (1977), 101-107.

[Car1] L. Carleson, *On convergence and growth of partial sums of*

Fourier series, Acta Mat., 116, (1966), 135-157.

[Car2] L. Carleson, *An explicit unconditional basis in H^1*, Bull. des Sciences Math., 104, (1980), 405-416.

[Chu] C.K. Chui, An Introduction to Wavelets, Academic Press, (1992).

[CS] C.K. Chui, X. Shi, *Inequalities of Littlewood-Paley type for frames and wavelets*, SIAM J. Math. Anal., Vol. 24, No. 1, (1993), 263-277.

[Co1] A. Cohen, *Ondelettes, analyses multirésolutions et filtres miroirs en quadrature*, Ann. Inst. Henri Poincaré, Anal. non linéaire, 7, No. 5, (1990), 439-459.

[Co2] A. Cohen, *Ondelettes et traitment numerique du signal*, Research Notes in Mathematics, Masson, Paris, (1992).

[Co3] A. Cohen, *Construction de bases d'ondelettes α-höldériennes*, Rev. Mat. Iberoamericana, 6, (1990), 91-108.

[CC] A. Cohen, J.P. Conze, *Régularité des bases d'ondelettes et mesures ergodiques*, Rev. Mat. Iberoamericana, 8, (1992), 351-365.

[CD] A. Cohen, I. Daubechies, *A stability criterion for biorthogonal wavelet bases and their related subband coding schemes*, Duke Math. J., 68, (1992), 313-335.

[CDF] A. Cohen, I. Daubechies, J.C. Feauveau, *Biorthogonal bases of compactly supported wavelets*, Comm. Pure Appl. Math., 45, (1992), 485-500.

[Coi] R. Coifman, *A real variable characterization of H^p*, Studia Math., 51, (1974), 269-274.

[CM1] R. Coifman, Y. Meyer, Ondelettes et opérateurs III: Opérateurs Multilinéaires, Hermann, Paris, (1991).

[CM2] R. Coifman, Y. Meyer, *Remarques sur l'analyse de Fourier à fenêtre*, C. R. Acad. Sci. Paris, Série I, 312, (1991), 259-261.

[CMQW] R. Coifman, Y. Meyer, S. Quake, M.V. Wickerhauser, *Signal processing and compression with wave packets*, in Wavelets and their Applications (J.S. Byrnes, Ed.), Kluwer Academic Publisher,

(1994), 363–378.

[CMW] R. Coifman, Y. Meyer, M.V. Wickerhauser, *Size properties of wavelet-packets*, in Wavelets and their Applications (Ruskai, M.B. et al., Eds.), Jones and Bartlett, (1992).

[CW1] R. Coifman, G. Weiss, *Review of Littlewood-Paley and multiplier theory*, Bull. Amer. Math. Soc., 84, (1977), 242-250.

[CW2] R. Coifman, G. Weiss, *Extensions of Hardy spaces and their use in analysis*, Bull. Amer. Math. Soc., 83, (1977), 569-645.

[CWi] R. Coifman, M.V. Wickerhauser, *Entropy-based algorithms for best basis selection*, IEEE Transations on Information Theory, Vol. 38, no. 2, March 1992, 713–718.

[Da1] I. Daubechies, Ten Lectures on Wavelets, CBS-NSF Regional Conferences in Applied Mathematics, 61, SIAM, (1992).

[Da2] I. Daubechies, *Orthonormal bases of compactly supported wavelets*, Comm. Pure Appl. Math., 41, (1988), 909-996.

[Da3] I. Daubechies, *The wavelet transform, time-frequency localization and signal analysis*, IEEE Trans. Inform. Theory, 36, (1990), 961-1005.

[DGM] I. Daubechies, A. Grossman, Y. Meyer, *Painless nonorthogonal expansions*, J. Math. Phys., 27(5), (1986), 1271-1283.

[DJJ] I. Daubechies, S. Jaffard, J.L. Journé, *A simple Wilson orthonormal basis with exponential decay*, SIAM J. Math. Anal., 22, (1991), 554-572.

[DMW] I. Daubechies, S. Mallat, A.S. Willsky (Guest Editors), Special issue on wavelet transforms and multiresolution signal analysis, IEEE Trans. Inform. Theory, Vol 38, No 2, March, 1992.

[DL1] I. Daubechies, J. Lagarias, *Two scale-difference equations I, Existence and global regularity of solutions*, SIAM J. Math. Anal., 22, (1991), 1388-1410.

[DL2] I. Daubechies, J. Lagarias, *Two scale-difference equations II, Local regularity, infinite products of matrices and fractals*, SIAM J. Math. Anal. 23, (1992), 1031-1079.

[Dav] G. David, *Wavelets and singular integrals on curves and surfaces*, Lecture Notes in Mathematics, 1465, Springer-Verlag, (1991).

[DS] R.J. Duffin, A.C. Shaffer, *A class of nonharmonic Fourier Series*, Trans. Amer. Math. Soc., 72, (1952), 341-366.

[DH] J. Dziubański, E. Hernández, *Band-limited wavelets with subexponential decay*, Preprint, Washington University in St. Louis, (1996).

[EG] R.E. Edwards, G.I. Gaudry, *Littlewood-Paley and multiplier theory*, Springer-Verlag, (1977).

[FW] X. Fang, X. Wang, *Construction of minimally supported frequency wavelets*, Journal of Fourier Analysis and Applications, Vol. 2, No. 4, (1996), 315-327.

[Fef] C. Fefferman, *Characterizations of bounded mean oscillation*, Bull. Amer. Math. Soc., 77, (1971), 587-588.

[FS1] C. Fefferman, E.M. Stein, H^p *spaces of several variables*, Acta Math., 129, (1972), 137-193.

[FS2] C. Fefferman, E.M. Stein, *Some maximal inequalities*, Amer. J. Math., 93, (1971), 107-115.

[Fra] Ph. Franklin, *A set of continuous orthogonal functions*, Math. Ann., 100, (1928), 522-529.

[FJ1] M. Frazier, B. Jawerth, *Decomposition of Besov spaces*, Indiana Univ. Math. J., 34, (1985), 777-799.

[FJ2] M. Frazier, B. Jawerth, *The φ-transform and applications to distribution spaces*, in Function Spaces And Applications (M. Cwikel, et al., Eds.), Springer Lecture Notes in Math., 1302, (1988), 223-246.

[FJ3] M. Frazier, B. Jawerth, *A discrete transform and decomposition of distribution spaces*, J. Func. Anal., 93, (1990), 34-170.

[FJW] M. Frazier, B. Jawerth, G. Weiss, Littlewood-Paley theory and the study of function spaces, CBMS – AMS, (1991).

[Gab] D. Gabor, *Theory of communication*, J. Inst. Electr. Eng., London, 93 (III), (1946), 429-457.

[GK1]   J. García-Cuerva, K. Kazarian, *Spline wavelet bases of weighted*
        L^p *spaces*, $1 \leq p < \infty$, Proc. Amer. Math. Soc., 123 (2), (1995),
        433-439.

[GK2]   J. García-Cuerva, K. Kazarian, *Calderón-Zygmund operators*
        and unconditional bases of weighted Hardy spaces, Studia Math.,
        109 (3), (1994), 257-276.

[GK3]   J. García-Cuerva, K. Kazarian, *Spline wavelet bases of weighted*
        spaces, in Fourier Analysis and Partial Differential Equations
        (J. García-Cuerva, E. Hernández, F. Soria, J.L. Torrea, Eds.),
        Studies in Advanced Mathematics, CRC Press, (1995).

[GR]   J. García-Cuerva, J.L. Rubio de Francia, Weighted norm inequal-
        ities and related topics, North-Holland, (1985).

[Gri1] G. Gripenberg, *A necessary and sufficient condition for the exis-*
        tence of a father wavelet, Studia Math., 114(3), (1995), 207-226.

[Gri2] G. Gripenberg, *Wavelet bases in* $L^p(\mathbb{R})$, Studia Math., 106(2),
        (1993), 175-187.

[Gri3] G. Gripenberg, *Unconditional bases of wavelets for Sobolev*
        spaces, SIAM J. Math. Anal., Vol. 24, No. 2, (1993), 1030-1042.

[HKLS] Y-H. Ha, H. Kang, J. Lee, J. Seo, *Unimodular wavelets for* L^2
        and the Hardy space H^2, Michigan Math. J., 41, (1994), 345-361.

[Haa]   A. Haar, *Zur theorie der orthogonalen funktionen systems*, Math.
        Ann., 69, (1910), 331-371.

[Hal]   P.R. Halmos, Introduction to Hilbert space, Chelsea Publishing
        Company, New York, (1951).

[Han]   B. Han, *Some applications of projection operators in wavelets*,
        Acta Math. Sinica, New Series, Vol. 11, No. 1, (1995), 105-112.

[HS]   Y. Han, E. Sawyer, *Para-accreative functions, the weak bound-*
        edness property and the Tb Theorem, Rev. Mat. Iberoamericana,
        Vol. 6, nos. 1, 2, (1990), 17-41.

[HL]   G.H. Hardy, J.E. Littlewood, *A maximal theorem with function-*
        theoretic applications, Acta Math., 54, (1930), 81-116.

[HWW1] E. Hernández, X. Wang, G. Weiss, *Smoothing minimally sup-*

ported frequency wavelets: part I, Journal of Fourier Analysis and Applications, Vol. 2, No. 4, (1996), 329-340.

[HWW2] E. Hernández, X. Wang, G. Weiss, *Smoothing minimally supported frequency (MSF) wavelets: part II*, Journal of Fourier Analysis and Applications, To appear.

[HWW3] E. Hernández, X. Wang, G. Weiss, *Characterization of wavelets, scaling functions and wavelets associated with multiresolution analyses*, Washington University in St. Louis, Preprint, (1995).

[Hun] R.A. Hunt, *On the convergence of Fourier series*, in Proc. Conf. Orthogonal expansions and continuous analogues (D.T. Haimo, Ed.), Southern Ill. Univ. Press, (1966), 235-255.

[Kat] Y. Katznelson, An introduction to harmonic analysis, John Wiley ans Sons, (1968).

[KKR1] S.E. Kelly, M.A. Kon, L.A. Raphael, *Pointwise convergence of wavelet expansions*, Bull. Amer. Math. Soc. (New Series), 30, no 1, (1994), 87-94.

[KKR2] S.E. Kelly, M.A. Kon, L.A. Raphael, *Local convergence for wavelet expansions*, J. of Funct. Analysis, 126, (1994), 102-138.

[KT] C. Kenig, P. Tomas, *Maximal operators defined by Fourier multipliers*, Studia Math., 68, (1980), 79-83.

[Lae] E. Laeng, *Une base orthonormal de $L^2(\mathbb{R})$ dont les éléments sont bien localisés dans l'espace de phase et leurs supports adaptés à toute partition symétrique de l'espace des fréquences*, C. R. Acad. Sci. Paris, Série 2, 311, (1990), 677-680.

[Law] W.M. Lawton, *Necessary and sufficient conditions for constructing orthonormal wavelet bases*, J. Math. Phys., 32-1, (1991), 57-61.

[Le1] P.G. Lemarié-Rieusset, *Analyse multi-schelles et ondelettes a support compact*, in Les ondelettes en 1989 (P.G. Lemarié, Ed.), Lecture Notes in Mathematics, 1438, Springer-Verlag, (1990), 26-38.

[Le2] P.G. Lemarié-Rieusset, *Existence de fonction-père pour les ondelettes à support compact*, C.R. Acad. Sci. Paris, Série I, 314, (1992), 17-19.

[Le3] P.G. Lemarié-Rieusset, *Sur l'existence des anlyses multi-résolutions en théorie des ondelettes*, Rev. Mat. Iberoamericana, Vol 8, no 3, (1992), 457-474.

[Le4] P.G. Lemarié-Rieusset, *Ondelettes à localisation exponentiells*, J. Math. Pure et Appl., 67, (1988), 227-236.

[LM]   P.G. Lemarié, Y. Meyer, *Ondelettes et bases hilbertiannes*, Rev. Mat. Iberoamericana, 2, (1986), 1-18.

[LP]   J.E. Littlewood, R.E.A.C. Paley, *Theorems on Fourier series and power series I and II*, J. London Math. Soc., 6, (1931), 230-233, and Proc. London Math. Soc., 42, (1936), 52-89.

[Low] F. Low, *Complete sets of wave packets*, A passion for Physics – Essays in Honor of Geoffrey Chew, World Scientific, Singapore, (1985), 17-22.

[Mad] W. R. Madych, *Some elementary properties of multiresolution analyses of $L^2(\mathbb{R}^n)$*, in Wavelets: A Tutorial in Theory and Applications (C. K. Chui, Ed.), Academic Press, (1992), 259-294.

[Mal1] S. Mallat, *Multiresolution approximations and wavelet orthonormal bases for $L^2(\mathbb{R})$*, Trans. of Amer. Math. Soc., 315, (1989), 69-87.

[Mal2] S. Mallat, *A theory of multiresolution signal decomposition: the wavelet representation*, IEEE Trans. Pattern Anal. Machine Intell., 11 (1989), 674-693.

[Malv] H. Malvar, *Lapped transforms for efficient transform /subband coding*, IEEE Trans. Acoustics Speech Signal and Processing, 38, (1990), 969-978.

[Mar] J.T. Marti, Introduction to the theory of bases, Springer-Verlag, Berlin, (1969).

[Mau] B. Maurey, *Isomorphismes entre espaces H^1*, Acta Math. 145, (1980), 79-120.

[Me1] Y. Meyer, Ondelettes et opérateurs. I: Ondelettes, Hermann, Paris, (1990). [English translation: Wavelets and operators, Cambridge University Press, (1992).]

[Me2] Y. Meyer, Ondelettes et opérateurs. II: Opérateurs de Calderón-

Zygmund, Hermann, Paris, (1990).

[Me3] Y. Meyer, *Ondelettes, fonctions splines et analyses graduées*, Lectures given at the University of Torino, Italy, (1986).

[Me4] Y. Meyer, *Wavelets and operators*, in Analysis at Urbana 1, London Math. Soc., Lecture Note Ser., (E. Berkson, N.T. Peck, J. Uhl, Eds.), Cambridge U. Press, (1989), 256-365.

[Me5] Y. Meyer, *Principe d'incertitude, bases hilbertiennes et algèbres d'opérateurs*, Séminare Bourbaki, 1985 - 1986, 38 année, no 662.

[Me6] Y. Meyer, Wavelets, algorithms and applications (translated by R.D. Ryan), SIAM (1993).

[Nür] G. Nürnberger, Approximation by spline functions, Springer-Verlag, New York, (1989).

[Pal] R.E.A.C. Paley, *A remarkable system of orthogonal functions*, Proc. London Math. Soc., 34 (1932), 241–279.

[PZ] R.E.A.C. Paley, A. Zygmund, *On some series of functions*, Proc. of the Cambridge Phil. Soc., 34 (1930), 337-357, 458-474 and 28, (1932), 190-205.

[PP] M. Plancherel, G. Pólya, *Functions entières et intégrales de Fourier multiples*, Comment. Math. Helv., 9, (1937), 224-248.

[Pe1] J. Peetre, *Sur les espaces de Besov*, C.R. Acad. Sci. Paris, Ser. A-B, 264, (1967), 281-283.

[Pe2] J. Peetre, *On spaces of Triebel-Lizorkin type*, Ark. Mat., 13, (1975), 123-130.

[Pe3] J. Peetre, New thoughts on Besov spaces, Duke Math. Series, Durham, N.C., (1976).

[Pol] D. Pollen, *Daubechies' scaling function on [0,3]*, in Wavelets: A tutorial in Theory and Applications (C.K. Chui, Ed.), Academic Press, (1992), 3-13.

[RS] F. Riesz, B. Sz-Nagy, Functional analysis, Frederick Ungar Publishing Co., New York, (1955).

[Rud] W. Rudin, Real and complex analysis, McGraw-Hill, (1966).

[Sch] I.J. Schoenberg, Cardinal spline interpolation CBMS–NSF Series in Applied Math., no 12, SIAM Publ., (1973).

[Sha] C.E. Shannon, *Communications in the present of noise*, Proc. of the Inst. of Radio Eng., 37, (1949), 10-21.

[Sin] I. Singer, Bases in Banach spaces, I, Springer-Verlag, Berlin, (1970).

[SB1] M.J. Smith, T.P. Barnwell, *Exact reconstruction techniques for tree-structured subband coders*, IEEE Trans. Acoust., Speech Signal Processing, 34, (1986), 434-441.

[SB2] M.J. Smith, T.P. Barnwell, *A new filter bank theory for time frequency representation*, IEEE Trans. Acoust., Speech Signal Processing, 35, (1987), 314-327.

[Ste1] E.M. Stein, Singular integrals and differentiability properties of functions, Princeton University Press, Princeton, New Jersey, (1970).

[Ste2] E.M. Stein, Harmonic Analysis: Real-Variable Methods, Orthogonality and Oscillatory Integrals, Princeton University Press, Princeton, New Jersey, (1993).

[SW] E.M. Stein, G. Weiss, Introduction to Fourier Analysis on Euclidean spaces, Princeton University Press, Princeton, New Jersey, (1971).

[SN] G. Strang, T. Nguyen, Wavelets and filter banks, Wellesley-Cambridge Press, (1996).

[St] R. Strichartz, *Construction of orthonormal wavelets*, in Wavelets, Mathematics and Applications (J.J. Benedetto and M.W. Frazier, Eds.), CRC Press, (1993), 23-50.

[Str] J.O. Strömberg, *A modified Franklin system and higher order spline systems on* $\mathbb{R}^n$ *as unconditional basis for Hardy spaces*, in Conference in honor of A. Zygmund (W. Beckner, Ed.), Vol. II, Wasdsword, (1981), 475-493.

[Sz-N] B. Sz-Nagy, *Expansion theorems of Paley-Wiener type*, Duke Math. J., 14, (1947), 975–978.

[Tai] M. Taibleson, *On the theory of Lipschitz spaces of distributions in Euclidean n-space I, II and III*, J. Math. Mec., 13, (1964), 407-480; 14, (1965), 821-840; and 15, (1966), 973-981.

[TW] M. Taibleson, G. Weiss, The molecular characterization of certain Hardy spaces, Asterisque, 77, (1980).

[Tr1] H. Triebel, *Spaces of distributions of Besov type on Euclidean n-space: duality, interpolation*, Arkiv. Mat., 11, (1973), 13-64.

[Tr2] H. Triebel, Interpolation Theory, Function Spaces, Differential Operators, North Holland, (1978).

[Tr3] H. Triebel, Theory of Function Spaces, Monographs in Mathematics, Vol. 78, Birkhauser Verlag, (1983).

[Tr4] H. Triebel, Theory of Function Spaces II, Monographs in Mathematics, Vol. 84, Birkhauser Verlag, (1992).

[VK] M. Vetterli, J. Kovacĕvić, Wavelets and subband coding, Prentice Hall, (1995).

[Walk] J. Walker, Fast Fourier transforms, Studies in Advanced Mathematics, CRC Press, (1991).

[Wal1] G.G. Walter, *Pointwise convergence for wavelet expansions*, J. of Aprox. Theory, 80, (1995), 108-118.

[Wal2] G.G. Walter, *A sampling theorem for wavelet subspaces*, IEEE Trans. on Infor. Theory, Vol. 38. No. 2, (1992), 881-884.

[Wan] X. Wang, *The study of wavelets from the properties of their Fourier transforms*, Ph.D. Thesis, Washington University in St. Louis, (1995).

[Wi1] M.V. Wickerhauser, *Smooth localized orthonormal bases*, C. R. Acad. Sci. Paris, 316, (1993), 423-427.

[Wi2] M.V. Wickerhauser, Adapted wavelet analysis from theory to software, Wellesley, MA, A.K. Peters (1994).

[Wi3] M.V. Wickerhauser, *Lectures on wavelet packet algorithms*, Washington University in St. Louis, (1991).

[Wil] K.G. Wilson, *Generalized Wannier functions*, Preprint, Cornell University, (1987).

[Wo1] P. Wojtaszczyk, *The Franklin system is an unconditional basis in H^1*, Arkiv für Mat., 20, no 2, (1982), 293-300.

[Wo2] P. Wojtaszczyk, Banach spaces for analysts, Cambridge University Press, (1991).

[You] R.M. Young, An introduction to nonharmonic Fourier series, Academic Press, New York, (1980).

[Zyg] A. Zygmund, Trigonometric series, Cambridge University Press, (1959).

Author index

Index